Building Energy Simulation Using DesignBuilder™

This book explains the key concepts of energy simulations, build simulation models and use of these models to integrate daylight and energy efficiency in their building designs. It covers the key concepts of energy simulation, construction of the building with respect to geometry and materials used, simulating the features of a building façade to optimize daylight and energy efficiency, followed by energy code compliance and the integration of life-cycle assessment (LCA) into the design process. It also does the following.

- Explains the building physics and related concepts behind energy performance of building without relying on high-level mathematics
- Covers the simulation of dynamic shading techniques and daylighting strategies, offering insights into optimizing building performance
- Explores adaptive controls in naturally ventilated buildings
- Looks into performing LCA and carbon emissions calculation
- Facilitates a systematic learning experience for readers at all levels of expertise

It is aimed at students, researchers in building energy (simulation)/building design courses, energy simulation professionals, architects, and so forth.

Building Energy Simulation Using DesignBuilder™

Envelope and Lighting

Third Edition

Vishal Garg, Jyotirmay Mathur and
Aviruch Bhatia

CRC Press
Taylor & Francis Group
Boca Raton London New York

CRC Press is an imprint of the
Taylor & Francis Group, an **informa** business

Designed cover image: Shutterstock

Third edition published 2027
by CRC Press
2385 NW Executive Center Drive, Suite 320, Boca Raton FL 33431

and by CRC Press
4 Park Square, Milton Park, Abingdon, Oxon, OX14 4RN

CRC Press is an imprint of Taylor & Francis Group, LLC

First edition published by CRC Press 2017
Second edition published by CRC Press 2020

ISBN: 978-1-032-91385-8 (hbk)
ISBN: 978-1-032-89601-4 (pbk)
ISBN: 978-1-003-56301-3 (ebk)

DOI: 10.1201/9781003563013

Typeset in Times
by Apex CoVantage, LLC

This book is dedicated to our beloved teacher, the late Professor N. K. Bansal, whose pioneering efforts in introducing building energy simulation in India laid the foundation for energy-efficient building design in the country.

Contents

Preface

The building sector accounts for nearly one-third of global energy consumption, making it a major contributor to climate change. To mitigate this impact, buildings must be designed for energy efficiency, and building energy simulation has emerged as a vital approach in achieving that goal. Over the past few decades, advances in high-level programming languages and powerful computing infrastructure have led to the development of sophisticated building energy simulation tools with user-friendly interfaces. Today, these tools are used not only in architectural design but also in the design of building services, life-cycle performance assessments, code compliance, green certifications, and real-time building operations. Simulation has become an integral part of modern building design and construction process.

This work, *Building Energy Simulation Using DesignBuilder™: Envelope and Lighting*, a tutorial-based training guide for energy simulation, originated from a series of training programs designed for participants from diverse professional backgrounds. Through these programs, the authors experimented with various instructional approaches and found that the most effective method was a tutorial-based format with step-by-step instructions and graphical illustrations. The simulations presented in this workbook were carried out using DesignBuilder to demonstrate the building energy simulation process. The workbook embraces a "learning by doing" pedagogical approach, helping readers grasp the fundamentals of building physics and building services while deepening their understanding of building energy performance.

The first edition of this book laid the groundwork and the second edition of the workbook, published in 2020, expanded the content to include topics such as thermal storage, renewable energy systems, cost analysis, and sensitivity assessments.

This third edition has several new features: First, entire text has been thoroughly revised to incorporate more aspects of building physics. Second, additional tutorials have been added based on the latest simulation capabilities. Finally, to make it more accessible to readers, the entire text has been reorganized into two volumes. The first and second volumes have been designed for architects and building service engineers as primary readers, respectively.

Building Energy Simulation Using DesignBuilder™: Envelope and Lighting, Volume 1, focuses on building envelope, passive design strategies (such as shading, insulation, cool and green roofs), and daylighting. It also covers interior lighting simulation with advanced control strategies. A newly added chapter on life-cycle assessment (LCA) introduces a quantitative approach to sustainability strategy, supporting informed design decisions. LCA simulations are now also important in earning green building certification credits. The daylighting section has been enhanced to include dynamic shading for internal windows and spatial daylight autonomy (sDA) simulations.

Building Energy Simulation Using DesignBuilder™: HVAC and Renewable Energy, Volume 2, is dedicated to air conditioning, photovoltaic systems, and costing. It includes topics on simulating heating, ventilation, and air conditioning (HVAC) components such as chillers, boilers, air-handling units, heat recovery systems, and HVAC controls. This volume also addresses advanced topics like integrating thermal storage for demand response, renewable energy integration, cost–benefit evaluations, sensitivity, and uncertainty analysis.

Along with these volumes, there are additional electronic resources which include three project exercises: simulating a small office, a single-family home, and a large office building.

Whether you are a student, architect, engineer, or sustainability practitioner, this workbook serves as a practical guide for understanding and applying building energy simulation.

Acknowledgments

Right at the start, we want to acknowledge the late Professor N. K. Bansal, who introduced us to the field of building science and remained a guiding influence throughout our journey in mastering both the subject and the pedagogy. With regards to the two volumes, our special thanks go to Madhan Kumar Kandasamy for running the simulations for all the tutorials and capturing screenshots. We also thank Sahil Chilana and Sraavani Gundepudi for their thorough review of the manuscript. This workbook would not have been possible without all those reviewers who took the time to patiently go through the contents and provided their valuable feedback.

This ten-year-long project would not have been possible without the foundational contributions of Surekha Tetali, Gaurav Chaudhary, Hema Rallapalli, Kopal Nihar, Kuntal Chattopadhyay, Naresh Arthem, Nishesh Jain, Shivraj Dhaka, and Vijay Singh, whose work on the previous edition laid the groundwork for this one.

Our heartfelt thanks go to DesignBuilder Software Ltd., for their technical assistance and ongoing support in responding to our queries throughout the development of this book.

We gratefully acknowledge the support of Indorama Ventures for their continued commitment toward sustainability, and we thank Indorama Ventures Center for Clean Energy, Plaksha University, for their generous support, which made this work possible.

Our thanks to all friends and family members for their support and encouragement. Without their support, this book would not have been possible.

Our sincere gratitude goes to the team at CRC Press/Taylor & Francis Group, especially Dr. Gagandeep Singh, for his trust in us and the numerous extensions to the timelines for delivering this workbook.

About the Authors

Vishal Garg is the University Chair Professor and Director of the Indorama Ventures Center for Clean Energy at Plaksha University. His research interests are in the areas of building energy simulations, smart energy homes, and cool roofs. He was awarded the inaugural Dr. Arthur Rosenfeld Urban Cooling Achievement Award and is a Fellow of the International Building Performance Simulation Association (IBPSA). He holds a Bachelor of Civil Engineering degree from MBM Engineering College, Jodhpur, and a Ph.D. from IIT Delhi.

Jyotirmay Mathur is a professor of mechanical engineering and the founding head of the Centre for Energy and Environment at Malaviya National Institute of Technology, Jaipur, India. He has done postgraduate work in energy studies at the Indian Institute of Technology, Delhi, India, and has received a doctorate in energy systems from the University of Essen, Germany. Dr. Mathur has published 100 research papers in refereed international journals and has presented more than 160 papers and talks at international seminars and conferences, besides writing five books. Dr. Mathur works in the field of energy modeling, codes and standards, energy conservation in buildings, passive cooling, adaptive thermal comfort, and building integrated photovoltaic systems.

Aviruch Bhatia is a senior research scientist at Plaksha University, Mohali, India. Before joining Plaksha, he served as an assistant professor at the TERI School of Advanced Studies in New Delhi. He holds a Ph.D. from the International Institute of Information Technology, Hyderabad. His research interests lie at the intersection of building physics, building performance simulation and analysis, and green buildings. Earlier in his career, he worked for three years as an assistant manager with the Sustainability Group at Spectral Consultants Pvt. Ltd. (an AECOM company), where he was involved in a range of green building projects.

1 Getting Started with Energy Simulation

This chapter is designed to acquaint you with building energy simulation. It begins by outlining the essential components and requirements for conducting a simulation and gradually progresses to hands-on modeling using a preconfigured example file. Through a series of 11 tutorials, you will explore the influence on overall energy consumption of critical input parameters such as lighting power density (LPD), building type, thermostat setpoints, window-to-wall ratio (WWR), building orientation, and fresh air supply rate. By the end of the chapter, you will be familiar with key simulation inputs, how to interpret output results, and the complete workflow of the simulation process.

BUILDING ENERGY SIMULATION

Building energy simulation uses computer-based tools to create a virtual representation of a building and perform physics-based calculations. Simulations can range in scale from individual components to entire buildings or clusters of buildings. To conduct an energy simulation, three key inputs are required: the building model, occupancy and usage patterns, and the weather data for the specific location. These inputs enable the estimation of outputs such as peak loads; heating, ventilation, and air conditioning (HVAC) system sizing; and energy consumption over a defined period. The resulting data can be used for estimating utility bills and for cost–benefit analysis of different design strategies.

Some of the uses for energy simulation tools include the following.

Early design decisions: During the initial stages of design, choices related to building orientation and spatial layout are made. Energy simulation can help in evaluating alternative strategies to inform these decisions. However, due to limited project details at this stage, simulations are typically high-level and less detailed.

Component or material selection: As the design progresses, simulation becomes a valuable tool for selecting specific components of the building envelope or mechanical systems. It is widely used for conducting cost–benefit analyses of different design options. Since decisions at this stage have direct impacts on performance, more accurate and detailed modeling is required.

Retrofitting decisions: For existing buildings, energy simulation is instrumental in identifying cost-effective retrofit strategies. To ensure reliability, simulations should be calibrated using actual performance data from the building, thereby enabling more precise evaluations of potential improvements.

WHAT IS NEEDED FOR ENERGY SIMULATION

Building energy simulation should follow a systematic approach, as it requires a substantial amount of input data. It is recommended to gather all necessary information before beginning the modeling process. The key inputs typically include the following.

- **Location and weather data:** Energy simulation tools generally require hourly climate data—including temperature, humidity, wind speed, and solar radiation—specific to the building's location. This data is provided through standardized weather files. If weather

DOI: 10.1201/9781003563013-1

data is unavailable for a particular location, a file from a nearby location with similar climatic conditions can be used.

- **Building geometry:** Accurate representations of the building's elevation and floor plans are essential for creating the geometry model. Although architectural drawings contain extensive detail, much of it may not be relevant for simulation. Simplifying these drawings by removing extraneous details and organizing the layout based on thermal zones greatly facilitates the modeling process.
- **Envelope components:** Detailed information about the construction of the building envelope is crucial. This includes the thickness and thermophysical properties of each material layer in walls, roofs, and floors. Additionally, the thermal and optical characteristics of windows, frames, and shading devices must be specified.
- **Building services:** Complete details of building systems such as HVAC and lighting are required. This includes equipment type, capacity, efficiency, and control strategies.
- **Operational schedules and internal gains:** Hourly schedules are needed for the following.
  - Occupancy
  - Lighting
  - Equipment
  - Thermostat setpoints
  - HVAC operation

Gathering this data accurately ensures reliable simulation results, enabling meaningful analysis and decision-making throughout the design or retrofit process.

How Simulation Software Works

Building energy simulation software integrates the geometric model of a building with external weather conditions, occupancy patterns, and building system operations to estimate energy loads on an hourly basis. These simulations are grounded in fundamental physics and energy balance equations. The software calculates the energy required to meet heating, cooling, lighting, and other loads, advancing hour by hour across the simulation period. Each calculation is based on the previous time step and contributes to a continuous simulation process. The results can be viewed as time-specific outputs or as aggregated summaries over days, months, or the full simulation year. Most tools also account for the transient behaviors of building components and systems, providing detailed insight into dynamic energy flows.

One of the main advantages of simulation over traditional static methods is the ability to reflect the real-time impact of material properties, weather conditions, and system performance. By modeling how the building responds to changing internal and external environments, simulations produce more accurate and insightful results. Despite their complexity, most energy simulation tools can run efficiently on standard desktop or laptop computers without the need for specialized hardware.

This chapter offers a series of foundational tutorials to help you build simple geometry models and explore how factors such as building orientation, WWR, internal heat gains, and ventilation rates influence energy consumption and HVAC system sizing. Each tutorial is followed by hands-on exercises to reinforce learning. While many simulation tools are available, this book uses **DesignBuilder v7.3.1.003**. You can download the latest version from https://designbuilder.co.uk/download. Please note that minor variations in the interface or results may occur between the latest release and the version referenced in this book.

TUTORIAL 1.1: OPENING AND SIMULATING AN EXAMPLE FILE

Goal

To evaluate the energy performance of a building model provided in an example file with DesignBuilder installation

What Are You Going to Learn?

- How to simulate an example file
- How to view energy consumption results based on utility type or fuel, such as electricity or gas
- How to view the end-use energy consumption
- How to view results on daily, monthly, and annual bases
- How to switch between metric (*système international* [SI]) and English (inch-pound [IP]) measurement units
- How to calculate energy use intensity (EUI)

Problem Statement

In this tutorial, you are going to simulate an existing template, **Courtyard with VAV Example**, from the DesignBuilder library and simulate it for a London location. You will learn how to view the daily, monthly, and annual energy consumption reports in a graphical format. You will also learn how to view the annual fuel breakdown and fuel total.

FUEL BREAKDOWN

The data for fuel consumption for building level broken down by end-use, as follows.

- **Heating:** total fuel consumption due to operation of heat generators such as boilers and heat pumps
- **Cooling:** total chiller fuel consumption plus any fuel consumption due to direct expansion (DX) cooling coils, air-cooled chiller condenser fans, and evaporative coolers
- **Heat rejection:** cooling tower fan electric consumption
- **Fans:** total electrical energy consumed by HVAC fans
- **Pumps:** total electrical energy consumed by HVAC pumps
- **Auxiliary energy:** total electrical energy consumed by HVAC fans and pumps
- **Preheat:** heating energy transferred from main plant preheat coils to HVAC system air when using detailed HVAC data
- **Lighting:** electricity consumed by general and task/display lights
- **Room electricity:** electricity consumed by room equipment other than lights (computers, equipment, process, etc.)
- **Room gas:** gas consumed by room equipment (miscellaneous, process, and catering)
- **Room other:** other fuel consumed by room equipment (miscellaneous, process, and catering)

Source: https://designbuilder.co.uk/helpv7.3/Content/Simulation_Detailed_Results.htm
https://designbuilder.co.uk/helpv7.3/Content/Reviewing_Simulation_Output.htm

FUEL TOTALS

Total fuel consumption for the building, as follows.

- **Electricity:** total building electricity consumption.
- **Gas:** total building gas consumption.
- **Oil:** total building or oil consumption.

- **Solid:** total building solid fuel consumption.
- **Other fuel consumption:** total consumption of all other fuels.

Source: https://designbuilder.co.uk/helpv7.3/Content/Simulation_Detailed_Results.htm
https://designbuilder.co.uk/helpv7.3/Content/Reviewing_Simulation_Output.htm

Solution

Step 1: Start DesignBuilder. The DesignBuilder main screen appears.

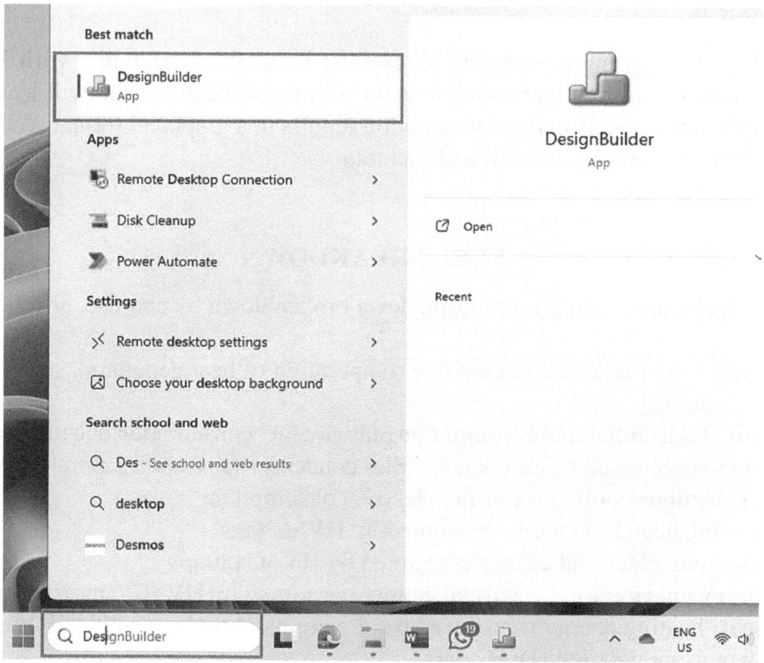

Step 2: Double-click **Courtyard with VAV Example** in the DesignBuilder templates under the **Recent Files** tab. A **New project** screen appears.

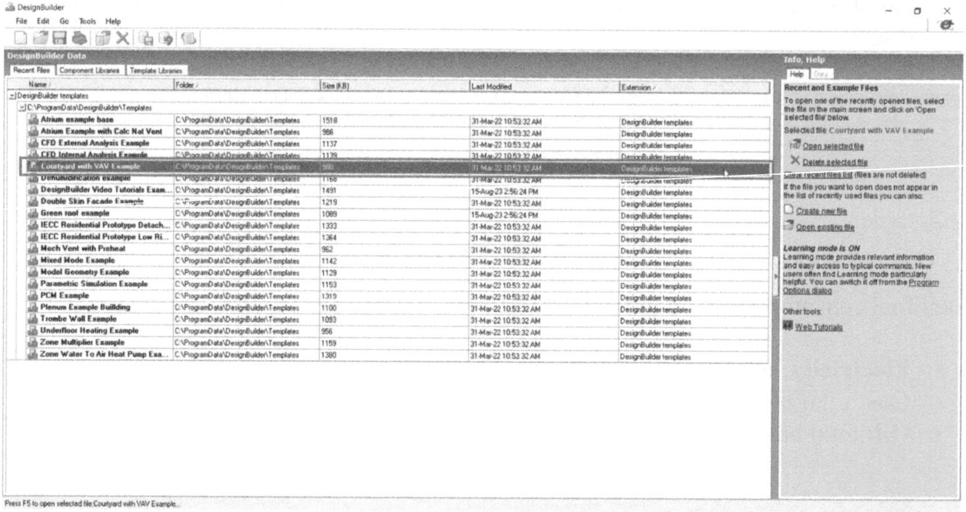

Courtyard with VAV Example is a DesignBuilder template file. A DesignBuilder template file helps to provide examples of various building typologies and systems such as geometry, HVAC system, and passive strategies. Relevant data from the example files can be exported and imported into other DesignBuilder models.

Step 3: To change location, click **CA-SAN FRANCISCO INTL**. Three dots (. . .) appear.

If you have installed DesignBuilder for the first time on your system, **CA-SAN FRANCISCO INTL** may appear as your location. If you have used DesignBuilder previously, you may get some other location based on your previous settings.

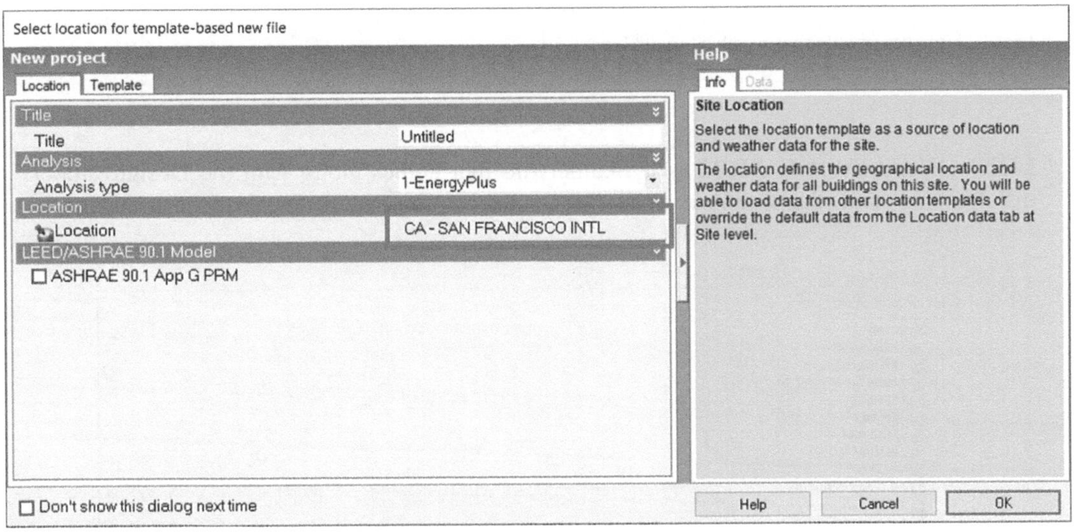

Step 4: Click the three dots (. . .). The **Select the location template** screen appears.

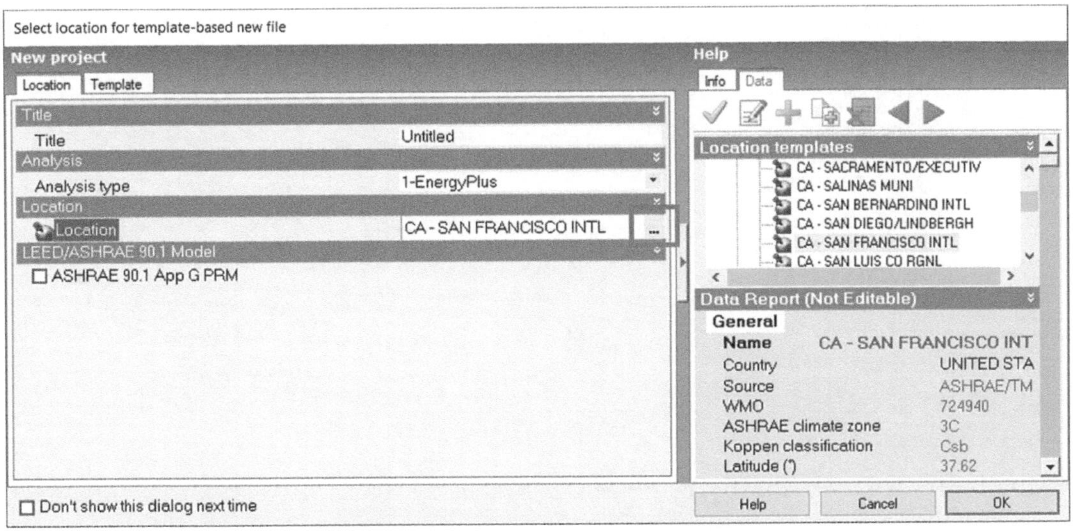

Step 5: Select **LONDON/GATWICK ARPT**. Click **OK**. The **Select location for template-based new file** screen appears.

DesignBuilder uses EnergyPlus Weather Format (EPW) for simulation. These files have hourly values of the following.

- Outside dry bulb temperature
- Outside dew point temperature
- Direct normal solar
- Diffuse horizontal solar
- Wind speed
- Wind direction
- Atmospheric pressure
- Solar altitude
- Solar azimuth

Source: https://energyplus.net/weather/simulation

LONDON/GATWICK ARPT is a weather file that comes along with the DesignBuilder installation. For all other locations, an internet connection is needed to download the weather files.

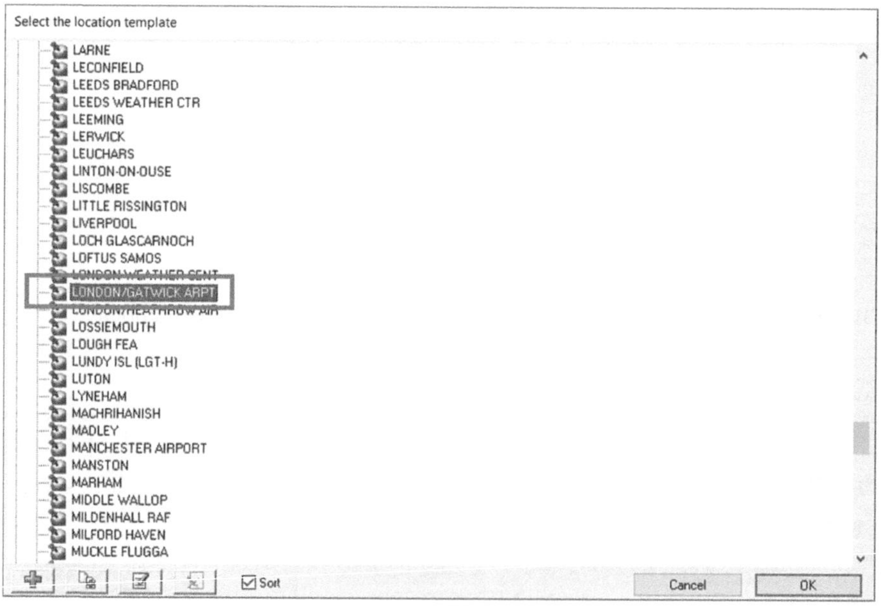

Step 6: Click **OK**.

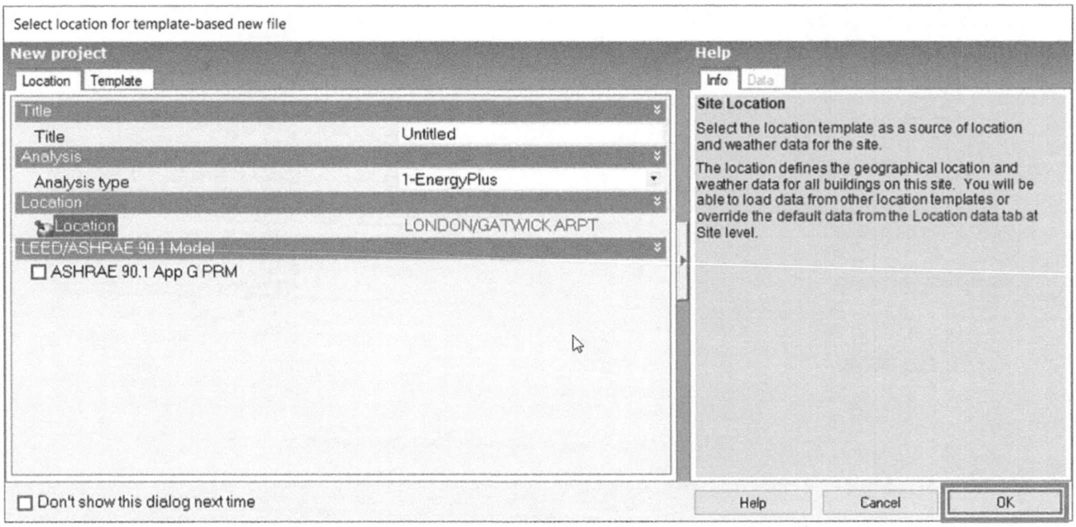

The building layout appears with the following tabs at the bottom of the central display screen.

- **Edit**
- **Visualize**
- **Heating design**
- **Cooling design**
- **Simulation**
- **CFD**
- **Daylighting**
- **Cost and carbon**

Step 7: Select the **Simulation** screen tab. Click **Update data**.

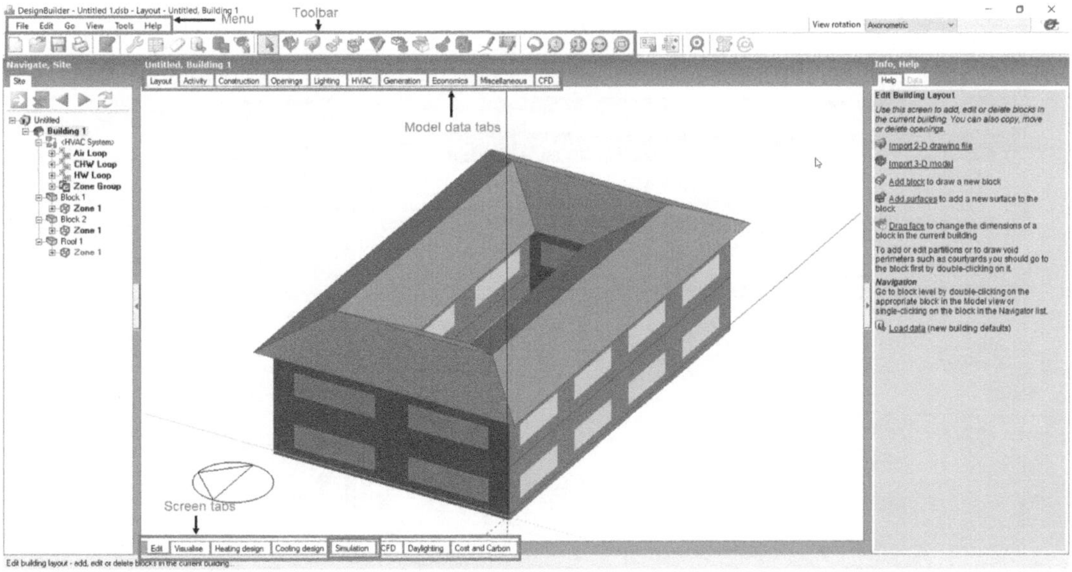

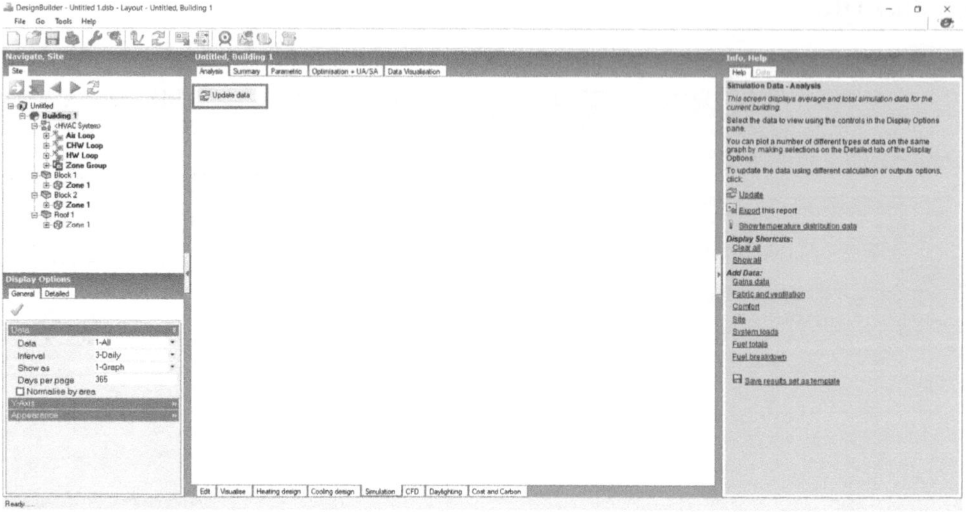

The **Edit Calculation Options** screen appears.

> **WHAT IS THE MEANT BY SIMULATING A MODEL?**
>
> **Simulate** a file?
>
> When you click on the **Simulate** Screen tab, the simulation engine EnergyPlus performs advanced dynamic thermal simulations. It provides detailed performance data, such as energy consumption, across various time intervals such as annual, monthly, daily, hourly, and even sub-hourly.
>
> *Source:* https://designbuilder.co.uk/simulation/

Step 8: Enter the **Simulation Period** start date as **1 Jan** and the end date as **31 Dec**. Click **OK** to start the simulation.

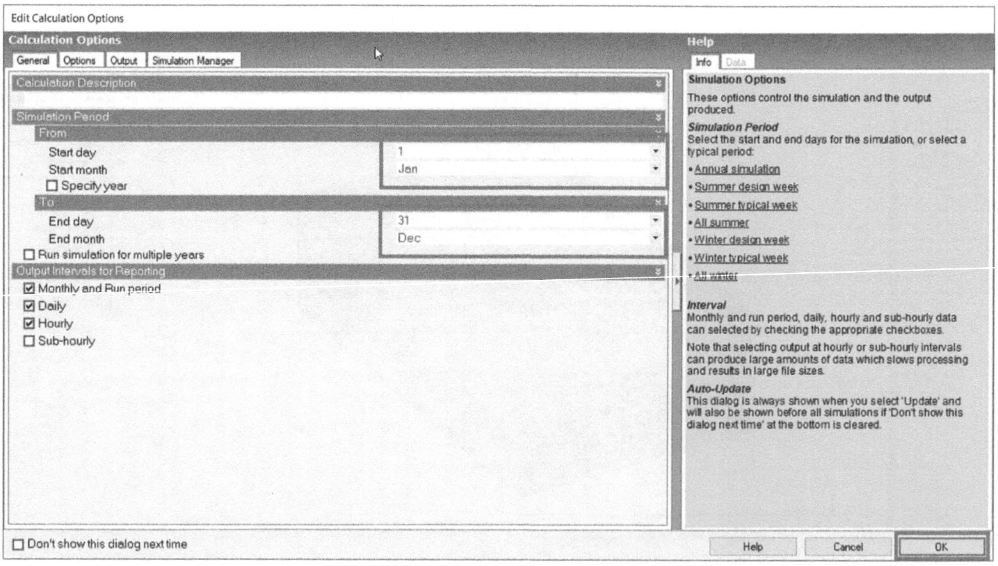

A screen appears showing the simulation progress. Depending on the configuration of the computer, it can take several minutes to complete the simulation.

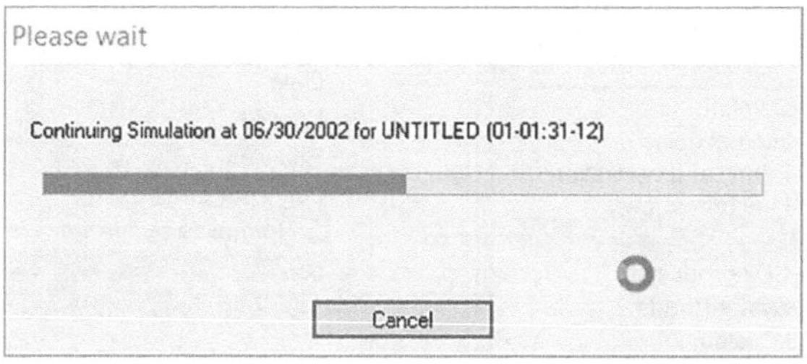

After the simulation is completed, results are displayed in the **Analysis** subtab of the **Simulation Screen** tab.

Step 9: Under **Display Options**, select **Fuel totals** from the **Data** drop-down list and **Run period** from the **Interval** drop-down list. A screen appears with a bar graph for annual fuel totals for electricity and gas.

Note: The results are either displayed in IP or SI units based on your configuration of DesignBuilder. Switching between SI and IP units is explained in Step 15.

After the simulation is complete and the output dataset is loaded, you can select different data intervals depending on the data generated during the simulation.

- **1-Run period:** show data for the entire simulation
- **2-Monthly:** show data broken down by month
- **3-Daily:** show data broken down by day
- **4-Hourly:** show hourly data; useful for checking operation of equipment and comfort conditions for particular design periods
- **5-Sub-hourly:** show data recorded at time step intervals (2 and 6 intervals per hour depending on the "Time steps per hour" calculation option); useful for detailed checking of building performance

Source: https://designbuilder.co.uk/helpv7.3/Content/Simulation_Period.htm
https://designbuilder.co.uk/helpv7.3/Content/Reviewing_Simulation_Output.htm

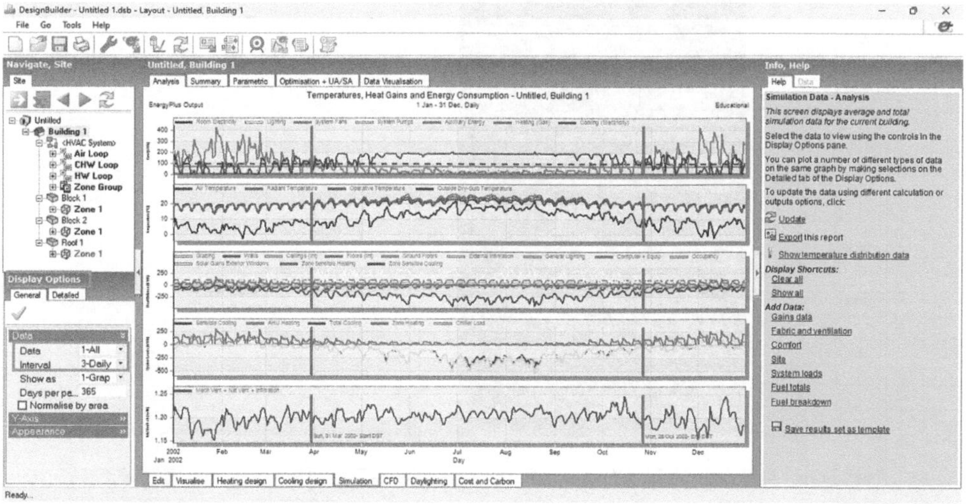

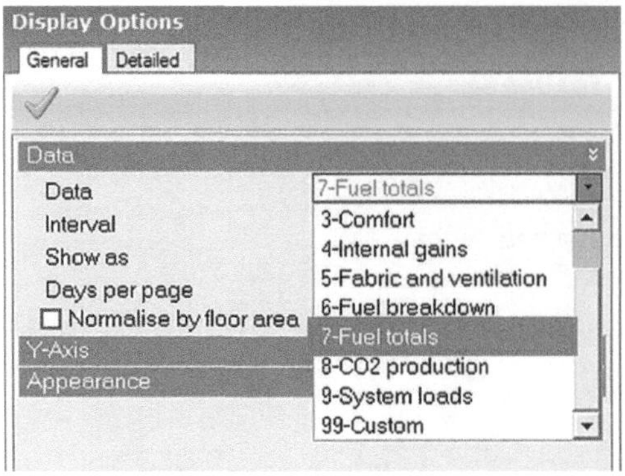

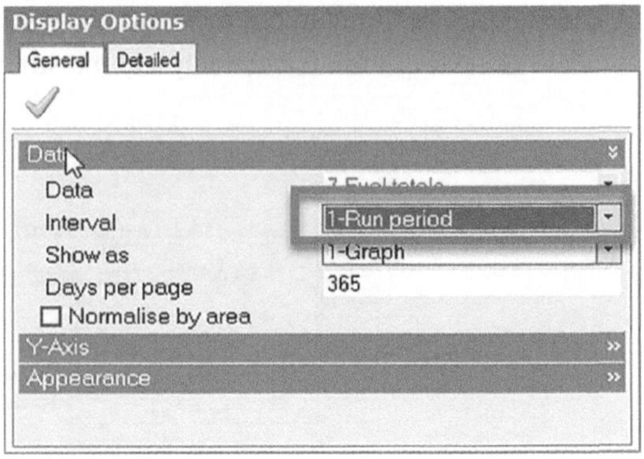

Step 10: Select the **Normalise by floor area** check box and **All floor area** from the **By** drop-down list to view the energy use intensity (EUI) of the building.

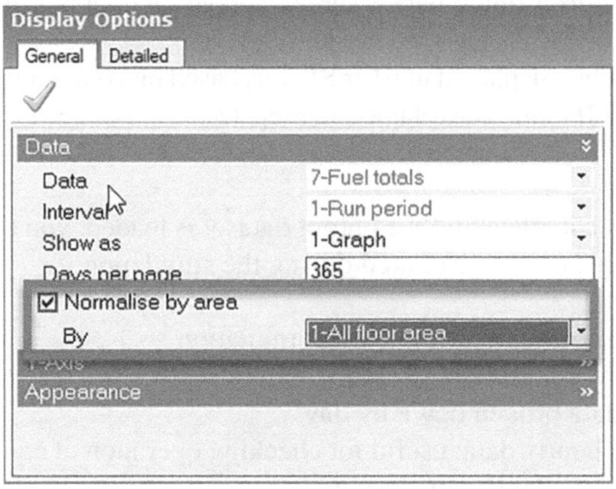

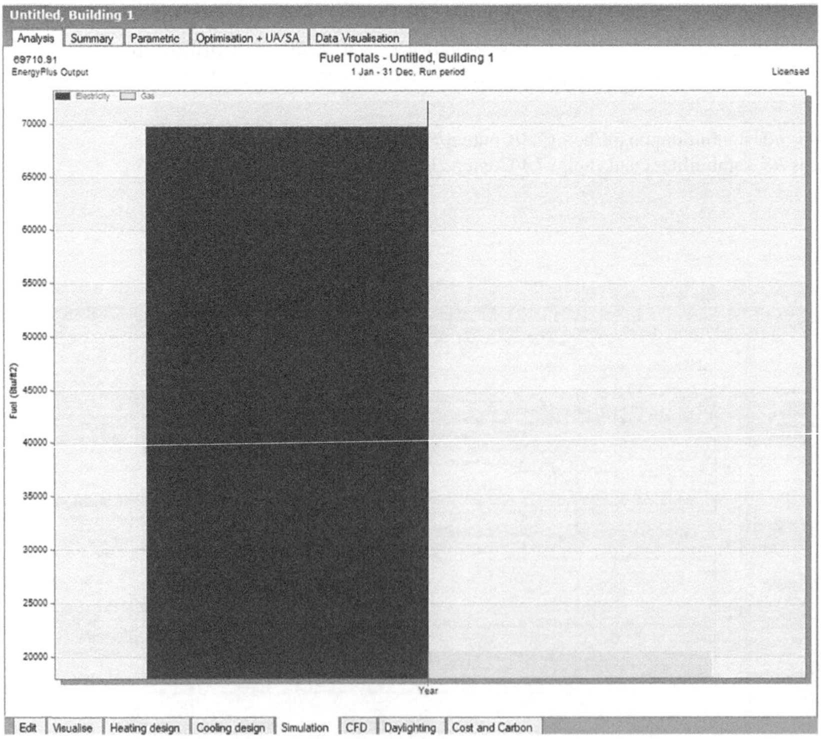

Units for the **Fuel totals** can be seen on the *y*-axis of the graph.

Record the results and fill in Table 1.1 under **Office building** column.

The term *energy use intensity* (EUI) refers to the energy consumption of the building per unit area per annum. The same term, in some countries, is also called the *energy performance index* (EPI). The EUI of a building is calculated by dividing the annual total energy consumption (all fuel types) of the building by its gross floor area. The unit of measurement is kWh/m^2 yr, kWh/ft^2 yr, or Btu/ft^2 yr. A lower EUI means a better energy performance.

The *gross floor area* of a building is the total built-up area, which includes all conditioned and unconditioned enclosed spaces of the building.

Step 11: Change the **Interval from Run** period to **Monthly** to view monthly results. Refer to Step 9.

A DesignBuilder message box appears. Click **Yes**. Click **Update Data**. Click **OK** to perform a simulation to update the results.

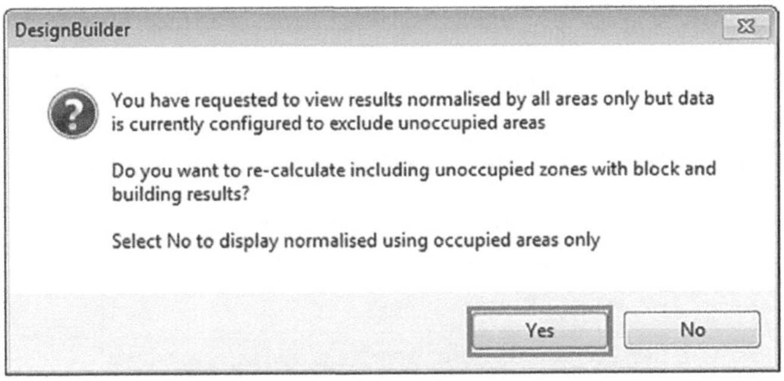

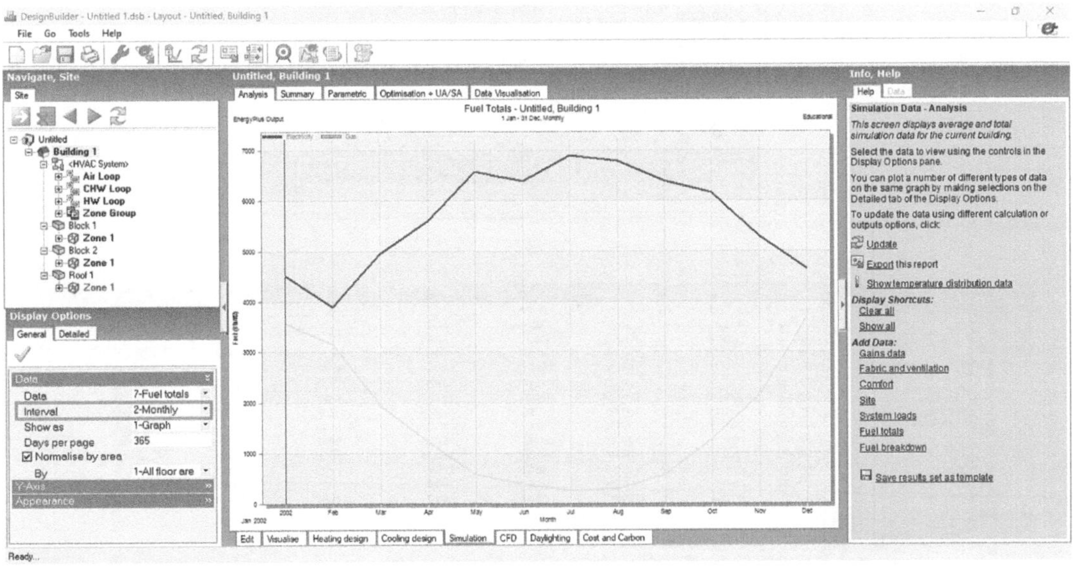

Step 12: Select **Grid** from the **Show as** drop-down list to view results in a grid format.

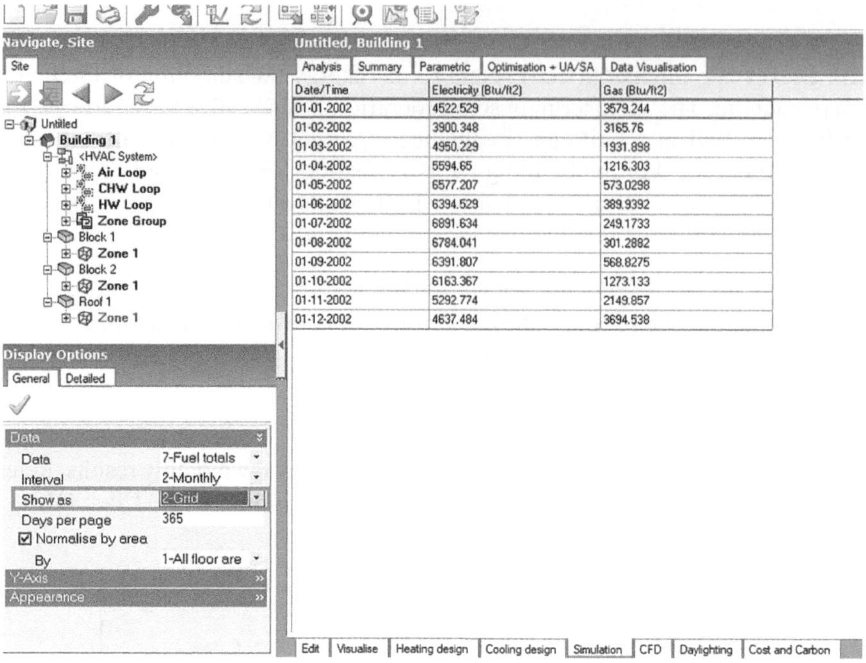

Step 13: Select **Fuel breakdown** from the **Data** drop-down list to view results for fuel breakdown.

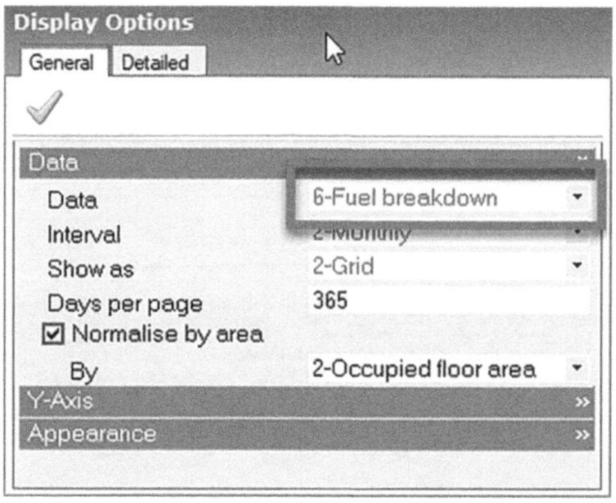

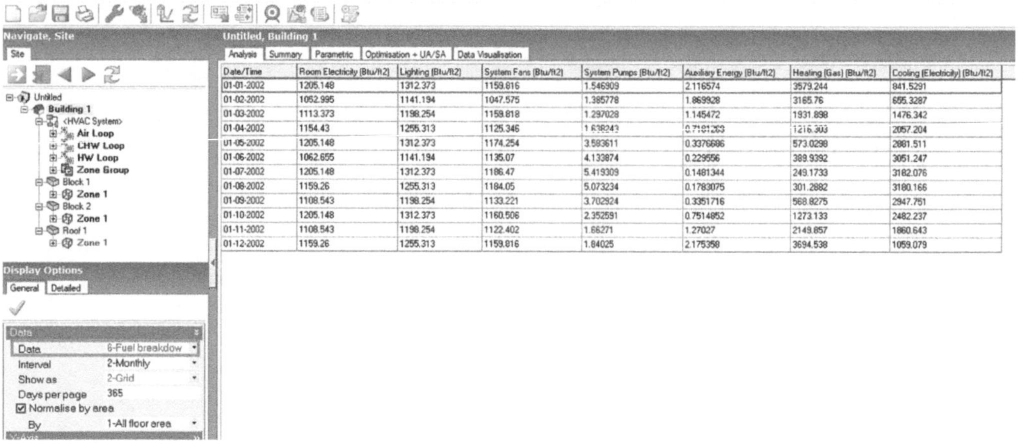

In all the preceding steps, the results are in SI or IP units based on the existing configuration of DesignBuilder in your system. To change the units, perform the following steps.

Step 14: From the **Tools** menu, select **Program options**. The **Edit Program Options** screen appears.

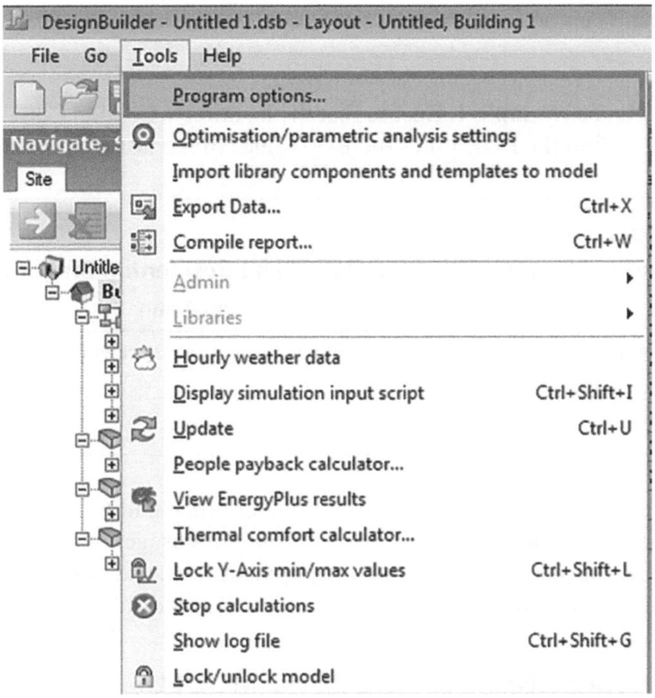

Step 15: Select the **International** tab and select **Units** as per your requirement. Click **OK**. DesignBuilder updates all the values in the selected units. SI units will be used in this workbook.

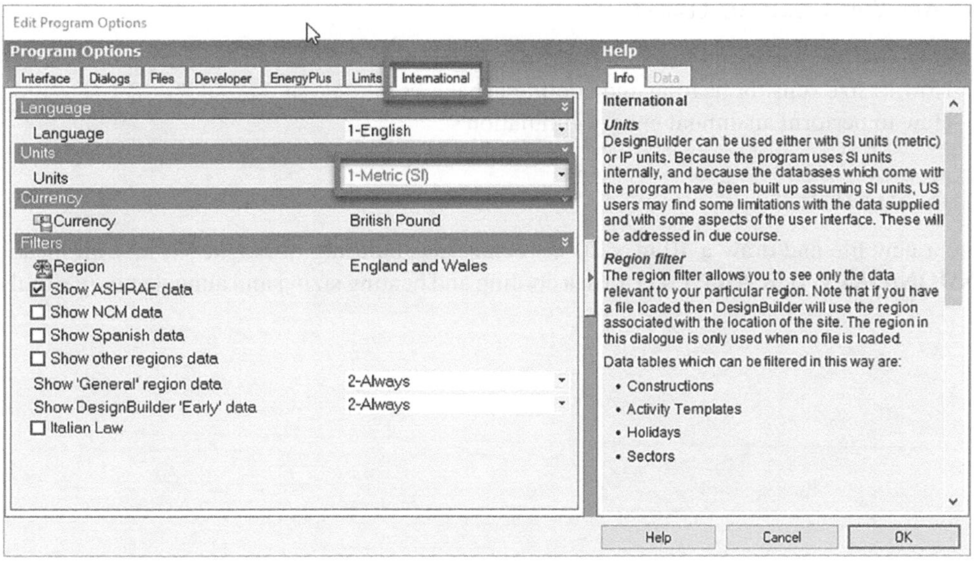

Step 16: From the **File** menu, select **Save** to save this model.

EXERCISE 1.1: COMPARE THE EUIS OF RESIDENTIAL AND OFFICE BUILDINGS

Courtyard with VAV Example is for an office space. Simulate a file with residential usage and compare the results.

> DesignBuilder can open only one file at a time.

Hint: Open the file **House example 1**. Ensure that the location is London Gatwick. Repeat the steps given in Tutorial 1.1 to find the EUI. Enter the EUI values in the last column of Table 1.1.

TABLE 1.1
Energy Use Intensities for Office and Residential Buildings

	EUI (kWh/m²)	
Energy Use	**Office Building**	**Residential Building**
Electricity		
Gas		

You can see that even with the same location, the EUIs of office buildings and residential buildings are different. This is mainly due to the difference in building usage, which includes differences in timing and duration of use, occupancy, nature of equipment used, and other factors. Office buildings usually have a higher EUI, as they have higher internal loads (see Table 1.1).

TUTORIAL 1.2: CREATING A SINGLE-ZONE MODEL

Goal

To create a single-zone model and find heating and cooling capacities

What Are You Going to Learn?

- How to draw, view, and render a building model
- How to size runs for heating and cooling capacities
- How to perform an annual energy simulation

Problem Statement

Create a new file and draw a 40 m × 20 m rectangular building of height 3.5 m with location as **LONDON/GATWICK ARPT**. Perform a cooling and heating sizing and annual energy simulation.

SOLUTION

Step 1: Click **File** and select **New file**. The **New file** screen appears.

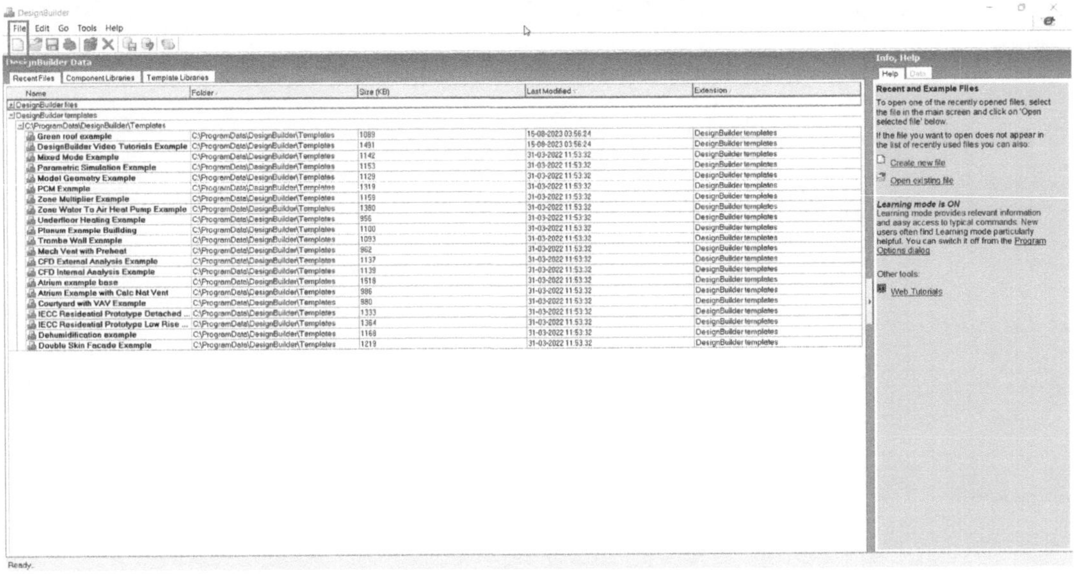

Step 2: Enter **First Building** in the **Title** text box and click **OK**. A blank building layout appears.

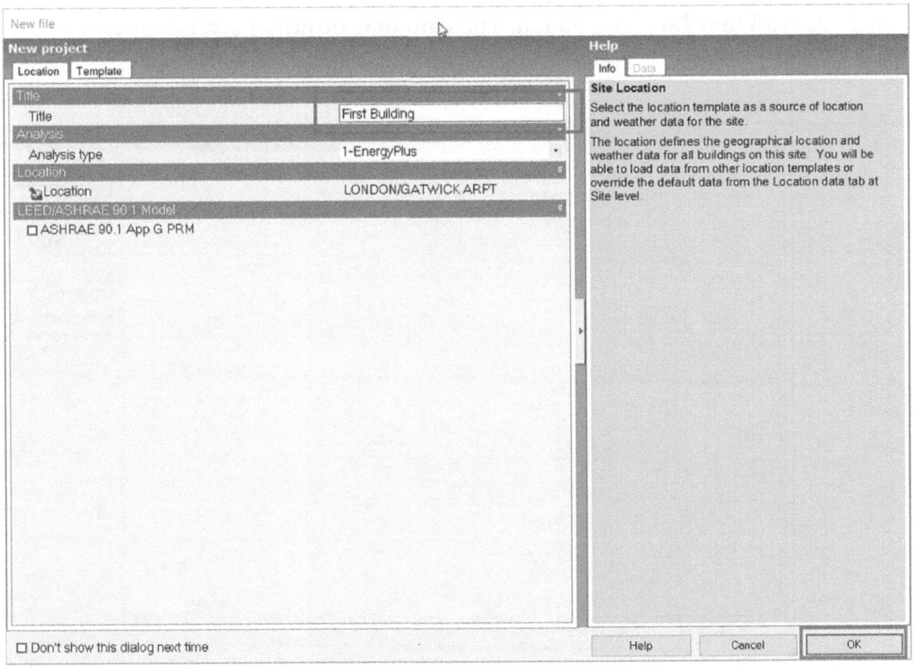

Step 3: Select **Plan** from the **View rotation** drop-down list.

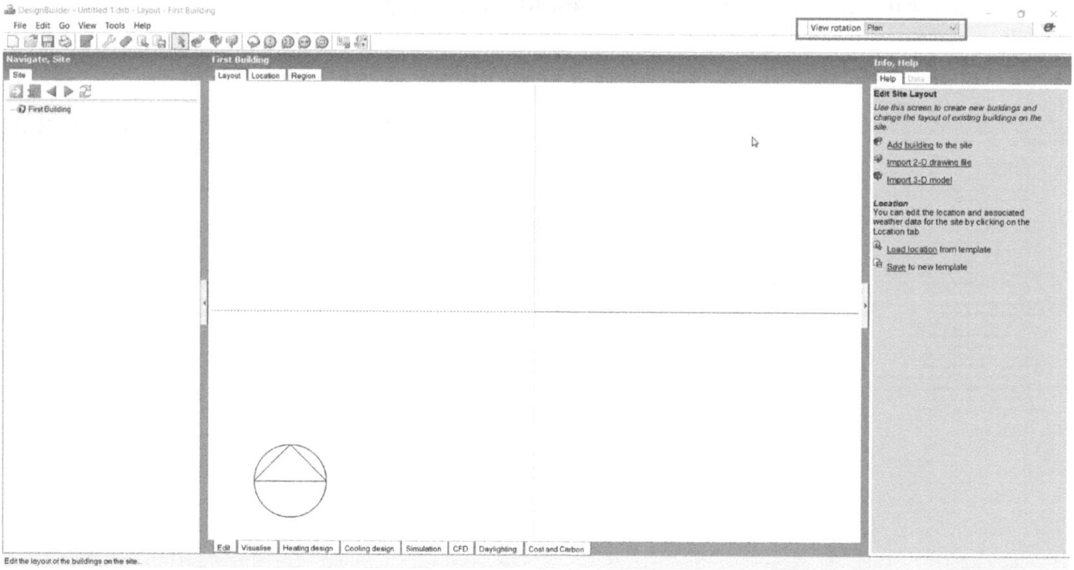

In the DesignBuilder display, the red line represents the x-axis, the green line the y-axis, and the blue line the z-axis.

Step 4: Click the **Add new building** button. The **Add new building** screen appears.

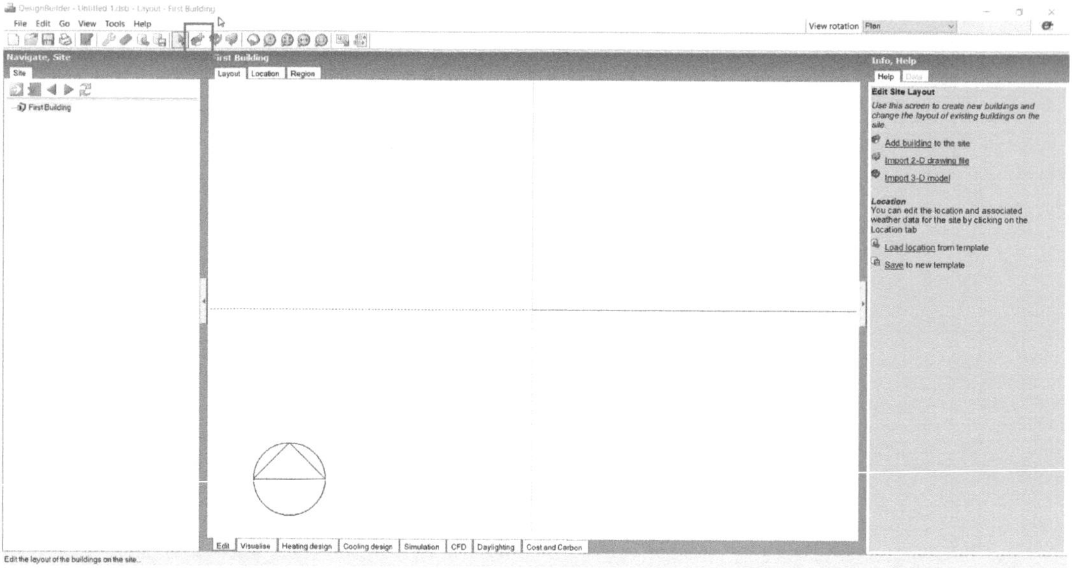

Step 5: Click **OK**. The screen appears with a cursor to draw a building.

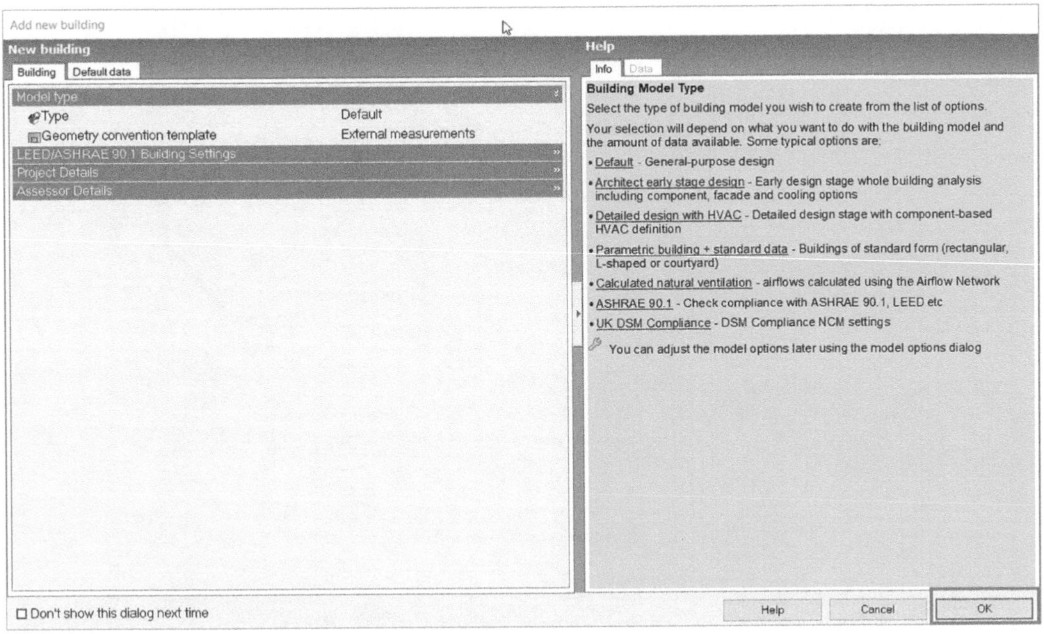

Step 6: Move the pointer to the **Origin**. Left-click and move on the positive x-axis. The length of the segment is shown. (In case of any mistake, you can choose to cancel the drawn line by pressing the **ESC** key and then trying again.)

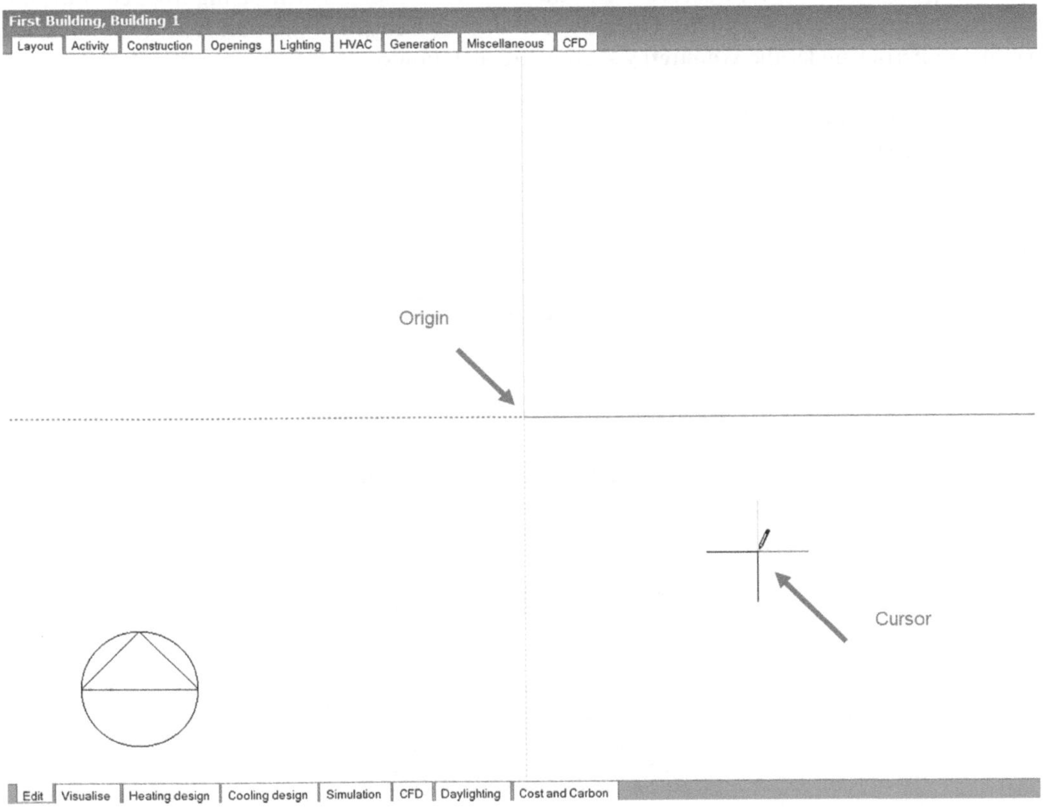

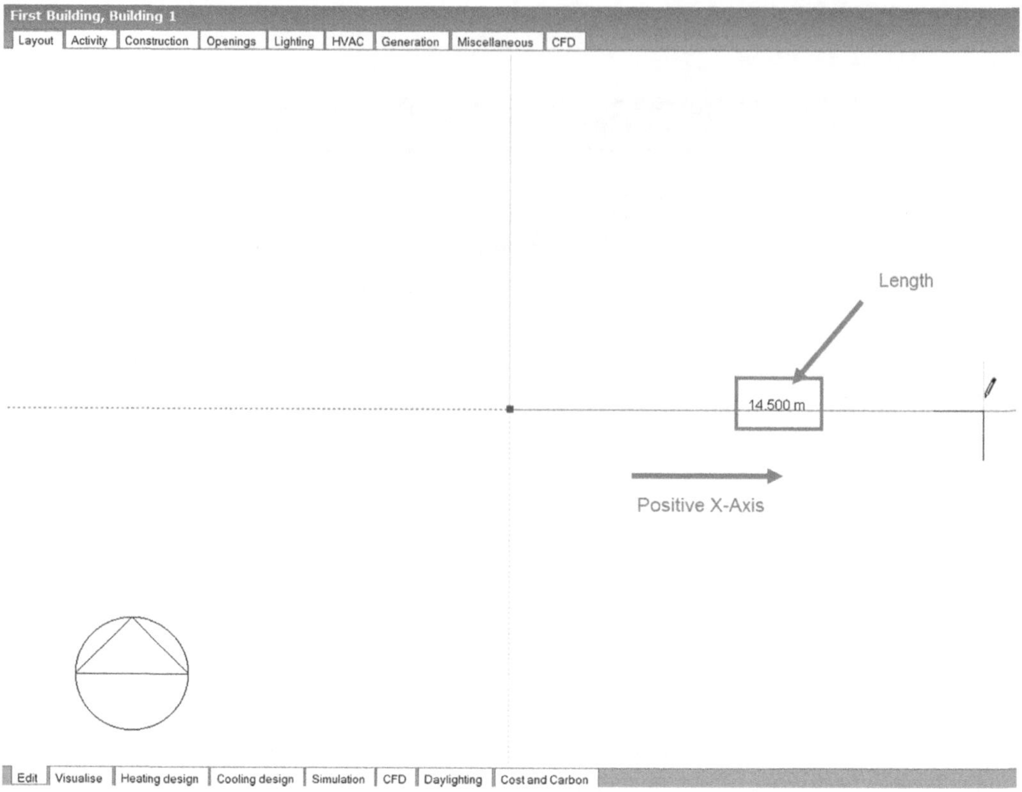

Step 7: Type **40**. You can type directly, and the text box for length will be highlighted. Fill in the value for the length in the **Drawing Options** section. The **Length** of the wall is set to **40** m. Ensure that the properties under the **Geometry** section are as follows:

- Block type: **Building Block**
- Form: **Extruded**
- Height: **3.5 m**

Press **Enter**.

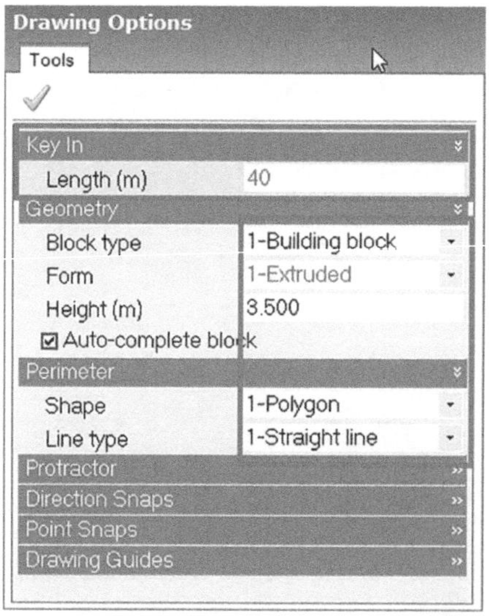

Step 8: Scroll the mouse wheel to zoom in or zoom out until you can see the complete line of 40 m in length. Move the pointer parallel to the positive *y*-axis. It snaps when a dashed vertical line appears. Type **20** to draw the second side of the rectangle. Press **Enter**.

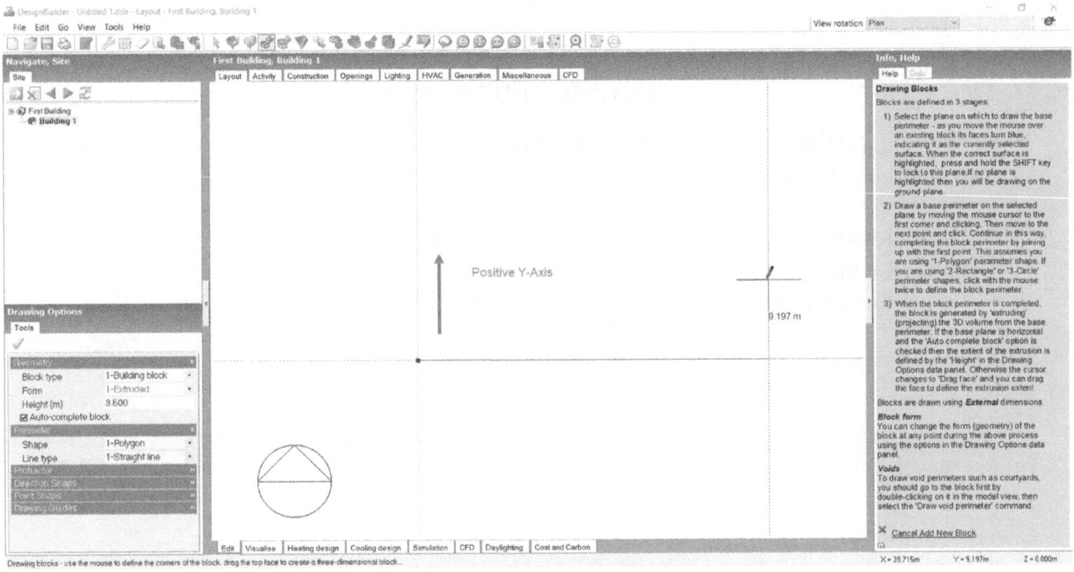

Step 9: Similarly draw other segments and complete the rectangular building block of dimensions 40 m × 20 m. The following screen shows the completed block.

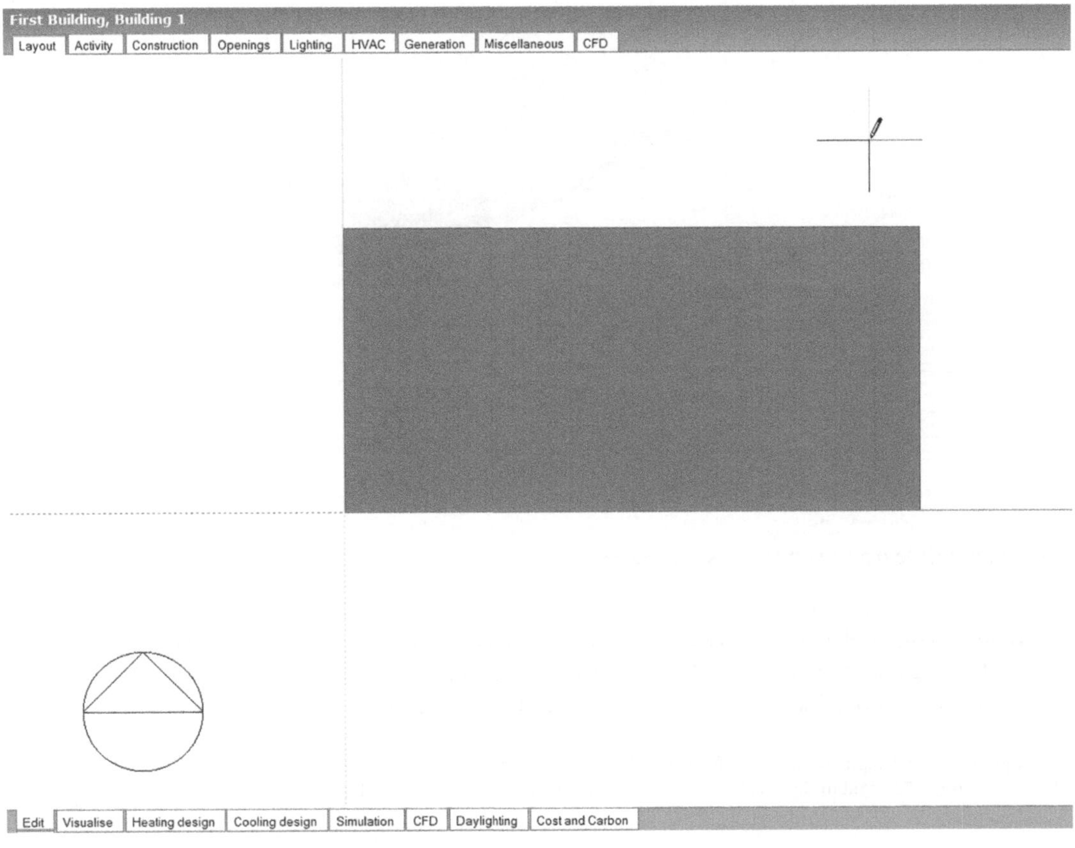

Step 10: Select the **Dynamic orbit** tab on the main menu bar. This shows the block in the orbit mode. Click and drag the mouse in the layout to view the block in the orbit mode.

DESIGNBUILDER SNAP

When drawing perimeter lines, block partition or surface opening direction snaps can be used to draw lines, as follows.

- Parallel with the active drawing plane axes (axis snap)
- Parallel with existing elements (parallel snap)
- Perpendicular to existing elements (normal snap)

Source: https://designbuilder.co.uk/helpv7.3/Content/Direction_snaps1.htm

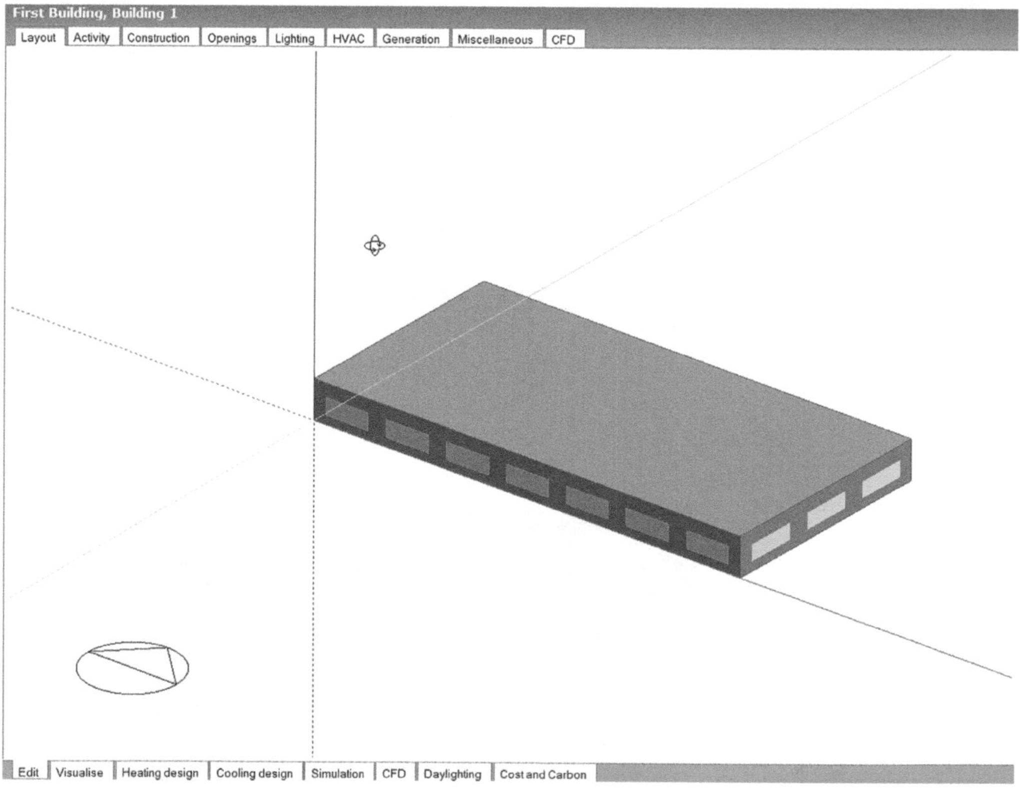

Now you will learn to run the system sizing.

System sizing calculations are carried out to determine the capacity of mechanical cooling and heating equipment required to meet the hottest summer design and coldest winter design weather conditions likely to be encountered at the site location.

Source: https://designbuilder.co.uk/helpv7.3/Content/_Heating_design_simulation.htm
https://designbuilder.co.uk/helpv7.0/Content/_Cooling_design_simulation.htm

Step 11: Select the **Heating design** screen tab at the bottom. Click **Update data**. The **Edit Calculation Options** screen appears.

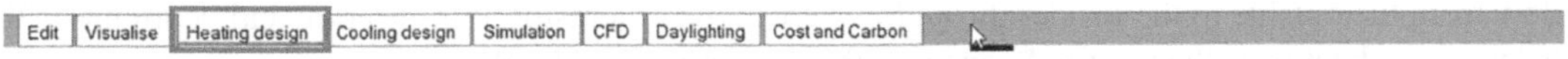

Heating design calculations show the heating system size in the worst conditions (with no internal loads and solar heat gains) for a winter design day. Steady-state simulations assume that the temperature across the envelope does not vary with time.

Worst conditions mean a peak for winter considering no internal gains and solar heat gains.

Source: https://designbuilder.co.uk/helpv7.3/Content/_Heating_design_simulation.htm

Step 12: Select **OK**. The screen appears with design data for the current building in graphical format.

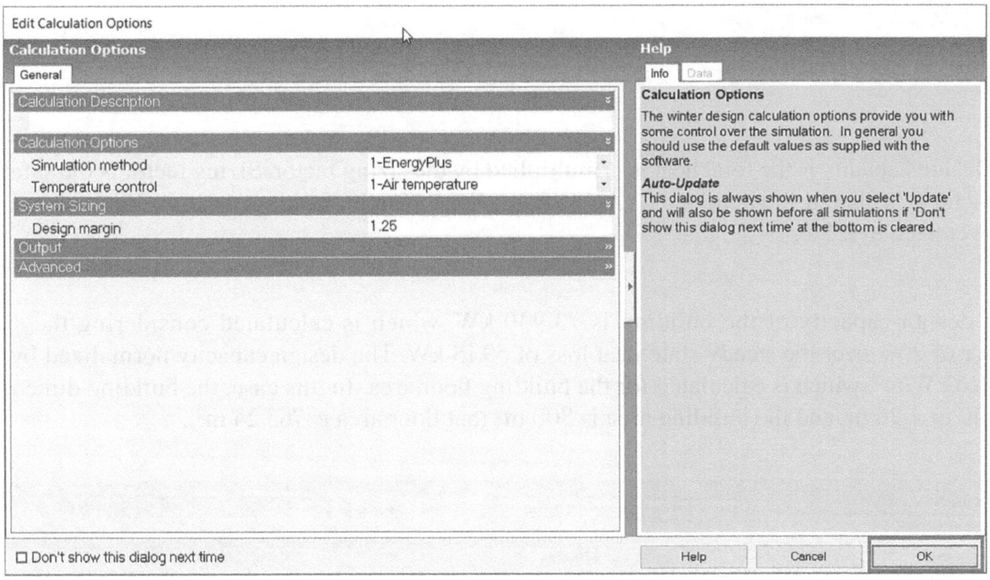

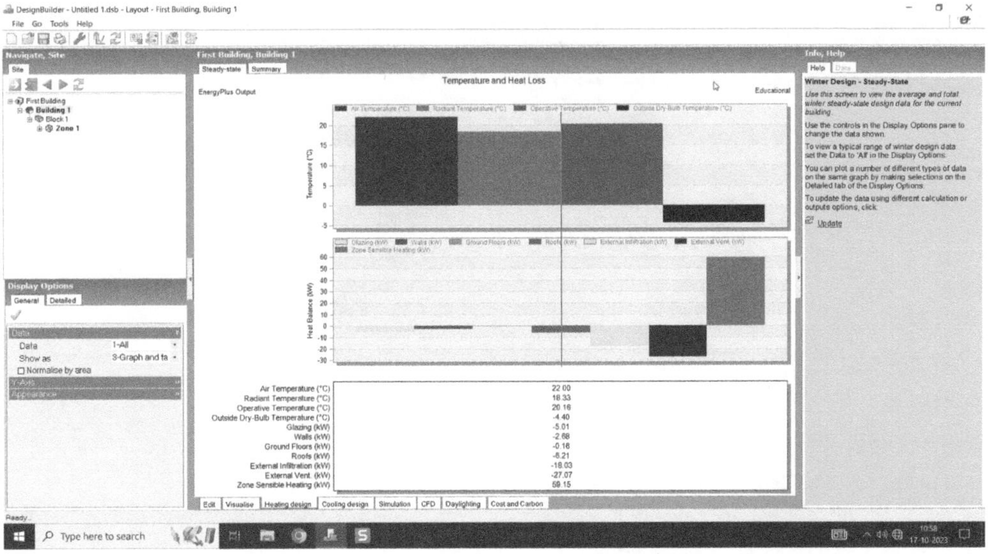

The graph in the preceding figure shows the heat balance to maintain an inside temperature of 22°C when the outdoor dry bulb temperature is –4.4°C. You can see that there is a heat flow from the inside to the outside (shown as negative numbers; the convention is that a positive value is used for a heat flow from the outside to the inside) through the envelope as a result of infiltration and ventilation. To balance this heat loss, the required zone sensible heating is 59.15 kW.

> Steady-state heat loss is the total heat loss from the building. In other words, it is the amount of heat needed to maintain the given indoor comfort temperature.

Step 13: Select the **Summary** tab. The heating design data summary appears.

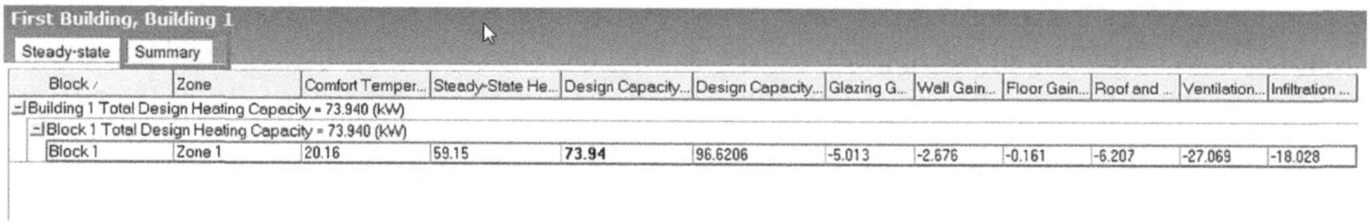
First Building, Building 1
Steady-state | Summary

Block	Zone	Comfort Temper...	Steady-State He...	Design Capacity...	Design Capacity...	Glazing G...	Wall Gain...	Floor Gain...	Roof and ...	Ventilation...	Infiltration ...
Building 1 Total Design Heating Capacity = 73.940 (kW)											
Block 1 Total Design Heating Capacity = 73.940 (kW)											
Block 1	Zone 1	20.16	59.15	**73.94**	96.6206	-5.013	-2.676	-0.161	-6.207	-27.069	-18.028

> Design capacity is the total heat loss multiplied by the sizing factor. Sizing factor is the safety factor considered in sizing an HVAC system. For example, heating systems are commonly oversized by 25%.

The design capacity of the building is 73.940 kW, which is calculated considering the sizing factor of 25% over the steady-state heat loss of 59.15 kW. The design capacity normalized by area is 96.62 W/m², which is calculated for the building floor area. In this case, the building dimensions are 40 m × 20 m, and the building area is 800 m² (net floor area is 765.24 m²).

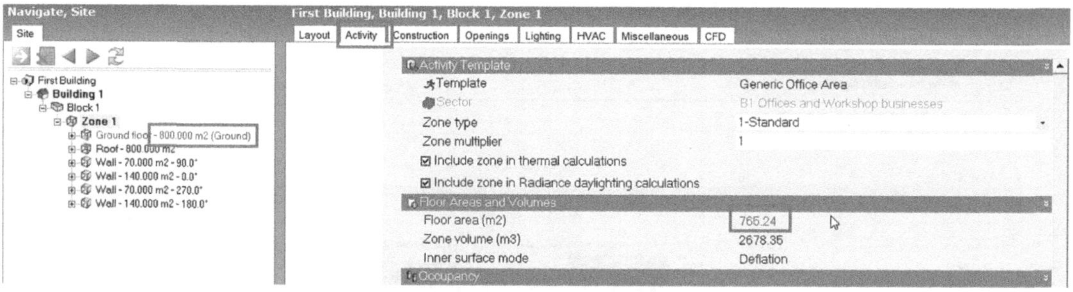

Step 14: Similarly select the **Cooling design** screen tab at the bottom. Click on **Update data**. The **Calculation Options—Building 1** screen appears. Click **OK**. The temperature, heat gains, and other parameters appear in graphical format.

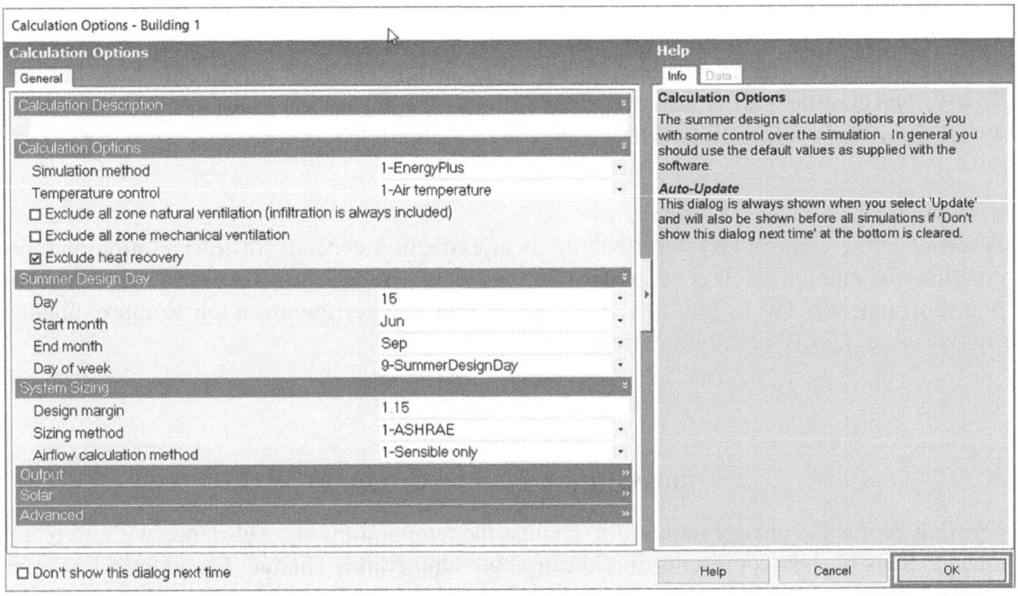

A cooling system is commonly oversized by 15%.

Step 15: Select the **Summary** tab. The screen appears with the cooling design data summary.

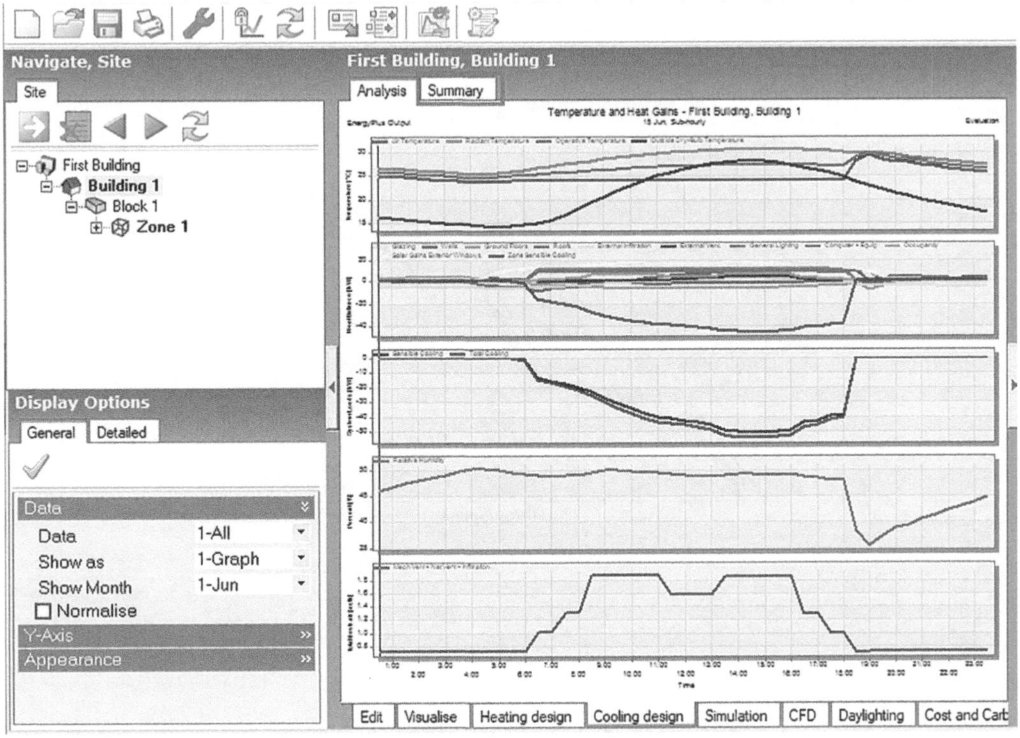

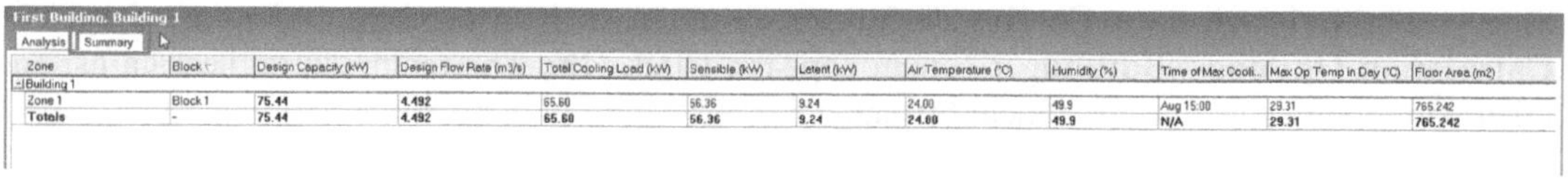

First Building, Building 1

Analysis | Summary

Zone	Block	Design Capacity (kW)	Design Flow Rate (m3/s)	Total Cooling Load (kW)	Sensible (kW)	Latent (kW)	Air Temperature (°C)	Humidity (%)	Time of Max Cooli...	Max Op Temp in Day (°C)	Floor Area (m2)
-Building 1											
Zone 1	Block 1	75.44	4.492	65.60	56.36	9.24	24.00	49.9	Aug 15:00	29.31	765.242
Totals	-	**75.44**	**4.492**	**65.60**	**56.36**	**9.24**	**24.00**	**49.9**	**N/A**	**29.31**	**765.242**

The screen data is provided in tabular format in Table 1.2. The load calculated in the building is 75.44 kW. The cooling load is also described in tons of refrigeration (TR). For this building, it is 21.49 TR. Enter the cooling and heating capacities in Table 1.4.

A *ton of refrigeration* (TR) describes the heat-extraction capacity of refrigeration and air-conditioning equipment. It is defined as the rate of heat transfer that results in the melting of one short ton (907 kg) of pure ice at 0°C in 24 hours. A refrigeration ton is approximately equivalent to 3.5 kW of cooling effect.

SENSIBLE VS. LATENT HEAT

Sensible heat is the energy required to change the temperature of a substance with no phase change. Sensible heat represents only the dry bulb temperature change. *Latent heat*, however, does not affect the temperature of a substance. The heat that causes a change of state with no change in temperature is called *latent heat*. A cooling system should be capable of removing both the sensible and latent heat from a building. Therefore, the total cooling capacity of a system will be (sensible heat load + latent heat load) × sizing factor. In this tutorial, sensible heat is 56.36 kW and latent heat is 9.24 kW. The sum of both sensible and latent heat is 65.6 kW. Therefore, the design capacity calculated by multiplying 65.6 by the sizing factor (1.15) is 75.44 kW.

You can divide this by 3.51 to convert it into the unit of tons of refrigeration: 21.49 TR

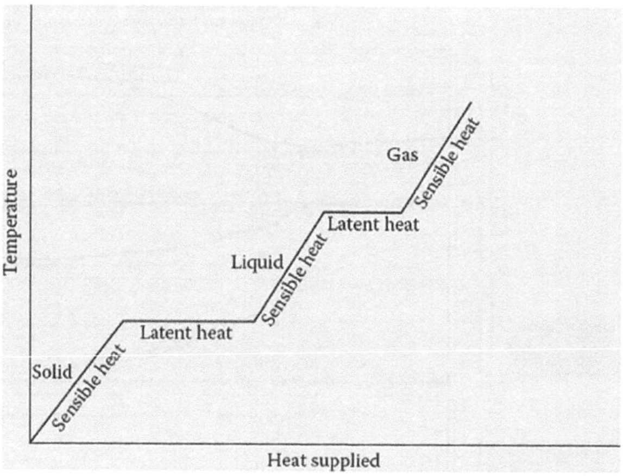

TABLE 1.2
Design Data for the Single-Zone Model

Zone	Block 1: Zone 1
Design capacity (kW)	75.44
Design flow rate (m^3/s)	4.49
Total cooling load (kW)	65.60
Sensible (kW)	56.36
Latent (kW)	9.24
Air temperature (°C)	24.0
Humidity (%)	49.9
Time of max cooling	August 15:00
Maximum operating temperature in day (°C)	29.31
Floor area (m^2)	765.3
Volume (m^3)	2,678.3
Flow/floor area [liters (l)/s-m^2]	5.87
Design cooling load per floor area (W/m^2)	98.588
Outside dry bulb temperature at time of peak cooling load (°C)	30.5

The **Summary** screen, **1-Zones** display option shows design cooling loads and flow rates for each zone plus other related data.

- **Design Capacity** (kW or kBtu/h), shown in bold, is the maximum value of **Total Cooling Load** and **Sensible** multiplied by the **Design margin** to give the design cooling capacity of the equipment. This value is written into the Cooling capacity Model data on the HVAC tab if the appropriate Plant sizing Model Option is set.
- **Design Flow Rate** (m^3/s or ft^3/min), shown in bold, is the supply air flow rate required to deliver the **Sensible** cooling load using the **Design supply temperature** and the zone air temperature at the time of the maximum load multiplied by the **Design margin**.
- **Total Cooling Load** (kW or kBtu/h) is the total maximum **Sensible + Latent** loads for the zone at the time of maximum Sensible Cooling load. If latent load is negative, then it is ignored here and the Total Cooling Load is the same as the Sensible load.
- **Sensible** (kW or kBtu/h) is the maximum Sensible Cooling load for the zone in the Design day.
- **Latent** (kW or kBtu/h) is the latent load for the zone at the time of maximum Sensible load, calculated as (Total Cooling – Sensible Cooling).

Source: https://designbuilder.co.uk/helpv7.3/Content/Summary_Cooling_Design_Table.htm

Step 16: Select the **Simulation** screen tab to perform the annual simulation. Select **Fuel breakdown** from the **Data** drop-down list, **Run period** from the **Interval** drop-down list, and **Grid** from the **Show as** drop-down list.

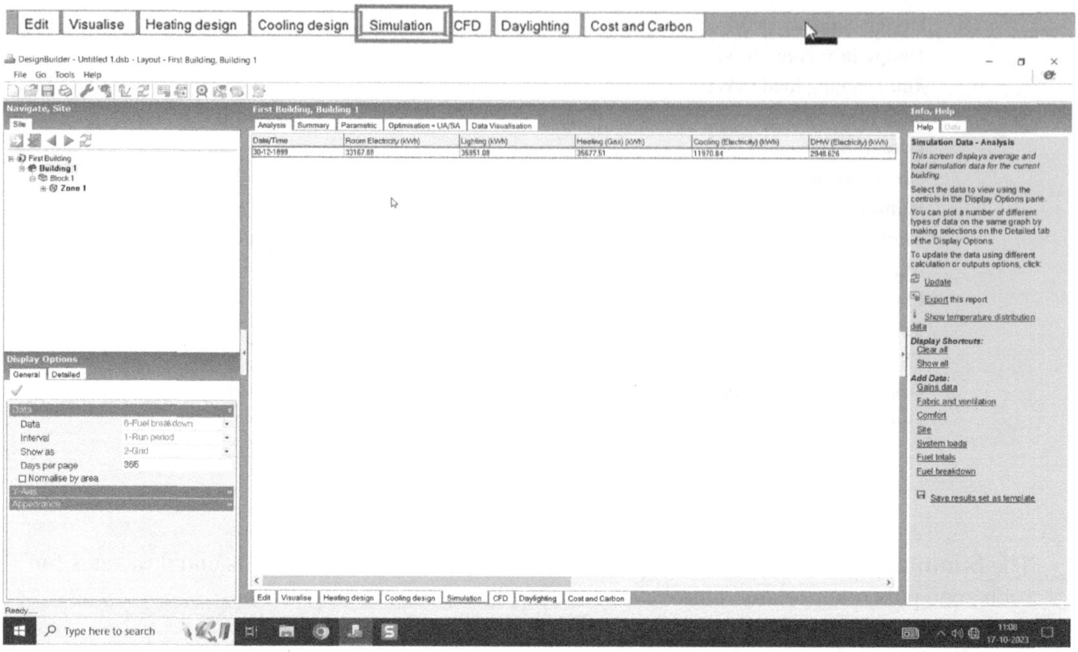

Record the results for use in subsequent tutorials.

Step 17: Select the **Normalise by floor area** checkbox and **All floor area** from the **By** drop-down list to view the EUI of the building. Also, select **Fuel totals** from the **Data** drop-down list. The DesignBuilder message screen appears. Click **Yes** to perform simulations for all floor area. Enter the electricity EUI value in Table 1.3.

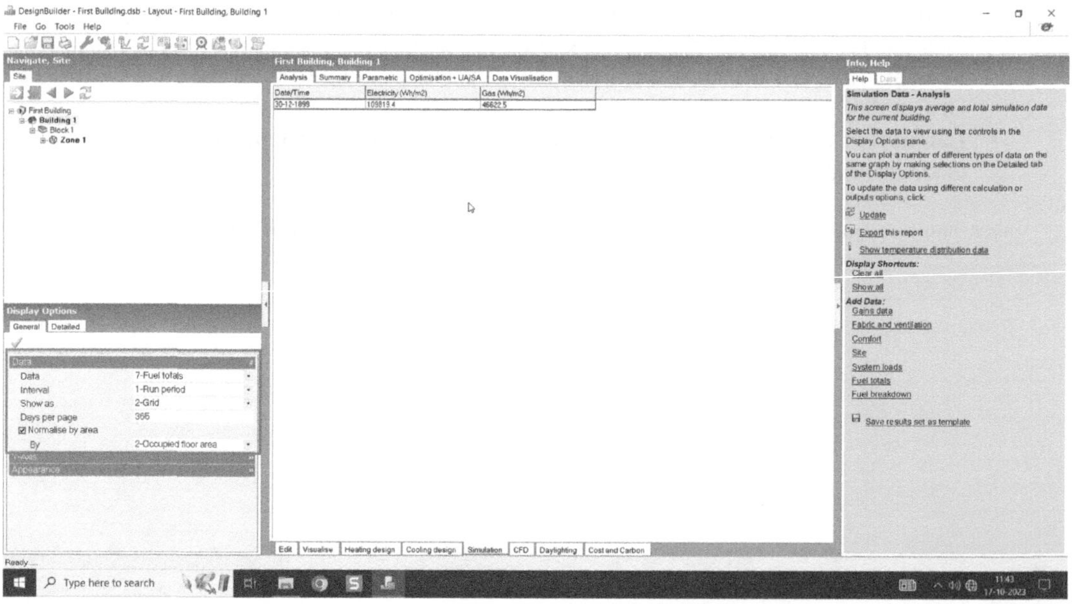

Step 18: Save the model as **First Building** on your desktop. You are going to use this model in forthcoming tutorials.

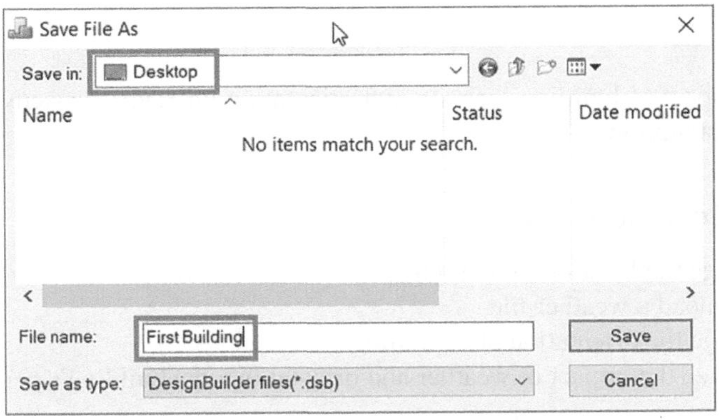

EXERCISE 1.2

Compare the EUI of a large building with that of a small building. Repeat the tutorial to create an 80 m × 40 m rectangular building. Compare the EUI of the larger building with that of the smaller building (the 40 m × 20 m building created in Tutorial 1.2), and enter the result in Table 1.3. Enter and compare the cooling and heating capacities of the larger building in Table 1.4. The method to calculate the EUI (kWh/m² yr) is explained in Tutorial 1.1.

TABLE 1.3
EUI for the Large and Small Buildings

	EUI (kWh/m² Year)	
Energy Use	**Small Building (40 m × 20 m)**	**Large Building (80 m × 40 m)**
Electricity		
Gas		

TABLE 1.4
Cooling and heating capacity for a large and small building

Capacity	**Small Building (40 m × 20 m)**	**Large Building (80 m × 40 m)**
Cooling equipment (kW)		
Heating equipment (kW)		
Cooling per unit area (kW/m²)		
Heating per unit area (kW/m²)		

The following observations can be made as a result of Exercise 1.2.

1. The EUIs of both buildings are different.
2. The cooling and heating equipment capacities are greater for the large building than for the small building. Larger buildings require higher-capacity equipment because of their greater envelope area, occupancy, and internal loads.
3. Cooling/heating equipment capacities per unit area are also different. A large building has a lower capacity per unit area. The main reason behind such a difference is that although most of the internal loads—such as occupancy, lighting, and equipment—vary linearly with the area, the heat gain/loss through the building envelope does not follow the same trend because of the nonlinear variation in the exposed surface area versus the floor or carpet area.

TUTORIAL 1.3: EVALUATING THE IMPACT OF BUILDING LOCATION AND ORIENTATION

GOAL

To evaluate the impact of building location and orientation on HVAC system sizing and building annual energy consumption

WHAT ARE YOU GOING TO LEARN?

- How to change the location of a building
- How to download a weather file
- How to change the orientation of a building
- How to analyze the impact of weather and orientation on a building's performance

PROBLEM STATEMENTS

1. In Tutorial 1.2, a 40 m × 20 m rectangular building was simulated for London. Simulate the same building model for New Delhi/Palam, and compare the HVAC sizing, monthly energy consumption, and annual energy consumption.
2. Rotate the building 90° clockwise, and simulate it for New Delhi. Compare the results of the two models (without rotation and with 90° of rotation).

SOLUTION

Step 1: Open the model saved in **Tutorial 1.2**.

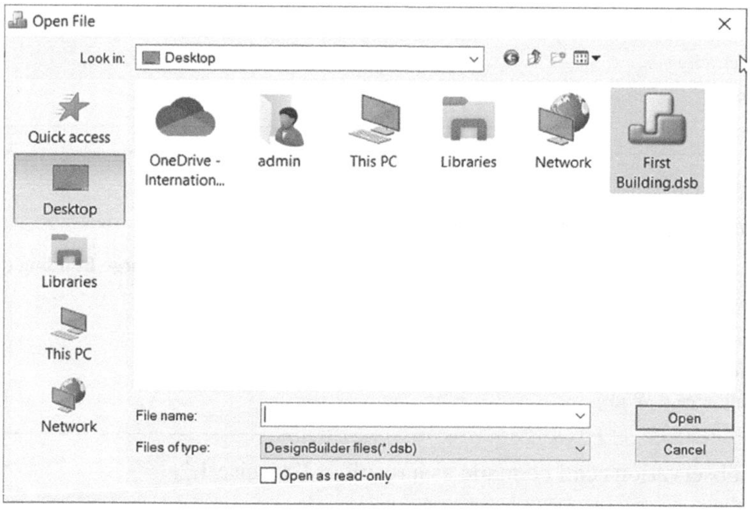

Step 2: Click **First Building**.

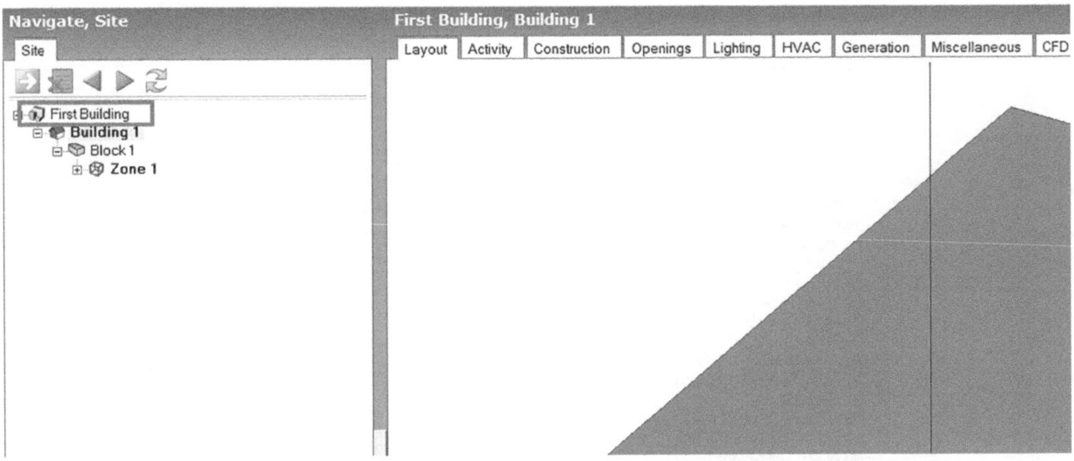

Step 3: Select the **Location** tab. The **Location Template** screen appears.

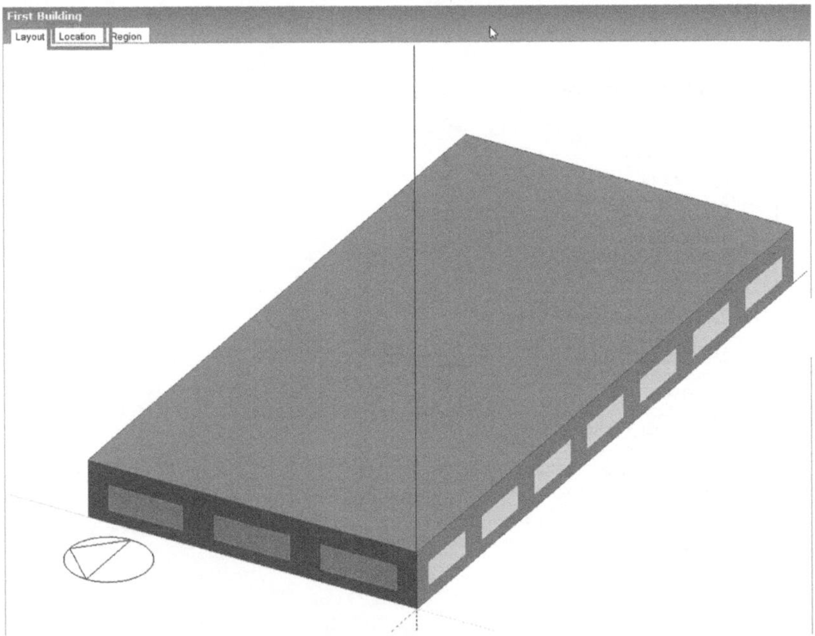

Step 4: Click **LONDON/GATWICK ARPT**. Three dots (. . .) appear. Click the three dots. The **Select the location template** screen appears.

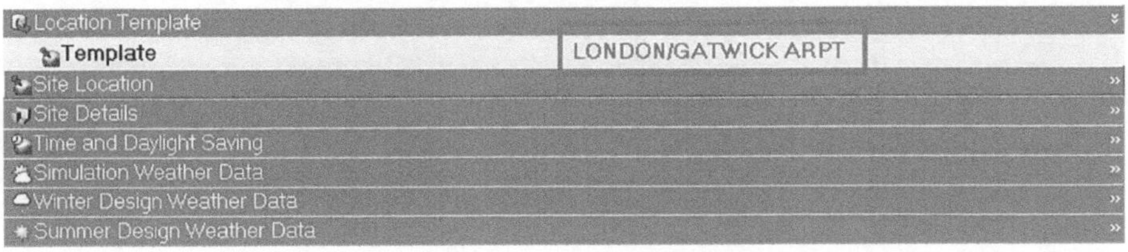

Step 5: Scroll to select **INDIA**. Double-click **INDIA,** and select **NEW DELHI/PALAM**. Click OK.

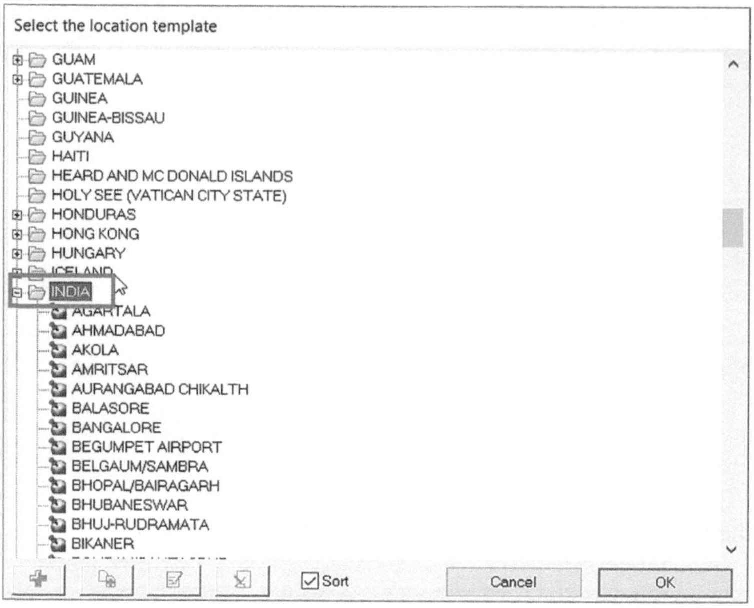

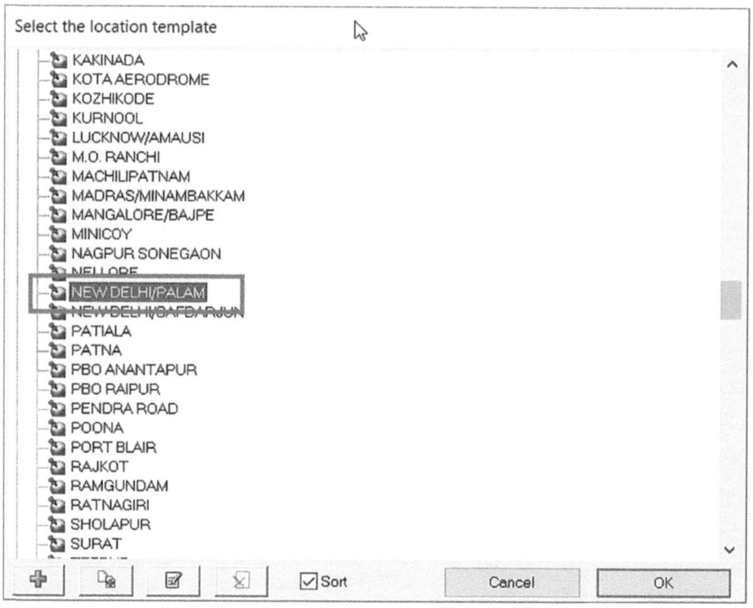

Step 6: Click **Building 1**.

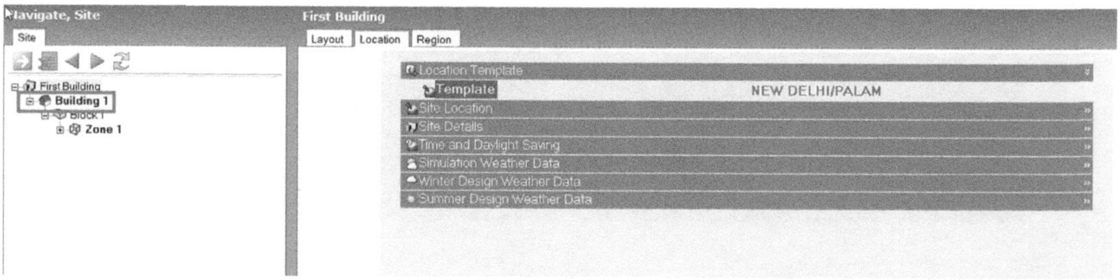

Step 7: Select the **Simulation tab**. Click on **Update data**. The **Edit Calculation Options** screen appears. Click **OK**. If your computer is connected to the internet, DesignBuilder directly downloads the weather file. If, for some reason, DesignBuilder is not able to download the weather file, you can manually download the weather file from https://energyplus.net/weather.

Place the downloaded weather file in the folder C:\ProgramData\DesignBuilder\WeatherData.

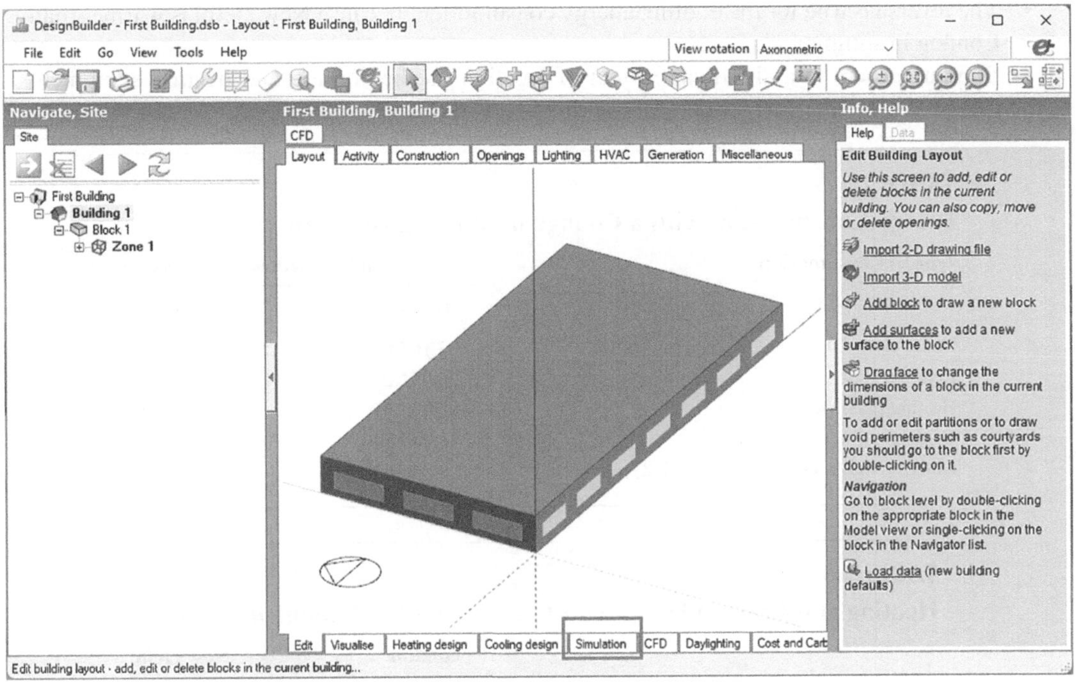

DesignBuilder uses hourly weather files with the .epw extension (weather files can be downloaded from the following link: https://energyplus.net/weather). Weather files that are downloaded directly from the website should be copied in the DesignBuilder weather data folder located at C:\ProgramData\DesignBuilder\Weather Data.

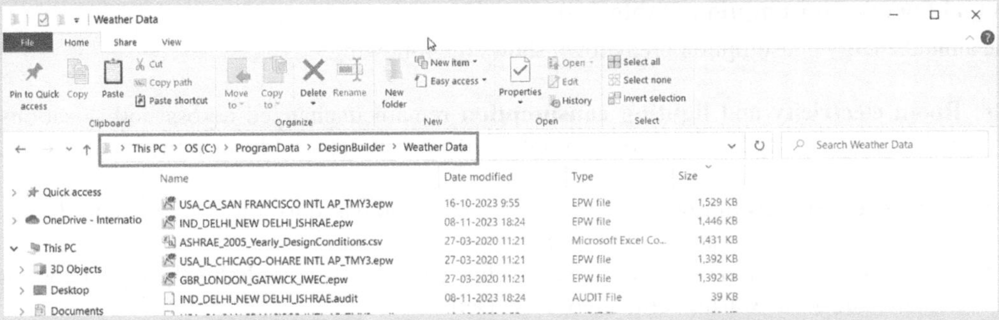

Step 8: Compare the simulation results for the London and New Delhi locations. Results for the London location can be obtained from Tutorial 1.2.

The impact of location on building energy consumption can be seen from the results reported in Table 1.5. You can see that there is no change in room electricity and lighting. This is because there was no change in internal loads; further, as there is no daylight-based control, the change in location does not change these values. From the results in Table 1.5 and Table 1.6, the following observations can be made.

- The heating energy consumption is higher in London than in New Delhi, because London has a colder climate than New Delhi
- The reverse is true for the cooling energy consumption, because New Delhi is warmer than London in summer
- Similarly, there is a significant effect of location on the cooling and heating equipment capacities

TABLE 1.5
Energy Consumption with a Change in Building Location

Type of Consumption	Annual Fuel Breakdown Data (kWh)	
	London	New Delhi
Room electricity	33,167.88	33,167.88
Lighting	35,951.08	35,951.08
Heating (gas)	35,677.51	1,680.90
Cooling	11,970.84	96,711.51
Domestic hot water (DHW, electricity)	2,948.63	2,948.63

TABLE 1.6
Heating and Cooling Equipment Capacity with a Change in Location

	London	New Delhi
Capacity of cooling equipment (kW)	75.44	115.47
Capacity of heating equipment (kW)	73.94	44.72

Results and Analysis

The impact of building location on HVAC system sizing and annual energy consumption is presented in Table 1.5 and Table 1.6.

Annual Energy Consumption (Table 1.5)

The annual energy consumption breakdown shows the following.

- **Room electricity** and **lighting consumption** remain unchanged across both locations. This is expected as these factors are influenced more by internal loads rather than external climate conditions.
- **Heating consumption (gas)** decreases in London compared to New Delhi due to the significantly warmer climate of New Delhi.
- **Cooling consumption** increases in New Delhi. This increase highlights the higher cooling demand due to the warmer climate of New Delhi.

HVAC Sizing (Table 1.6)

The HVAC sizing breakdown shows the following.

- **Cooling equipment capacity** increases in New Delhi, reflecting the much higher cooling demand required to maintain indoor comfort in a warmer climate

- **Heating equipment capacity**, conversely, is lower in New Delhi because the coldest days are not as severe, resulting in a significantly reduced peak heating demand compared to London

Analysis

The results demonstrate that building location significantly impacts HVAC system sizing and energy consumption. Relocating the building from London to New Delhi leads to a decrease in heating requirements, while increasing cooling requirements considerably. This change is due to the contrasting climate conditions of the two locations.

The following steps show how to change the orientation.

Step 9: Ensure that you are at the **First Building** level in the **Navigation** panel. Select the Location tab (refer to Steps 2 and 3 of this tutorial), and click **Site Details**. Enter **90** in the **Site Orientation** text box. Select the **Layout** tab. You can observe the change in the north arrow direction before and after changing the site orientation.

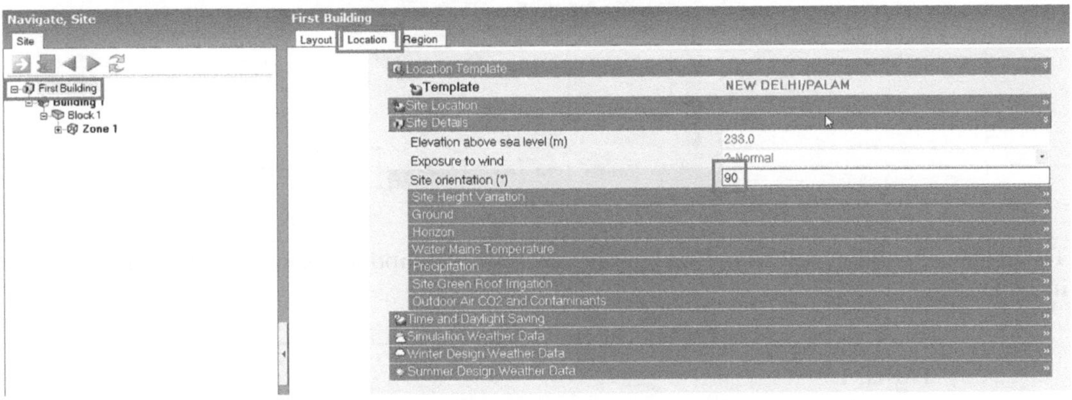

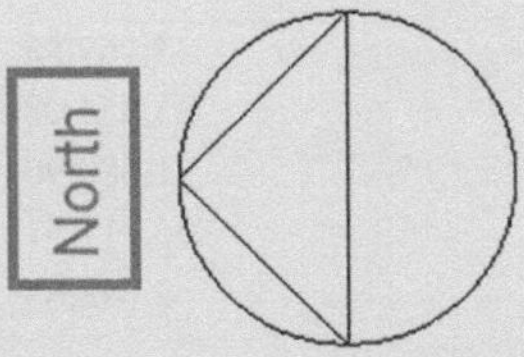

North is indicated by the direction of the north arrow in the sketch plan view. Working at the "untitled" level (here First Building) of the tree means working at the site level. For example, if there are multiple buildings on a site that are to be modeled in DesignBuilder, after creating one building, return to the site level in the tree to start creating the next building. Therefore, all site-specific information such as location and orientation can be assigned to the model only when the site level is selected in the navigation tree.

Site orientation (°) in DesignBuilder represents the alignment of the building(s) with respect to true north. However, because there could be multiple buildings on the site, this is called *site orientation* instead of *building orientation*.

Step 10: Click **Building 1** then select the **Simulation** screen tab to get the results.

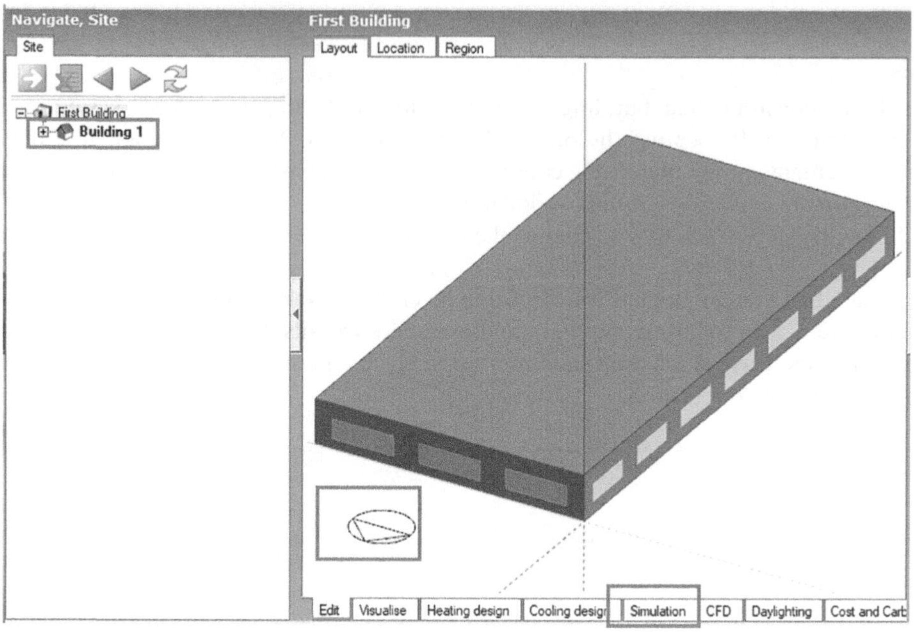

The impact of building orientation on building energy consumption is shown in the results recorded in Table 1.7.

TABLE 1.7
Annual Energy Consumption with a Change in Building Orientation

Type of Consumption	0°	90°
Room electricity (kWh)	33,167.88	33,167.88
Lighting (kWh)	35,951.08	35,951.08
Heating (gas) (kWh)	1,680.90	1,843.63
Cooling (electricity) (kWh)	96,711.51	98,434.28
DHW (electricity) (kWh)	2,948.63	2,948.63

Changing the orientation of a building changes the heating and cooling energy consumption, as well as the size of the system. You may note that heating capacity is not affected by the change in orientation because of the fact that the peak requirement for heating occurs in non-sunshine hours. Hence, any orientation would require the same amount of heating (Table 1.8).

TABLE 1.8
Heating and Cooling Equipment Capacity with a Change in Building Orientation

Type of equipment	0°	90°
Cooling (kW)	115.47	121.73
Heating (kW)	44.72	44.72

Impact of Building Orientation (Table 1.7 and Table 1.8)

The impact of changing the building's orientation by 90° clockwise is evaluated for New Delhi, and the results are summarized as follows.

- **Room electricity, lighting, and DHW consumption** remain unchanged, as these components are unaffected by the change in orientation
- **Heating energy consumption** slightly increases from 1,680.90 kWh (0°) to 1,843.63 kWh (90°), due to minor changes in the building's thermal exposure, but is generally not a significant increase
- **Cooling energy consumption** increases from 96,711.51 kWh (0°) to 98,434.28 kWh (90°), indicating that the building orientation adjustment has a noticeable effect on cooling demand

Equipment Sizing

- **Cooling equipment capacity** increases from 115.47 kW (0°) to 121.73 kW (90°), which indicates that the building's cooling requirement rises when it is rotated by 90°
- **Heating equipment capacity** remains the same at 44.72 kW, as the peak heating requirement typically occurs during non-sunshine hours, making orientation irrelevant

Analysis of Orientation Impact

The results show that changing the building's orientation has a more noticeable impact on cooling energy consumption than on heating. This is because cooling loads are heavily influenced by solar gain, which is directly affected by the building's orientation. Heating loads, however, are generally unaffected, since peak heating demands occur during nighttime or overcast conditions.

The findings highlight the importance of considering building orientation during simulation to ensure optimal cooling system sizing and accurate estimation of cooling energy consumption.

EXERCISE 1.3A

For New Delhi, run simulations for three more orientations: 45, 180 and 270°. Compare the results (Table 1.9 and Table 1.10).

TABLE 1.9
Energy Consumption with a Change in Building Orientation

	Annual Fuel Breakdown Data (kWh)				
Type of Consumption	**0°**	**45°**	**90°**	**180°**	**270°**
Room electricity					
Lighting					
Heating (gas)					
Cooling (electricity)					
DHW (electricity)					

TABLE 1.10
Heating and Cooling Equipment Capacity with a Change in Building Orientation

Type of equipment	0°	45°	90°	180°	270°
Heating capacity (kW)					
Cooling capacity (kW)					

Compare the energy consumption for 0° and 180°. What do you observe? It can be seen that the results for 0° and 180° are the same, and those for 90° and 270° are the same. This is due to symmetry in the shape and window distribution of the building such that solar exposure for these two cases becomes similar. A comparison of the energy consumption for 0° and 90° shows a difference in the cooling capacity, as well as energy consumption owing to the change in solar exposure.

EXERCISE 1.3B

Change the weather locations of the model to New York and Singapore and enter the values in Table 1.11 and Table 1.12.

TABLE 1.11
Energy Consumption with a Change in Building Location

	Annual Fuel Breakdown Data (kWh)	
Type of Consumption	**New York**	**Singapore**
Room electricity		
Lighting		
Heating (gas)		
Cooling (electricity)		
DHW (electricity)		

TABLE 1.12
Heating and Cooling Equipment Capacity with a Change in Building Location

Type of Equipment	New York	Singapore
Heating (kW)		
Cooling (kW)		

Observe and write why energy consumption and system sizing changed. It can be seen that the extent of variation is significantly less for a change in orientation versus a change in location. The change in energy consumption due to orientation may be more pronounced in buildings that have a higher aspect ratio (the ratio of longer and shorter sides) because of the change in solar exposure of the building that influences heat gain.

The change in energy consumption with location is largely due to the change in the harshness of the climatic conditions. This can be proportional to the *cooling degree days* (CDDs) for cooling energy consumption and *heating degree days* (HDDs) for heating energy consumption. Values for CDDs and HDDs can be found in weather files, design data books, and references such as American Society of Heating, Refrigerating and Air-Conditioning Engineers (ASHRAE) handbooks.

HDDs and **CDDs** are common measures used to interpret the heating and cooling needs of a location. Degree days are the summation of the product of the difference in temperature (ΔT) between the average outdoor and the hypothetical average indoor temperatures and the number of days the outdoor temperature is above or below the hypothetical average indoor temperature.

The degree days are calculated in reference to a baseline temperature that is commonly 18°C. Temperatures above 18°C need to be cooled, and temperatures below 18°C need to be heated.

Example to calculate CDDs for two consecutive days:

Baseline temperature: 18°C
Daily average outdoor dry bulb temperature (DBT) on day 1: 20°C
Daily average outdoor DBT on day 2: 17°C
CDD on day 1: 20 – 18 = 2
CDD on day 2: 17 – 18 = –1
Total CDD: 2 (sum for days that have positive values only)

The unit of measurement for CDDs and HDDs is degree days.

Reference: Mayank Bhatnagar, Jyotirmay Mathur, Vishal Garg (2018) Determining base temperature for heating and cooling degree days for India, Journal of Building Engineering.

TUTORIAL 1.4: EVALUATING THE IMPACT OF OPAQUE ENVELOPE COMPONENTS

Goal

To understand the impact of the thermal properties of opaque envelope components (the external wall and roof) on HVAC system sizing and energy consumption

What Are You Going to Learn?

- How to change wall and roof construction

Problem Statement

For the **Courtyard with VAV Example**, change the external wall from **Best practice Wall, Medium weight** to **Brickwork single leaf construction dense plaster,** and record its impact on the building energy performance for the London weather file.

Best practice Wall, Medium weight wall is an insulated wall, and its cross section is as follows.

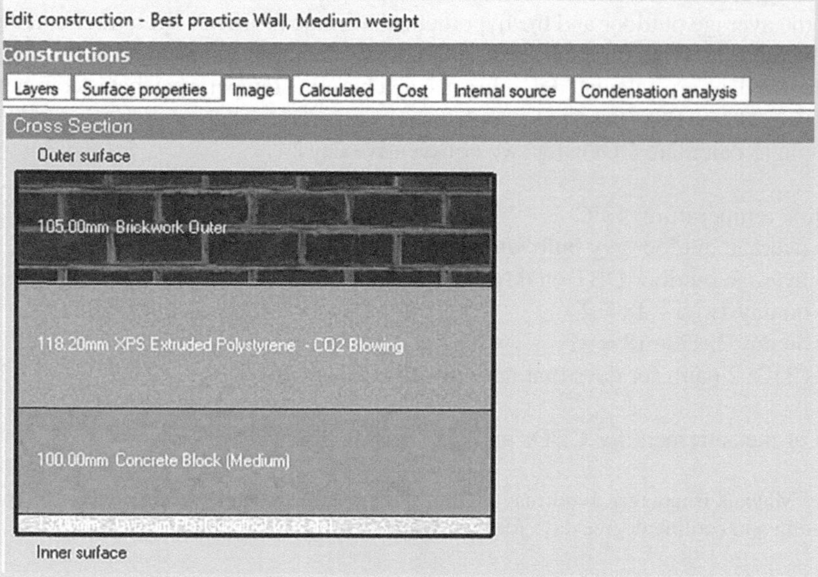

Brickwork single leaf construction dense plaster is an uninsulated wall, and its cross section is as follows.

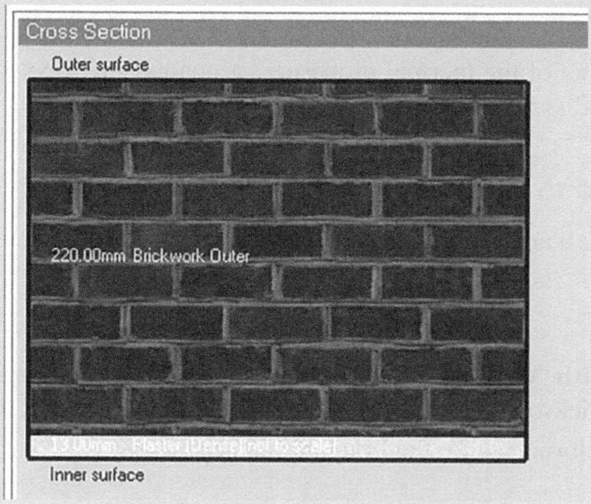

SOLUTION

Step 1: Open the **Courtyard with VAV Example** model from the template.

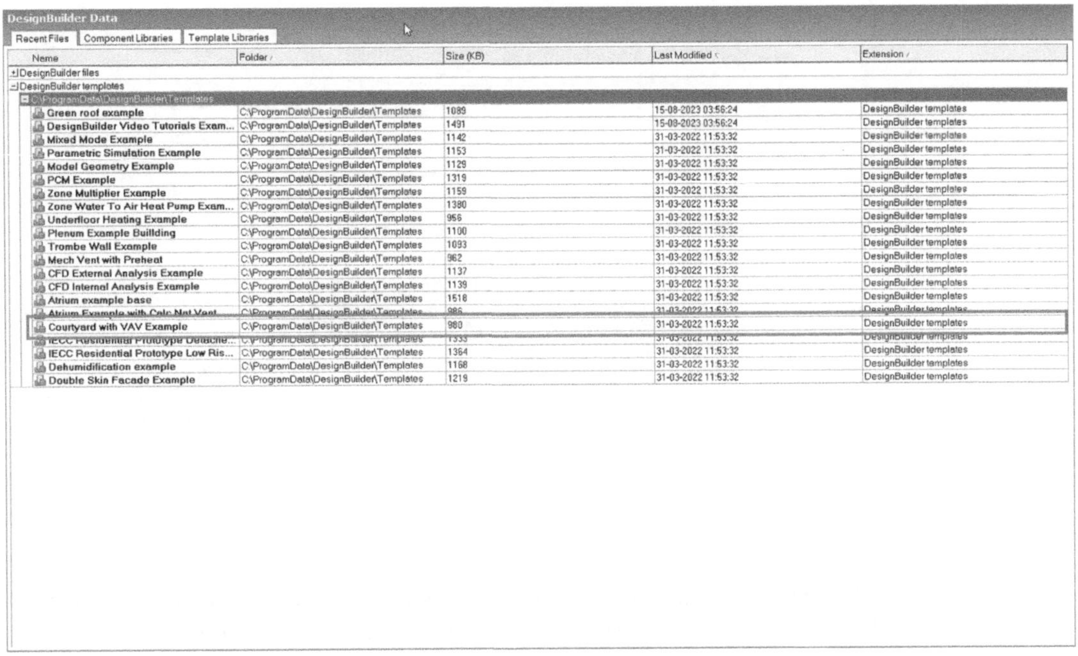

Step 2: Select the **Construction** tab.

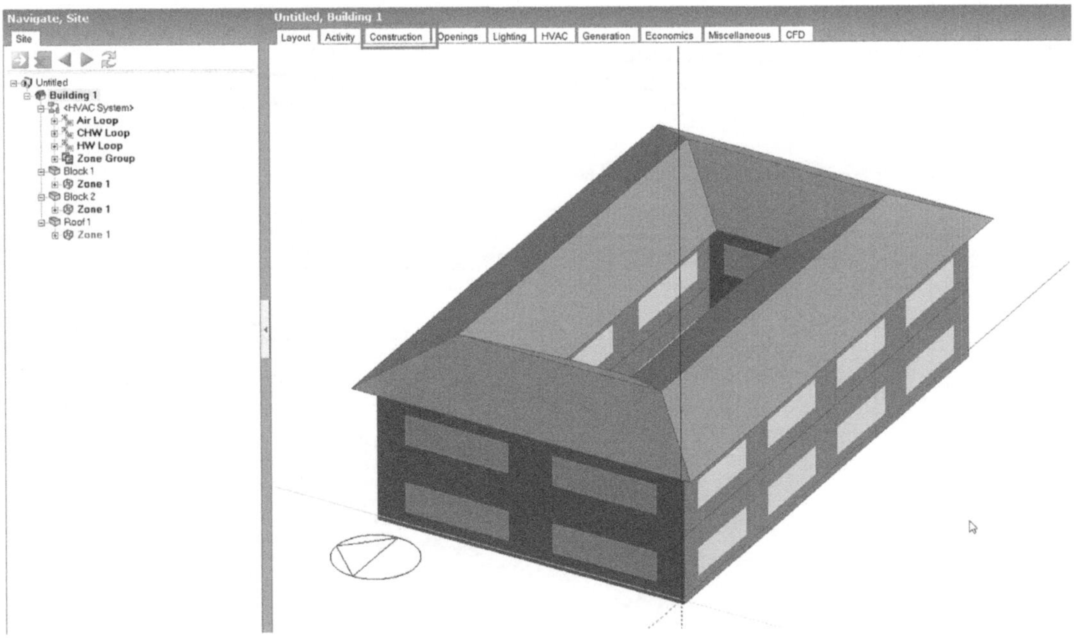

Step 3: Make sure that **Best practice Wall, Medium weight** is selected as External walls.

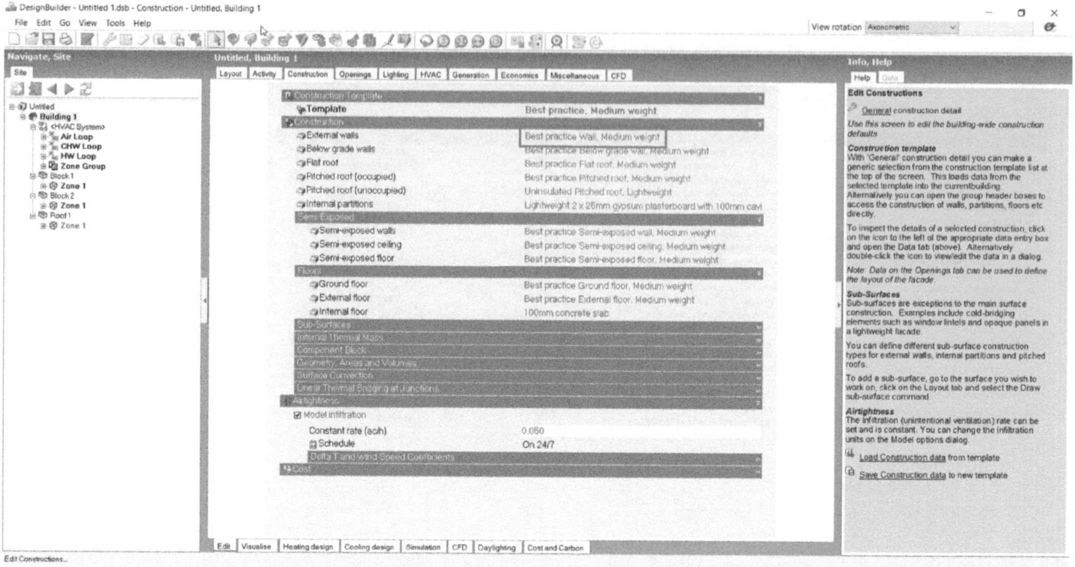

Step 4: Select the **Simulation** screen tab. Click on **Update data**. The **Edit Calculation Options** screen appears. Click the **Annual simulation** link to set the annual simulation period.

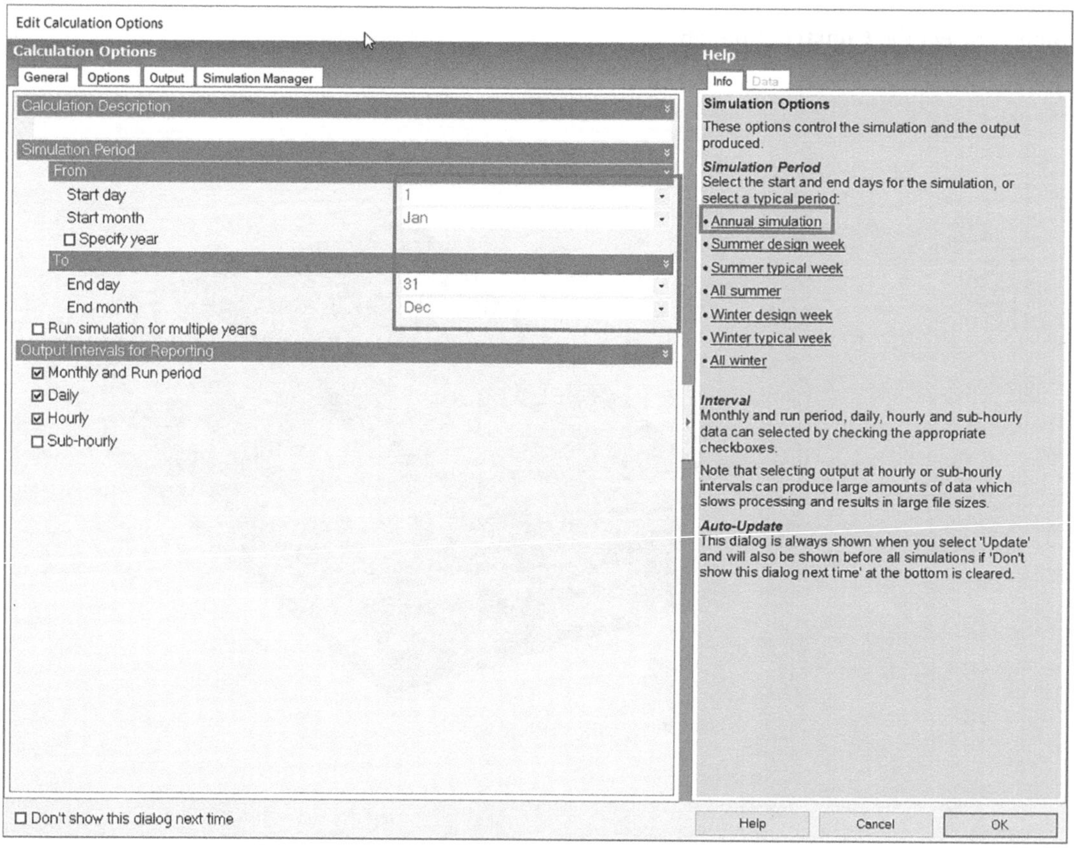

Step 5: Record the energy simulation results.

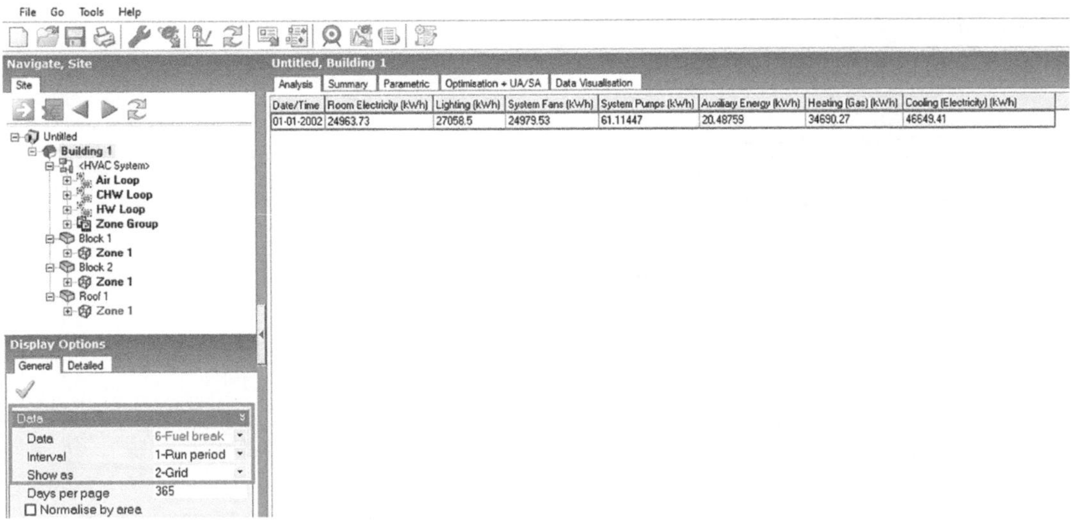

In the following steps, you are going to change the external wall construction.

Step 6: Click **the Edit** screen**. The Construction Template** screen appears.

Step 7: Click **External walls** under the **Construction** section. Three dots (. . .) appear. Click the three dots. The **Select the construction** screen appears.

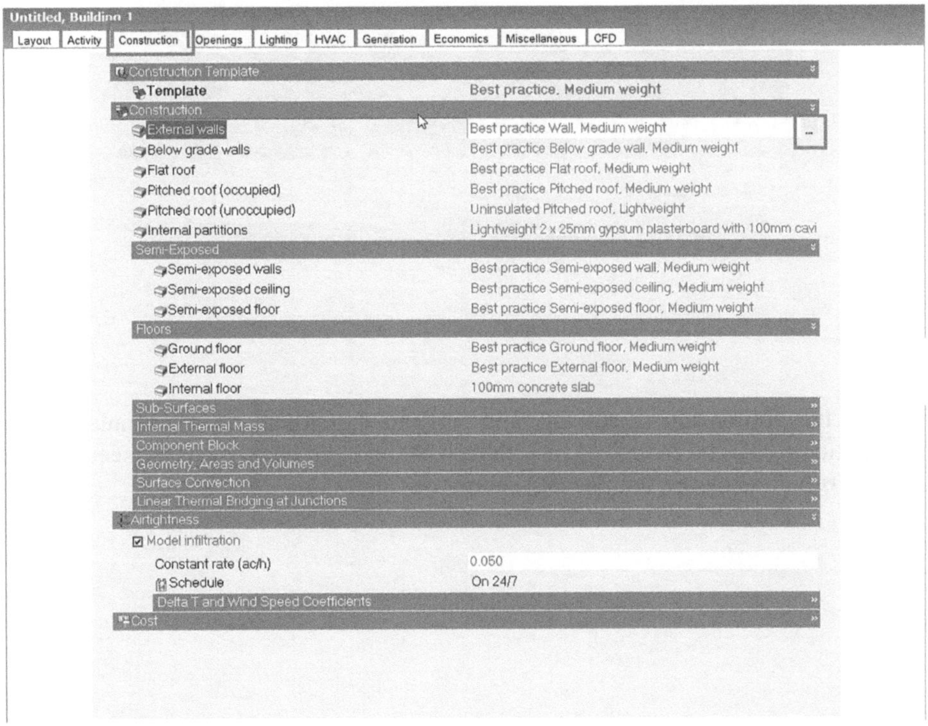

Step 8: Select **Brickwork single leaf construction dense plaster**.

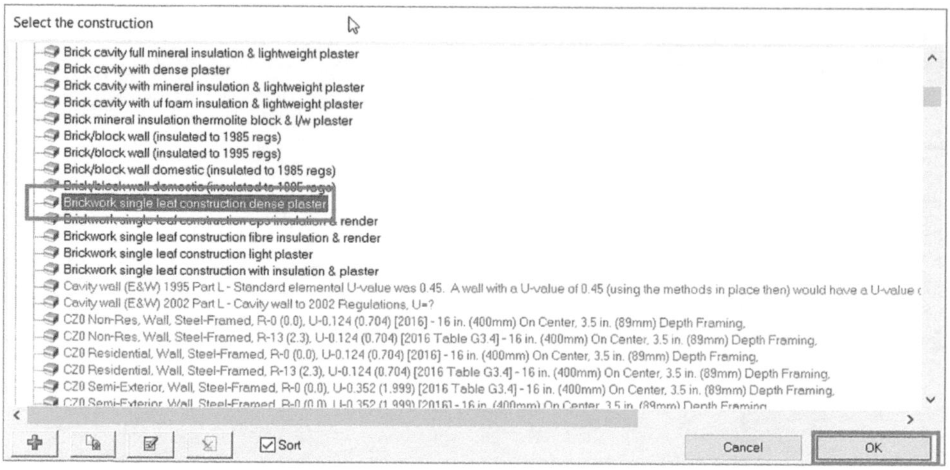

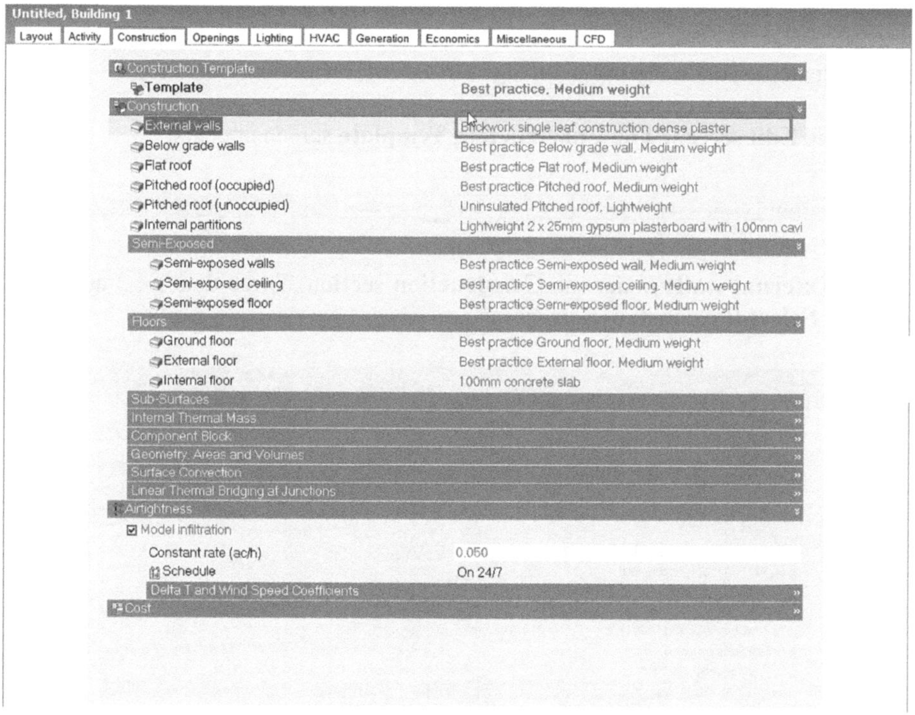

Step 9: Select the **Simulation** screen tab, and perform the annual energy simulation. View the energy simulation results in Table 1.13 and Table 1.14, and compare. It can be seen that changing the wall impacts both equipment sizing and energy consumption.

TABLE 1.13
Energy Consumption with a Change in External Wall Construction

	Annual Fuel Breakdown Data (kWh)	
Type of Consumption	**Best practice Wall, Medium weight**	**Brickwork single leaf construction dense plaster**
Room electricity	24964	24,964
Lighting	27059	27,059
System fans	24980	25,685
System pumps	61	62
Auxiliary energy	20	25
Heating (gas)	34690	72,266
Cooling (electricity)	46649	44,281

Heating Energy Consumption

- **Uninsulated Building:** In the absence of insulation, heat escapes easily through the building envelope. During cold weather, the heating system must work significantly harder to maintain comfortable indoor temperatures, resulting in higher heating energy consumption.
- **Insulated Building:** Insulation acts as a barrier, reducing heat transfer between the interior and exterior of the building. As a result, heat loss is minimized, and the heating system requires much less energy to keep the indoor environment warm and comfortable.

Cooling Energy Consumption

- **Uninsulated Building:** Without insulation, heat from the external environment can also penetrate the building more easily. However, in London's climate, which is predominantly cool and mild, the effect of external heat gain is less significant. In fact, the lack of insulation allows the building to lose some of the indoor heat to the cooler outdoor environment, slightly reducing cooling energy consumption.
- **Insulated Building:** Insulation also slows down heat gains during warmer periods. However, given London's generally cool climate, the cooling demand is typically low to begin with. The slight reduction in cooling energy for the uninsulated building is mainly due to the ability of heat to escape, rather than accumulate inside the building.

TABLE 1.14
Heating and Cooling Equipment Capacity with a Change in External Wall Construction

Type of Capacity	Best practice Wall, Medium weight	Brickwork single leaf construction dense plaster
Cooling (kW)	70.38	71.64
Heating (kW)	55.04	89.95

EXERCISE 1.4

Open the model prepared in Tutorial 1.2. Assign flat roof construction as shown in Table 1.15 and Table 1.16. Simulate and compare the results for London.

TABLE 1.15
Energy Consumption with a Change in Roof Construction

	Annual Fuel Breakdown Data (kWh)	
Type of Consumption	**Roof, Insulated Entirely above Deck, R-24 (4.2), U–0.040 (0.23)**	**Uninsulated Flat Roof, Medium Weight**
Room electricity		
Lighting		
System fans		
System pumps		
Heating (electricity)		
Heating (gas)		
Cooling (electricity)		

TABLE 1.16
Heating and Cooling Sizing with a Change in Roof Construction

	Heating and Cooling Sizing (kW)	
Type	**Roof, Insulated Entirely above Deck, R-24 (4.2), U–0.040 (0.23)**	**Uninsulated Flat Roof, Medium Weight**
Heating		
Cooling		

You can refer to the following screenshots for help.

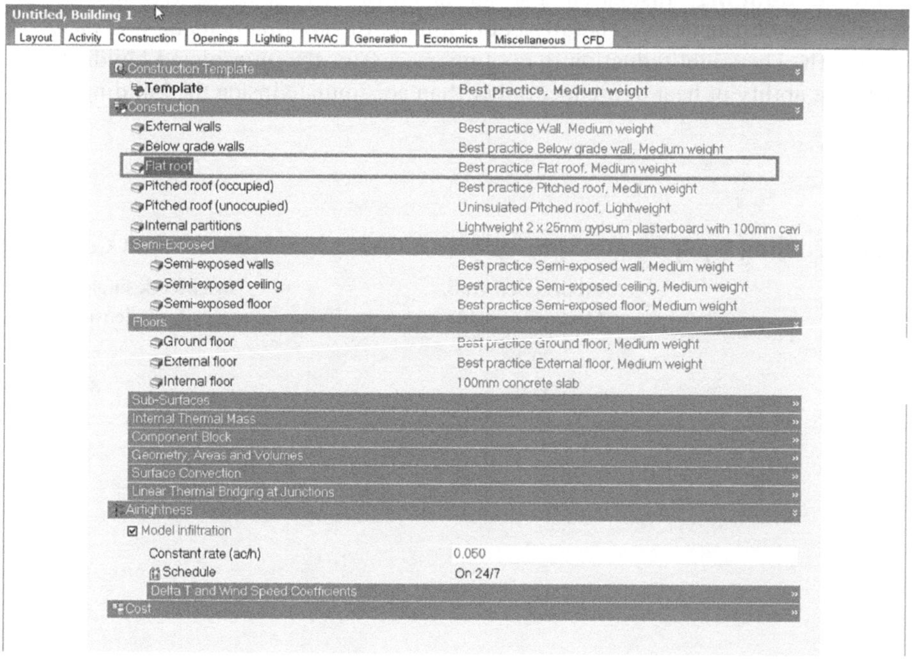

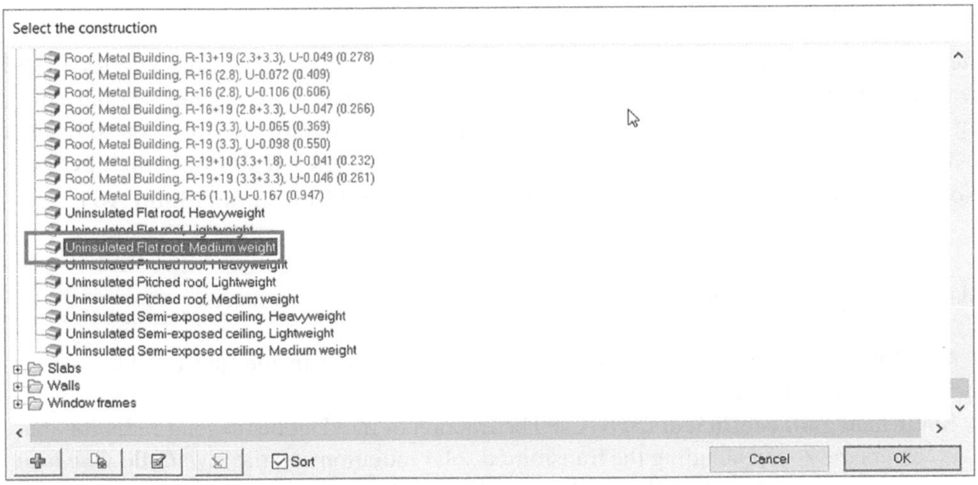

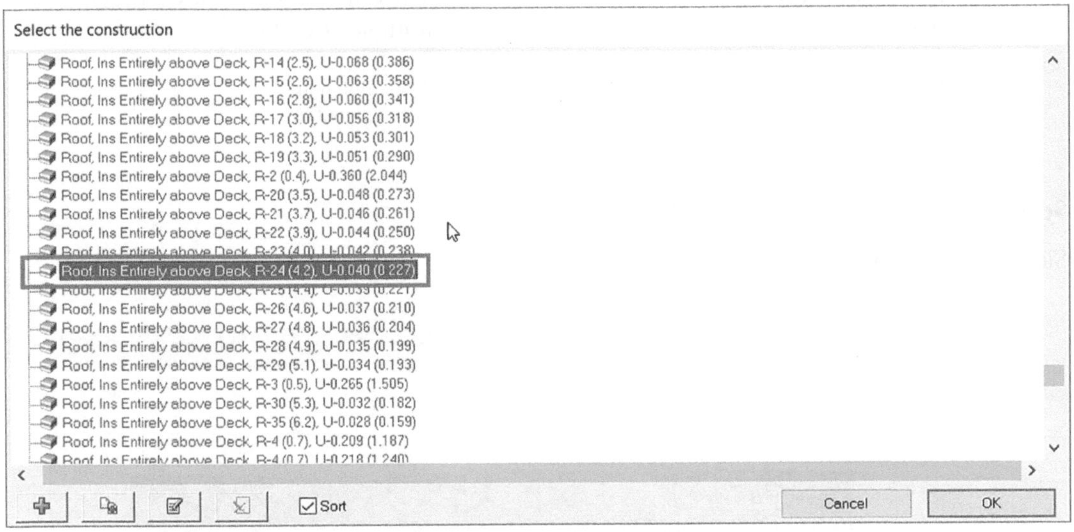

TUTORIAL 1.5: EVALUATING THE IMPACT OF WWR AND GLASS TYPE

Goal

To evaluate the impact of glazing area and glazing properties on the energy performance of a building

In the field of building energy simulations, glazing area is most commonly quantified in terms of *window-to-wall ratio* (WWR). The WWR is the ratio of the total glazing area to the total external wall area in conditioned zones.

What Are You Going to Learn?

- How to create zones in a building
- How to change the WWR
- How to change glazing type

Problem Statement

Create a 60 m × 40 m five-zone single-story building model for London. Take WWRs of 30% and 80%, and evaluate the building energy performance. Compare cooling and heating equipment sizing of the north and south zones. Similarly, compare sizing in east and west zones. Evaluate the impact of glazing by changing the single-pane clear glazing (**Sgl Clr 3mm**) to high-performance, low solar heat gain coefficient (SHGC) dual-pane glazing (**Dbl Blue 6mm/13mm Air**).

Glazing properties are mainly defined by the following three parameters.

U-value: The overall heat transfer coefficient, including inside and outside surface convective and radiant resistances

Solar heat gain coefficient (SHGC): The fraction of incident beam solar radiation that enters the zone, including the transmitted solar radiation and the inward flowing heat from the solar radiation that is absorbed by the glazing

Visible transmittance (VT): A measure of how much visible light passes through a window or glass, also known as visible light transmittance (VLT)

Sources: https://designbuilder.co.uk/helpv7.3/Content/SimpleGlazingDefinition.htm
https://designbuilder.co.uk/helpv7.3/Content/_Glazing_calculated_data.htm

Solution

Step 1: Open a new project, and draw a building with a size of 60 m × 40 m.

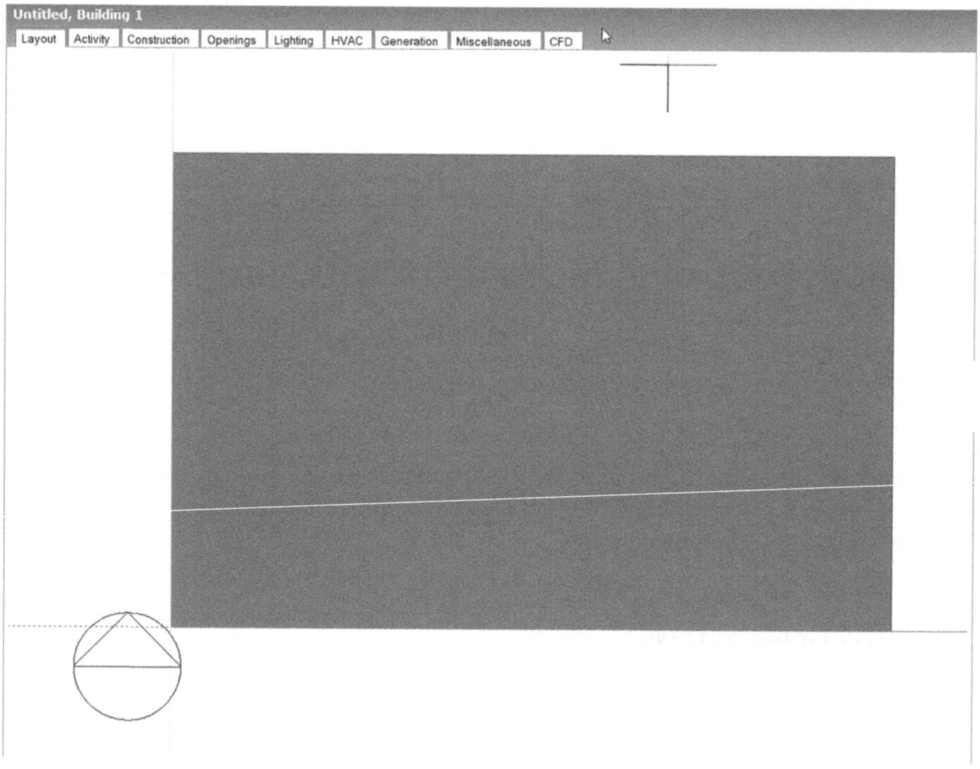

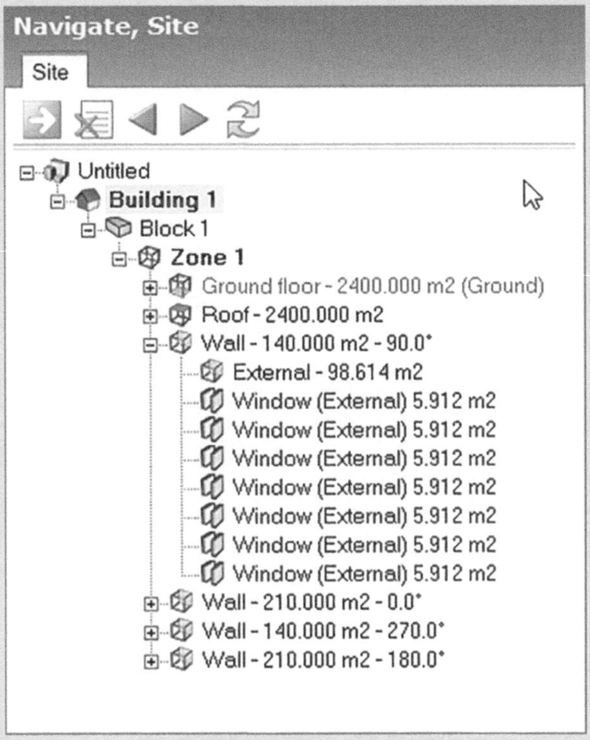

DATA STRUCTURE HIERARCHY IN DESIGNBUILDER

Site → Building → Block → Zone → Surface → Opening

DesignBuilder models are organized in a simple hierarchy. Default data is inherited from the level above in the hierarchy, so block data is inherited from building level, zone data is inherited from block data, and surface data is inherited from zone data. This arrangement allows you to make settings at building level which can becomes active throughout the whole building or make settings at block level to change data for all zones/surfaces in the block.

Assigning a template to the upper-level branch will result in the same template for all the sub-branches. For example, if a construction template named "Framed Construction" is assigned to the building, all floors and zones will have the same template: "Framed Construction". When a particular sub-branch is edited, it is separated from the main branch, resulting in a need for a separate edit from the main branch.

For example, if a particular zone (such as zone 1) of the first floor of the building is selected and the construction template is changed from "Framed Construction" to "Mass Construction", and later the whole building is selected and the construction template is changed from "Framed Construction" to "Steel Construction", the construction template of zone 1 still retains "Mass Construction".

Source: https://designbuilder.co.uk/helpv7.0/#_Model_data_hierarchy_and_data_inheritance.htm

Step 2: Click **Block 1** in navigation tree. The display changes to **Zone 1**.

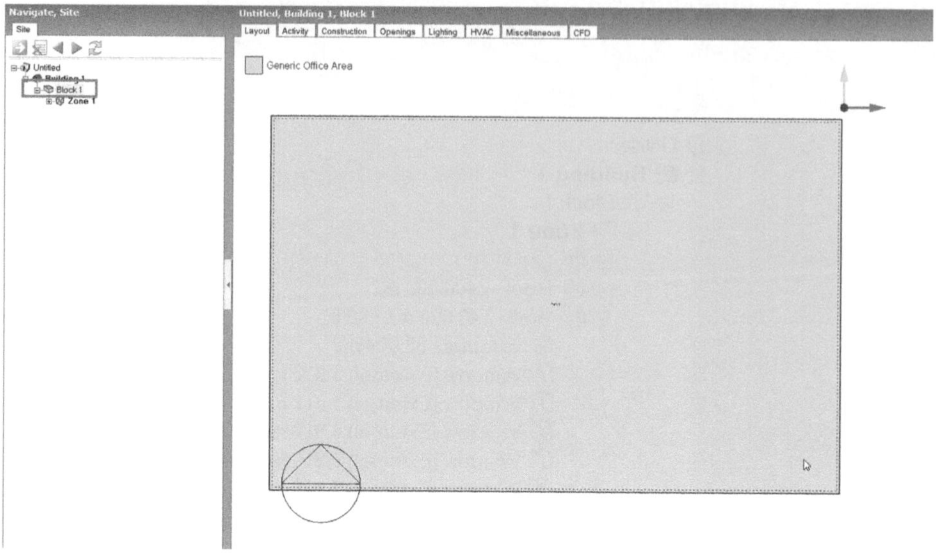

Step 3: Click the **Draw construction line** icon to mark lines on the zone. You need to draw an internal partition with an offset of 10 m.

Similar to other drafting programs, DesignBuilder provides important drafting options such as snap points, gridlines, and direction snap. Make sure to select the relevant options while creating the custom geometry.

Step 4: Place the pointer near the top-left corner. The pointer snaps to the corner (green square appears at the center of the crosshairs). Left-click once the pointer has snapped, and then move the mouse in the right direction over the north wall of the building and type **10**. This draws a 10 m construction line from the top-left corner on the north wall.

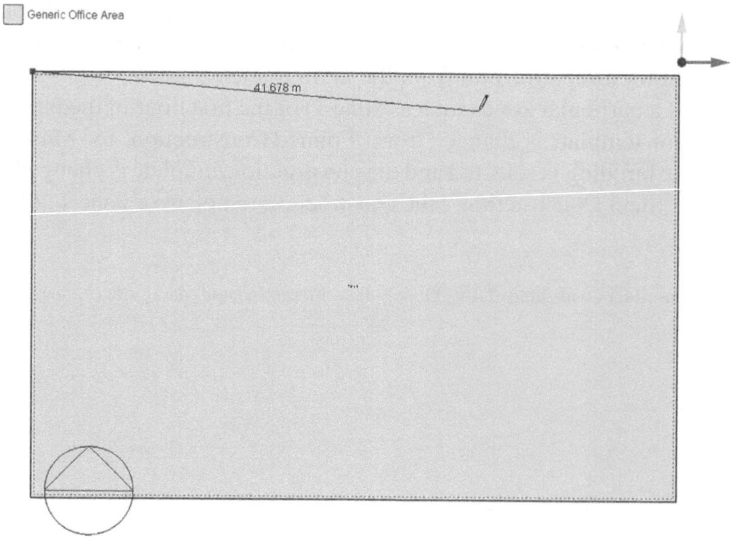

Step 5: Click the right end of the construction line. Now move the mouse vertically down parallel to the west wall of the building with an offset of 10 m (as achieved in the preceding step) and snap to the south wall (this time the crosshairs will change to the red square because it will be edge snapping) then click.

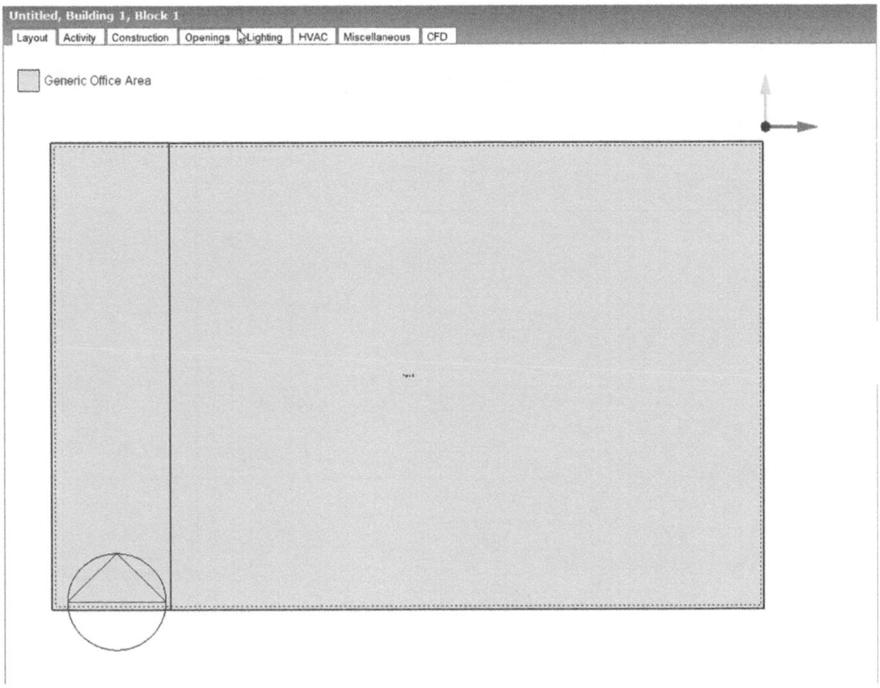

Step 6: Continue drawing all construction lines to mark core and perimeter zones.

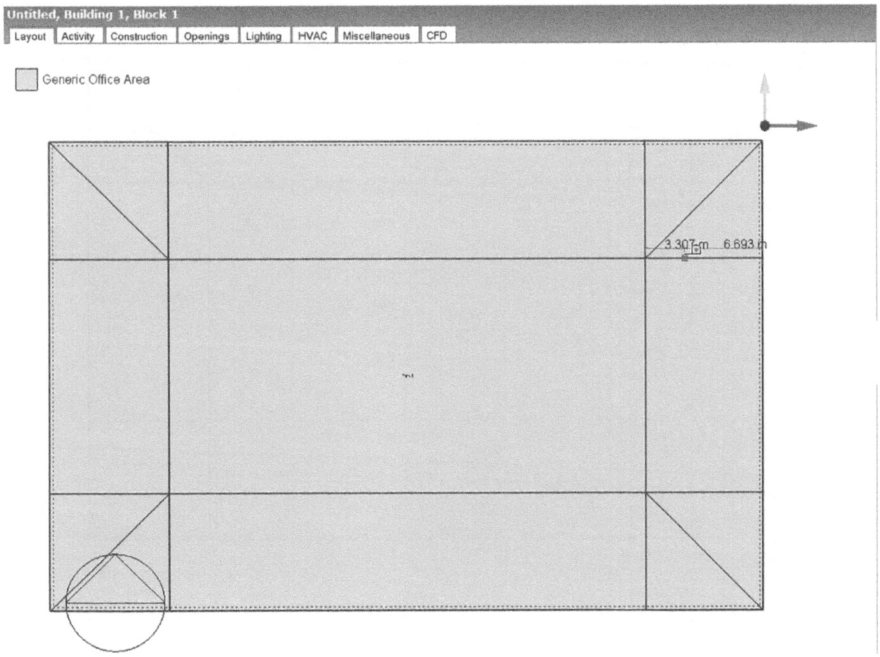

Step 7: Click the **Draw partition** icon. The screen appears with a cursor to draw the internal partitions.

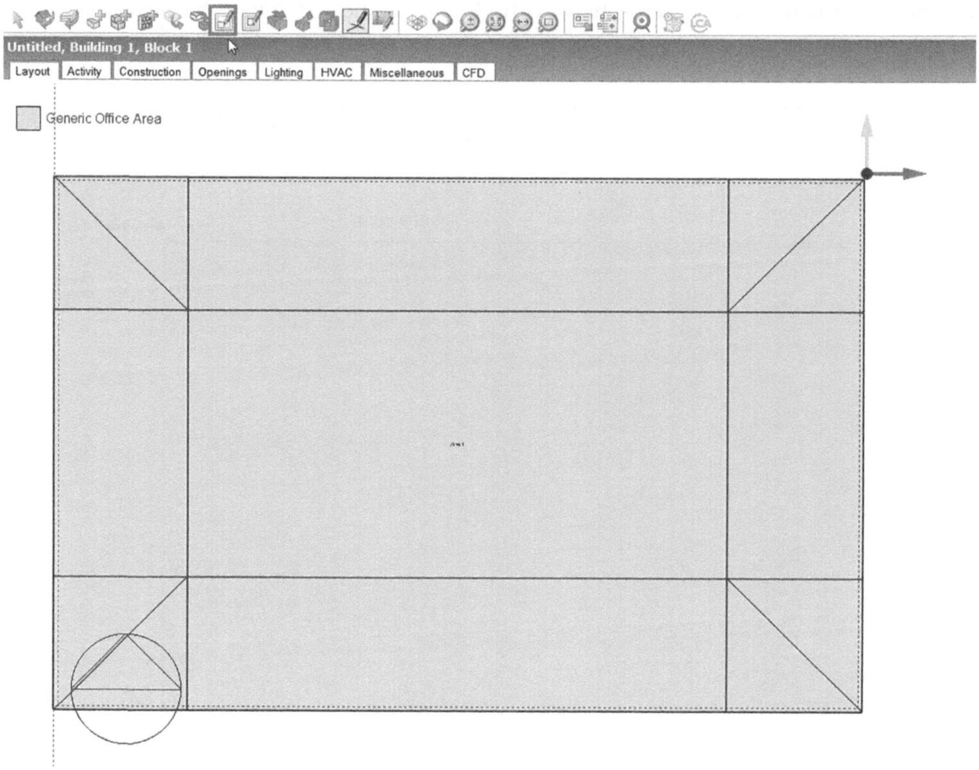

Step 8: Trace the construction line to draw a perimeter zone on the north.

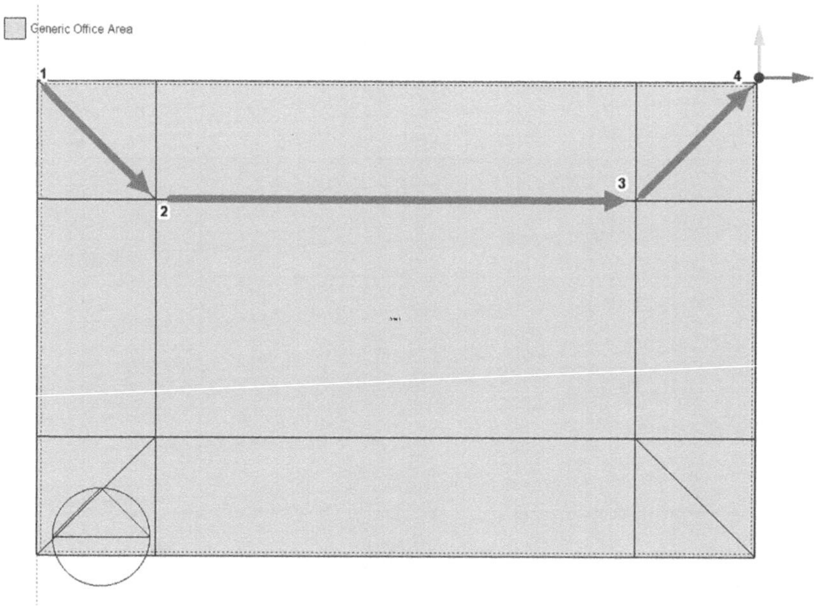

Step 9: Repeat Step 8 to draw the other three zones. This step results in a total of five zones, four perimeters and one core, as shown below with **Axonometric** view. (*Note:* There is no need to redraw the partition drawn in the preceding steps.)

Perimeter and core zoning is common in building energy simulations, especially when the internal layout of the building that is being modeled is not designed. Perimeter and core zoning is also practiced while modeling large open floor plans.

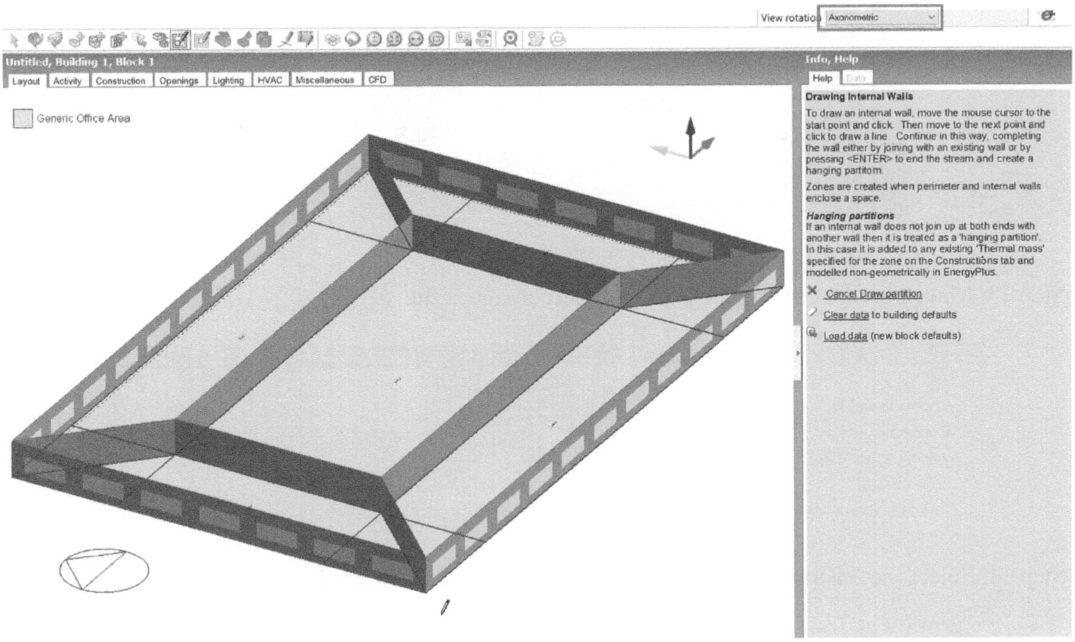

Step 10: Select and rename each zone based on its orientation. (For renaming, you need to single-click the zone name and wait for about one second and then click again. A text box appears to enter a new name.)

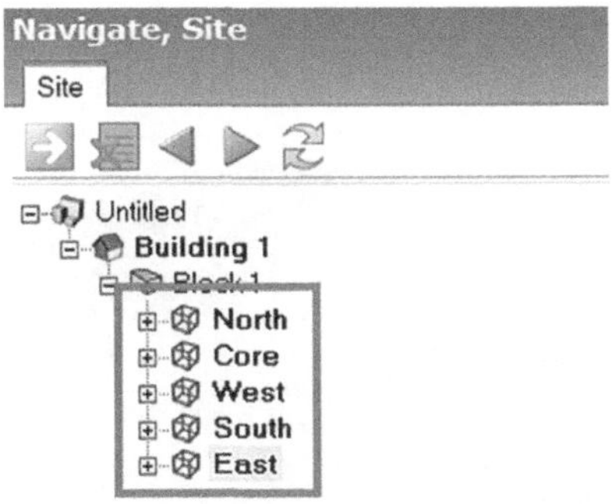

Step 11: Click **Building 1** in the navigation tree. Now you are going to set the WWR for the whole building.

Step 12: Select the **Openings** tab. The **Glazing Template** screen appears.

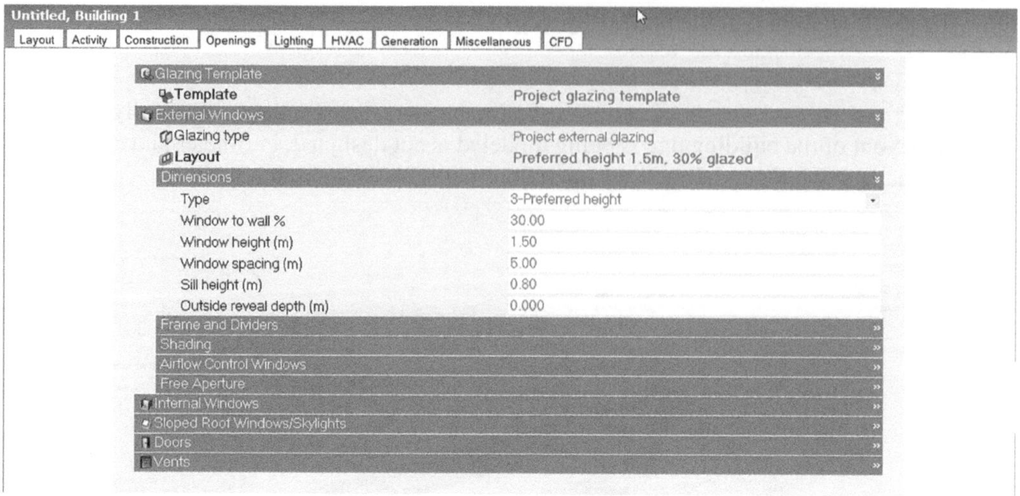

Step 13: Make sure that the window-to-wall percentage is **30**.

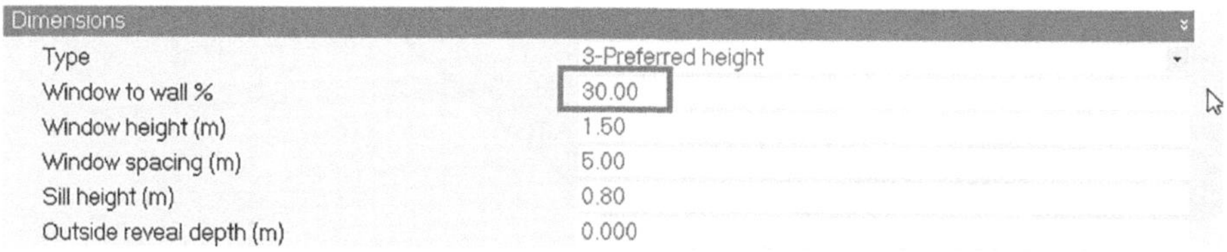

Step 14: Select the **Visualise** screen tab. Use the orbit tool for three-dimensional (3D) visuals.

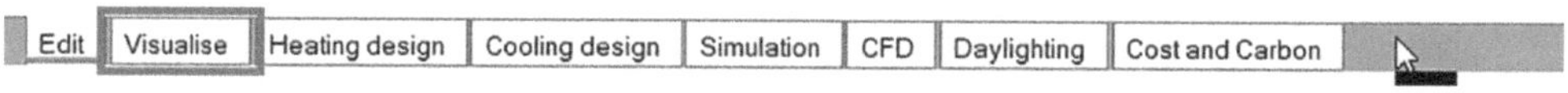

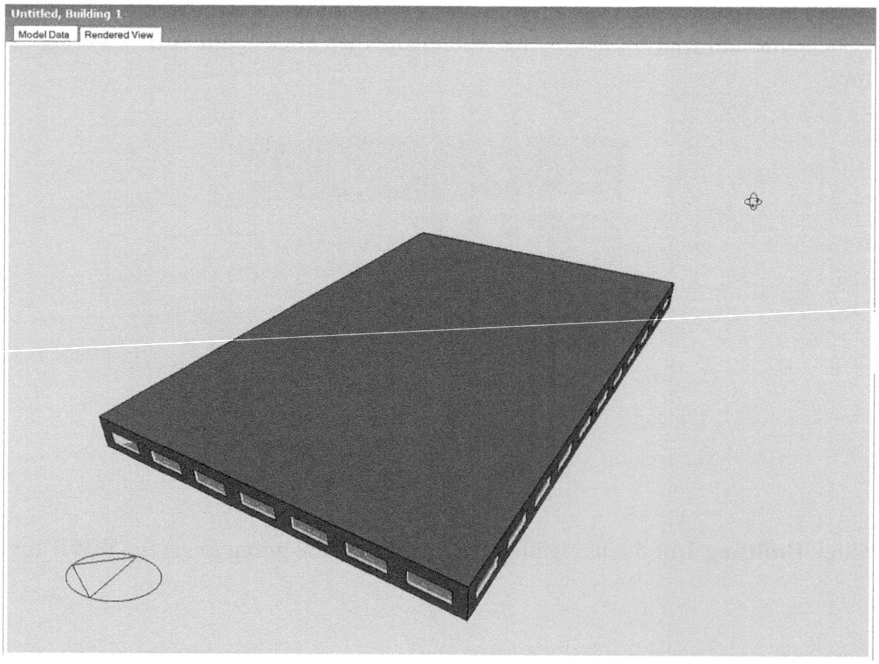

Step 15: Select the **Heating design** screen tab. Click on **Update data**. The **Calculation Options** screen appears.

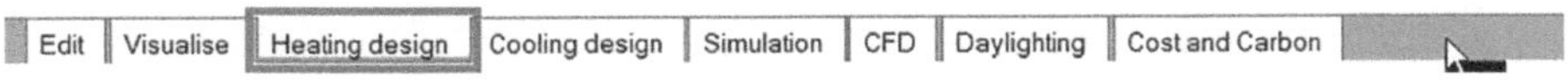

Step 16: Ensure that **Air temperature** is selected in the **Temperature control** drop-down list. Click **OK**.

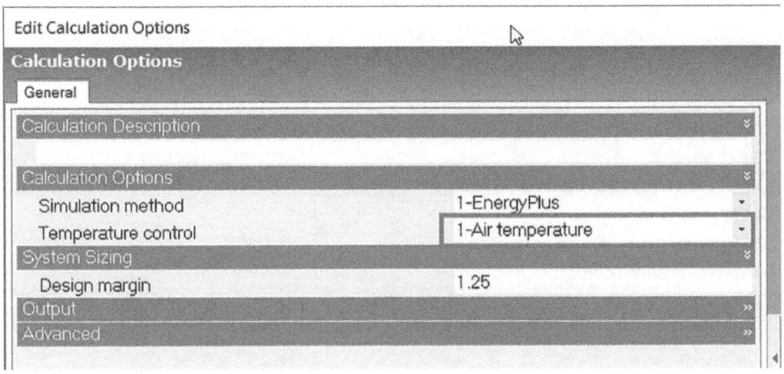

Air temperature control means controlling the mean air temperature of the zone to the assigned setpoint temperatures. Other control types mainly include operative temperature and adjusting zone radiant temperature control fraction. These control types are mainly used in advanced research and comfort analysis studies.

Step 17: Select the **Summary** tab. The results are displayed in a grid view. For each zone, the heating capacity is shown in the table.

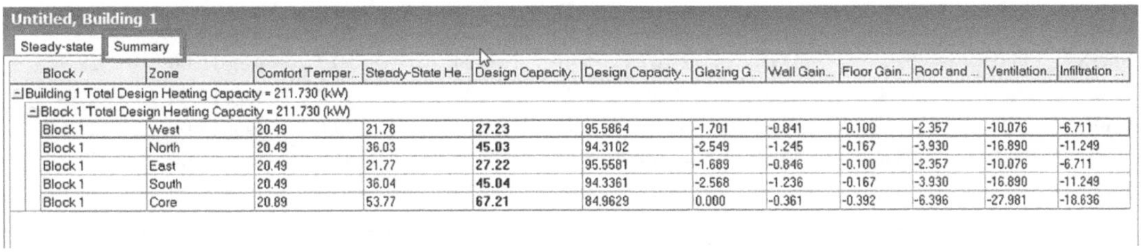

Untitled, Building 1

Steady-state | Summary

Block	Zone	Comfort Temper...	Steady-State He...	Design Capacity...	Design Capacity...	Glazing G...	Wall Gain...	Floor Gain...	Roof and ...	Ventilation...	Infiltration ...
Building 1 Total Design Heating Capacity = 211.730 (kW)											
Block 1 Total Design Heating Capacity = 211.730 (kW)											
Block 1	West	20.49	21.78	**27.23**	95.5864	-1.701	-0.841	-0.100	-2.357	-10.076	-6.711
Block 1	North	20.49	36.03	**45.03**	94.3102	-2.549	-1.245	-0.167	-3.930	-16.890	-11.249
Block 1	East	20.49	21.77	**27.22**	95.5581	-1.689	-0.846	-0.100	-2.357	-10.076	-6.711
Block 1	South	20.49	36.04	**45.04**	94.3361	-2.568	-1.236	-0.167	-3.930	-16.890	-11.249
Block 1	Core	20.89	53.77	**67.21**	84.9629	0.000	-0.361	-0.392	-6.396	-27.981	-18.636

Step 18: Select the **Cooling design** screen tab. Click on **Update data**. The **Calculation Options** screen appears. Ensure that **Air temperature** is selected as temperature control. Click **OK**.

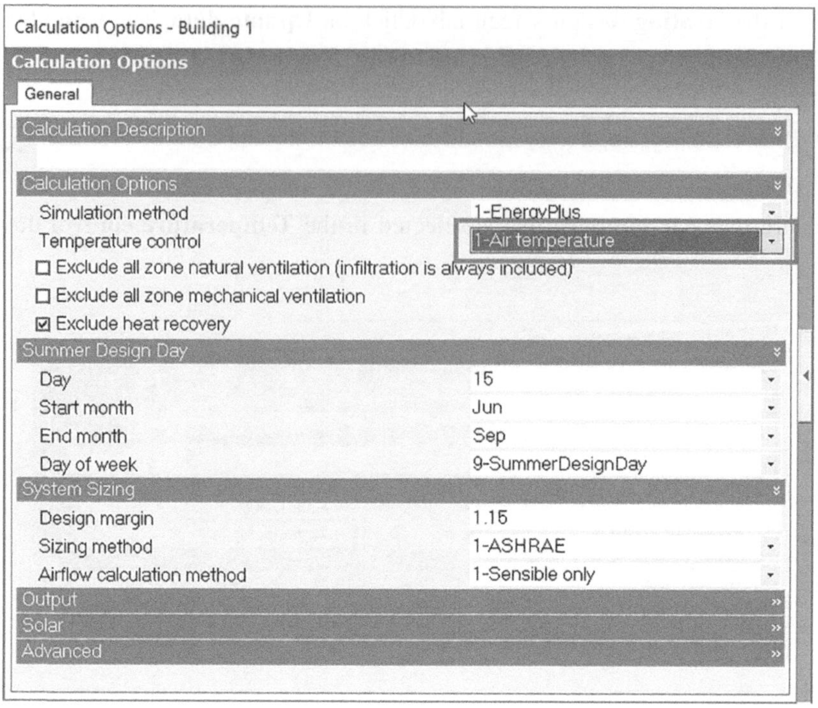

Step 19: Select the **Summary** tab to view the cooling system sizing results of each zone (Table 1.17).

TABLE 1.17
Cooling Design Data for 30% WWR

Zone	Design Capacity (kW)	Design Flow Rate (m3/s)	Total Cooling Load (kW)	Sensible (kW)	Latent (kW)	Air Temperature (°C)	Humidity (%)	Time of Max Cooling	Max Operating Temperature in Day (°C)
West	33.46	2.027	29.09	25.43	3.66	24	49.4	Aug 16:00	30.19
North	38.03	2.205	33.07	27.67	5.4	24	50.7	Aug 15:00	27.92
East	26.23	1.54	22.81	19.33	3.48	24	49.9	Aug 10:00	28.4
South	52.02	3.144	45.23	39.45	5.78	24	49.6	Aug 14:00	28.74
Core	58.26	3.346	50.66	41.98	8.68	24	51	Aug 15:00	27.91
Total	207.99	12.263	180.86	153.87	26.99	24	50.3	N/A	30.19

It can be seen that the peak cooling for the east zone is at 10:00 and for the west zone at 16:00. Further, note that the cooling capacity for the west zone is more than that for the east zone. Also, the design capacity for the south zone is more than that for the north zone. This is because London is located at 51.5° north latitude, resulting in more solar radiation on the south façade.

Step 20: Run the energy simulation and record the results for fuel breakdown. Save the model to use in forthcoming tutorials.

Step 21: Repeat the preceding steps and set the WWR to **80**.

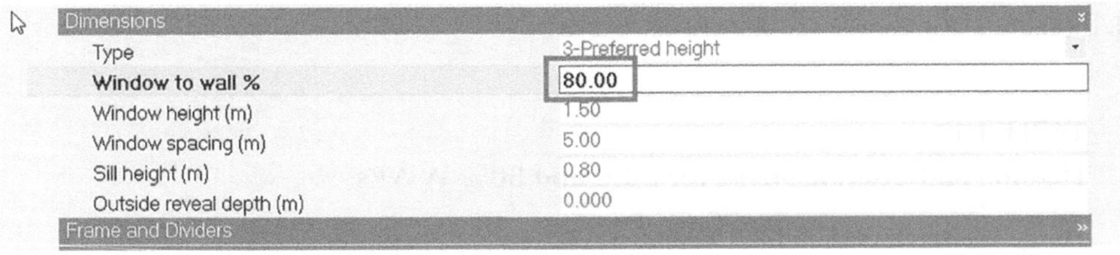

Step 22: Select the **Visualise** screen tab. Use the orbit tool for 3D visuals.

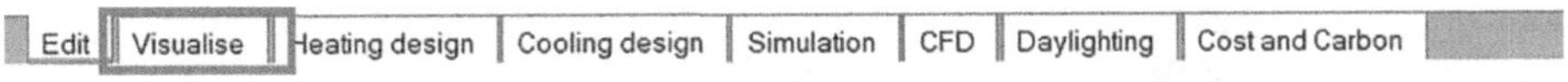

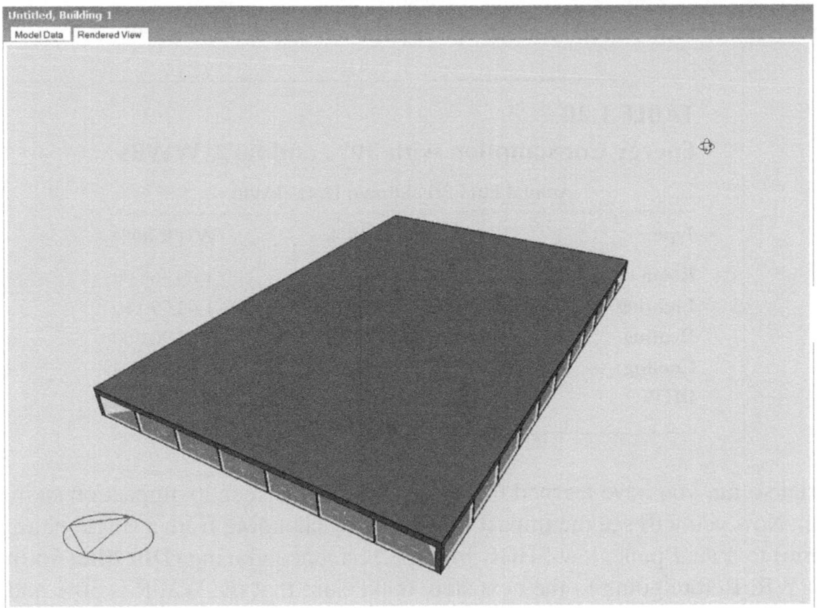

Step 23: Record the cooling design data and fuel breakdown (Tables 1.18 and 1.20).

TABLE 1.18
Cooling Design Data for 80% WWR

Zone	Design Capacity (kW)	Design Flow Rate (m3/s)	Total Cooling Load (kW)	Sensible (kW)	Latent (kW)	Air Temperature (°C)	Humidity (%)	Time of Max Cooling	Max Operating Temperature in Day (°C)
West	57.42	3.658	49.93	45.9	4.03	24	48.3	Aug 16:00	34.55
North	44.58	2.637	38.76	33.08	5.68	24	50.1	Aug 15:00	28.77
East	48.09	3.024	41.82	37.94	3.88	24	48.5	Aug 09:30	30.39
South	83.75	5.326	72.82	66.82	6	24	48.5	Aug 13:30	30.49
Core	60.52	3.491	52.62	43.81	8.82	24	50.9	Aug 15:00	28.21
Total	294.35	18.135	255.96	227.55	28.4	24	49.6	N/A	34.55

Now compare the results for each zone to gain a better understanding of the impact of WWR on sizing. The comparison is shown in Table 1.19. Also, compare the energy consumption of the two models (Table 1.20).

TABLE 1.19
Heating and Cooling Sizing for 30% and 80% WWRs

	Heating Sizing (kW)		Cooling Sizing (kW)	
Zone	**WWR 30%**	**WWR 80%**	**WWR 30%**	**WWR 80%**
North	27.23	29.6	33.46	57.42
South	45.03	48.56	38.03	44.58
East	27.22	29.57	26.23	48.09
West	45.04	48.61	52.02	83.75
Core	67.21	67.66	58.26	60.52

TABLE 1.20
Energy Consumption with 30% and 80% WWRs

	Annual Fuel Breakdown Data (kWh)	
Type	**WWR 30%**	**WWR 80%**
Room electricity	1,00,368.00	1,00,368.00
Lighting	1,08,790.10	1,08,790.10
Heating	1,05,850.30	1,04,901.50
Cooling	29,358.01	55,477.89
DHW	8,922.72	8,922.72

In this tutorial so far, you have learned to change WWR and seen its impact on sizing and energy consumption. Now you will see the impact of glazing by changing from a single-pane clear glazing (**Sgl Clr 3mm**) to a dual-pane, low SHGC high-performance glazing (**Dbl Blue 6mm/13mm Air**) with 30% WWR. Before going to the next step, make sure that the WWR is 30% and that you are at the building level in the navigation tree.

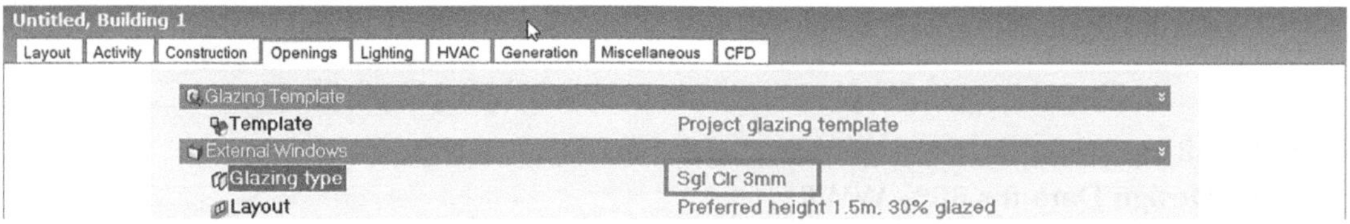

Step 24: Select the **Openings** tab, and choose **Sgl Clr 3mm** from the library, and run the annual simulation.

Step 25: Now select **Dbl Blue 6mm/13mm Air** from the library, and run the annual simulation.

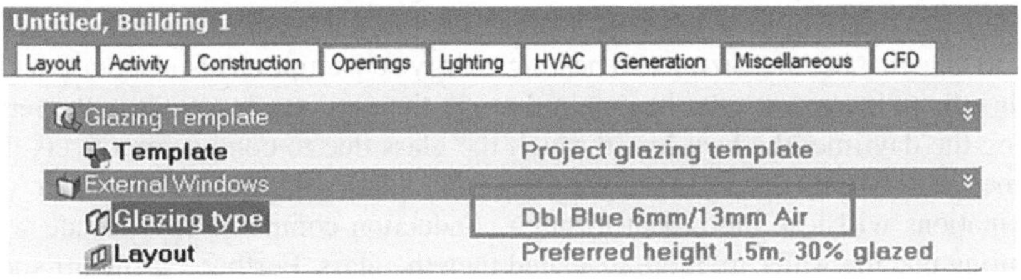

When you compare the results by changing the glazing from single to double glazing, notice that there is a change in the consumption of electricity for cooling and gas for heating (Table 1.21).

TABLE 1.21
Annual Energy Consumption with a Change in Glass Type

	Annual Fuel Breakdown Data	
Type	**Single Glazing (Sgl Clr 3 mm)**	**Double Glazing (Dbl Blue 6 mm/13 mm Air)**
Room electricity (kWh)	100368	100368
Lighting (kWh)	108790.1	108790.1
Heating (gas) (kWh)	127055.5	117427.7
Cooling (kWh)	29230.9	24344.5
DHW (kWh)	8922.7	8922.7

EXERCISE 1.5

Repeat this tutorial for Sydney, Australia.

1. Observe the effect when WWR is changed from 30% to 80% (Table 1.22 and Table 1.23).

TABLE 1.22
Energy Consumption with 30% and 80% WWRs

	Annual Fuel Breakdown Data	
Type	**WWR (30%)**	**WWR (80%)**
Room electricity		
Lighting		
Heating (electricity)		
Heating (gas)		
Cooling (electricity)		

TABLE 1.23
Heating and Cooling Sizing Capacity with 30% and 80% WWRs

	Heating Sizing (kW)		Cooling Sizing (kW)	
Zone	**WWR 30%**	**WWR 80%**	**WWR 30%**	**WWR 80%**
North				
South				
East				
West				
Core				

It can be seen that cooling sizing and cooling energy consumption increase with an increase in the WWR. This is because a larger glass area results in a larger solar gain. Further, as the U-value of the window is inferior to the U-value of the wall, a larger window results in a larger heat gain in summer. It is interesting to note that heating sizing and heating energy consumption also increase with an increase in WWR. Though an increased solar heat gain through the glass tends to reduce the requirement for heating during the daytime, the heat loss through the glass due to conduction offsets this effect. At any given time, glass of only one or two orientations will allow solar radiation to enter, whereas glass of other orientations will lose more heat through conduction compared to a façade with a smaller WWR, assuming that the walls are more insulated than the glass. Further, during off-sunshine hours, the entire glazed area results in more heat loss if the U-value of the glass is inferior to that of the wall.

However, in subsequent chapters, it will be seen that this effect is combined with a reduction in lighting energy consumption if artificial light is simulated with a dimming feature.

> What is U-Value?
> U-value is the measure of the overall rate of heat transfer through a particular section of construction. U-value is also known as thermal transmittance.

2. Observe the effect with the change in glass type for 30% WWR (Table 1.24 and Table 1.25).

TABLE 1.24
Energy Consumption with a Change in Glass Type

	Annual Fuel Breakdown Data	
Type	**Single Glazing (Sgl Clr 3 mm)**	**Double Glazing (Dbl Blue 6 mm/13 mm Air)**
Room electricity		
Lighting		
Heating (electricity)		
Heating (gas)		
Cooling (electricity)		

TABLE 1.25
Heating and Cooling Sizing Capacity with a Change in Glass Type

	Cooling and Heating System Sizing	
Type	**Single Glazing (Sgl Clr 3 mm)**	**Double Glazing (Dbl Blue 6 mm/13 mm Air)**
Heating		
Cooling		

TUTORIAL 1.6: EVALUATING THE IMPACT OF OCCUPANCY DENSITY

Goal

To evaluate the impact of occupancy density on cooling and heating loads and whole-building energy consumption

What Are You Going to Learn?

- How to change occupancy density

Problem Statement

Use the model created in Tutorial 1.5 (60 m × 40 m building with five zones). Set the WWR to 30%, minimum fresh air (l/s-person) to 2.5, mechanical ventilation per area (l/s-m^2) to 0.3, and model infiltration [air changes (ac)/h] to 0.20. Run simulations with occupancy density (people/m^2) set to 0.07 and 0.10. Analyze the change in energy consumption. Use the **AZ-PHOENIX/SKY HARBOR, USA** weather file.

Solution

Step 1: Select the **Activity** tab, and under the **Occupancy** section, set the occupancy density (people/m^2) to 0.111. Under the **Minimum Fresh Air** section, set Fresh air (l/s-person) to **2.5** and mechanical vent per area (l/s-m^2) to **0.3**.

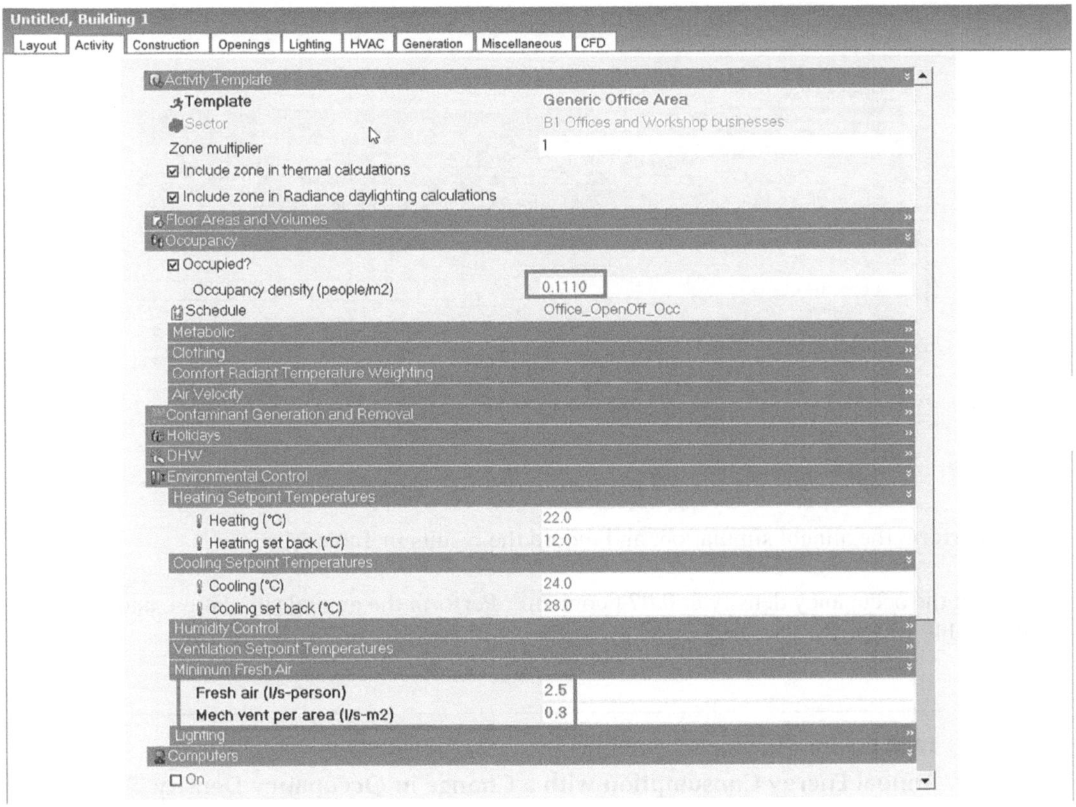

Step 2: Select the **Construction** tab. Under the **Airtightness** section, set **Model infiltration, Constant** rate as 0.200 ac/h.

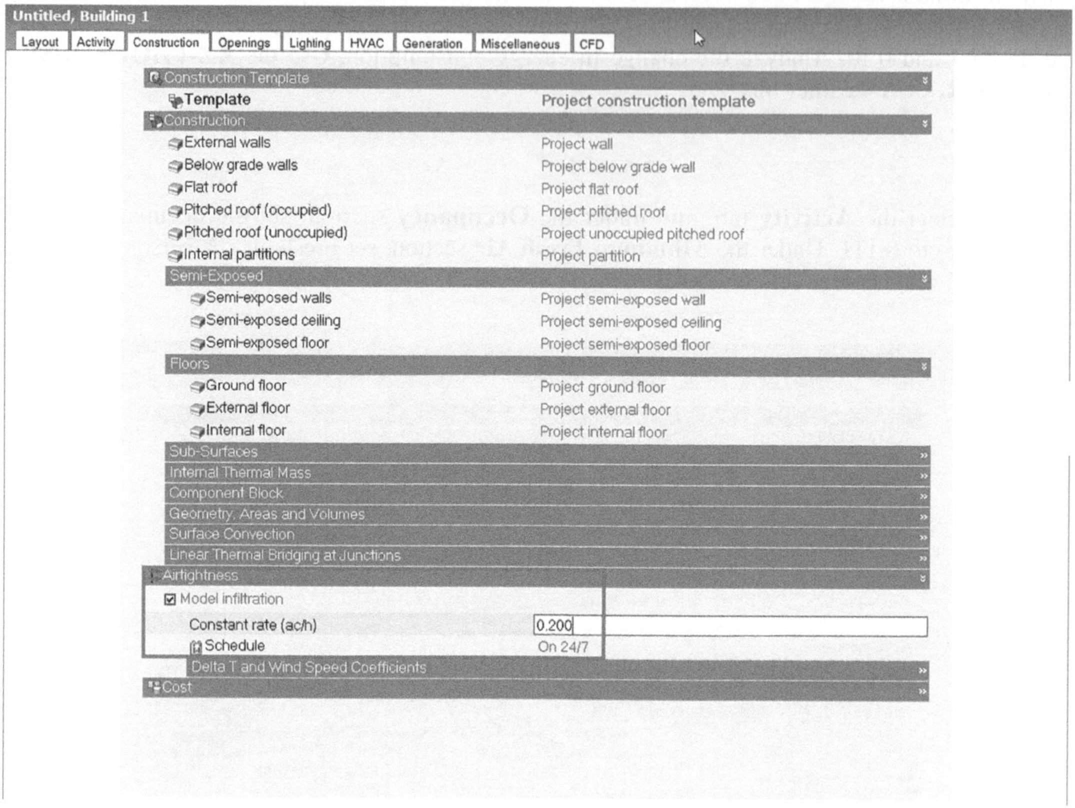

Step 3: Perform the annual simulation, and record the results in Table 1.26.

Step 4: Set the occupancy density to **0.07** people/m^2. Perform the annual simulation, and record the results (Table 1.26).

TABLE 1.26
Annual Energy Consumption with a Change in Occupancy Density

Type	0.111 People/m^2	0.07 People/m^2
Room electricity (kWh)	100,368	100,368
Lighting (kWh)	108,790	108,790
Heating (gas) (kWh)	1,283	1,349
Cooling (electricity) (kWh)	185,668	175,450
DHW (electricity) (kWh)	8,923	8,923

Analysis

The analysis of the simulation results reveals the following impacts of occupancy density on energy consumption.

Cooling energy consumption: Increasing the occupancy density from 0.07 to 0.111 people/m² results in an increase in cooling energy consumption. This is attributed to the higher internal load due to the presence of more occupants, which increases both sensible and latent heat loads in the zones.

Heating energy consumption: The heating energy consumption decreases when the occupancy density is increased. This reduction is due to the heat generated by occupants, which contributes to the indoor temperature and reduces the heating load; however, the increase in fresh air intake to maintain indoor air quality for higher occupancy levels partially offsets this reduction, leading to higher heating energy consumption associated with conditioning the fresh air. Additionally, the increase in fan power required to supply more air should also be considered.

Lighting and equipment energy consumption: The lighting and room electricity consumption remain constant for both occupancy densities. This indicates that the lighting and equipment usage are independent of occupancy density in the current simulation setup.

Conclusion

The simulation results demonstrate that increasing occupancy density primarily affects cooling and heating energy consumption. Designers should carefully consider occupancy density during the design phase to accurately estimate cooling and heating loads, especially when planning for buildings with high occupancy densities.

Two important parameters are used to show the occupancy of a zone: *occupancy density* and *schedule*. As seen in this tutorial, occupancy density is the maximum number of people in a zone. Schedule of occupancy defines when a zone is occupied or unoccupied and by how many people. *Metabolic activity* is the amount of heat given out by people, and it depends on the activity they engage in. For example, a person who is exercising gives out more heat than a person who is sleeping.

EXERCISE 1.6

For the same tutorial, observe the effect on cooling and heating equipment sizing (Table 1.27).

TABLE 1.27
Heating and Cooling Sizing Capacity with a Change in Occupancy Density

Type	0.1 People/m²	0.07 People/m²
Cooling (kW)		
Heating (kW)		

TUTORIAL 1.7: EVALUATING THE IMPACT OF SPACE ACTIVITY

Goal

To understand the impact of space activity on cooling and heating loads and energy consumption

What Are You Going to Learn?

- How to understand various activity types
- How to change space activity
- How to understand the impact of activity type on sizing and energy consumption

Problem Statement

Create a 20 m × 15 m single-zone model. Set the **Activity Template** to **Office** and **Restaurant** and compare. Study the effect on sizing and energy consumption for London.

Solution

Step 1: Create a 20 m × 15 m single-zone model. You can refer to Tutorial 1.2 to create a single-zone model.

Step 2: Select the **Activity** tab. The **Activity Template** appears.

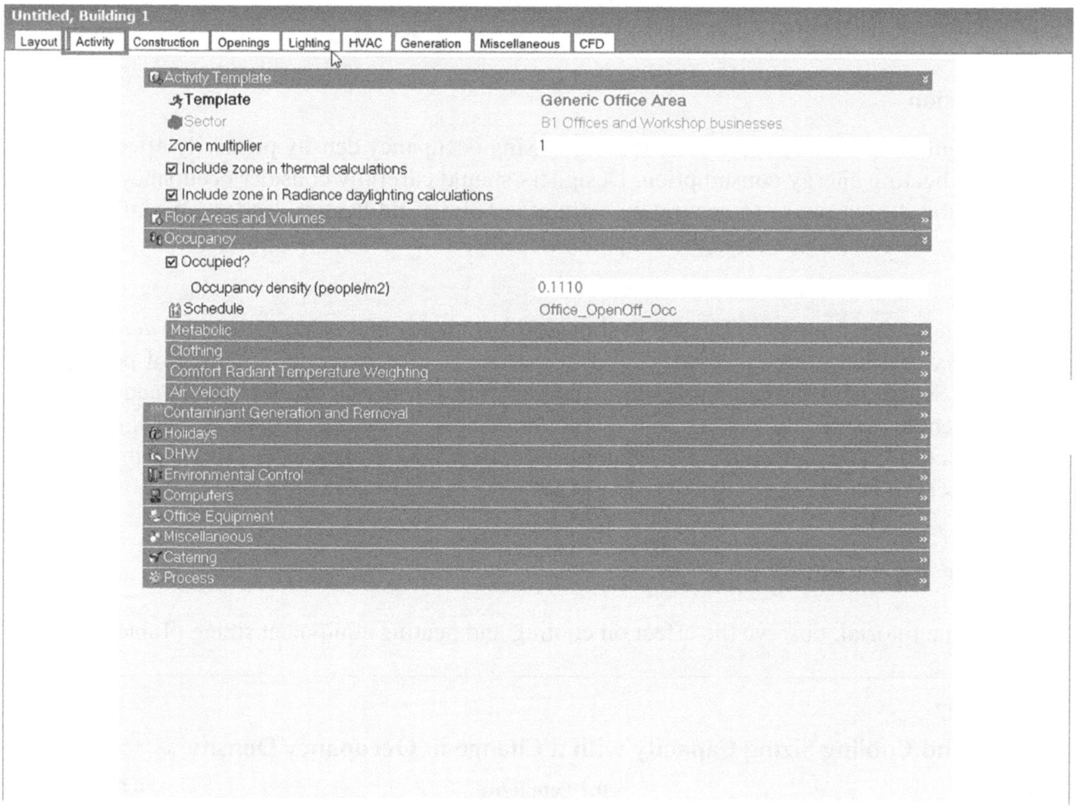

Step 3: Simulate the model, and record the results.

Step 4: On the **Activity Template**, select the **Eating/drinking area** under **Restaurants/Cafes/Drinking Establishments/Hot Food takeaways**. This changes internal gains and schedules of the space.

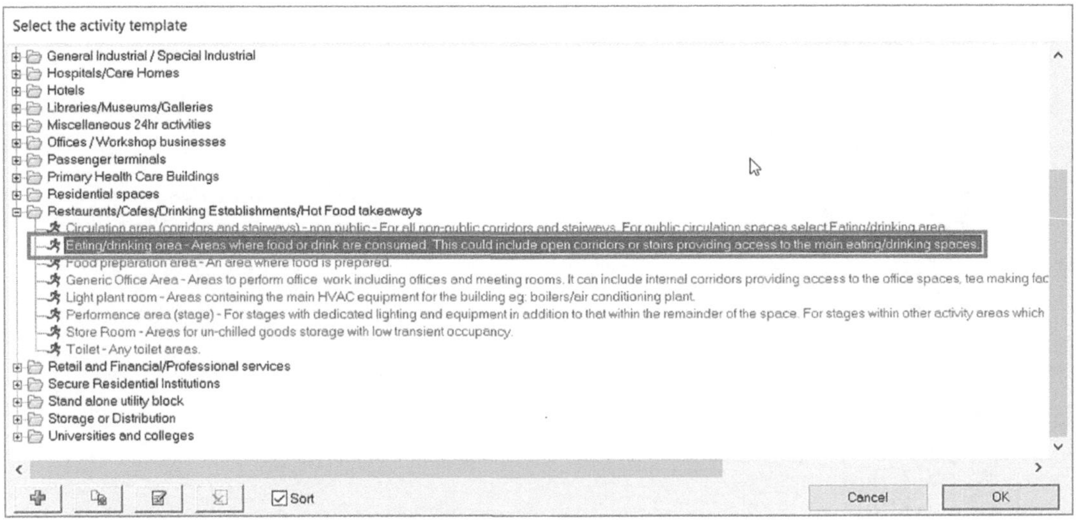

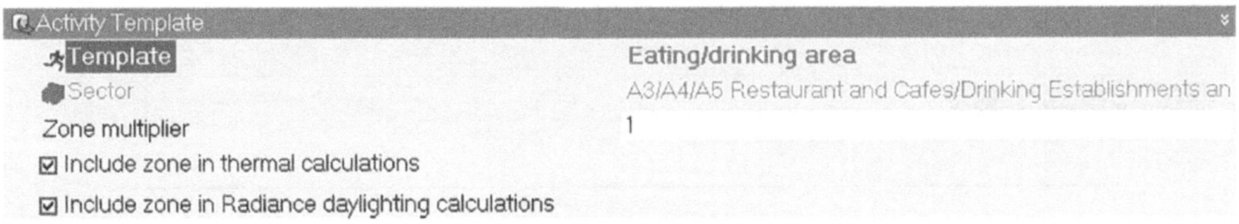

Step 5: Right-click on the **RestPub_EatDrink_Occ** schedule under the **Occupancy** tab and then click on **edit**. The **Edit schedule-RestPub_EatDrink_Occ** screen appears. (You can note the occupancy schedule for comparison with the other **Activity Template** options, as shown in Table 1.28.)

SCHEDULES

Schedules are used in DesignBuilder to define with respect to time the following.

- Occupancy
- Equipment, lighting, and HVAC system operation
- Heating and cooling temperature setpoints
- Transparency of component blocks (usually seasonal)

Occupancy, equipment and lighting schedule are defined by a fraction (0–1). The maximum gain values (e.g., people/m^2) are multiplied by the values in the schedule to obtain the actual value to use at each time step in the simulation.

Source: https://designbuilder.co.uk/helpv7.3/Content/Schedules_-_EnergyPlus_Compact_Schedules.htm

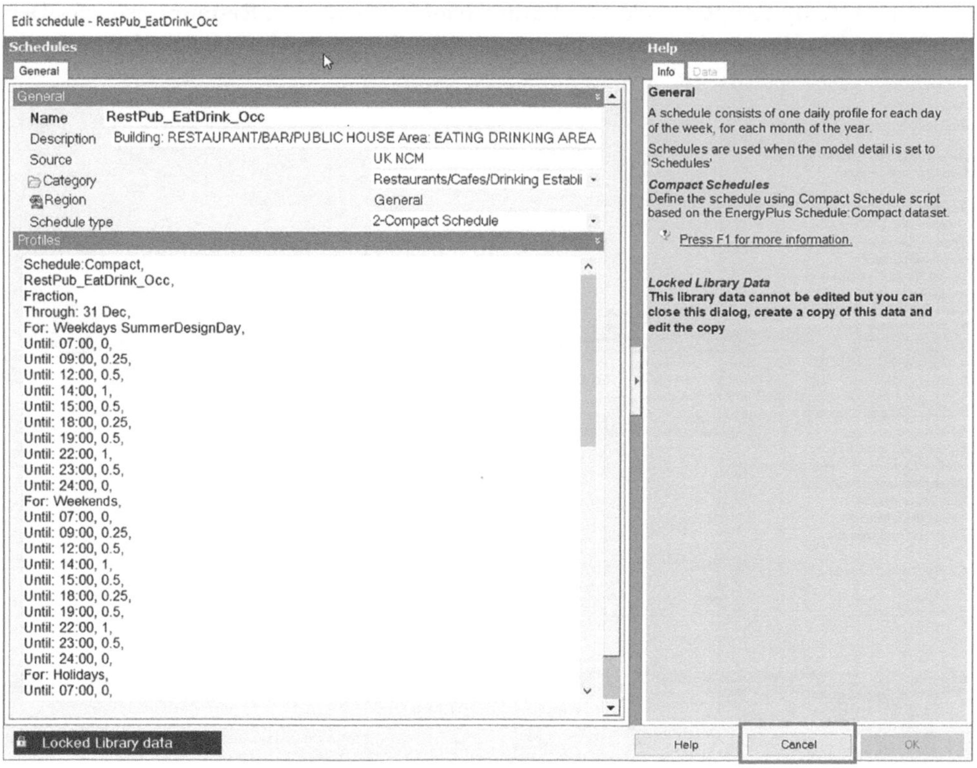

TABLE 1.28
Workday Schedules for a Generic Office and Eating/Drinking Area

Generic Office Area Office_OpenOff_Occ	Eating/Drinking Area RestPub_EatDrink_Occ
Schedule: Compact,	Schedule:Compact,
Office_OpenOff_Occ,	RestPub_EatDrink_Occ,
Fraction,	Fraction,
Through: 31 Dec,	Through: 31 Dec,
For: Weekdays	For: Weekdays SummerDesignDay,
SummerDesignDay,	Until: 07:00, 0,
Until: 07:00, 0,	Until: 09:00, 0.25,
Until: 08:00, 0.25,	Until: 12:00, 0.5,
Until: 09:00, 0.5,	Until: 14:00, 1,
Until: 12:00, 1,	Until: 15:00, 0.5,
Until: 14:00, 0.75,	Until: 18:00, 0.25,
Until: 17:00, 1,	Until: 19:00, 0.5,
Until: 18:00, 0.5,	Until: 22:00, 1,
Until: 19:00, 0.25,	Until: 23:00, 0.5,
Until: 24:00, 0,	Until: 24:00, 0,
For: Weekends,	For: Weekends,
Until: 24:00, 0,	Until: 07:00, 0,
For: Holidays,	Until: 09:00, 0.25,
Until: 24:00, 0,	Until: 12:00, 0.5,
For: WinterDesignDay	Until: 14:00, 1,
AllOtherDays,	Until: 15:00, 0.5,
Until: 24:00, 0;	Until: 18:00, 0.25,

TABLE 1.28 *(Continued)*
Workday Schedules for a Generic Office and Eating/Drinking Area

Generic Office Area Office_OpenOff_Occ	Eating/Drinking Area RestPub_EatDrink_Occ
	Until: 19:00, 0.5, Until: 22:00, 1, Until: 23:00, 0.5, Until: 24:00, 0, For: Holidays, Until: 07:00, 0, Until: 09:00, 0.25, Until: 12:00, 0.5, Until: 14:00, 1, Until: 15:00, 0.5, Until: 18:00, 0.25, Until: 19:00, 0.5, Until: 22:00, 1, Until: 23:00, 0.5, Until: 24:00, 0, For: WinterDesignDay AllOtherDays, Until: 24:00, 0;

Step 6: Click **Cancel**. The screen closes.

Step 7: Select the **Simulation** screen tab to get energy consumption results. Compare both cases (Table 1.29).

TABLE 1.29
Annual Energy Consumption with a Change in Space Activity

Type	Generic office area	Eating/drinking area
Room electricity (kWh)	12,130.28	32,222.99
Lighting (kWh)	13,148.16	16,344.24
Heating (gas) (kWh)	14,489.3	2,2016.83
Cooling (electricity) (kWh)	5,632.9	9,437.9
DHW (electricity) (kWh)	1,078.3	46,048.32

Analysis

The results show the impact of activity type on energy consumption. From the model, you can determine the interior load parameters (Table 1.30). You can get these data from the **Activity** and **Lighting** tabs.

The analysis of the simulation results highlights the following key findings.

Room electricity consumption: The Eating/Drinking area (Restaurant) shows a much higher electricity consumption compared to the Office area. This difference is primarily due to higher internal loads associated with kitchen equipment, food preparation, and increased operational hours typical of restaurant spaces.

Lighting energy consumption: The lighting consumption is slightly higher for the Eating/Drinking area compared to the Office area. The increase is due to a higher lighting density requirement typical for restaurant environments, providing enhanced visibility and ambience.

Heating energy consumption: The heating demand is significantly higher for the Eating/Drinking area compared to the Office area. This is likely due to higher ventilation requirements in restaurants to maintain indoor air quality, resulting in additional heating load to condition the fresh air intake.

Cooling energy consumption: Cooling energy consumption is higher for the Eating/Drinking area compared to the Office area. The increase in cooling load can be attributed to higher internal heat gains from occupants, kitchen equipment, and lighting in restaurants.

Domestic hot water (DHW) consumption: The DHW consumption in the Eating/Drinking area is higher than in the Office area, which is expected due to the greater overall demand for hot water in such spaces.

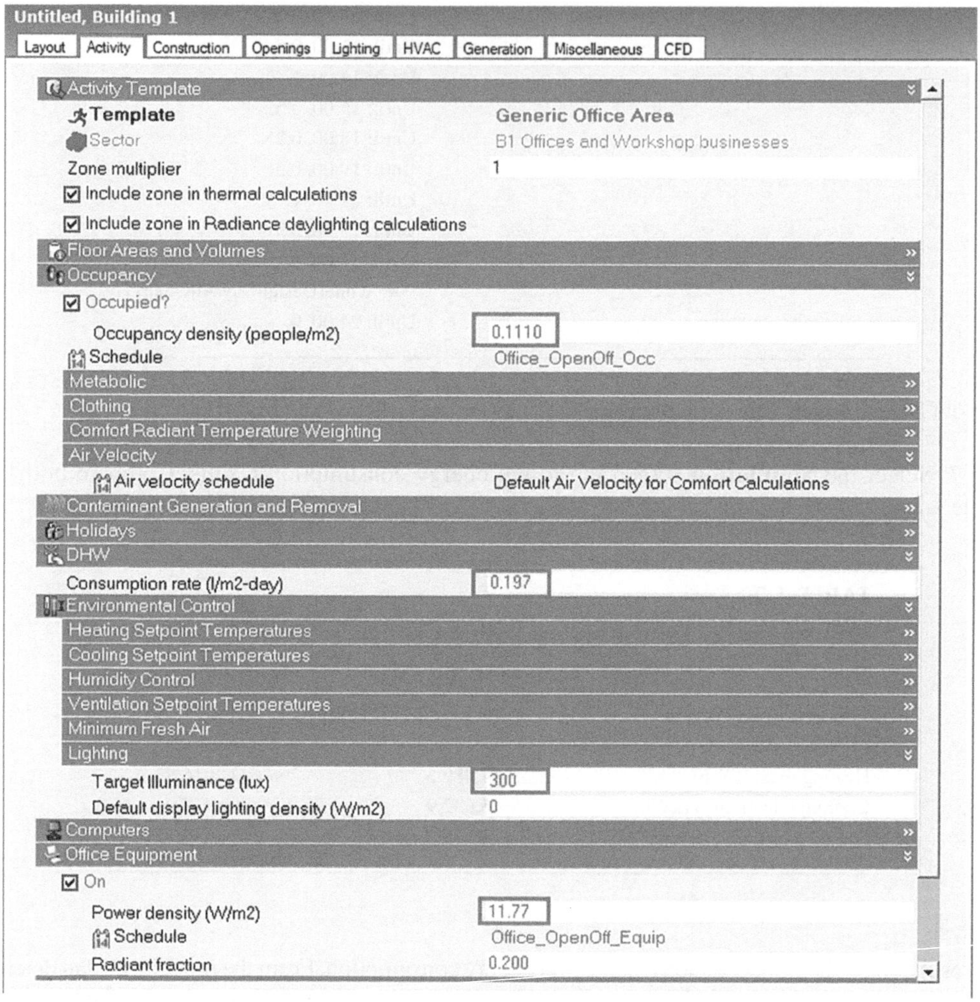

TABLE 1.30
Internal Load Data for Generic Office and Eating/Drinking Area

Interior Load	Unit	Generic Office Area	Eating/Drinking Area
Occupancy	People/m^2	0.11	0.20
Target illuminance	lux	400	150
Interior light	W/m^2 per 100 lux	5	5
Equipment	W/m^2	11.77	18.88
DHW	l/s-day	0.20	5.69

EXERCISE 1.7

Compare the cooling and heating energy consumption for a generic office area and classroom (Table 1.31).

Activity type: **Classroom** can be found under the **Universities and college** category.

TABLE 1.31
Annual Energy Consumption for a Generic Office and Eating/Drinking Area

Type	Generic Office Area (kWh)	Classroom Area (kWh)
Room electricity		
Lighting		
Heating (gas)		
Cooling (electricity)		
DHW (electricity)		

From the model, you can find the change in the internal load parameter (Table 1.32).

TABLE 1.32
Internal Load for a Generic Office and Eating/Drinking Area

Interior load	Unit	Generic office area	Classroom area
Occupancy	People/m^2		
Target illuminance	lux		
Interior light	W/m^2 per 100 lux		
Equipment	W/m^2		
DHW	l/s-day		

Fill in and Compare office and classroom schedules in Table 1.33.

TABLE 1.33
Workday Schedules for a Generic Office and Classroom Area

Generic Office Area Office_OpenOff_Occ	Classroom Area

TUTORIAL 1.8: EVALUATING THE IMPACT OF LIGHTING AND EQUIPMENT POWER

Goal

To understand the impact of lighting and equipment power density on HVAC system sizing and energy consumption

What Are You Going to Learn?

- How to change lighting power density (LPD)
- How to change equipment power density (EPD)

Problem Statement

Compare the annual energy consumption when you set the LPD to 13.13 W/m² and to 10.0 W/m². Use the **Courtyard with VAV Example** template for London Gatwick Airport.

Solution

Step 1: Select the **Lighting** tab, and under the **General Lighting** section, set the LPD shown as **Lighting energy (W/m²)** to 13.13.

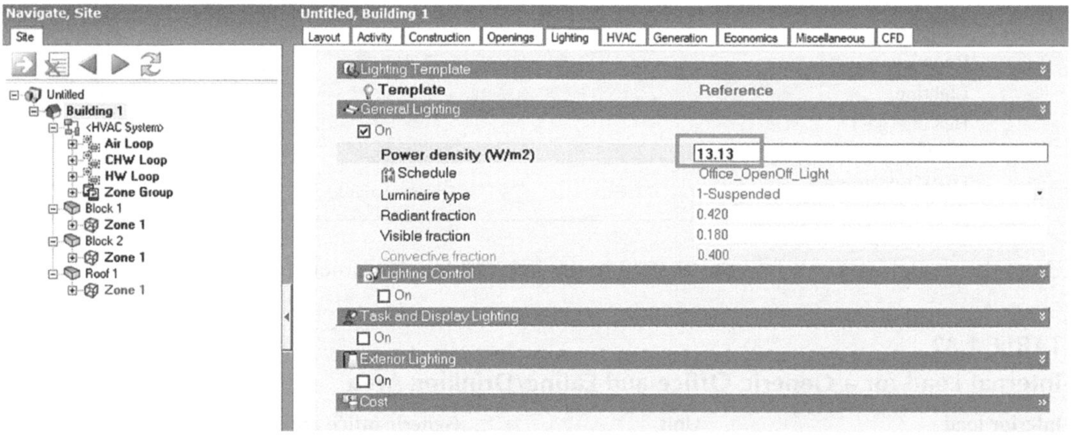

Lighting power density (LPD) is the lighting power per unit area of a building, space, or outdoor area expressed in W/m².

Tip: Does your lighting energy have units as W/m² per 100 lux rather than W/m²?
If yes, change this to W/m² in the following way.

go to Edit → Model Options Data → Data → Gain Data → Lighting Gains Units, and change the lighting gain units to W/m².

Step 2: Click the **Simulation** screen. Click on **Update data**. Click the **Annual simulation** link to perform annual simulation.

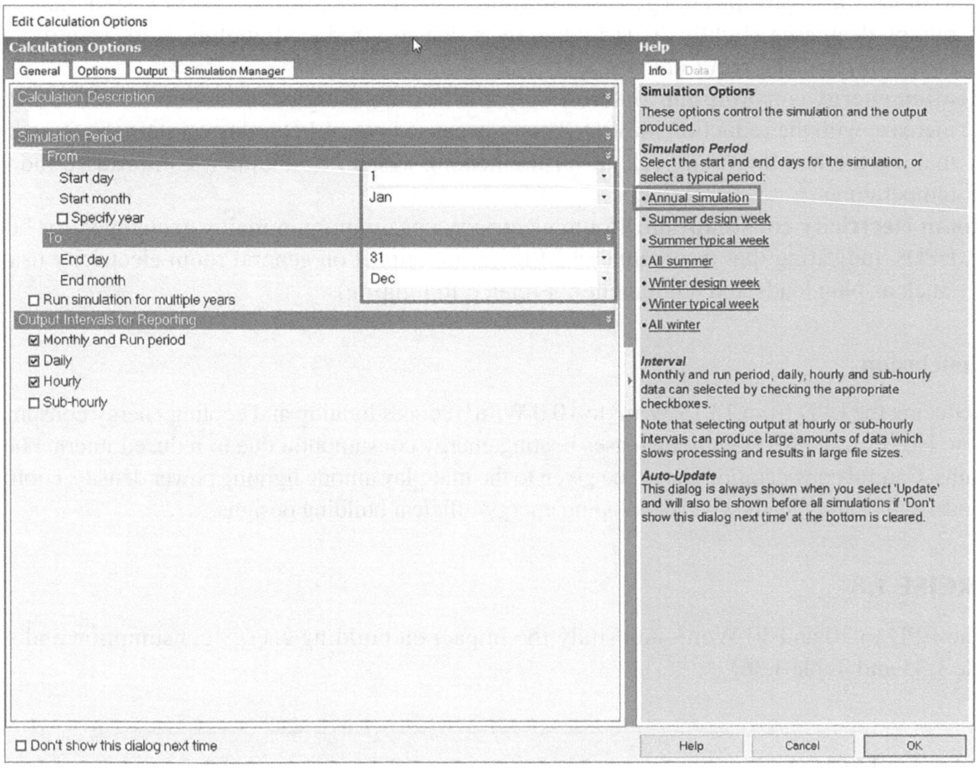

Step 3: Repeat the steps to set the LPD to 10.00. Perform annual energy simulation. Compare the end-use energy consumption of both cases (Table 1.34).

TABLE 1.34
Annual Fuel Breakdown with a Change in LPD

Type	LPD: 13.13 W/m^2	LPD: 10.0 W/m^2
Room electricity (kWh)	24,964	24,964
Lighting (kWh)	23,685	18,039
System fans (kWh)	24,413	23,143
System pumps	59	54
Heating (gas) (kWh)	21	23
Cooling (electricity) (kWh)	44,863	40,808

Analysis

The analysis of the simulation results reveals the following insights.

Lighting energy consumption: The reduction in LPD from 13.13 W/m^2 to 10.0 W/m^2 results in a decrease in lighting energy consumption. This reduction is a direct result of the lower LPD value, which reduces the amount of electricity required for lighting while maintaining the same operational schedule.

Cooling energy consumption: Cooling energy consumption is reduced when LPD is decreased. This reduction occurs because lower lighting power density reduces internal heat gains, which consequently lowers the cooling load.

System fans and pumps energy consumption: The energy consumption of system fans and pumps decreases slightly. This reduction is due to a reduced cooling load, resulting in lower air-handling requirements.

Heating energy consumption: Interestingly, the heating energy consumption shows a slight increase with the reduction in LPD. This is because lower LPD reduces internal heat gains, making the building more reliant on the heating system to maintain comfortable indoor temperatures.

Room electricity consumption: Room electricity consumption remains unchanged for both LPDs, indicating that the change in LPD has no impact on general room electricity usage (such as plug loads and equipment not related to lighting).

Conclusion

Reducing the LPD from 13.13 W/m² to 10.0 W/m² reduces lighting and cooling energy consumption. However, it also slightly increases heating energy consumption due to reduced internal heat gains. Careful consideration should be given to the interplay among lighting power density, cooling loads, and heating loads when developing energy-efficient building designs.

EXERCISE 1.8

Set the EPD to 20 and 10 W/m², and study the impact on building energy consumption and sizing (Table 1.35 and Table 1.36).

TABLE 1.35
Energy Consumption with a Change in Equipment Power Density

Annual Fuel Breakdown Data		
Type	**EPD (20 W/m²)**	**EPD (10 W/m²)**
Room electricity		
Lighting		
System fans		
System pumps		
Auxiliary energy		
Heating (gas)		
Cooling (electricity)		

TABLE 1.36
Heating and Cooling Sizing Capacity with a Change in Equipment Power Density

Type	EPD (20 W/m²)	EPD (10 W/m²)
Heating		
Cooling		

EQUIPMENT POWER DENSITY (EPD)

In DesignBuilder, under the **Activity** tab, you can set **Equipment gain** units. This option provides control over the method used to enter **Internal gains** such as computers, office equipment, miscellaneous, catering, and process.

1. *Power density*, where zone equipment gains are entered as a power density; that is, as a power per zone floor area in W/m^2 or W/ft^2.
2. *Absolute zone power*, where zone equipment gains are entered as a power value in watts (W). This option is usually used when detailed zone-by-zone survey or design data are available. Unless you need to enter specific known lighting equipment gains zone by zone, this option is best avoided, as it increases the amount of work involved because of the inheritance mechanism not working effectively for gains that are not normalized by floor area.

Source: https://designbuilder.co.uk/helpv7.3/Content/_Gains_data_model_detail.htm#Equipmen%20gain

It can be clearly seen from the results that reducing the EPD not only reduces the energy consumption for equipment, respectively, but it also leads to a reduction in the cooling energy. This is because the energy consumed by equipment ultimately gets converted into heat and usually gets added to the space. This additional heat increases the energy consumption of the fans and pumps and also increases system sizing. In cases of heating energy consumption, the reverse effect is observed because equipment adds additional heat to the space, thereby reducing energy requirements for heating.

TUTORIAL 1.9: EVALUATING THE IMPACT OF DAYLIGHT CONTROLS

GOAL

To evaluate the impact of daylight controls on energy consumption

WHAT ARE YOU GOING TO LEARN?

- How to specify daylight controls in a building model
- How to evaluate the impact of daylight controls on energy consumption

PROBLEM STATEMENT

Create a 60 m × 40 m model with a core and four perimeter zones. Consider a perimeter depth of 5 m. (Refer Tutorial 1.5 for steps to create the model.) Set the WWR to 60%, and use **Sgl Grey 3mm** glass (VLT ~ 60%). Perform an annual energy simulation without daylight sensors, and then install daylight control in the north zone. Compare the lighting energy consumption with and without daylight control for the London Gatwick Airport location, United Kingdom.

Tip: For single-pane glass with 60% visible light transmittance, select **Sgl Grey 3mm** with a light transmission value of 0.61 under the **Single** category of the glazing library.

Solution

Step 1: Create a five-zone core and perimeter model of size 60 m × 40 m. Set the WWR to 60%, and select **Sgl Grey 3mm** glass.

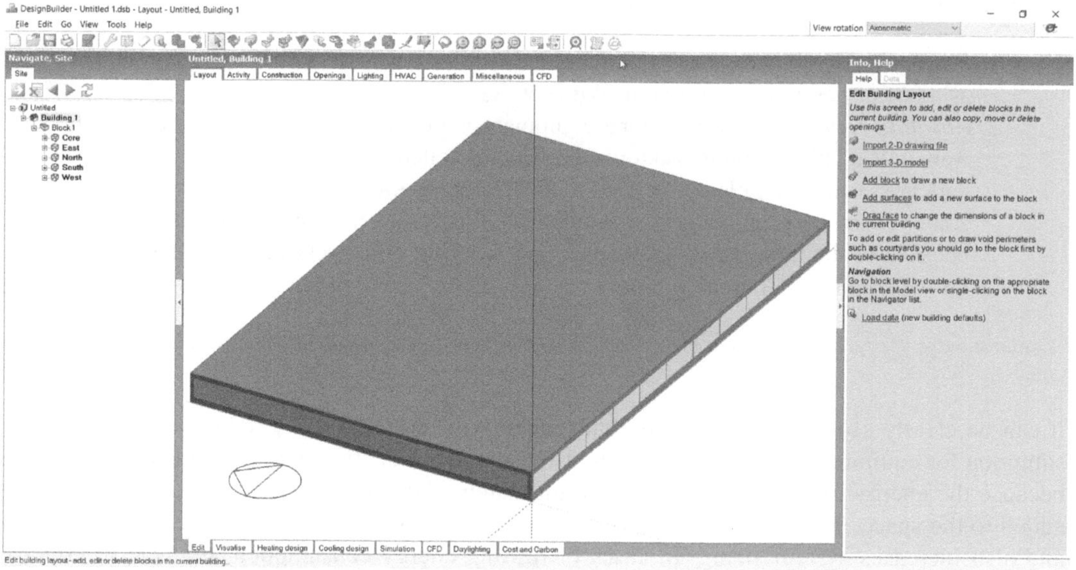

Step 2: Click the **Simulation** screen tab. Click on **Update data**. The **Edit Calculation Options** screen appears. Select the **Hourly** checkbox under **Output Intervals for Reporting**, and then click **OK**. After the simulation, hourly results appear under the **Analysis** tab.

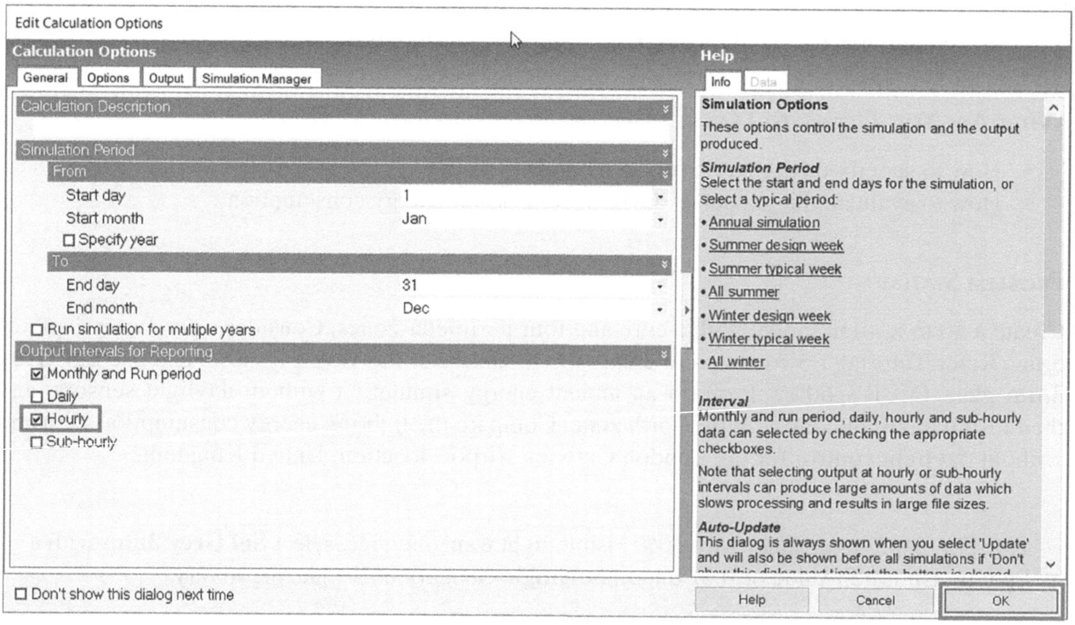

Step 3: Click the **Export data** icon to export the hourly energy consumption in the spreadsheet. (Ensure that **Fuel breakdown**, **Hourly interval**, and **Grid** options are selected.)

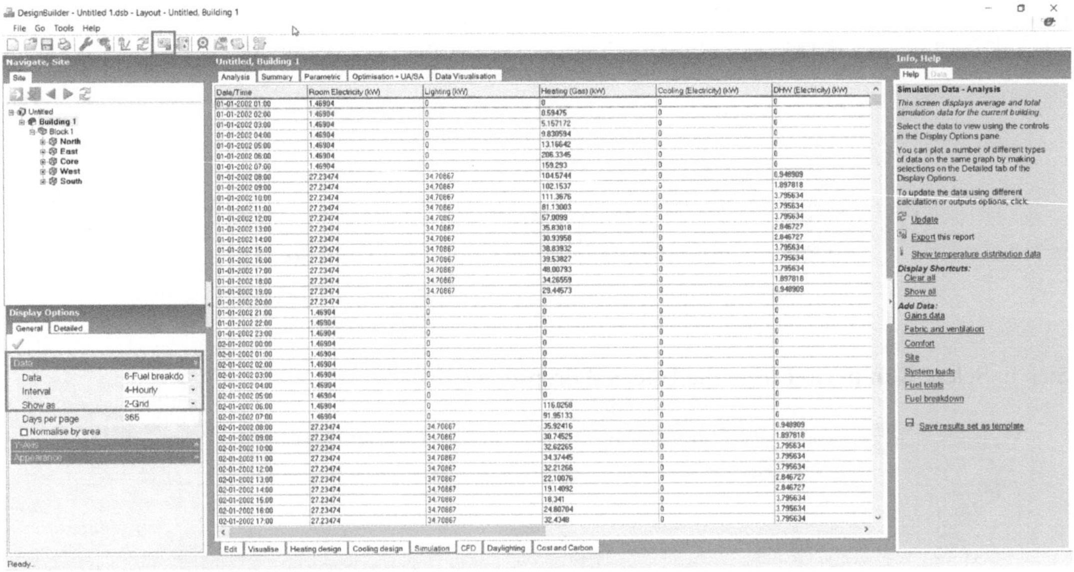

Step 4: Select **File** from the **Export to** drop-down list. Select **CSV spreadsheet** from the **Format** drop-down list. Click **OK**. The **Export CSV file** screen appears.

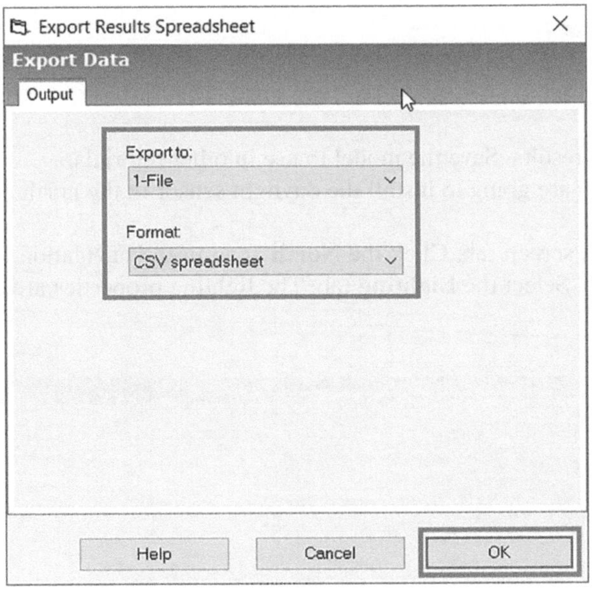

Step 5: Enter the file name **Without daylight sensor**, and save the file to **Desktop**.

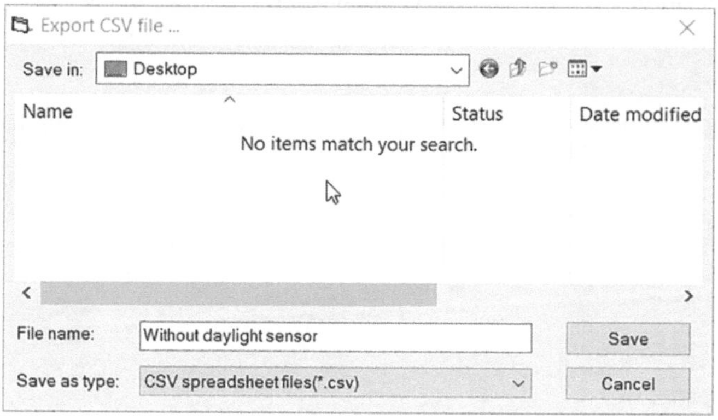

Step 6: Select **Run period** from the **Interval** drop-down list.

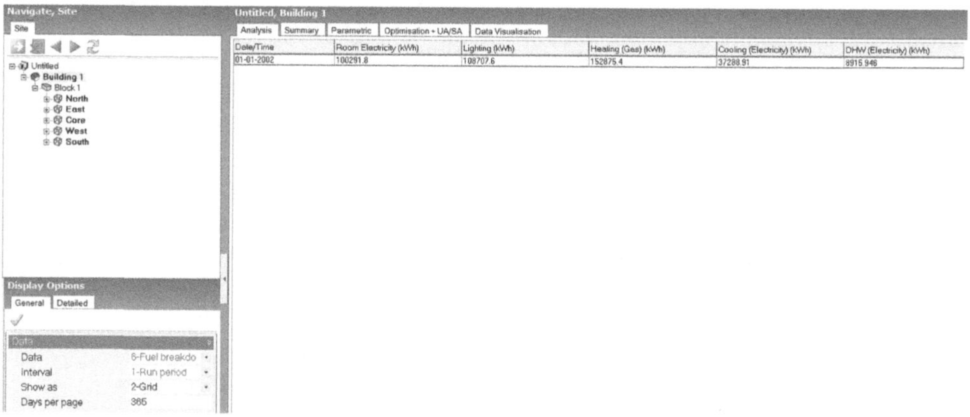

Record the simulation results. Save the model to use in other tutorials.

In the next step, you are going to install the daylight sensor in the north zone.

Step 7: Select the **Edit** screen tab. Click the **North** zone in the navigation tree. It shows the north zone in the **Layout** tab. Select the **Lighting** tab. The lighting properties are displayed.

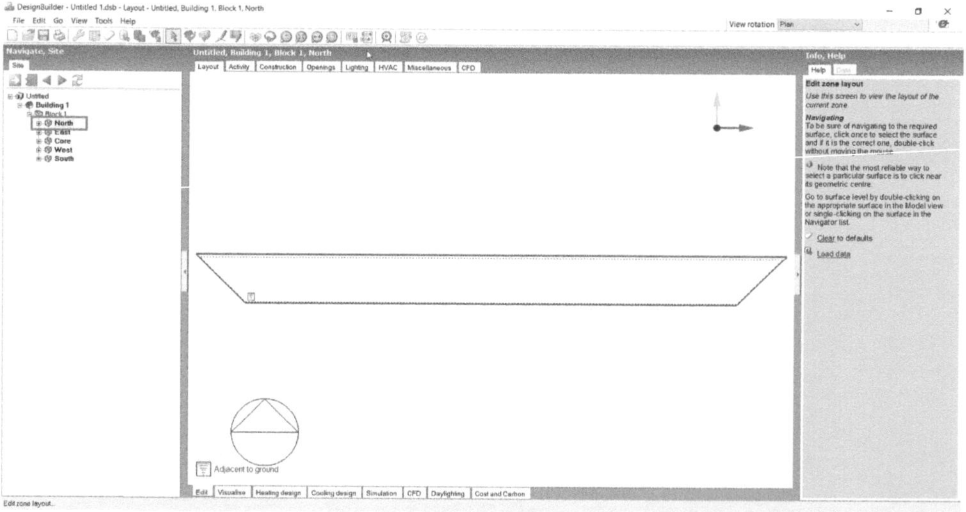

Step 8: Expand the **Lighting control** section. Select the **On** checkbox. Select **Linear** from the **Control type** drop-down list.

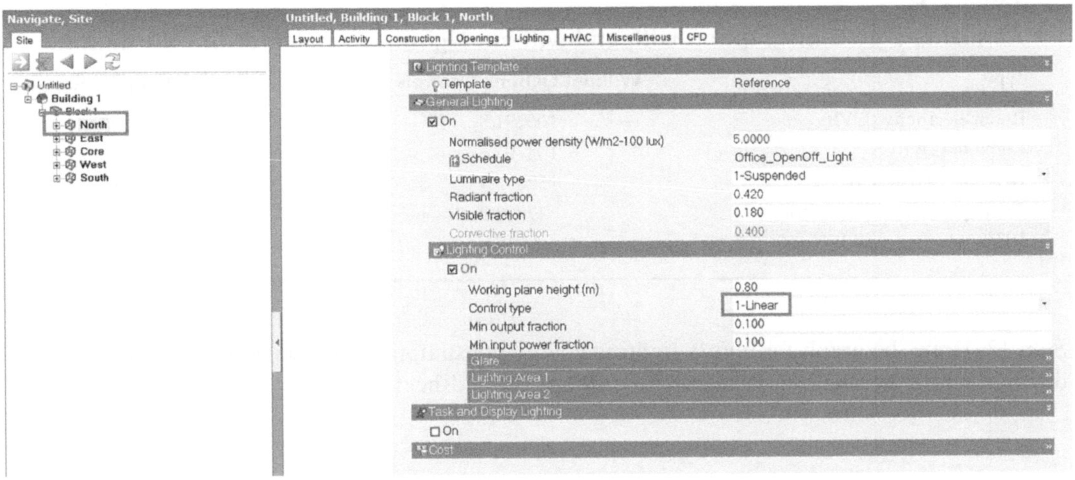

Different types of lighting control options that exist in DesignBuilder are as follows.

1 Linear
2 Linear/off
3 Stepped

Linear control, also called as *continuous control*, reduces the power input of the luminaire continuously, thereby decreasing the output light from the lamp until it reaches the minimum input and output fraction provided as a user input. This is possible with continuously dimmable fixtures and lamps. The decrease in input power (proportionally the lighting output) depends on the daylight illuminance requirement in the space. As the daylight illuminance increases, the input power decreases until it reaches the minimum input power and light output fraction and remains at the minimum specified ratio with a further increase in daylight illuminance.

Step 9: Select the **Layout** tab. The **Daylight Sensor** is displayed.

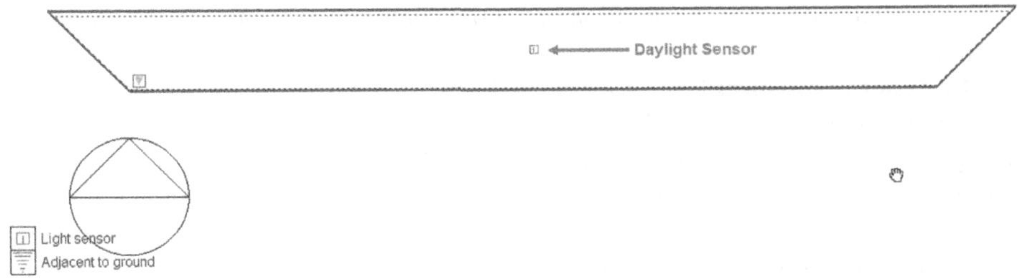

Step 10: Perform the simulation for the entire building, and record the annual and hourly results of lighting energy consumption.

Step 11: Compare the results with and without lighting controls (Table 1.37).

Table 1.37
Annual Energy Consumption with and without Lighting Control for the North Zone

Type	Without Lighting Controls	With Lighting Controls
Room electricity (kWh)	100291.8	100291.8
Lighting (kWh)	108707.6	99734.4
Heating (gas) (kWh)	152875.4	157343.8
Cooling (electricity) (kWh)	37288.9	35865.5
DHW (electricity) (kWh)	8915.9	8915.94

Step 12: Draw the graph for hourly lighting energy consumption for 1 January with and without a daylight sensor by using the data exported using a spreadsheet program.

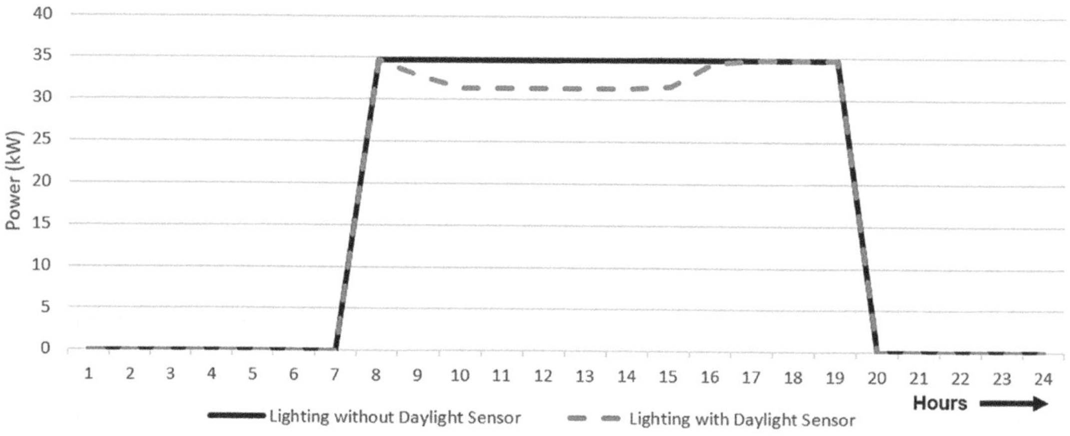

The preceding figure shows the hourly lighting energy consumption for 1 January. It can be observed that there is a reduction in energy consumption with the use of a daylight sensor and a dimmable luminaire.

Key Observations

1. **Room electricity consumption:** The room electricity consumption remains constant in both cases, indicating that the addition of daylight controls does not affect room electricity use in the building.
2. **Lighting energy consumption:** With the inclusion of daylight controls, the lighting energy consumption is reduced. This demonstrates that daylight controls are effective in reducing lighting energy demand by dimming or switching off artificial lights when sufficient natural daylight is available. The lighting loads are reduced to the extent that the combined lux levels of artificial light and daylight equal the setpoint lux level. This reduction in LPD reduces lighting energy consumption. Different zones would have different daylight extents and timing of availability, and hence would have different energy savings.
3. **Heating energy consumption:** Heating energy consumption increases when daylight controls are used. This increase of about 2.93% is attributed to reduced internal heat gains from artificial lighting, thereby increasing the heating demand to maintain comfort levels.

4. **Cooling energy consumption:** The cooling energy consumption decreases with daylight controls. Reduced heat generated from artificial lighting due to daylight control contributes to lower cooling loads.

Analysis

The implementation of daylight controls reduces lighting energy consumption, contributing to lower overall electricity consumption. Cooling energy consumption is also reduced as a beneficial side effect.

EXERCISE 1.9

Add daylight control in the south zone of the model and study the effect on energy consumption (Table 1.38).

TABLE 1.38
Annual Energy Consumption with and without Lighting Control for the South Zone

Type	Without Daylight Controls	With Daylight Controls
Room electricity		
Lighting		
System fans		
System pumps		
Heating (electricity)		
Heating (gas)		
Cooling (electricity)		

It can be seen that the deployment of lighting control increases the heating energy consumption and decreases the cooling energy consumption.

TUTORIAL 1.10: EVALUATING THE IMPACT OF SETPOINT TEMPERATURE

Goal

To evaluate the impact of setpoint temperature on sizing and energy consumption

What Are You Going to Learn?

- How to change the setpoint for heating and cooling

Problem Statement

Open the model created in Tutorial 1.9 (60 m × 40 m model with a core and four perimeter zones; consider a perimeter depth of 5 m) without a daylight sensor. Set the heating setpoint to 20°C and 22°C. Simulate the model with the **PARIS-AEROPORT CHAR** weather file. Analyze the effect of setpoint on energy consumption and the HVAC system sizing.

Solution

Step 1: Create a five-zone core and perimeter model of size **60 m × 40 m**.

Step 2: Select the **Activity** tab, and go to **Heating Setpoint Temperatures**. Enter **20** in **Heating** (°C) box.

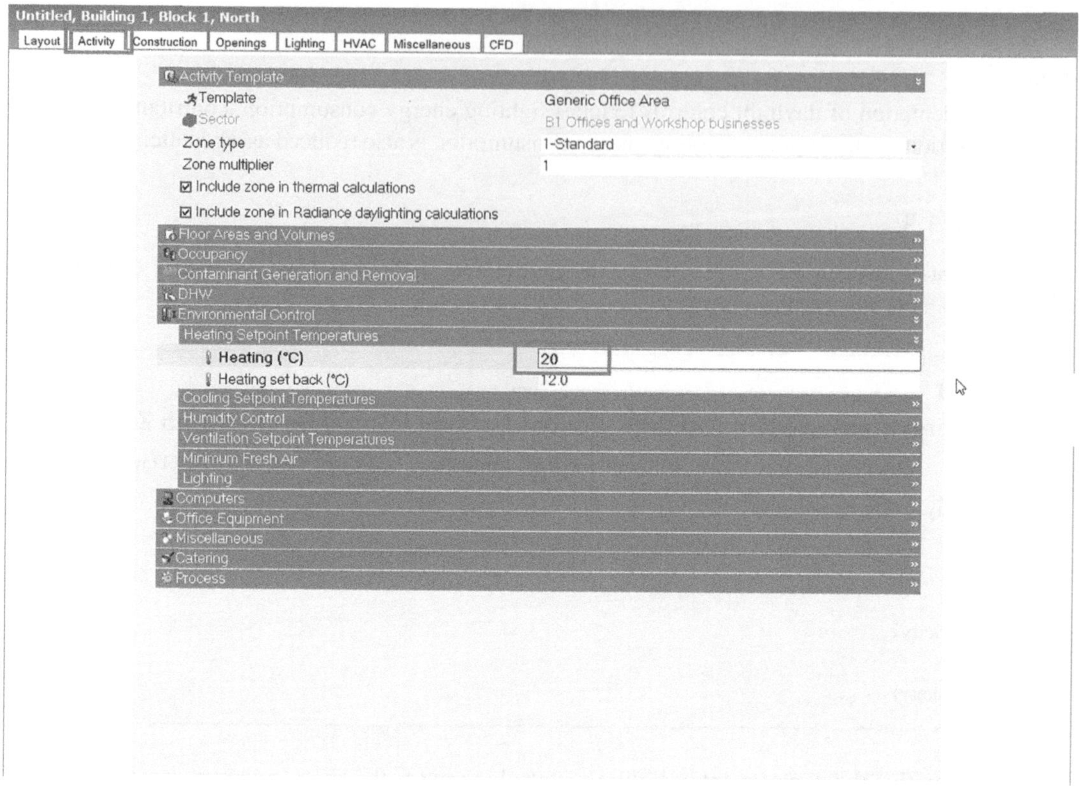

Step 3: Perform the annual simulation and record all end-use energy consumption.

Step 4: Repeat the steps to set the heating temperature to 22°C. Simulate the model.

Step 5: Compare results for both cases (Table 1.39 and Table 1.40). You can see that increasing the heating setpoints increases the heating energy consumption and heating sizing (Table 1.40).

TABLE 1.39
Energy Consumption with a Change in Heating Setpoint Temperature

	Annual Fuel Breakdown Data	
Type	**Heating Setpoint Temperature 20°C**	**Heating Setpoint Temperature 22°C**
Room electricity (kWh)	100,291.80	100,291.80
Lighting (kWh)	108,707.60	108,707.60
Heating (gas) (kWh)	112,341.60	153,152.40
Cooling (electricity) (kWh)	52,696.47	53,953.71
DHW (electricity) (kWh)	8,915.94	8,915.94

TABLE 1.40
Heating Sizing Capacity with a Change in Heating Setpoint Temperature

Type	With Heating Setpoint Temperature 20°C	With Heating Setpoint Temperature 22°C
Heating sizing (kW)	237.39	257.15

Key Observations

Heating energy consumption: Increasing the heating setpoint temperature from 20°C to 22°C results in a rise in heating energy consumption. This increase is due to the greater amount of heat required to maintain a higher indoor temperature, which also increases the heating system's capacity requirement.

Room electricity and lighting: The room electricity and lighting electricity consumption remain unchanged by the change in heating setpoint temperature. This is expected, as these components are independent of the indoor air temperature control.

System sizing: As the heating setpoint increases, the heating system sizing also increases. The higher heating setpoint demands a greater rate of heat addition to counterbalance the increased heat loss, thereby raising the system's capacity.

Overall Analysis

Increasing the heating setpoint temperature leads to higher energy consumption for heating, resulting in increased energy demand and system capacity. Maintaining a lower heating setpoint temperature is more energy-efficient, but comfort requirements must be balanced against energy-saving objectives.

EXERCISE 1.10

Repeat the tutorial for a change in the cooling setpoint from 24°C to 25°C. Compare the results for cooling sizing and energy consumption. Set the weather location as Brisbane, Australia (Table 1.41 and Table 1.42).

TABLE 1.41
Energy Consumption with a Change in Cooling Setpoint Temperature

	Annual Fuel Breakdown Data	
Type	With Cooling Setpoint 24°C	With Cooling Setpoint 25°C
Room electricity		
Lighting		
System fans		
System pumps		
Heating (electricity)		
Heating (gas)		
Cooling (electricity)		

TABLE 1.42
Cooling Sizing Capacity with Change in Cooling Setpoint Temperature

Type	With Cooling Setpoint 24°C	With Cooling Setpoint 25°C
Cooling (kW)		

TUTORIAL 1.11: EVALUATING THE IMPACT OF FRESH AIR SUPPLY RATE

Goal

To evaluate the impact of the fresh air supply quantity on energy consumption

What Are You Going to Learn?

- How to change the fresh air flow rate

Problem Statement

Create a single-zone 20 m × 15 m model. Set the fresh air supply rate (l/s-person) to 5 and 7.5, and compare. Study the effect on energy consumption for the **AZ-PHOENIX DEER VALLEY, USA** location.

Solution

Step 1: Create a single-zone 20 m × 15 m model. Select the **Activity** tab.

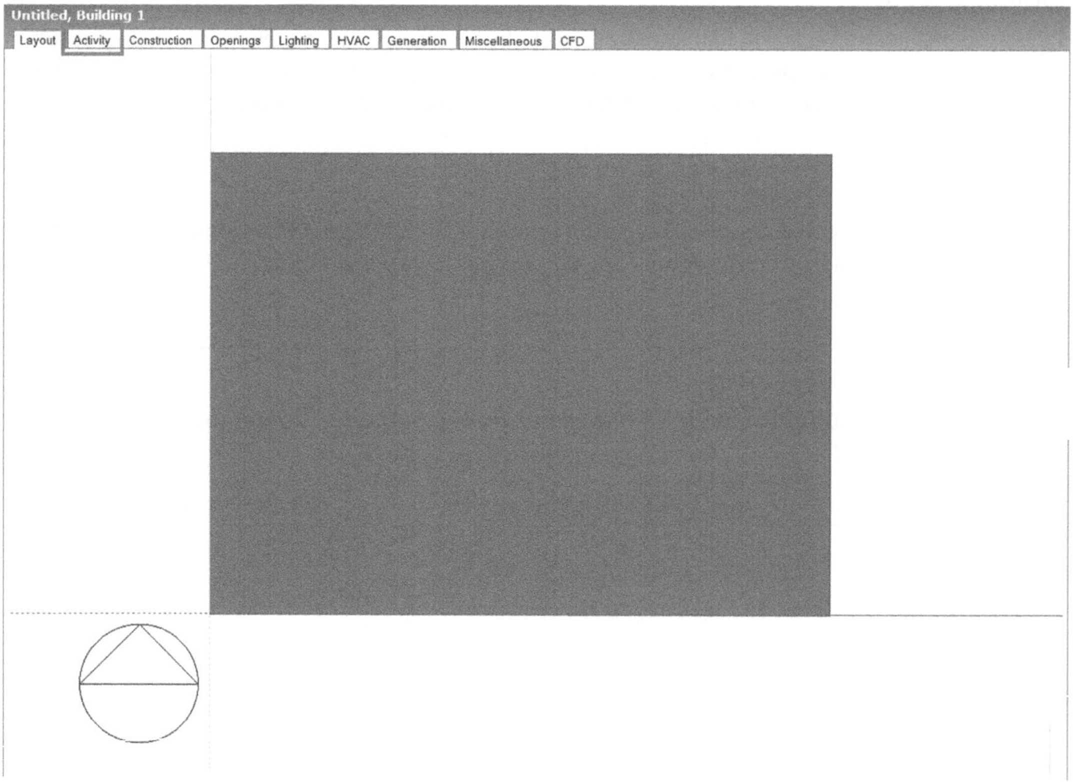

Step 2: Expand the **Minimum Fresh Air** section, and enter **5** in the **Fresh air (l/s-person)** section.

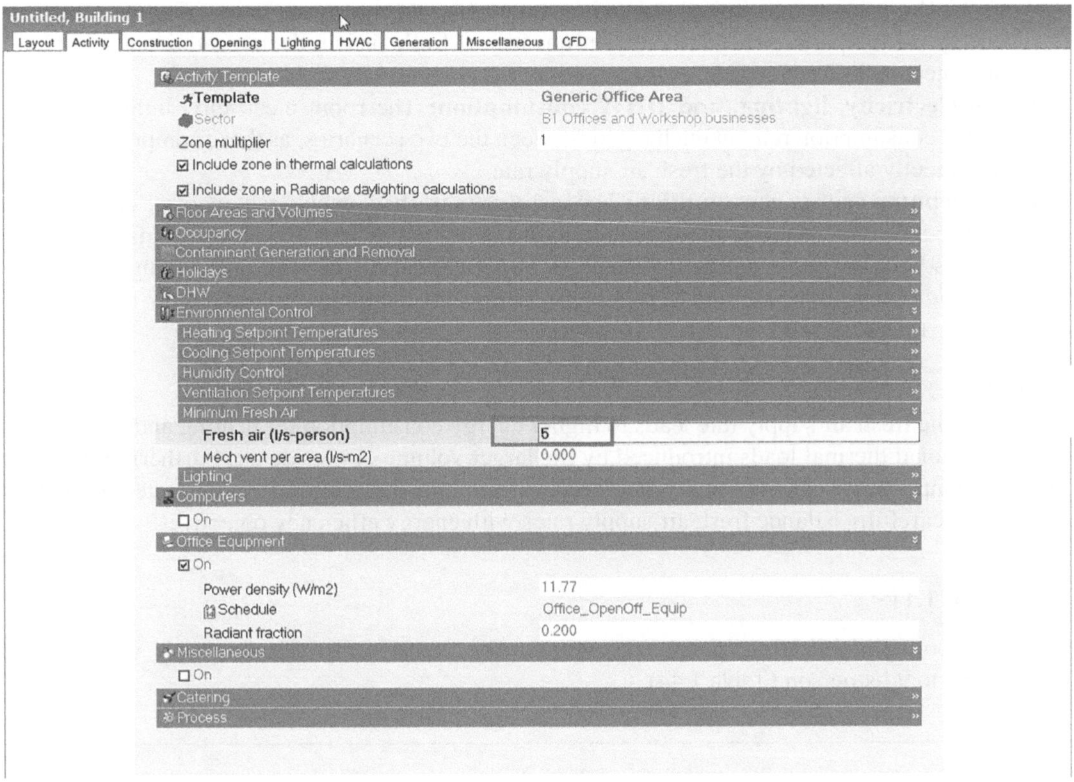

Step 3: Simulate the model, and record the results. Compare the end-use energy consumption results for the fresh air supply of 5 l/s-person and 7.5 l/s-person (Table 1.43).

TABLE 1.43
Energy Consumption with a Change in Fresh Air Supply Quantity

	Annual Fuel Breakdown Data	
Type	**Fresh Air Supply of 5 l/s-Person**	**Fresh Air Supply of 7.5 l/s-Person**
Room electricity (kWh)	12,130	12,130
Lighting (kWh)	13,148	13,148
Heating (gas) (kWh)	1,723	1,786
Cooling (electricity) (kWh)	29,801	30,177
DHW (electricity) (kWh)	1,078	1,078

Key Observations

Heating and cooling energy consumption: Increasing the fresh air supply rate from 5 l/s-person to 7.5 l/s-person results in higher heating and cooling energy consumption. The increase in energy consumption is due to the higher quantity of fresh air introducing additional sensible and latent loads from outside.

Room electricity, lighting, and DHW consumption: The room electricity, lighting, and DHW consumption remain unchanged between the two scenarios, as these components are not directly affected by the fresh air supply rate.

Fan and pump energy consumption: The increase in fresh air supply rate increases the volumetric flow rate, resulting in higher energy consumption by fans and pumps. This increase occurs because more supply air needs to be circulated for heating and cooling, requiring additional fan power and higher pump ratings for the increased water flow needed to handle the increased load.

Analysis

Increasing the fresh air supply rate leads to higher energy consumption for heating and cooling due to the additional thermal loads introduced by the larger volume of outside air. Furthermore, higher fan and pump energy consumption is required to maintain the required airflow rates, making it essential to carefully balance fresh air supply rates with energy efficiency objectives.

EXERCISE 1.11

Compare cooling and heating energy consumption for a change in fresh air supply volume from 5 l/s-person to 9 l/s-person (Table 1.44).

TABLE 1.44
Heating and Cooling Energy Consumption by Changing Fresh Air Supply Quantity

	Cooling Energy Consumption, Electricity (kWh)		Heating Energy Consumption, Gas (kWh)	
Month	5 l/s-Person	9 l/s-Person	5 l/s-Person	9 l/s-Person
Jan				
Feb				
Mar				
Apr				
May				
Jun				
Jul				
Aug				
Sept				
Oct				
Nov				
Dec				

2 Geometry of Buildings

In this chapter, we discuss geometrical aspects such as thermal zoning, aspect ratio, floor multiplier, and surface adjacency. Usually, buildings have several rooms; however, from a modeling perspective, there may not be a requirement to model each room separately. If adjacent spaces have the same specifications—such as schedule, occupancy, and cooling and heating temperature setpoint—you can combine these spaces and model the result as a single zone. By doing this, the complexity of the model and its simulation run time can be reduced without affecting the energy simulation results very much. Similarly, in cases of multi-storied buildings with typical floors, the model can be simplified by modeling only three floors: ground, typical intermediate, and top. One of the important parameters that need to be considered while designing the building is the *aspect ratio*, which is the ratio of the floor length to the width of the building. The *aspect ratio* affects the envelope area, which refers to the total surface area of all external components, such as walls and the roof. As the envelope area increases, the building experiences higher heat gains and losses. On the other hand, a higher *aspect ratio* helps with better distribution of daylight and more access to windows for occupants. The impact of these aspects on building performance is explained through four tutorials in this chapter.

TUTORIAL 2.1: DEFINING THERMAL ZONING FOR A BUILDING

Goal

To evaluate the effect of architectural and thermal zoning on the end-use energy consumption and simulation run time

What Are You Going to Learn?

- How to define thermal zoning
- How to add internal mass

Problem Statement

Create a multi-zone building (15 zones, as shown in the following figure) with a rectangular footprint of 100 m × 50 m. Find its energy performance and simulation run time. Create another model by combining similar spaces into thermal zones and compare the energy performance and run-time duration with those of the first model. Use the **FRANKFURT MAIN ARPT**, Germany, weather file.

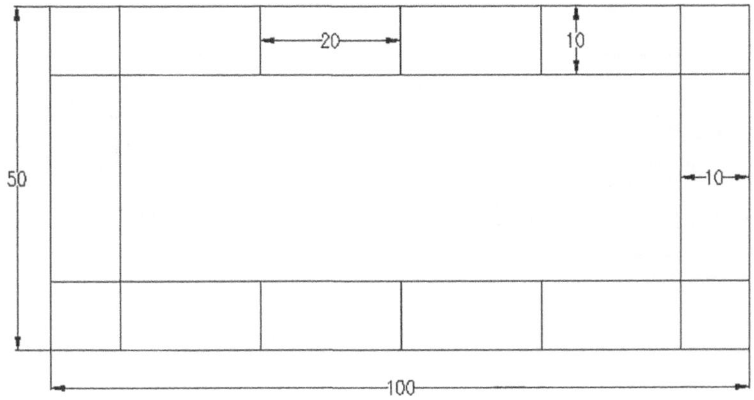

DOI: 10.1201/9781003563013-2

Solution

Step 1: Open a **New Project** and create a **100 m × 50 m** block with internal partitions. (Use construction lines to facilitate easy snapping while creating the partitions.)

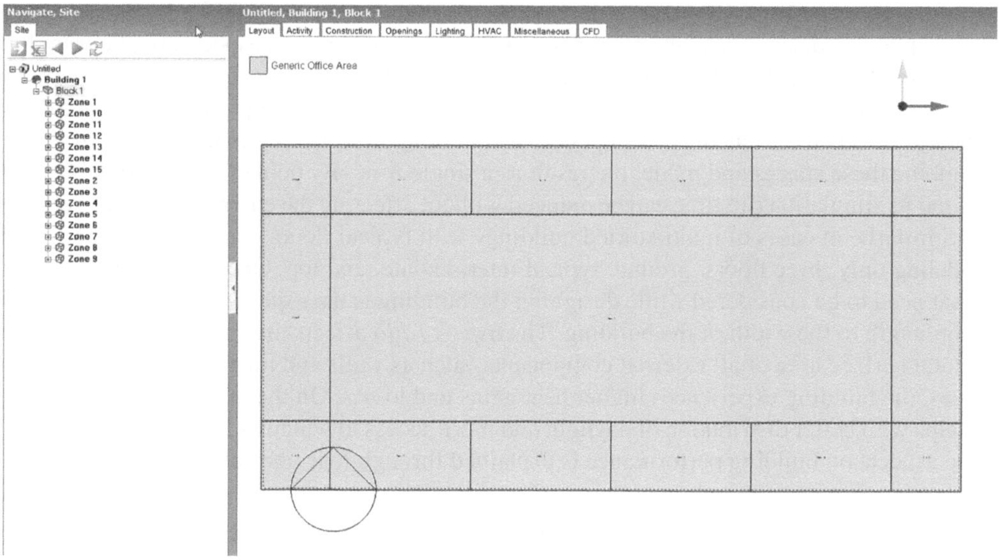

Step 2: Select the **Construction** tab and select **Outer volume** in the **Zone floor area calculation method** drop-down list.

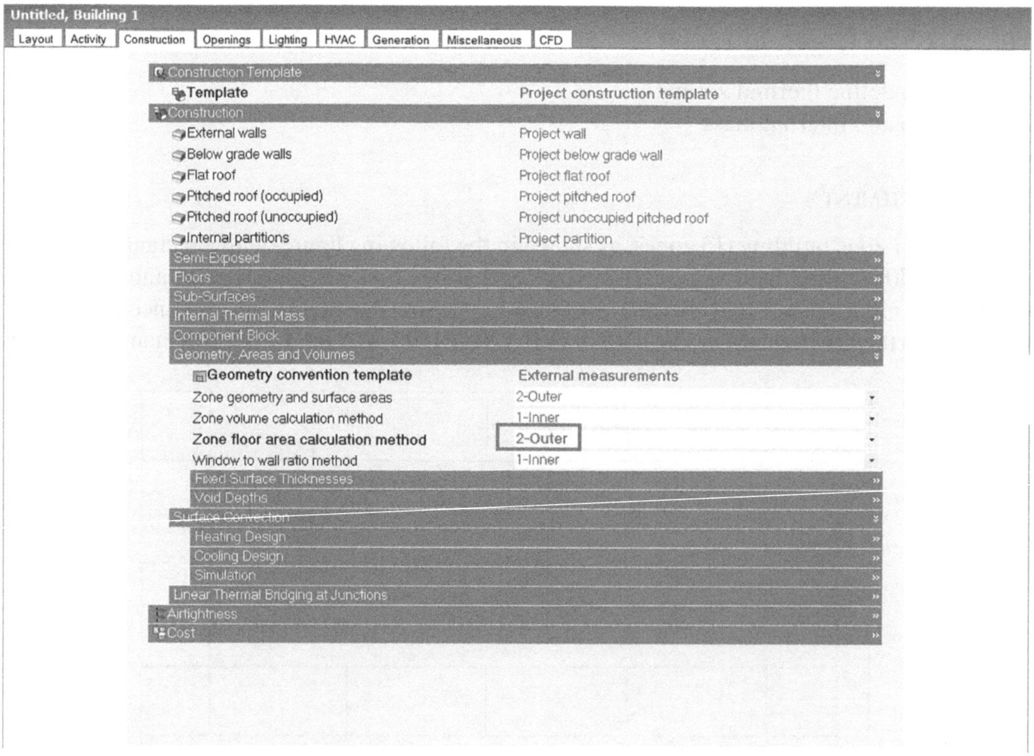

ZONE FLOOR AREA CALCULATION METHOD

This setting dictates whether internal or external measurements (including external wall thickness inside the zone boundary, or not) are used to calculate the zone floor area. This is required to calculate per square-meter values such as occupancy and other internal gains, as well as floor area values for general reporting.

Inner: Where the zone volume used in thermal calculations is derived from the zone inner geometry.

Outer: Where the zone volume used in thermal calculations is derived from the zone outer geometry.

Source: http://designbuilder.co.uk/helpv7.3/Content/GeometryAreasAndVolumes.htm

Step 3: Perform hourly simulation for the whole year and record the energy consumption and run time.

RECORDING A RUN TIME

After the simulation is complete, type the following path to open the eplusout.err file. You can use any text editor to open this file.

C:\Users\User\AppData\Local\DesignBuilder\EnergyPlus\eplusout.err

The actual folder name on a computer depends on the language setting and your Windows username. In this path given, the Windows username is "User". Alternatively, you can open the EnergyPlus folder from the DesignBuilder file option. At the end of the file, you can find Elapsed Time. You need to record the elapsed time.

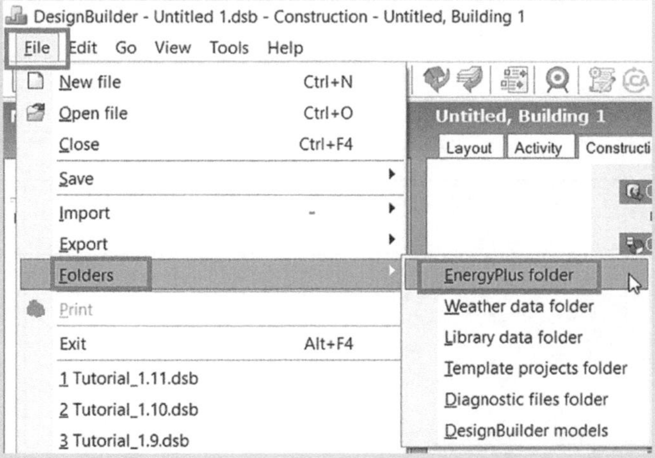

Source: https://designbuilder.co.uk/helpv7.3/Content/_DesignBuilder_files_location_and_extensions.htm

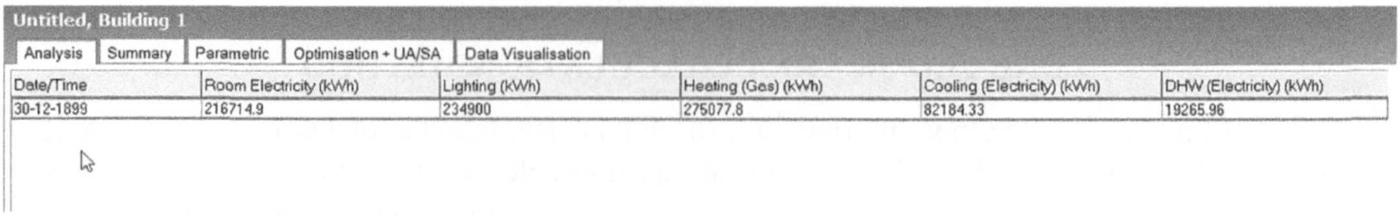

Date/Time	Room Electricity (kWh)	Lighting (kWh)	Heating (Gas) (kWh)	Cooling (Electricity) (kWh)	DHW (Electricity) (kWh)
30-12-1899	216714.9	234900	275077.8	82184.33	19265.96

Step 4: Repeat Step 1 while combining zones for spaces with similar activity, schedules, and set-points. In this example, let us assume that the north zones (zones 7–10), as shown in the following figure, are similar and can be combined into a single thermal zone. Similarly, the south zones (zones 12–15) can be combined into a single thermal zone. Do note that zone numbers in your model might differ because zones are numbered in the sequence in which they were created.

You can select an internal partition and delete an internal partition by pressing the **Delete** key. (Please ensure that you are at the block level.)

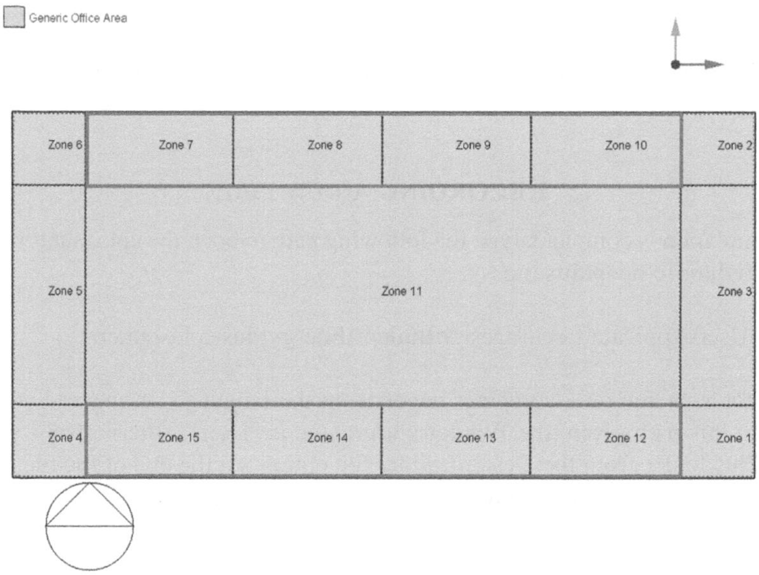

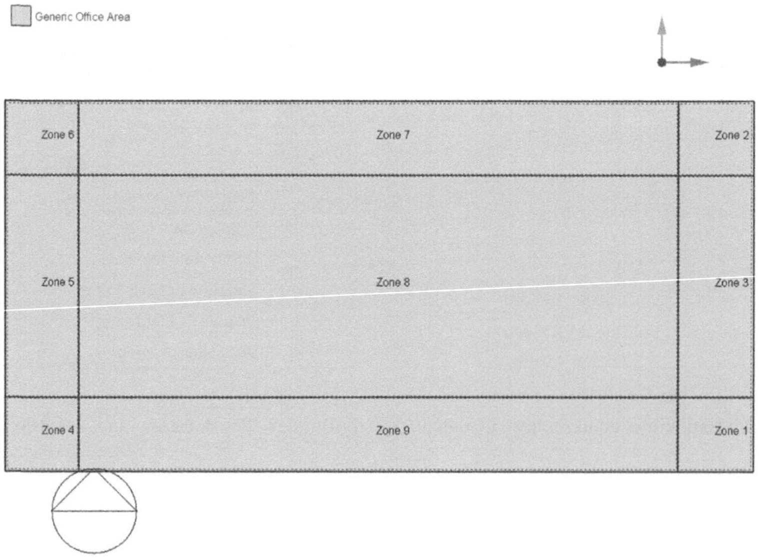

A *thermal zone*, usually termed simply *zone*, is a virtual or real segment of a building that has a homogeneous enclosed volume of air. In a simple approach, each physical space can be treated as one zone. However, to simplify the modeler's work and to reduce the calculation time, areas having similar thermal and usage conditions such as occupancy, setpoint, and solar exposure, and those which are serviced by common mechanical equipment, can be clubbed to create one zone.

Step 5: Perform annual simulation and record the energy consumption and the run time.
Compare the energy simulation results and the simulation run time (Table 2.1 and Table 2.2).

TABLE 2.1
Annual Fuel Breakdown Energy with Architectural Zoning and Lumped Thermal Zones

Type	With architectural zoning	With lumped thermal zones
Room electricity (kWh)	2,16,715	2,16,715
Lighting (kWh)	2,34,900	2,34,900
Heating (gas) (kWh)	2,75,078	2,74,407
Cooling (electricity) (kWh)	82,184	81,945
Domestic hot water (DHW, electricity) (kWh)	19,266	19,266

TABLE 2.2
Simulation Run Time* with Architectural Zoning and Lumped Thermal Zones

With Architectural Zoning	With Lumped Thermal Zones
48.04 seconds	31.40 seconds

*Simulation run times may vary from the values given in the preceding table based on your computer configuration.

Analysis

The results demonstrate the impact of thermal zoning on simulation performance and energy consumption.

Simulation Run Time

By combining architectural physical zones to form virtual thermal zones, the simulation run time decreased from 48.04 seconds to 31.40 seconds, representing an approximate 35% reduction in computation time.

Energy Performance

The energy consumption results for various categories—such as Room electricity and Lighting—remain unchanged between the models. This is because these loads are typically independent of thermal zoning and are more related to internal gains.

Slight differences are observed in heating and cooling energy consumption between the two models due to changes in internal thermal mass.

Similarly, slight differences in system capacity can also be observed due to differences in peak load occurrence.

EXERCISE 2.1

Create thermal zoning for the plan shown in the following figure. Compare the simulation run time and energy consumption for the models with architectural and thermal zoning (Table 2.3 and Table 2.4).

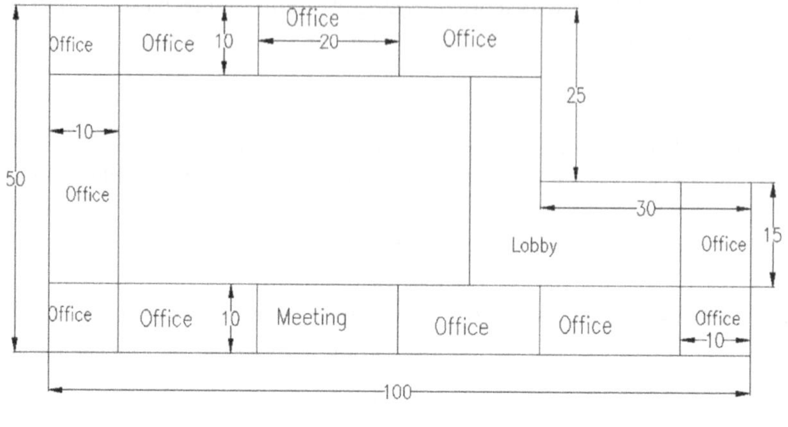

TABLE 2.3
Annual Fuel Breakdown Energy with Architectural Zoning and Lumped Thermal Zones

Type	With Architectural Zoning	With Lumped Thermal Zones
Room electricity (kWh)		
Lighting (kWh)		
Heating (gas) (kWh)		
Cooling (electricity) (kWh)		
DHW (electricity) (kWh)		

TABLE 2.4
Simulation Run Time with Architectural Zoning and Lumped Thermal Zones

With Architectural Zoning (Seconds)	With Lumped Thermal Zones (Seconds)

TUTORIAL 2.2: EVALUATING THE EFFECT OF A ZONE MULTIPLIER

Goal

To evaluate the effect of a zone multiplier on the energy consumption and simulation run time

What Are You Going to Learn?

- How to use a zone multiplier

Problem Statement

Use the **Zone Multiplier Example** template file to evaluate the impact of a floor multiplier on a building's energy consumption and simulation run time. This file contains two buildings, with and without the floor multiplier. You need to select one building at a time and simulate for London Gatwick Airport, United Kingdom.

CONCEPT OF ZONE MULTIPLIER

The zone multiplier data allow you to reduce the size of your model in cases where there are similar zones by specifying that certain zones are repeated and so only need to be simulated once. A typical use is for multi-story buildings with identical (or very similar) floors. The concept of zone multiplier when applied at the floor level multiplies all the zones on the given floor, effectively working as a floor multiplier. Hence, in a building where there are several identical floors, you can model one floor and use a zone multiplier on that floor.

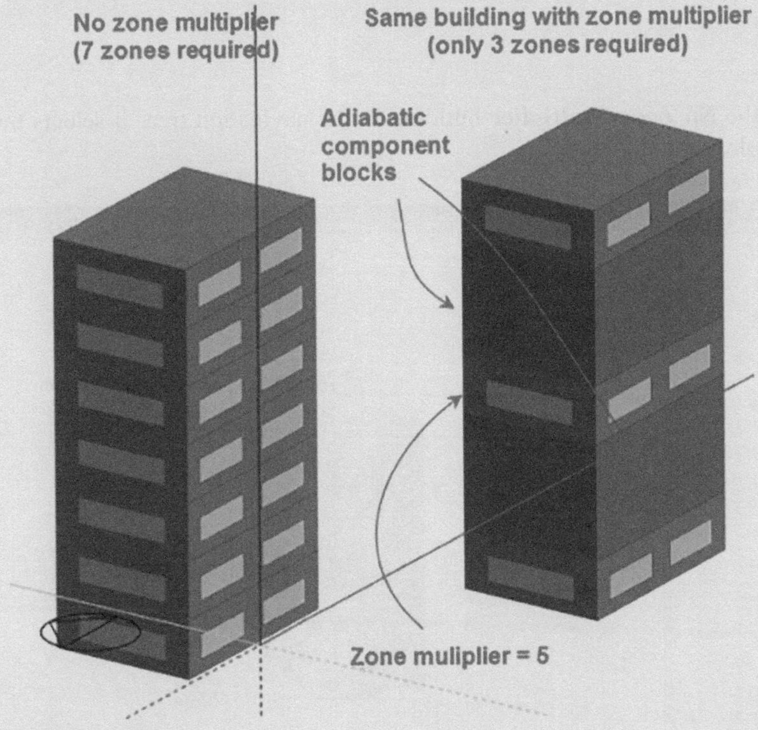

Source: https://designbuilder.co.uk/helpv7.0/Content/Zone%20Multiplier.htm

Solution

Step 1: Open the **Zone Multiplier Example** template from DesignBuilder templates. The layout appears with two buildings.

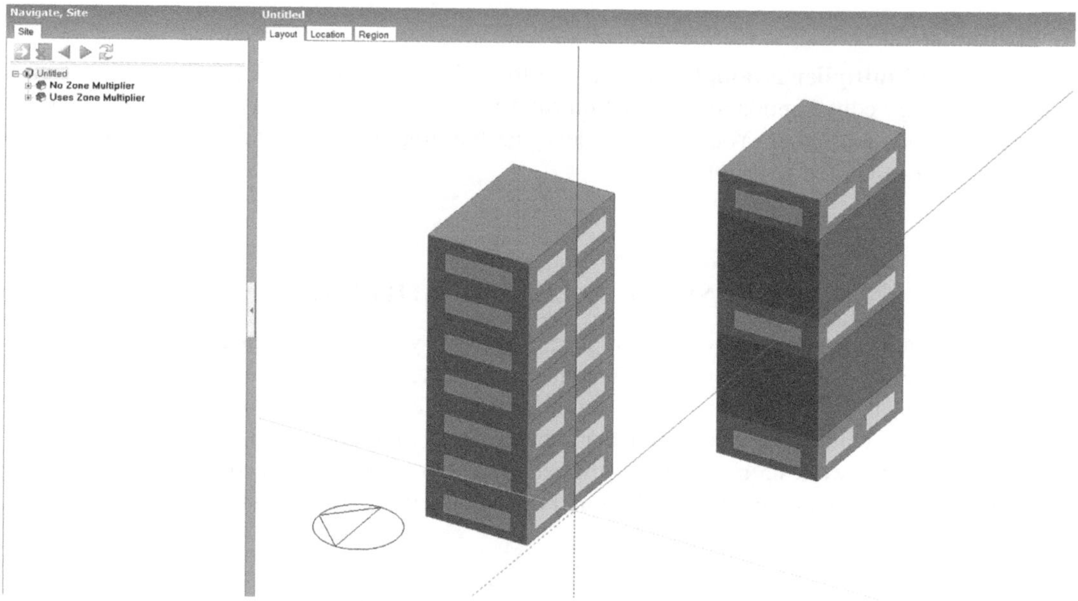

Step 2: Click the **No Zone Multiplier** building in the navigation tree. It selects the building that does not use a floor multiplier.

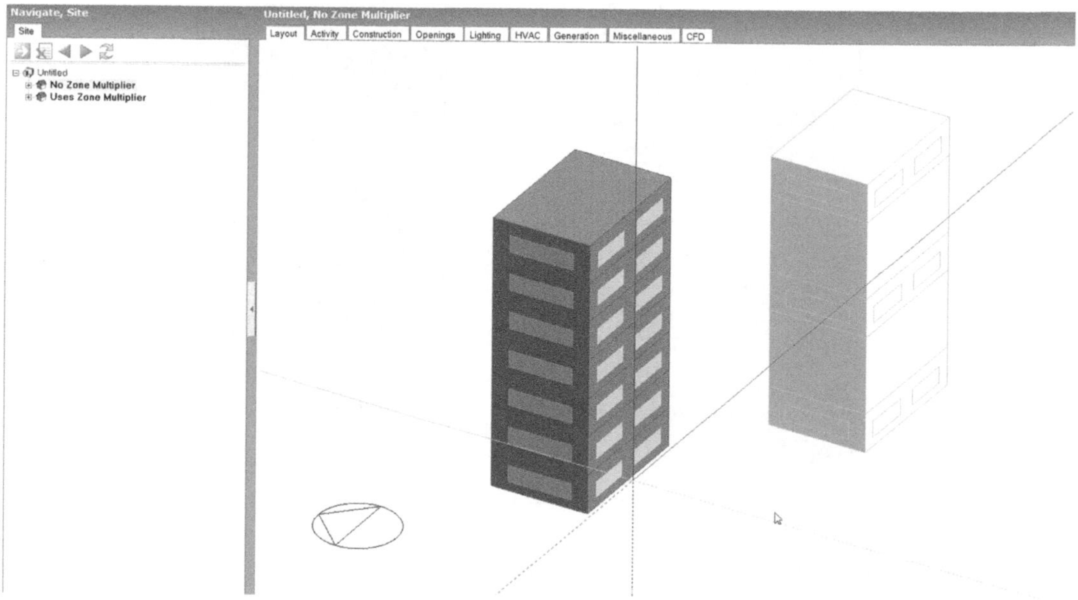

Step 3: Perform annual simulation. Record the simulation run time. The results appear. Record the energy simulation results.

Step 4: Select the **Edit** tab. Click the **Uses Zone Multiplier** building in the navigation tree. Expand the **Uses Zone Multiplier** option.

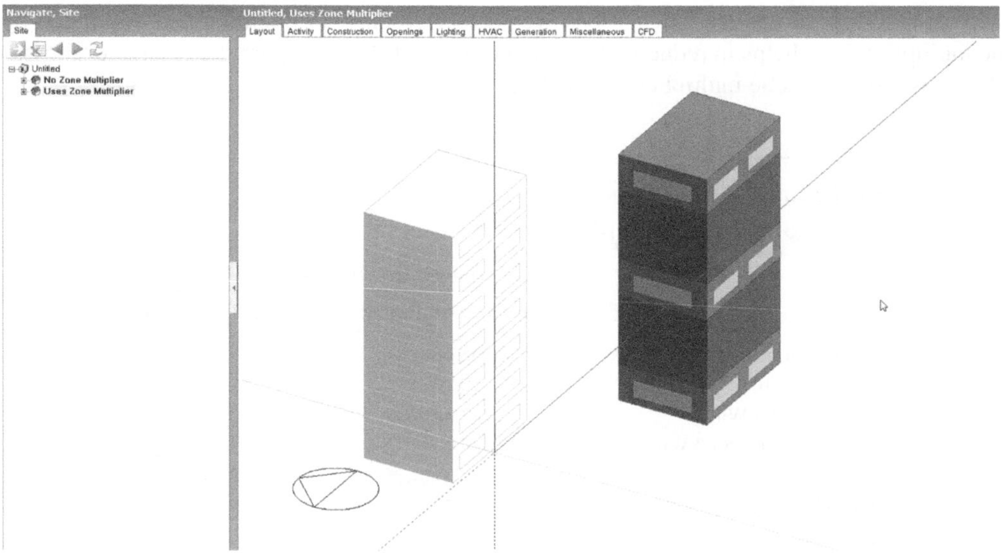

Step 5: Under **Mid**, click **Zone 1**.

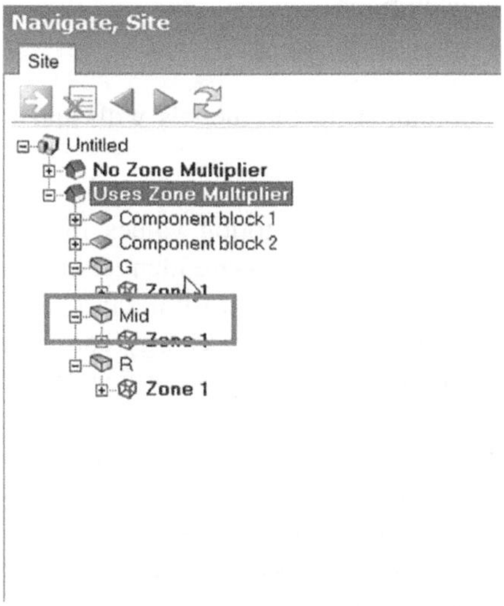

Step 6: Select the **Activity** tab. Because the middle five floors are typical floors, the **Zone Multiplier** for the whole floor is set to **5**.

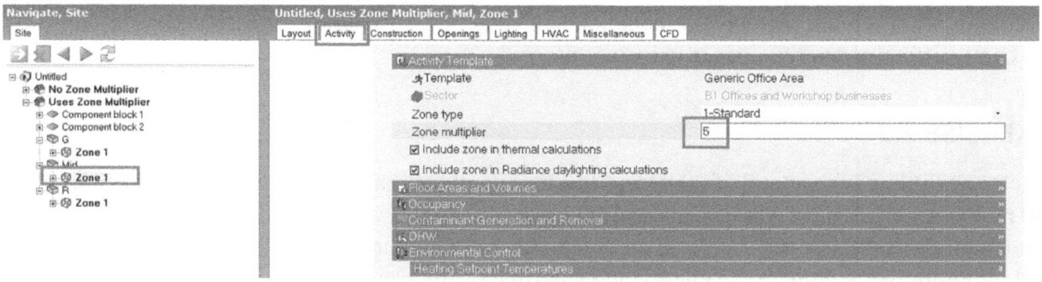

When using a floor multiplier, calculations are performed only for one floor and multiplied by the zone multiplier. This helps in reducing the simulation run time. Compare the results for the models with and without the zone multiplier (Table 2.5).

TABLE 2.5
Impact of a Zone Multiplier

	Annual Fuel Breakdown Data	
Type	Without Floor Multiplier	With Floor Multiplier
Room electricity (kWh)	29,238.68	29,238.68
Lighting (kWh)	52,820.29	52,820.29
Heating (gas) (kWh)	32,782.83	32,768.4
Cooling (electricity) (kWh)	22,459.23	21,946.7
DHW (gas) (kWh)	2,998.37	2,998.37
Simulation time (seconds)	8.74	7.67

The results show that using a zone multiplier reduces the simulation run time. However, this has a slight impact on the energy consumption.

Results and Analysis

The results from the simulation comparing the models with and without the floor multiplier are summarized in Table 2.5.

Energy consumption: The energy consumption values for various categories—including room electricity, lighting, and DHW (gas)—are identical for both models. The differences in heating and cooling energy are minimal, with the model using a floor multiplier showing a slight reduction in heating and cooling energy consumption. This small difference indicates that using a floor multiplier does not significantly affect the overall energy consumption of the building. The variations observed in heating and cooling loads may be due to minor calculation differences introduced by the multiplier mechanism.

Simulation run time: The simulation run time is reduced when using the floor multiplier. The model without the floor multiplier takes 8.74 seconds to complete, whereas the model with the floor multiplier completes the simulation in 7.67 seconds. This improvement in simulation speed highlights the efficiency advantage of using the floor multiplier when simulating large or repetitive floors within a building.

Conclusion

Using a zone multiplier proves beneficial in reducing simulation run time without significantly affecting energy consumption results.

EXERCISE 2.2

Create a building with a total of 20 floors. The first ten floors have a floor plate of dimensions 50 m × 50 m. Floors 11–20 have a floor plate of dimensions 25 m × 25 m. Each floor has a height of 3 m. All floors are centrally aligned. Perform an annual energy simulation of the building with and without a floor multiplier. Compare the energy and run time of the simulation.

TUTORIAL 2.3: EVALUATING THE IMPACT OF THE ASPECT RATIO

Goal

To evaluate the impact of building aspect ratio on energy performance with and without daylight sensors

What Are You Going to Learn?

- How to model a building with different aspect ratios but with the same floor area

Problem Statement

In this tutorial, you are going to analyze the impact of the aspect ratio of a building on the building's energy performance. You need to create models with different aspect ratios for floor plates of areas 64 m^2 and 625 m^2. Simulate each of the following cases and compare their energy consumption for the London Gatwick weather location. Make sure that the WWR is 30%.

1. For a floor area of 64 m^2, Table 2.6 gives the length and breadth values for various aspect ratios
2. For a floor area of 625 m^2, Table 2.7 gives the length and breadth values for various aspect ratios

For a rectangular building, the *aspect ratio* is the ratio of the longest dimension of the building footprint to the narrowest dimension. An aspect ratio of 1.0 represents a square building footprint.

TABLE 2.6
Different Aspect Ratios for a 64 m² Floor Area

Serial No.	Length l (m)	Breadth b (m)	Aspect Ratio l/b	Floor Area $l \times b$ (m^2)	Façade Area (Window + Wall) (m^2)	Façade Area/ Floor Area
1	8.00	8.00	1	64	112.00	1.75
2	11.31	5.65	2	64	118.79	1.86
3	13.85	4.62	3	64	129.33	2.02
4	16.00	4.00	4	64	140.00	2.19

TABLE 2.7
Different Aspect Ratios for a 625 m² Floor Area

Serial No.	Length l (m)	Breadth b (m)	Aspect Ratio l/b	Floor Area $l \times b$ (m^2)	Façade Area (Window + Wall) (m^2)	Façade Area/ Floor Area
1	25.0	25.0	1	625	350.00	0.56
2	43.3	14.4	3	625	404.15	0.65
3	55.9	11.2	5	625	469.57	0.75
4	66.1	9.4	7	625	529.15	0.85

Solution

Step 1: Open a new project file.

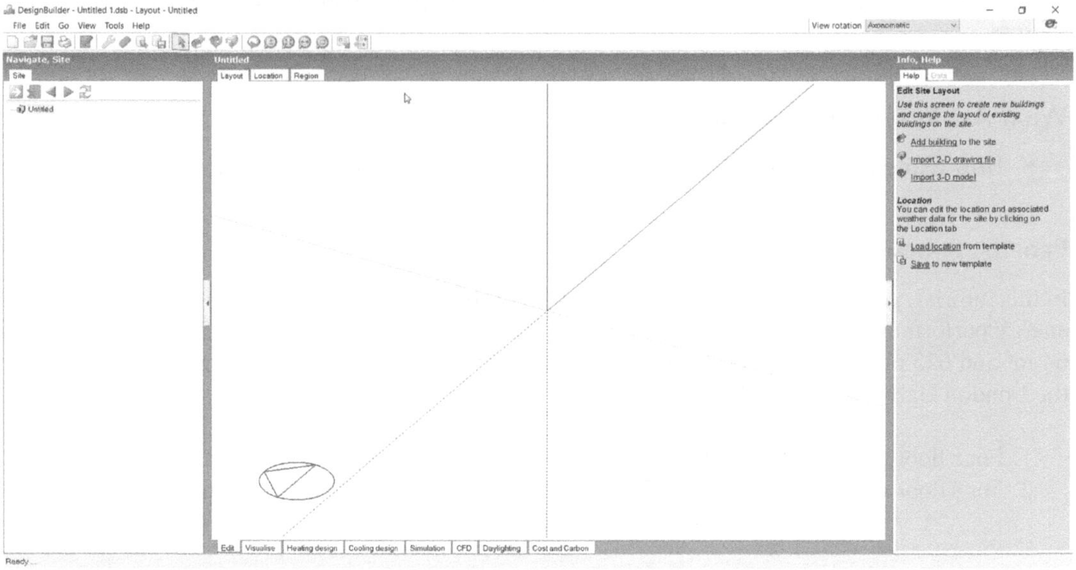

Step 2: Create an **8 m × 8 m** building (aspect ratio 1).

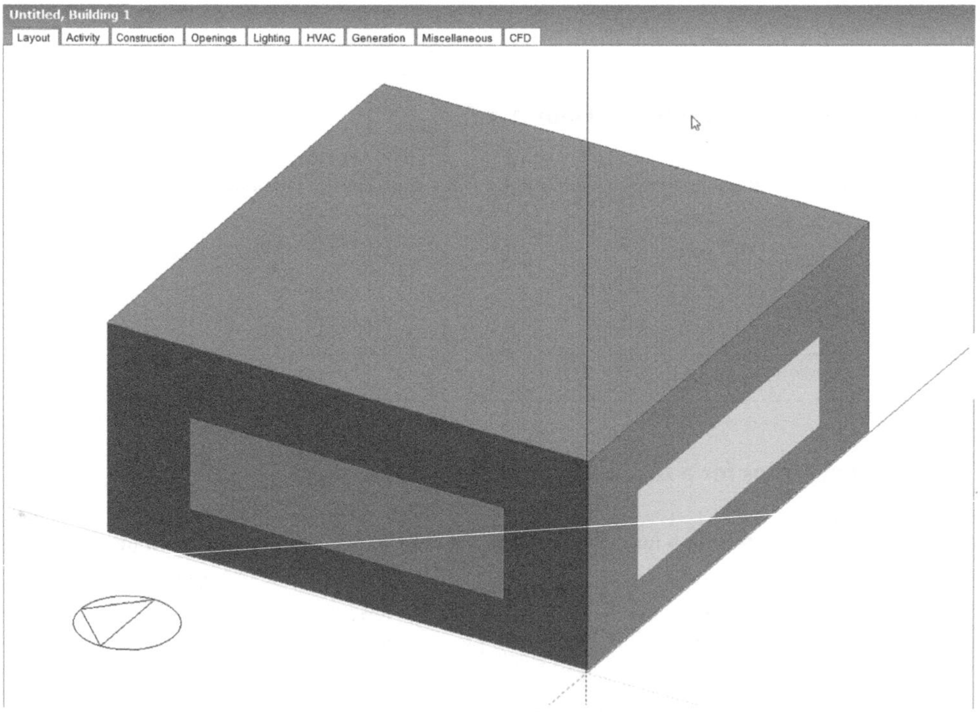

Step 3: Simulate the model and store the results normalized by area (Table 2.8).

TABLE 2.8
Energy Simulation Results for an 8 m × 8 m Model

Room Electricity (Wh/m²)	Lighting (Wh/m²)	Heating (Gas) (Wh/m²)	Cooling (Electricity) (Wh/m²)	DHW (Electricity) (Wh/m²)
43,343	46,980	71,719	34,514	3,853

Step 4: Repeat Steps 1–3 for aspect ratios 2–4, respectively. Compare the energy simulation results for all cases (Table 2.9).

TABLE 2.9
Energy Consumption with Different Aspect Ratios (Without a Daylight Sensor)

Aspect Ratio	1 (8 m × 8 m)	2 (11.31 m × 5.65 m)	3 (13.85 m × 4.62 m)	4 (16 m × 4 m)
Room electricity (Wh/m²)	43,343	43,343	43,343	43,343
Lighting (Wh/m²)	46,980	46,980	46,980	46,980
Heating (gas) (Wh/m²)	71,719	72,613	75,504	79,182
Cooling (electricity) (Wh/m²)	34,514	35,167	37,699	40,191
DHW (electricity) (Wh/m²)	3,853	3,853	3,853	3,853

The following figure shows the energy consumption with aspect ratio when daylight sensors are not modeled.

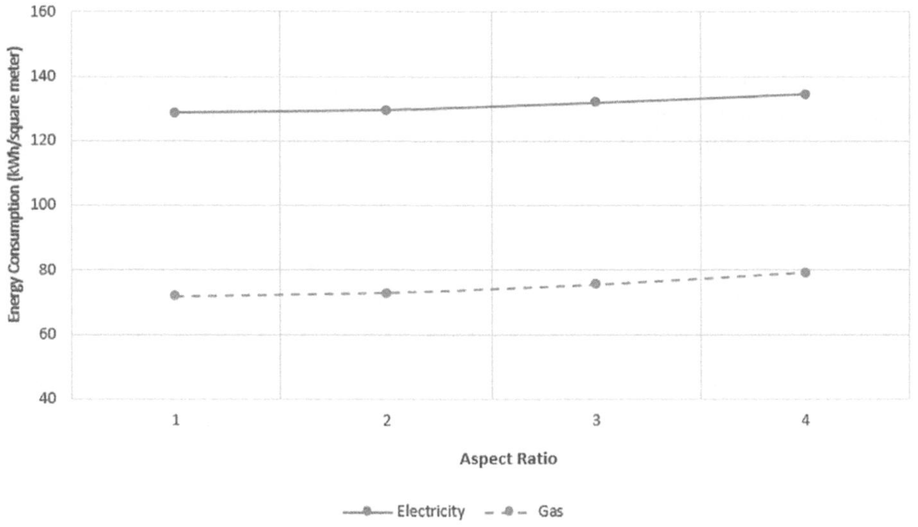

Step 5: Repeat all the preceding steps with a daylight sensor (Refer to Tutorial 1.9 for modeling of a daylight sensor) (Table 2.10).

TABLE 2.10
Energy Consumption with Varying Aspect Ratios (With a Daylight Sensor)

Aspect Ratio	**1 (8 m × 8 m)**	**2 (11.31 m × 5.65 m)**	**3 (13.85 m × 4.62 m)**	**4 (16 m × 4 m)**
Room electricity (Wh/m^2)	43,343	43,343	43,343	43,343
Lighting (Wh/m^2)	11,476	11,469	11,363	10,711
Heating (gas) (Wh/m^2)	85,405	86,083	88,789	92,819
Cooling (electricity) (Wh/m^2)	26,495	27,135	29,559	31,967
DHW (electricity) (Wh/m^2)	3,853	3,853	3,853	3,853

The following figure shows energy consumption with varying aspect ratios when a daylight sensor is modeled.

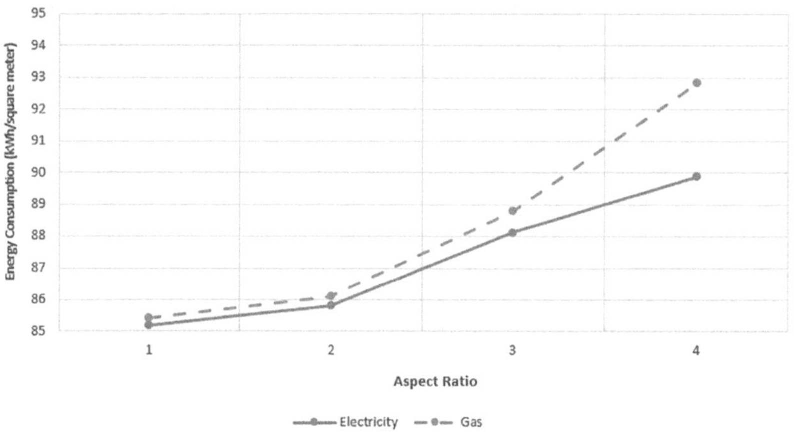

Step 6: Repeat all the preceding steps for a floor area of 625 m^2 (Table 2.11and Table 2.12). You need to draw the model with the help of plans as shown in the following figures with aspect ratios of 1, 3, 5, and 7. Perimeter depth for all plans is 600 mm.

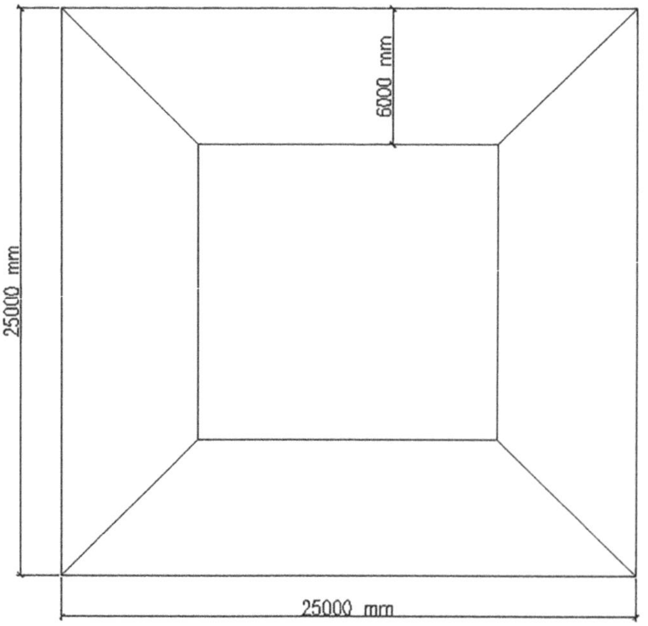

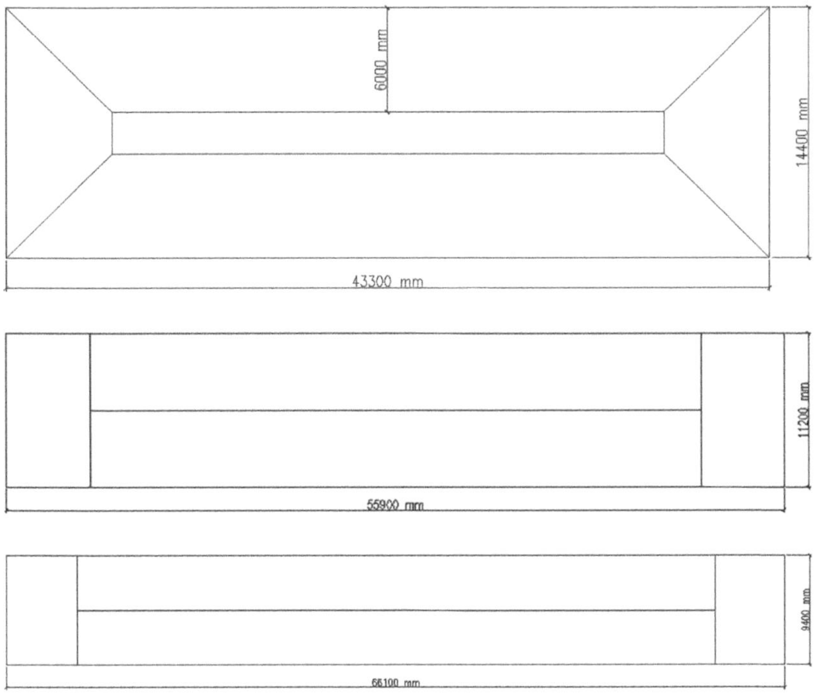

TABLE 2.11
Energy Consumption for Different Aspect Ratios (Without a Daylight Sensor)

Aspect ratio	1 (25 m × 25 m)	3 (43.3 m × 14.4 m)	5 (55.9 m × 11.2 m)	7 (66.1 m × 9.4 m)
Room electricity (Wh/m^2)	43,343	43,343	43,343	43,343
Lighting (Wh/m^2)	46,980	46,980	46,980	46,980
Heating (gas) (Wh/m^2)	51,720	52,823	52,771	54,669
Cooling (electricity) (Wh/m^2)	17,507	18,705	19,963	21,861
DHW (electricity) (Wh/m^2)	3,853	3,853	3,853	3,853

TABLE 2.12
Energy Consumption for Different Aspect Ratios (With Daylight Sensor)

Aspect Ratio	1 (25 m × 25 m)	3 (43.3 m × 14.4 m)	5 (55.9 m × 11.2 m)	7 (66.1 m × 9.4 m)
Room electricity (Wh/m^2)	43,343	43,343	43,343	43,343
Lighting (Wh/m^2)	23,406	18,106	13,704	12,973
Heating (gas) (Wh/m^2)	61,419	64,712	66,354	68,523
Cooling (electricity) (Wh/m^2)	12,134	12,146	12,469	14,120
DHW (electricity) (Wh/m^2)	3,853	3,853	3,853	3,853

The first of the following figures shows the energy consumption with aspect ratio when a daylight sensor is not modeled, and the second figure shows energy consumption with varying aspect ratios when a daylight sensor is modeled.

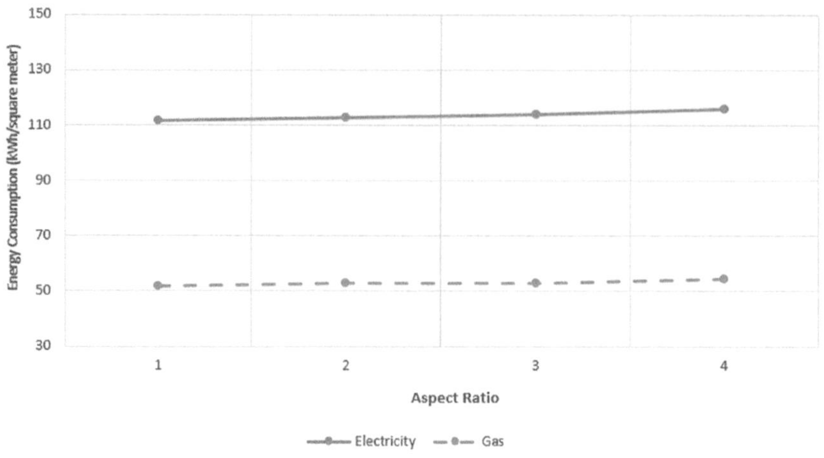

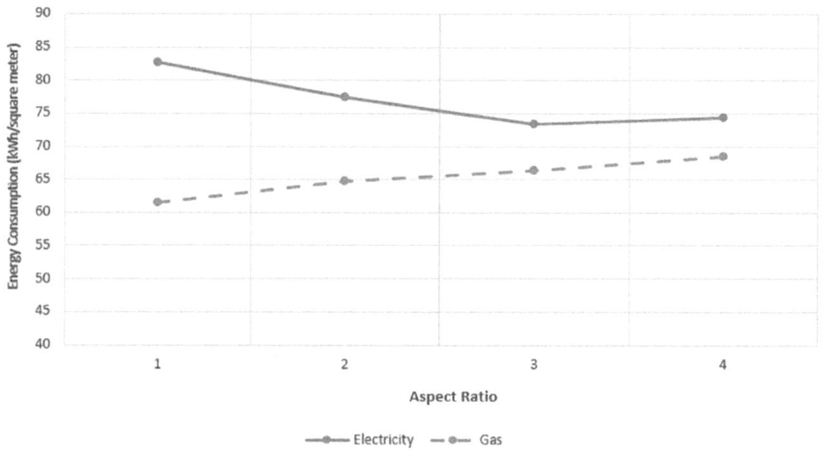

Results and Analysis

The results of the simulation comparing different aspect ratios (1, 3, 5, and 7) for a floor area of 625 m², both with and without daylight sensors, are provided in Tables 2.11 and 2.12.

Energy consumption (without daylight sensors): It is evident from Table 2.11 that increasing the aspect ratio of the building results in higher heating and cooling energy consumption. The increase in aspect ratio leads to greater heat gains and losses due to the larger exposed surface area, contributing to higher cooling and heating demands. However, lighting consumption remains constant regardless of the aspect ratio, as it is independent of change in aspect ratio in the absence of daylight sensors.

Energy consumption (with daylight sensors): It is evident from Table 2.12 that the use of daylight sensors leads to a significant reduction in lighting energy consumption as the aspect ratio increases. This is due to the improved availability of natural light for elongated floor plates.

Conclusion

The results indicate that increasing the aspect ratio of a building can lead to higher heating and cooling energy consumption, due to increased heat gains and losses from larger surface areas. This effect is more pronounced if the orientation of larger surfaces is sun-facing for a longer duration. However, incorporating daylight sensors substantially reduces lighting energy consumption, especially for buildings with elongated floor plates. Designers should balance the trade-off between aspect ratio and energy performance by optimizing natural lighting strategies and insulation.

EXERCISE 2.3

1. Analyze the effect of aspect ratio for all cases described in this tutorial for the 90° orientation
2. Repeat the tutorial for a hot and dry climate (**UAE**) and observe the energy simulation results

TUTORIAL 2.4: EVALUATING THE IMPACT OF ADJACENCY OF THE SURFACE

Goal

To evaluate the impact of the ground surface on energy performance

What Are You Going to Learn?

- How to assign surface adjacency
- How to make a surface adiabatic

Adiabatic surface means that heat is not transferred across its external surface.

Source: https://designbuilder.co.uk/helpv7.3/#Adjacency.htm

Problem Statement

In this tutorial, you are going to create a 50 m × 25 m single-zone model. Assign ground floor construction to a 200 mm aerated concrete slab. Set the ground floor surface adjacency to **Auto** and as **Adiabatic**. Compare the effect of the surface adjacency on the energy consumption for **QC—Montreal/Mirabel INT'L A, Canada**.

Solution

Step 1: Create a **50 m × 25 m single zone** model. Expand **Zone 1** in the navigation tree.

Step 2: Click **Ground Floor**.

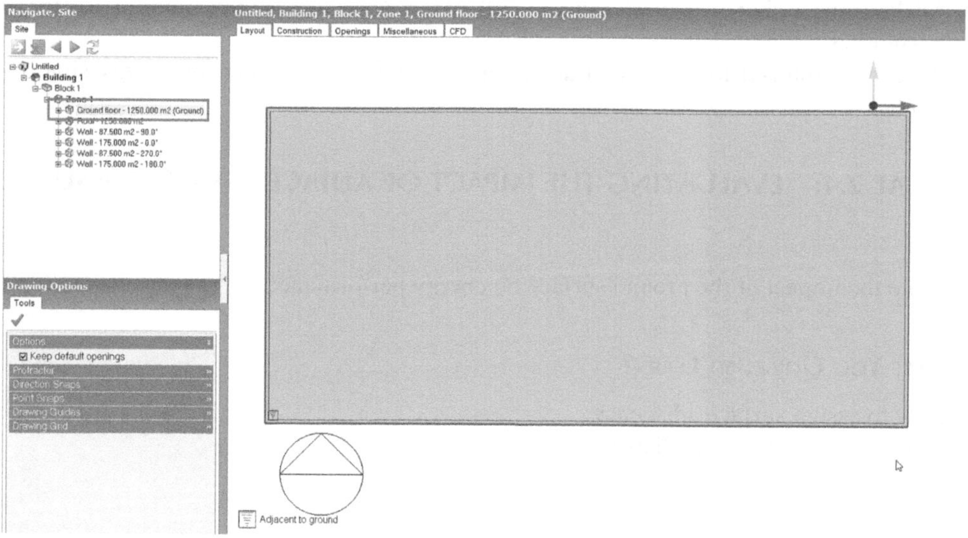

Step 3: Select the **Construction** tab.

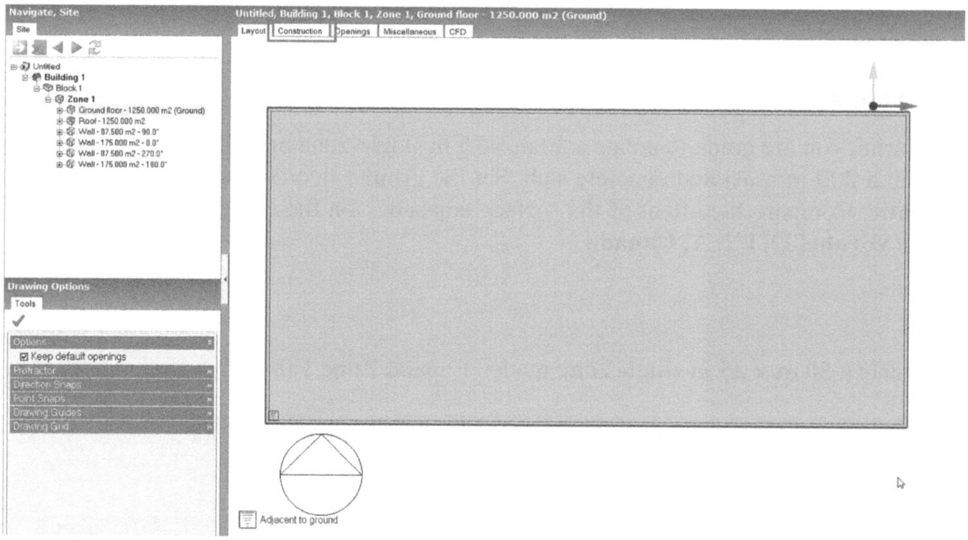

Step 4: Under the **Floors** section, click **Ground floor**. Three dots (. . .) appear. Click the dots. The **Select the Construction** screen appears.

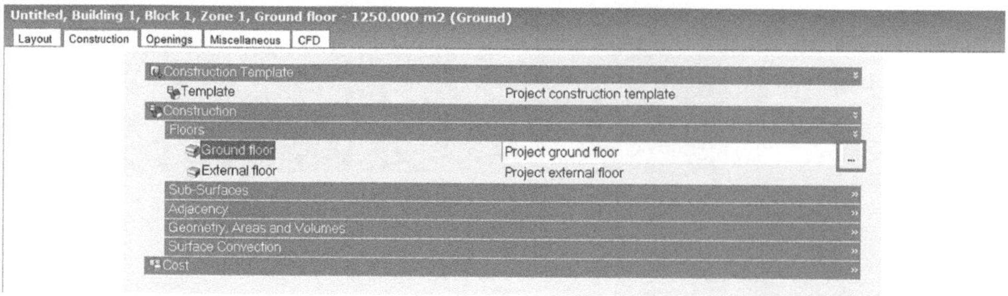

Step 5: Click the **Add new data** button to create a new construction.

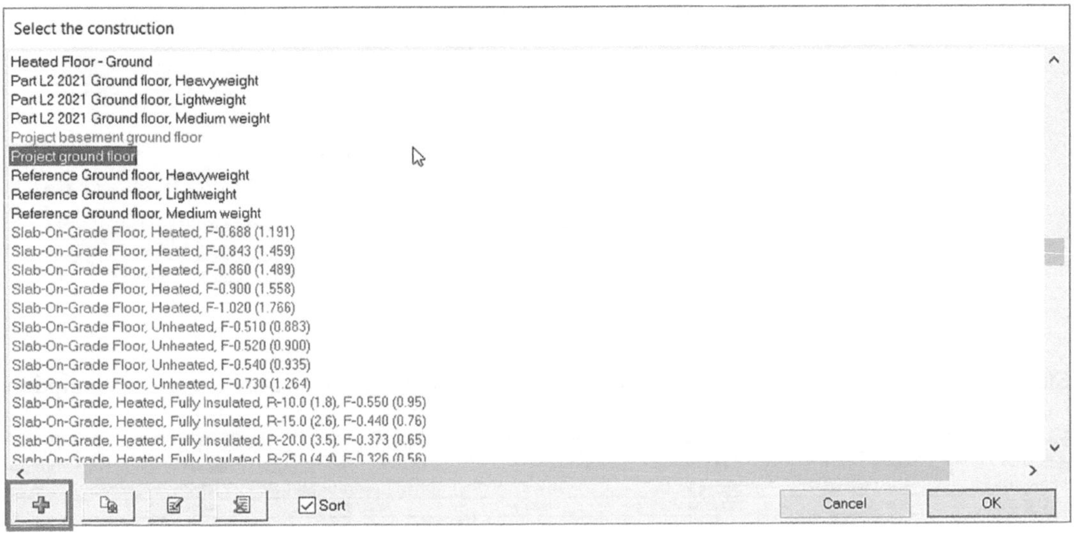

Step 6: Go to the **Layers** section, and make sure that **Aerated Concrete Slab** is the material and that the thickness is **0.2000** m. Click **OK**.

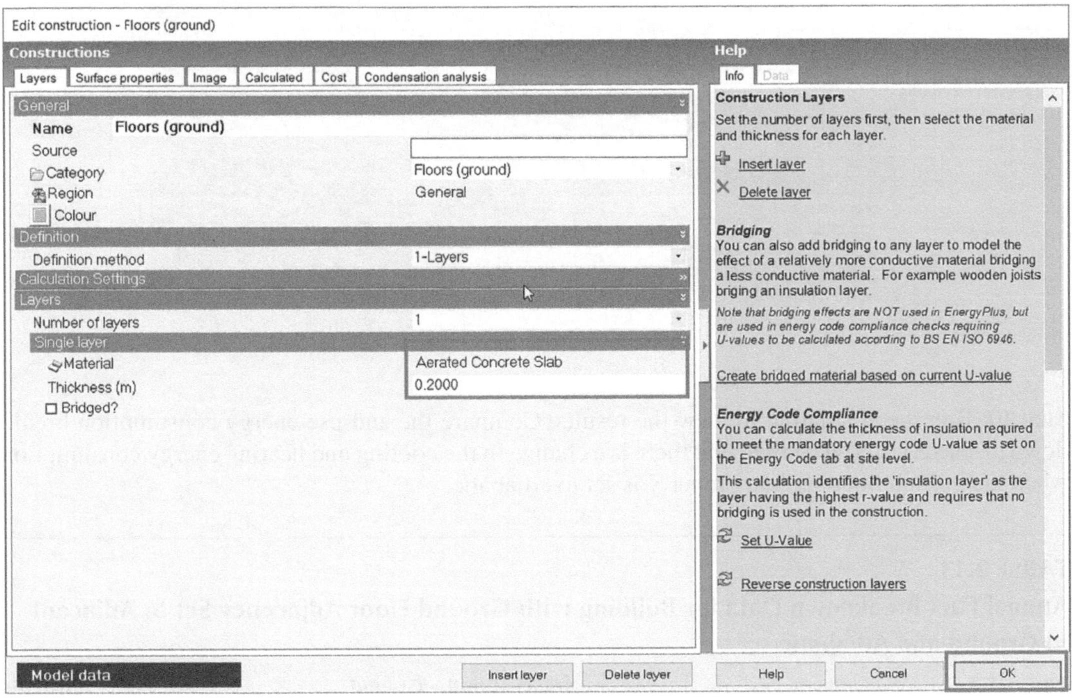

Step 7: Click the **Adjacency** section. It displays the adjacency property of the selected surface. Ensure that **Auto** is selected in the **Adjacency** drop-down list.

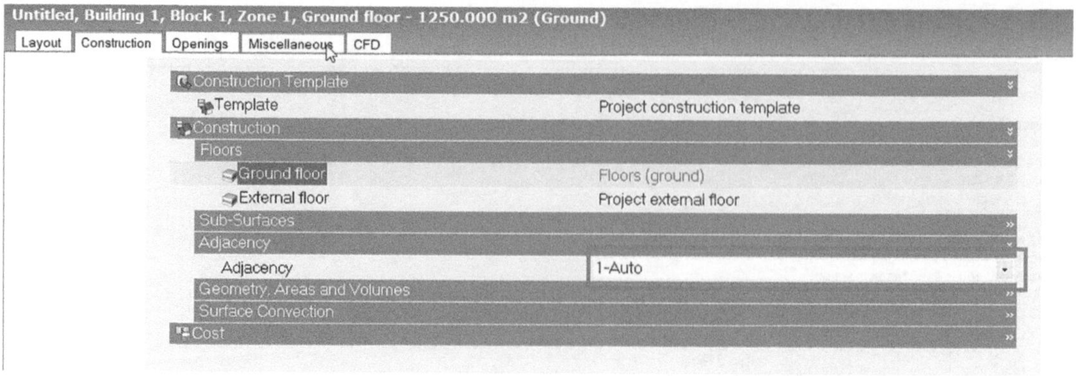

Step 8: Perform an annual energy simulation. Record the results for the end-use energy consumption.

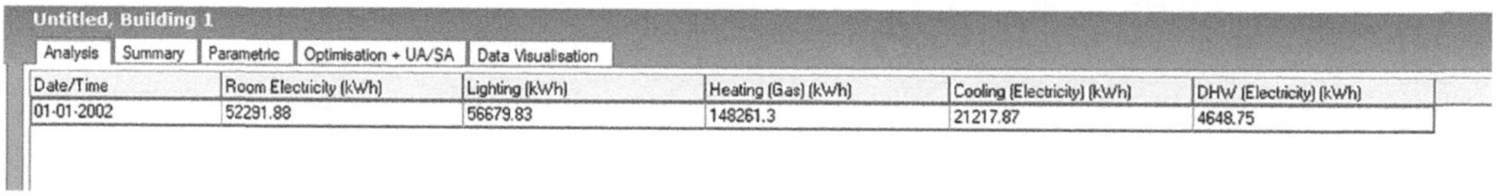

Step 9: Repeat Step 6 to select **Adiabatic** from the **Adjacency** drop-down list.

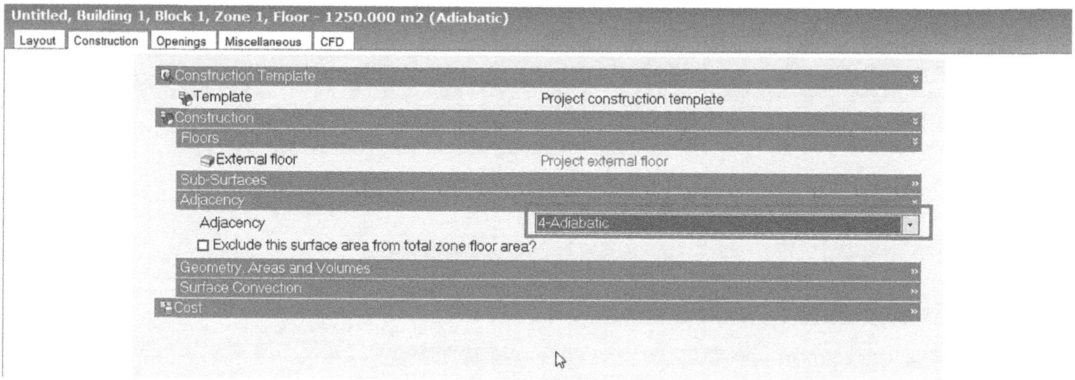

Step 10: Run the simulation to view the results. Compare the end-use energy consumption breakdown (Table 2.13). You can see that there is a change in the cooling and heating energy consumption when the ground floor surface property is set to adiabatic.

TABLE 2.13
Annual Fuel Breakdown Data for Building with Ground Floor Adjacency Set to Adjacent to Ground and Adiabatic

Type	Adjacent to the Ground	Adiabatic
Room electricity (kWh)	52,292	52,292
Lighting (kWh)	56,680	56,680
Heating (gas) (kWh)	1,48,261	1,44,954
Cooling (electricity) (kWh)	21,218	25,151
DHW (electricity) (kWh)	4,649	4,649

Results and Analysis

The results of the simulation comparing the models with ground floor adjacency set to Adjacent to the ground and Adiabatic are summarized in Table 2.13.

Energy Consumption

- **Heating (gas):** The heating energy demand is slightly lower for the **Adiabatic** case compared to the **Adjacent to the ground** case. This reduction is because the adiabatic condition prevents heat loss to the ground, thus retaining heat within the zone.
- **Cooling (electricity):** The cooling energy demand is higher for the **Adiabatic** case compared to the **Adjacent to the ground** case. The absence of heat dissipation to the ground results in higher cooling requirements.
- **Room electricity, lighting, and DHW:** These remain constant across both conditions, as expected.

Conclusion

The results demonstrate that setting the ground floor surface adjacency to **Adiabatic** decreases heating energy consumption but increases cooling energy consumption. This outcome is expected, as the adiabatic condition eliminates heat exchange between the ground and the building.

The EnergyPlus weather file provides ground temperatures for undisturbed sites. However, you should not use the ground temperatures in the EnergyPlus weather file header because they are for undisturbed sites.

The temperature beneath a building is significantly affected by the building itself—the EnergyPlus documentation recommends using a ground temperature of 2°C below average internal temperatures for large commercial buildings (where the perimeter heat loss is relatively less important). Note that this temperature should be applied directly below the slab and should not include ground material; thus, if you use this approach to ground temperature definition, you should switch off the use of the ground construction at the site level.

EnergyPlus cannot model very thick constructions, so it is necessary to use less thick constructions (2 m or less) combined with some assumptions about temperatures at about half a meter below the floor.

Source: www.designbuilder.co.uk/helpv7.3/Content/Ground_Modelling.htm

Many modelers prefer to define the ground temperature just below the slab and exclude the earth layers from the model. This has the advantage of simplicity and clarity, and this is the approach recommended by the EnergyPlus developers.

Note: The default ground temperatures provided in DesignBuilder assume that an earth layer is included in the constructions adjacent to the ground. If an earth layer is not included, then you should increase the default site ground temperatures to values closer to those typically found just below the ground slab.

Source: www.designbuilder.co.uk/helpv7.3/Content/Ground_Modelling.htm

3 Material and Construction

This chapter delves into the critical role that materials and constructions play in shaping the thermal and energy performance of buildings. It guides readers through the process of defining and assigning materials within a building energy model, and demonstrates how different construction choices directly impact energy consumption and occupant comfort.

The chapter opens with a tutorial that investigates the concept of thermal mass by comparing lightweight and heavyweight external wall assemblies. This comparison highlights how material selection affects heat storage and temperature regulation, especially in climates with significant day/night temperature variations. Following this, the chapter explores the influence of roof insulation placement—distinguishing between overdeck and underdeck applications—and examines how these choices affect heat transfer through the roof.

Further tutorials introduce the concepts of cool roofs and radiant barriers, emphasizing their ability to reflect solar radiation and reduce cooling loads, particularly in hot climates. These passive design strategies demonstrate how even small changes in surface properties can yield substantial energy savings.

Materials form the foundation of any building, not only structurally but also thermally. Selecting appropriate construction materials is essential, not merely for managing construction costs but for ensuring long-term energy efficiency and resilience. Key properties such as thermal conductivity, density, specific heat capacity, solar reflectance, and emissivity significantly influence how heat is absorbed, stored, and transmitted through building elements. By understanding and strategically applying these properties, designers and engineers can optimize building performance and support broader sustainability goals.

TUTORIAL 3.1: EVALUATING THE EFFECT OF LIGHTWEIGHT AND HEAVYWEIGHT CONSTRUCTION

Goal

To evaluate the effect of thermal mass by modeling and comparing lightweight and heavyweight external wall construction on the thermal performance of the building

What Are You Going to Learn?

- How to assign lightweight and heavyweight construction
- How to get zone temperatures

Problem Statement

In this tutorial, you are going to use a 50 m × 25 m model with night purge and set external wall construction to lightweight and heavyweight. Determine the air temperature inside the zone for both cases. You need to analyze it for the **DUBAI INTERNATIONAL, United Arab Emirates** weather location.

You are going to use the following construction for external walls.

- Uninsulated wall, lightweight (wood derivative—plywood 15 mm)
- Uninsulated wall, heavyweight (stone—granite, 450 mm)

DOI: 10.1201/9781003563013-3

Solution

Step 1: Open a new project file. Create a **50 m × 25 m** building.

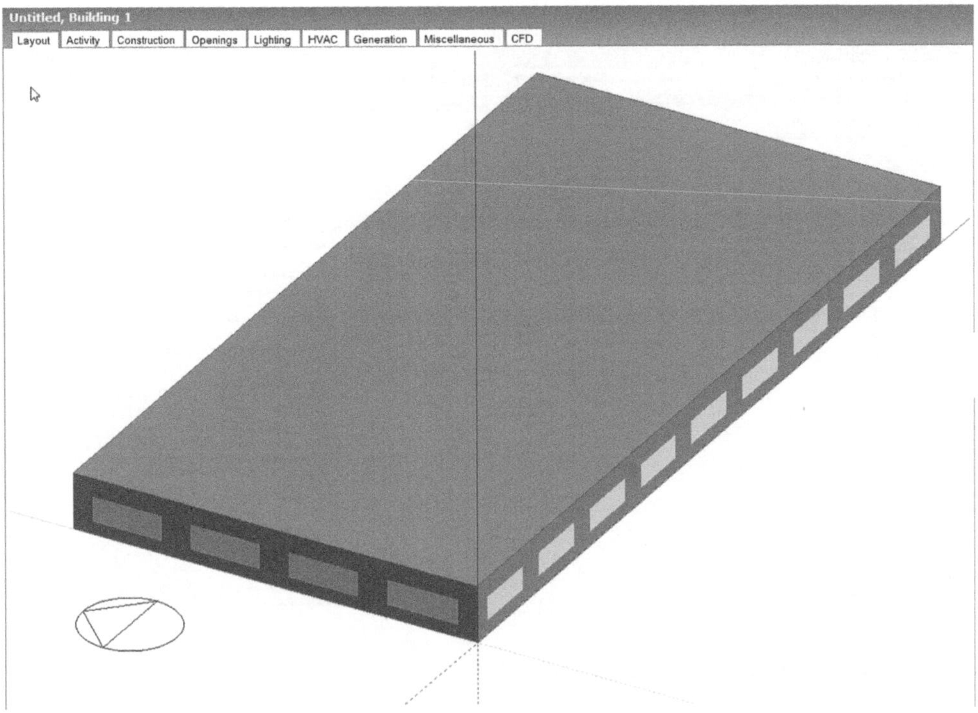

Step 2: Select the **Construction** tab.

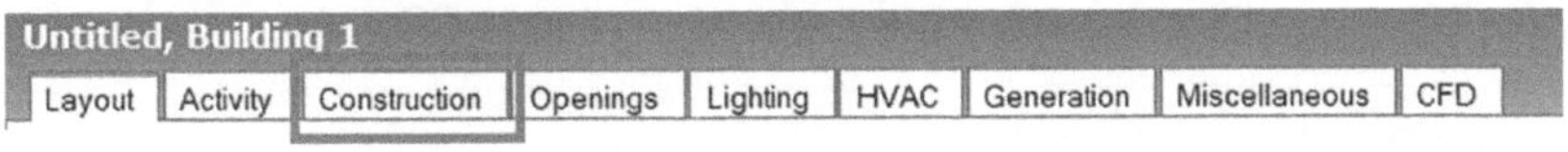

Step 3: In the **Construction** section, click **Project wall. External walls** get highlighted. Click the **Add new item** icon. The **Edit construction—Wall** screen appears.

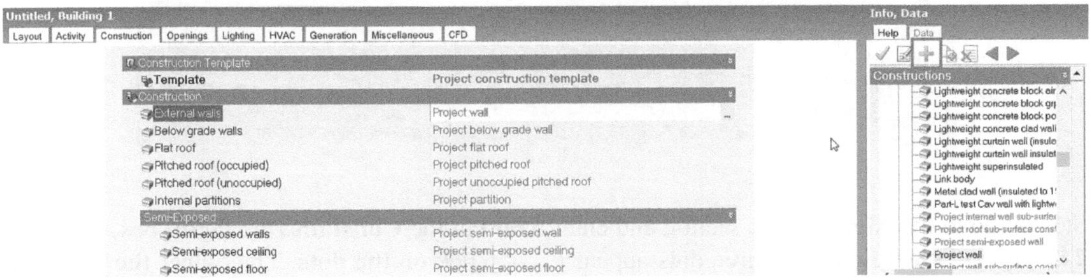

Step 4: Enter **Light weight** as the **Name**. Go to the **Layers** section and select **Plywood (Lightweight)** as the material (you can find it under the **Wood** branch). Set the thickness to **0.025**. Click **OK**. A message box appears.

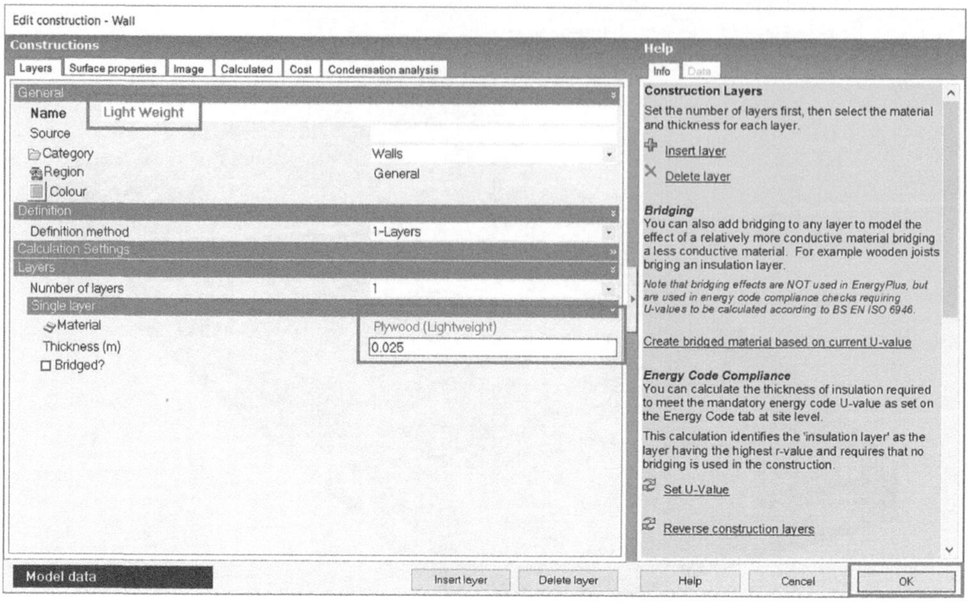

Step 5: Select **Light Weight** as the external wall.

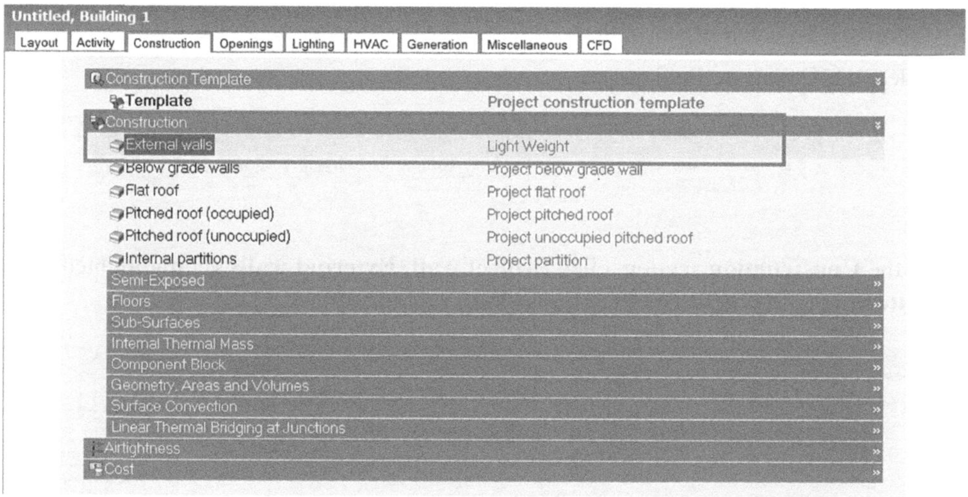

Step 6: Go to the **Airtightness** section and enter **3.000** in the **Constant rate (ac/h)** box. Click **On 24/7** in the **Schedule** field. Three dots appear (. . .). Click on the **dots**. The **Select the Schedule** screen appears.

Step 7: Click the **Create copy of highlighted item** icon. Make sure that the copy of the item is highlighted.

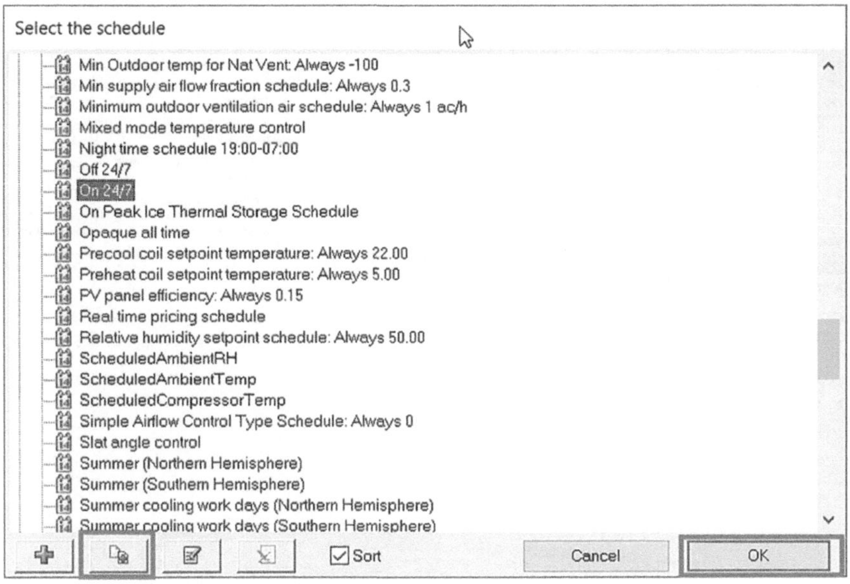

Step 8: Click the **Edit selected data** icon. The **Edit Schedule** screen appears.

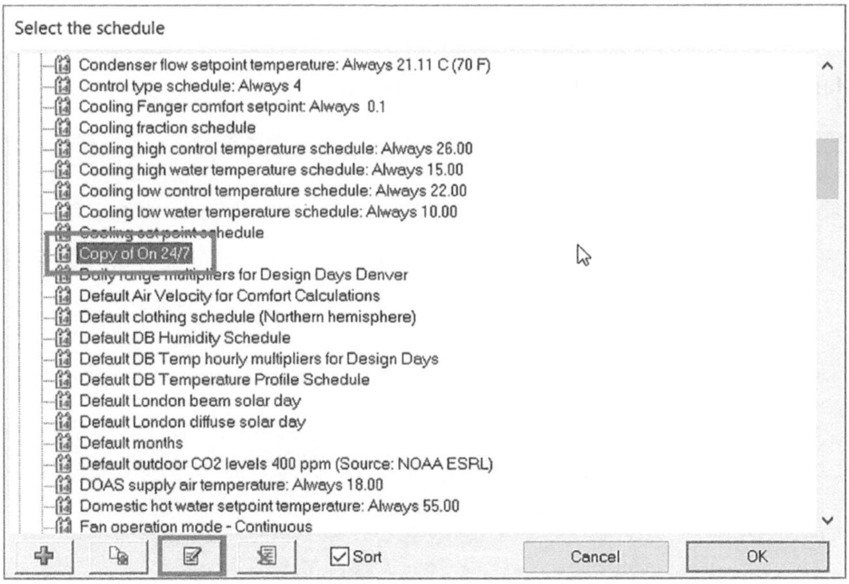

Step 9: In the **General** section, enter **Night Purge** in the **Name** box. Under the **Profiles** section, edit the schedule as shown below. Click **OK**.

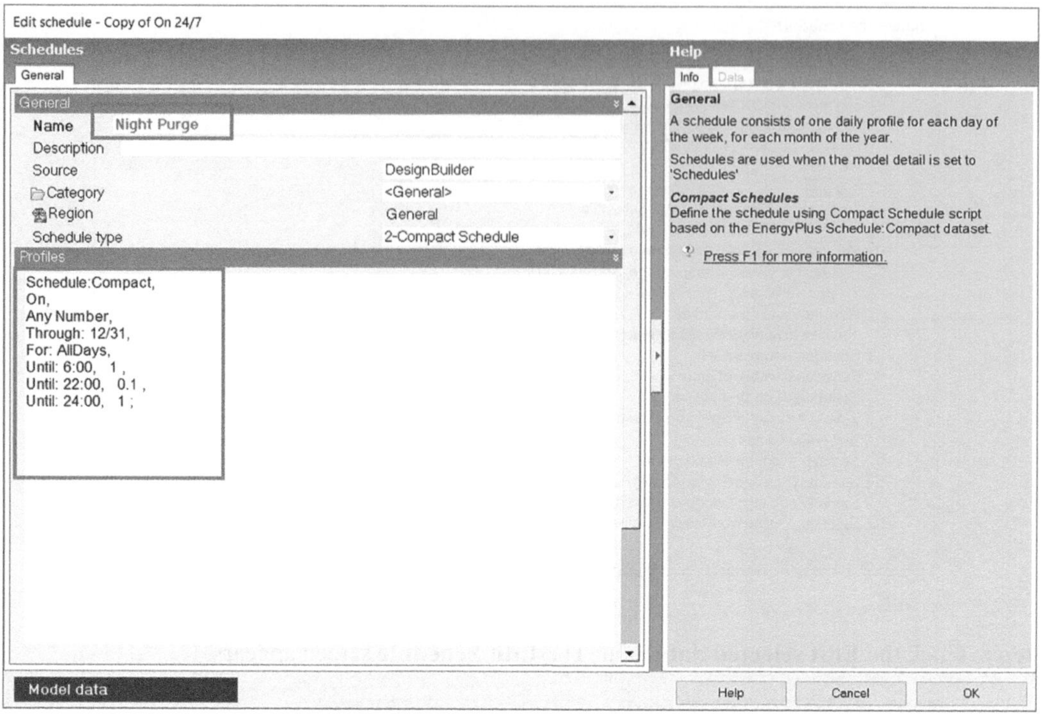

Step 10: Make sure that the **Night Purge** schedule is set.

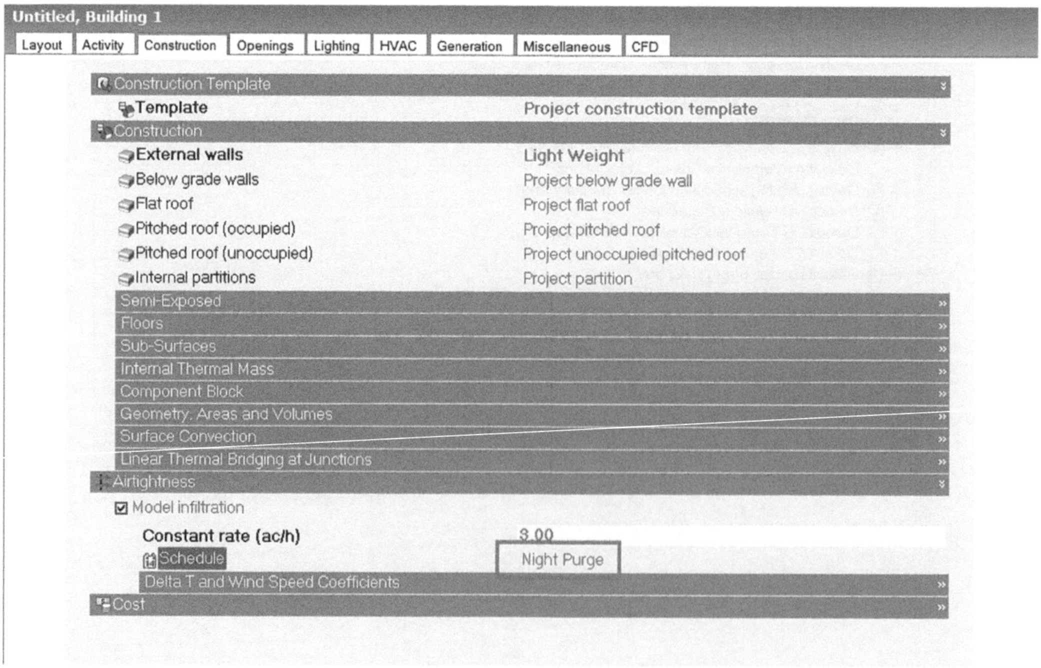

Night purge is a technique used in conditioned buildings by which cool ambient air is passed through the building during unoccupied night hours to flush out the heat released/accumulated in the building.

Step 11: Click **Flat roof element—1250.000 m²** in the navigation tree, and select **Adiabatic** from the **Adjacency** drop-down list. Click **Building 1** in the navigation tree.

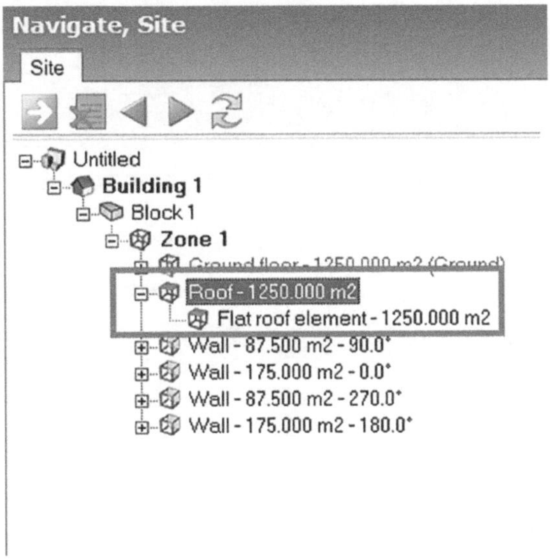

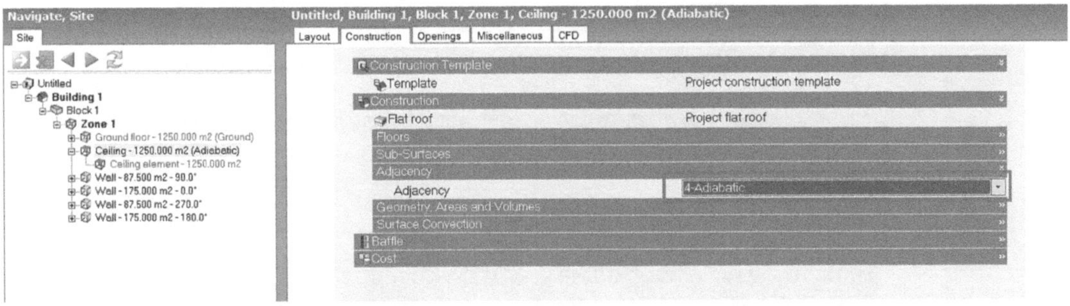

To understand the effect of the thermal mass of the external wall, we want to remove the heat gains and losses from the roof. To achieve this, the roof has been made adiabatic.Step 12: Go to the **Building 1** level. Select the **Activity** tab and select **Density (people/m²)** as **0.06** and **Office Equipment Gain (W/m²)** as **1.00**. (The internal load has been reduced so that the effect of the thermal mass is clearly visible. If the internal loads are higher, then the inside zone temperature is dominated by the internal load, and you cannot observe the effect of the thermal mass of the envelope.)

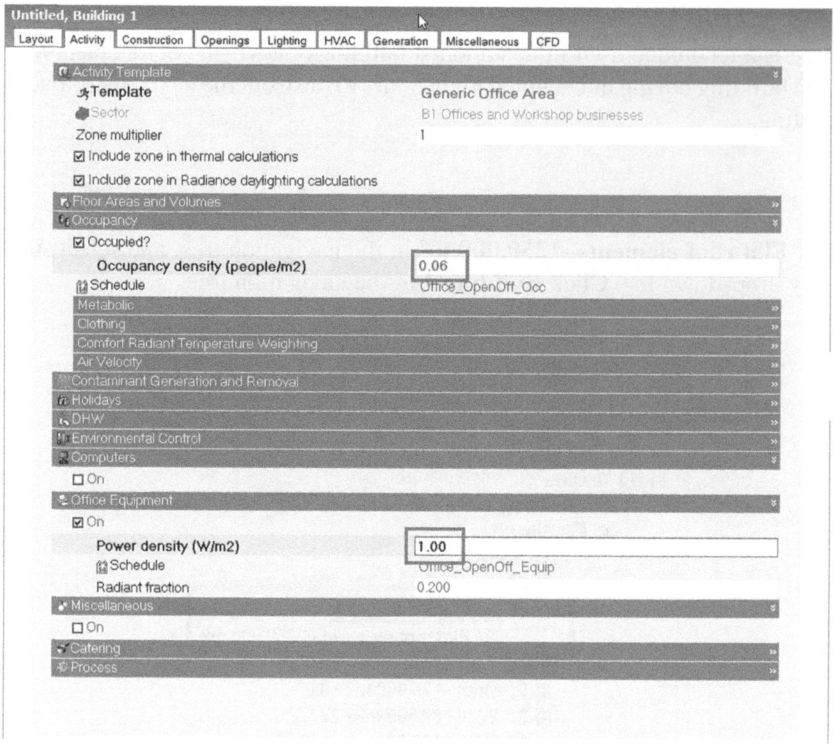

Step 13: Select the **Openings** tab and ensure that **Preferred height** is selected, and **Window-to-wall percentage** is set as **0.00**.

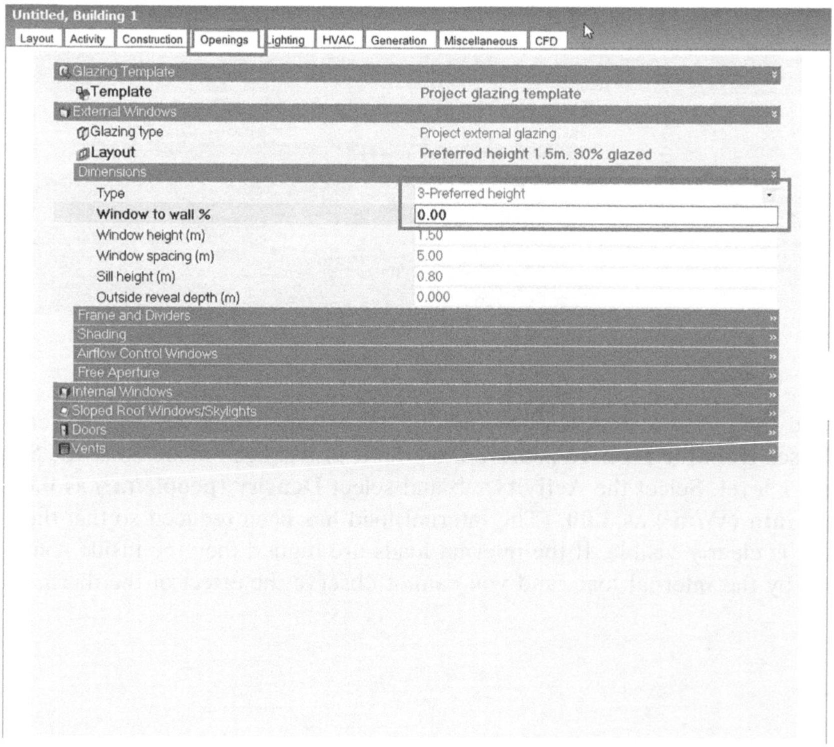

Step 14: Select the **Lighting** tab and set the **Lighting energy (W/m²-100 lux)** as **0.25**.

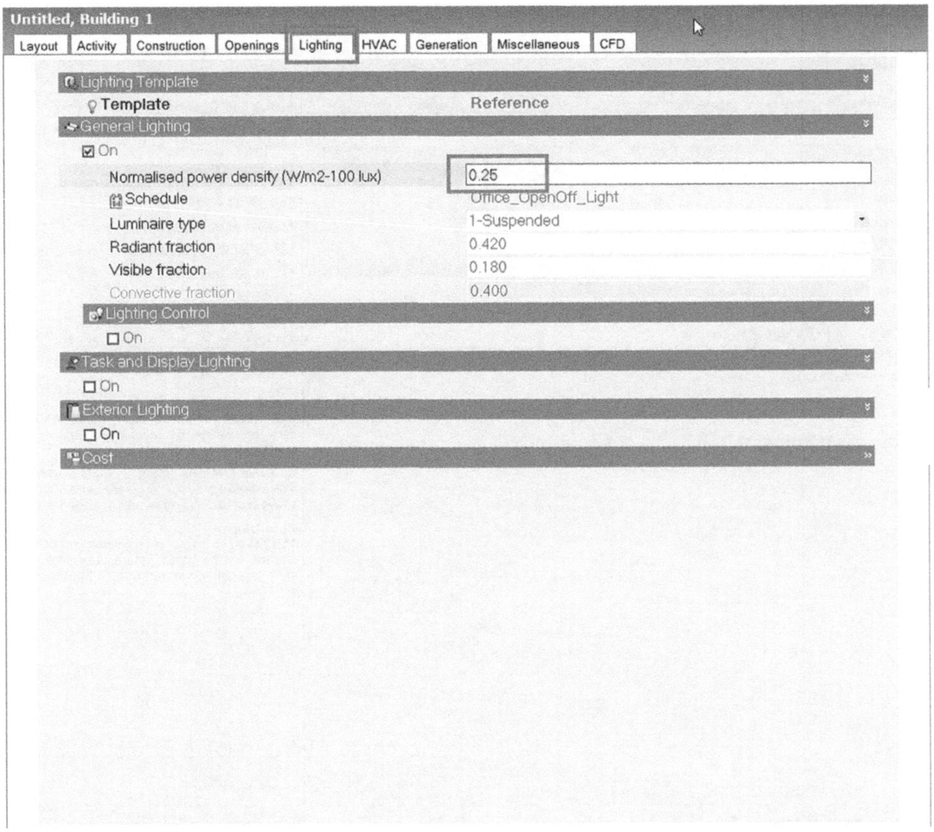

Step 15: Select the **HVAC** tab and set the **template** as **<None>**. Also, clear all checkboxes including **DHW** and **Natural Ventilation**.

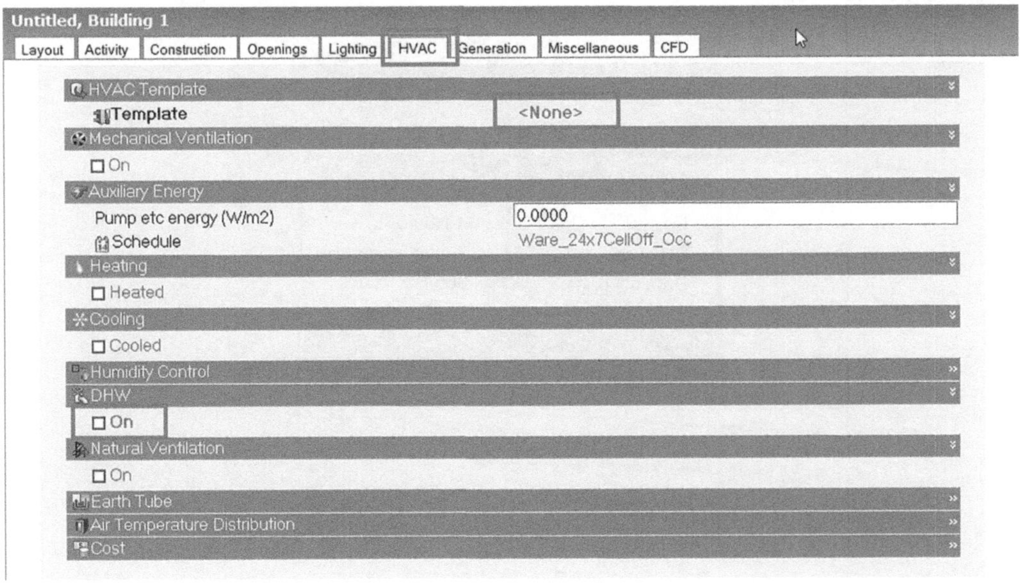

Step 16: Simulate the model for hourly interval reporting.

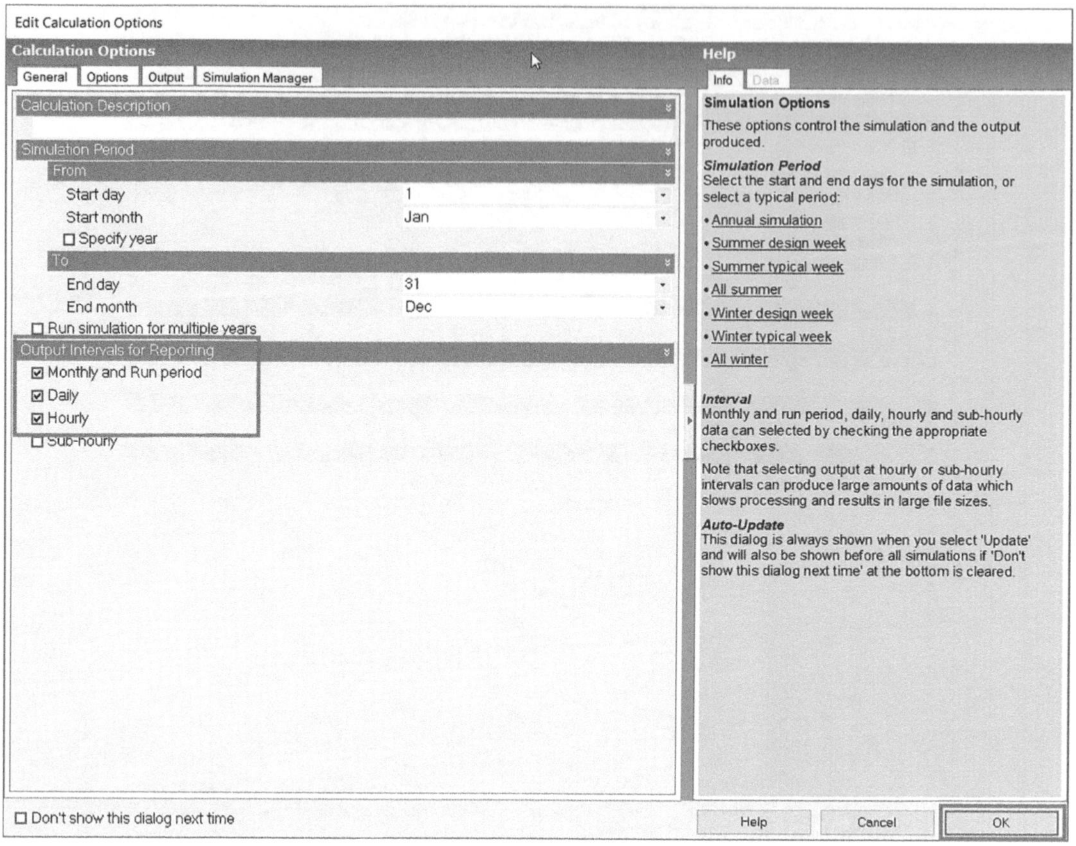

Step 17: Click **OK**. The results are displayed in the grid. You need to click on **Zone 1** to get the results at the zone level.

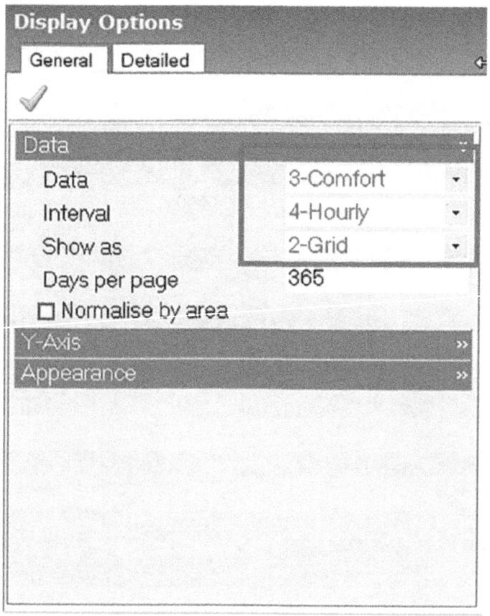

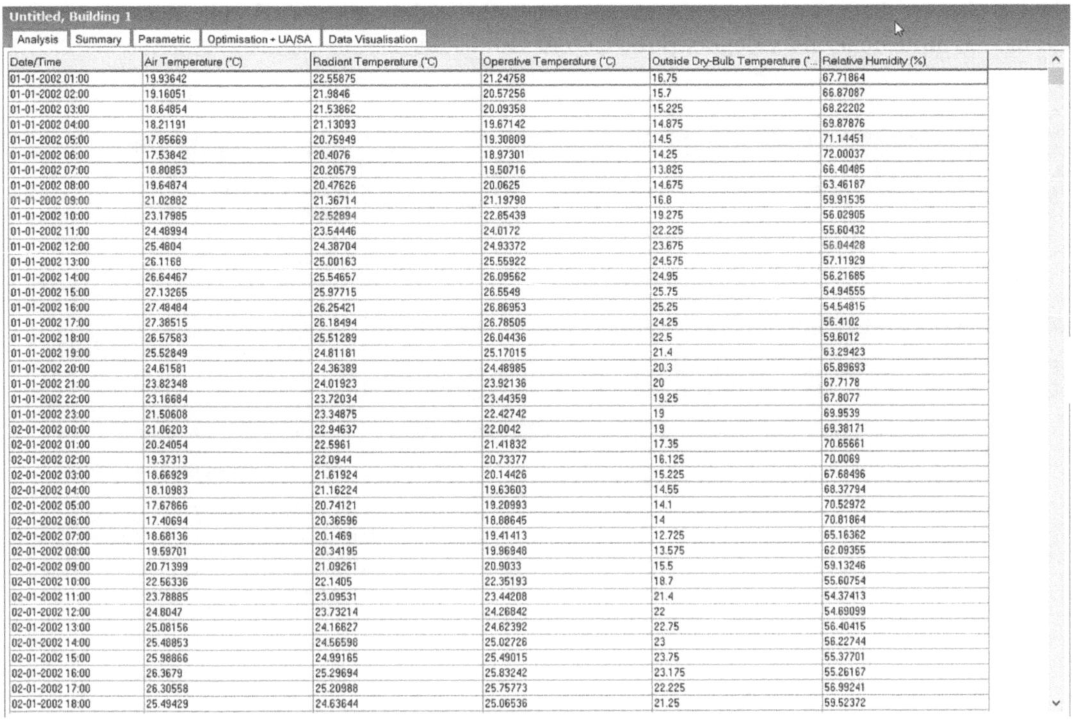

Date/Time	Air Temperature (°C)	Radiant Temperature (°C)	Operative Temperature (°C)	Outside Dry-Bulb Temperature (°...	Relative Humidity (%)
01-01-2002 01:00	19.93642	22.55875	21.24758	16.75	67.71864
01-01-2002 02:00	19.16051	21.9846	20.57256	15.7	66.87087
01-01-2002 03:00	18.64854	21.53862	20.09358	15.225	68.22202
01-01-2002 04:00	18.21191	21.13093	19.67142	14.875	69.87876
01-01-2002 05:00	17.85669	20.75949	19.30809	14.5	71.14451
01-01-2002 06:00	17.53842	20.4076	18.97301	14.25	72.00037
01-01-2002 07:00	18.80853	20.20579	19.50716	13.825	66.40485
01-01-2002 08:00	19.64874	20.47626	20.0625	14.675	63.46187
01-01-2002 09:00	21.02882	21.36714	21.19798	16.8	59.91535
01-01-2002 10:00	23.17985	22.52894	22.85439	19.275	56.02905
01-01-2002 11:00	24.48994	23.54446	24.0172	22.225	55.60432
01-01-2002 12:00	25.4804	24.38704	24.93372	23.675	56.04428
01-01-2002 13:00	26.1168	25.00163	25.55922	24.575	57.11929
01-01-2002 14:00	26.64467	25.54657	26.09562	24.95	56.21685
01-01-2002 15:00	27.13265	25.97715	26.5549	25.75	54.94555
01-01-2002 16:00	27.48484	26.25421	26.86953	25.25	54.54815
01-01-2002 17:00	27.38515	26.18494	26.78505	24.25	56.4102
01-01-2002 18:00	26.57583	25.51289	26.04436	22.5	59.6012
01-01-2002 19:00	25.52849	24.81181	25.17015	21.4	63.29423
01-01-2002 20:00	24.61581	24.36389	24.48985	20.3	65.89693
01-01-2002 21:00	23.82348	24.01923	23.92136	20	67.7178
01-01-2002 22:00	23.16684	23.72034	23.44359	19.25	67.8077
01-01-2002 23:00	21.50608	23.34875	22.42742	19	69.9539
02-01-2002 00:00	21.06203	22.94637	22.0042	19	69.38171
02-01-2002 01:00	20.24054	22.5961	21.41832	17.35	70.65661
02-01-2002 02:00	19.37313	22.0944	20.73377	16.125	70.0069
02-01-2002 03:00	18.66929	21.61924	20.14426	15.225	67.68496
02-01-2002 04:00	18.10983	21.16224	19.63603	14.55	68.37794
02-01-2002 05:00	17.67866	20.74121	19.20993	14.1	70.52972
02-01-2002 06:00	17.40694	20.36596	18.88645	14	70.81864
02-01-2002 07:00	18.68136	20.1469	19.41413	12.725	65.16362
02-01-2002 08:00	19.59701	20.34195	19.96948	13.575	62.09355
02-01-2002 09:00	20.71399	21.09261	20.9033	15.5	59.13246
02-01-2002 10:00	22.56336	22.1405	22.35193	18.7	55.60754
02-01-2002 11:00	23.78885	23.09531	23.44208	21.4	54.37413
02-01-2002 12:00	24.8047	23.73214	24.26842	22	54.69099
02-01-2002 13:00	25.08156	24.16627	24.62392	22.75	56.40415
02-01-2002 14:00	25.48853	24.56598	25.02726	23	56.22744
02-01-2002 15:00	25.98866	24.99165	25.49015	23.75	55.37701
02-01-2002 16:00	26.3679	25.29694	25.83242	23.175	55.26167
02-01-2002 17:00	26.30558	25.20988	25.75773	22.225	56.99241
02-01-2002 18:00	25.49429	24.63644	25.06536	21.25	59.52372

Step 18: Click the **Export Data** icon. The **Export Results Spreadsheet** dialog box appears. You can save this results file on your desktop to retrieve it easily.

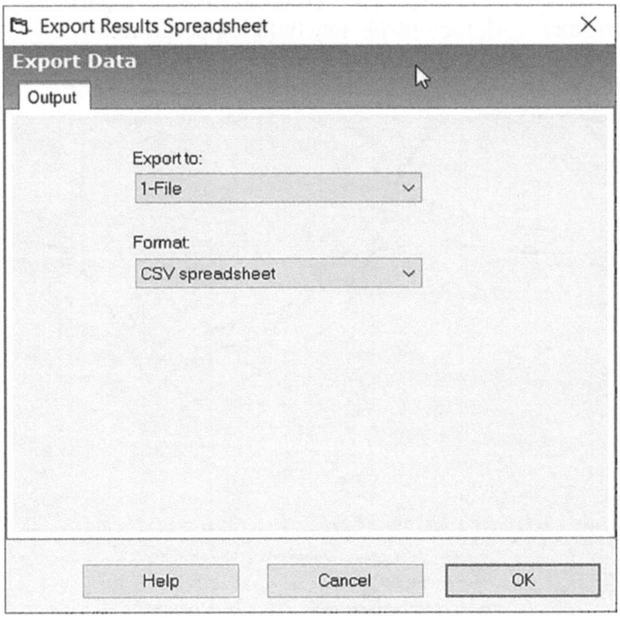

Step 19: Repeat the preceding steps to create a high–thermal-mass external wall. Select material as **Stone—granite** under the category **Sands, stones and soils**.

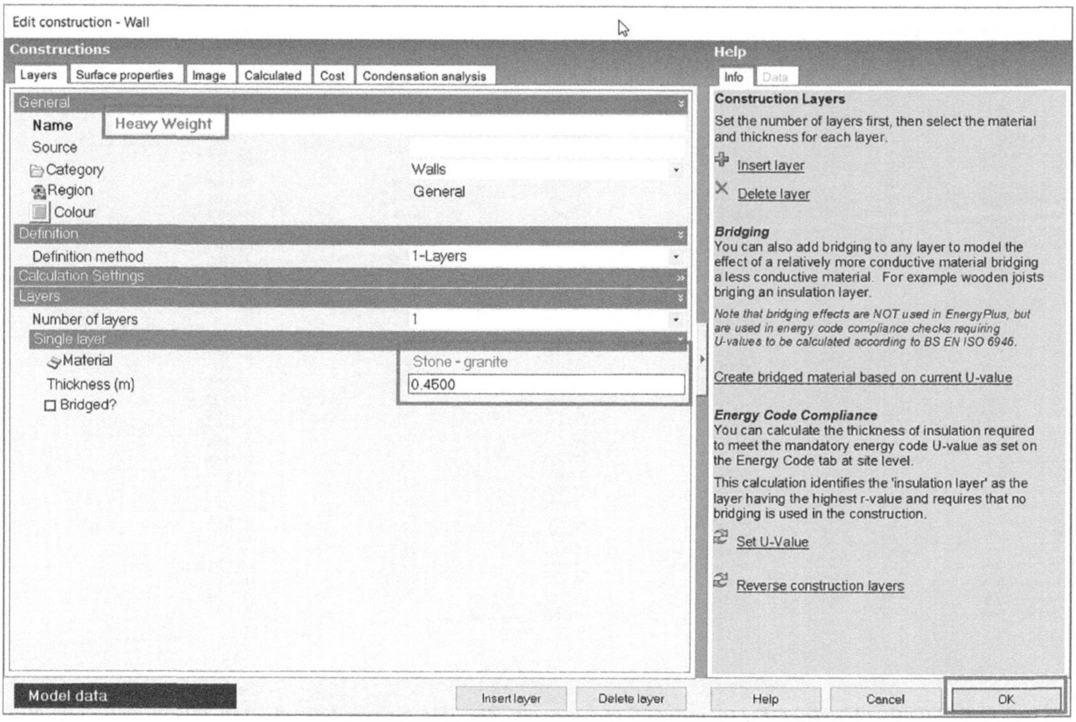

Step 20: Perform an hourly simulation and record the results. Compare the indoor air temperature for both cases (5 and 6 June) with the outside dry bulb temperature.

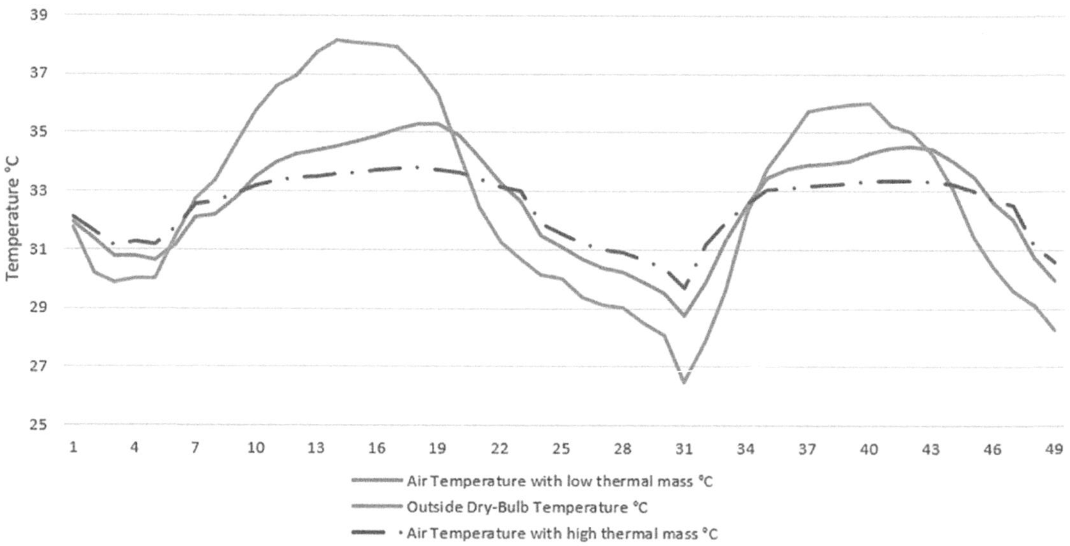

Conclusions

- High–thermal-mass buildings (heavyweight construction) exhibit better thermal stability, with reduced temperature swings and a noticeable delay in peak temperatures compared to low–thermal-mass buildings (lightweight construction)
- The thermal lag effect observed in heavyweight construction demonstrates its ability to delay heat transfer, enhancing indoor thermal comfort during periods of high outdoor temperatures
- These two effects also get translated to reduction of HVAC system capacity with high thermal mass due to lesser intense peak of cooling load

These findings highlight the benefits of using high–thermal-mass materials for enhancing indoor thermal performance, especially in hot climates where temperature fluctuations are extreme.

EXERCISE 3.1

Repeat the tutorial with insulation on the lightweight external wall. Compare the results with and without insulation on the lightweight wall.

TUTORIAL 3.2: EVALUATING THE IMPACT OF ROOF INSULATION

Goal

To study the effect of roof insulation on building energy consumption

What Are You Going to Learn?

- How to create roof construction
- How to set the U-value of a roof

Problem Statement

In this tutorial, you are going to use a 50 m × 25 m model. Construct a roof with a 100 mm aerated concrete slab and glass fiber slab insulation of varying thicknesses. Achieve the U-values given in Table 3.1 by varying the insulation thickness. Find out the energy consumption for each thickness. Use the weather file for **WIEN/SCHWECHAT-FLUG, AUSTRIA**.

TABLE 3.1
U-Values and R-Values of a Roof

Serial No.	U-Value (W/m^2-K)	R-Value (m^2-K/W)	S. No.	U-Value (W/m^2-K)	R-Value (m^2-K/W)
1	1	1	**10**	0.31	3.25
2	0.8	1.25	**11**	0.29	3.5
3	0.67	1.50	**12**	0.27	3.75
4	0.57	1.75	**13**	0.25	4.00
5	0.50	2.00	**14**	0.24	4.25
6	0.44	2.25	**15**	0.22	4.50
7	0.40	2.50	**16**	0.21	4.75
8	0.36	2.75	**17**	0.20	5.00
9	0.33	3.00			

R-VALUE OR THERMAL RESISTANCE R

The R-value of any section having one or more layers with parallel surfaces is an indication of the resistance offered by the section to heat flow. It is the reciprocal of thermal conductance. For a structure having plane parallel faces, the thermal resistance is equal to the thickness *L* of the structure divided by the thermal conductivity *k*:

$$R = \frac{L}{k} \tag{3.1}$$

The R-value of individual layers can be added to arrive at the total R-value of the section. It is also expressed as the ratio of the temperature difference across an insulator and the heat flux (the heat transfer per unit area per unit time).

THERMAL TRANSMITTANCE (U-FACTOR)—THERMAL TRANSMITTANCE (*U*)

The thermal transmission in the unit time through the unit area of the given section is divided by the temperature difference between the fluids on either side of the building unit in steady-state conditions. This is also called the *U-value*. Its unit is W/m²-K. It can be treated as a measure of the heat loss through the unit area of a building section such as a wall, floor, or roof. A low U-value generally indicates high levels of insulation.

Solution

Step 1: Open a new project and create a **50 m × 25 m** building.

Step 2: Select the **Construction** tab.

Step 3: Go to the **Construction** section, click **Flat roof**, and then click the **Add new item** icon. The **Edit construction—Roof** screen appears.

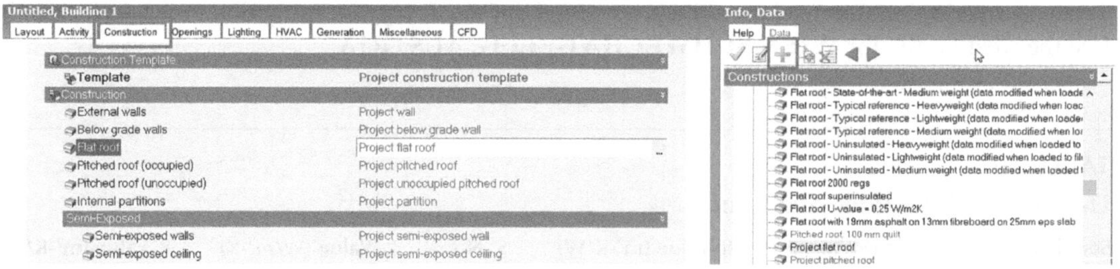

Step 4: Select **2** from the **Number of layers** drop-down list.

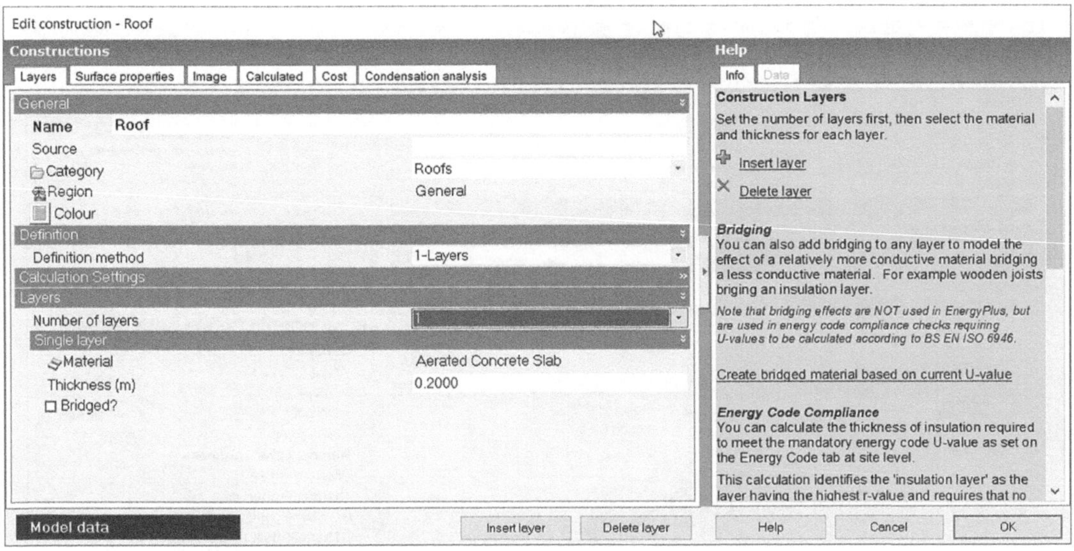

Step 5: Select **Aerated Concrete Slab** as the **Material and** enter **0.1000** as the **Thickness (m)** in the **Outermost layer** section. Select **Glass Fibre Slab** from **Insulating materials** and enter **0.2000** as the **Thickness (m)** for the innermost layer.

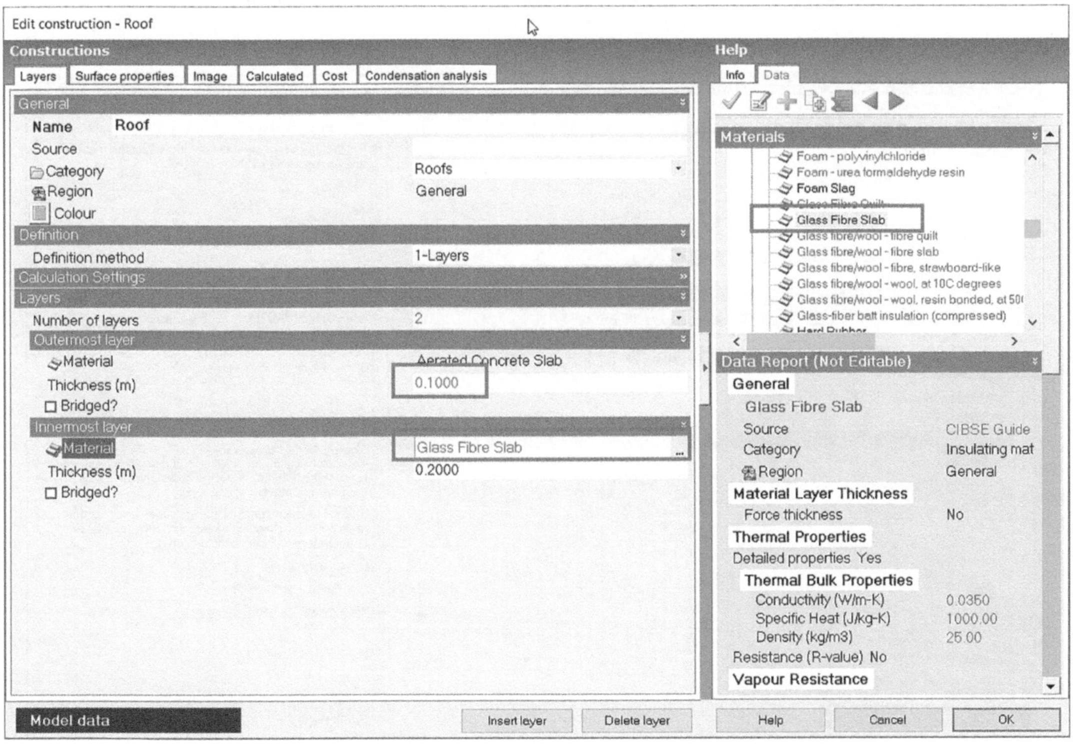

Step 6: Click anywhere in the blank space under the **Innermost layer** section to update the help section.

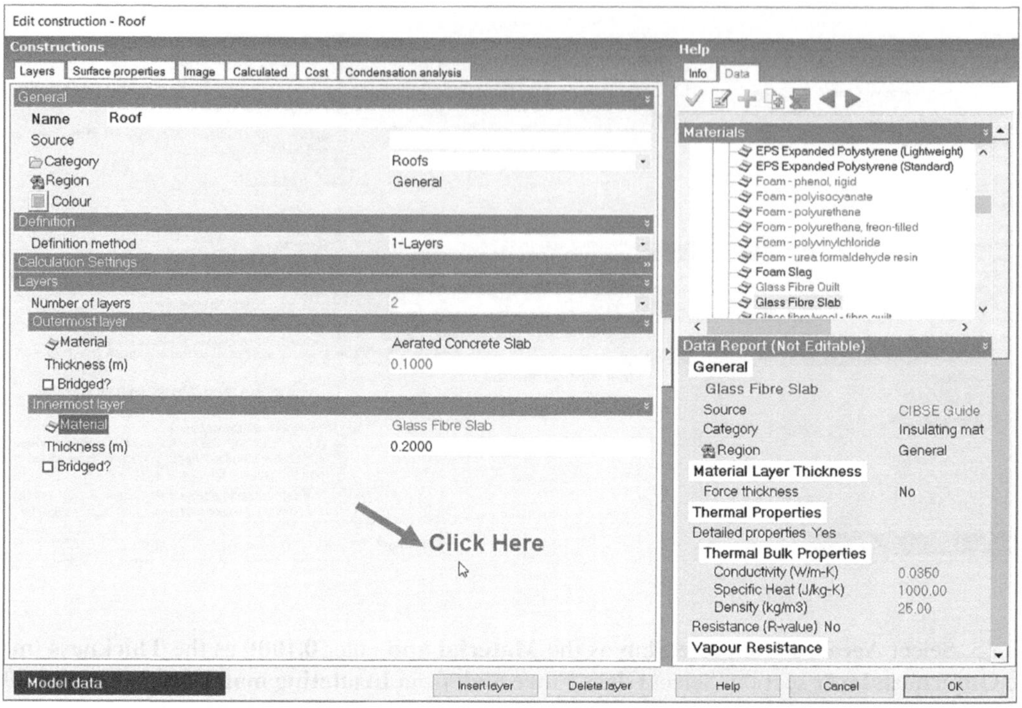

Step 7: Click the **Set U-value** link. The **Set Construction U-value** screen appears.

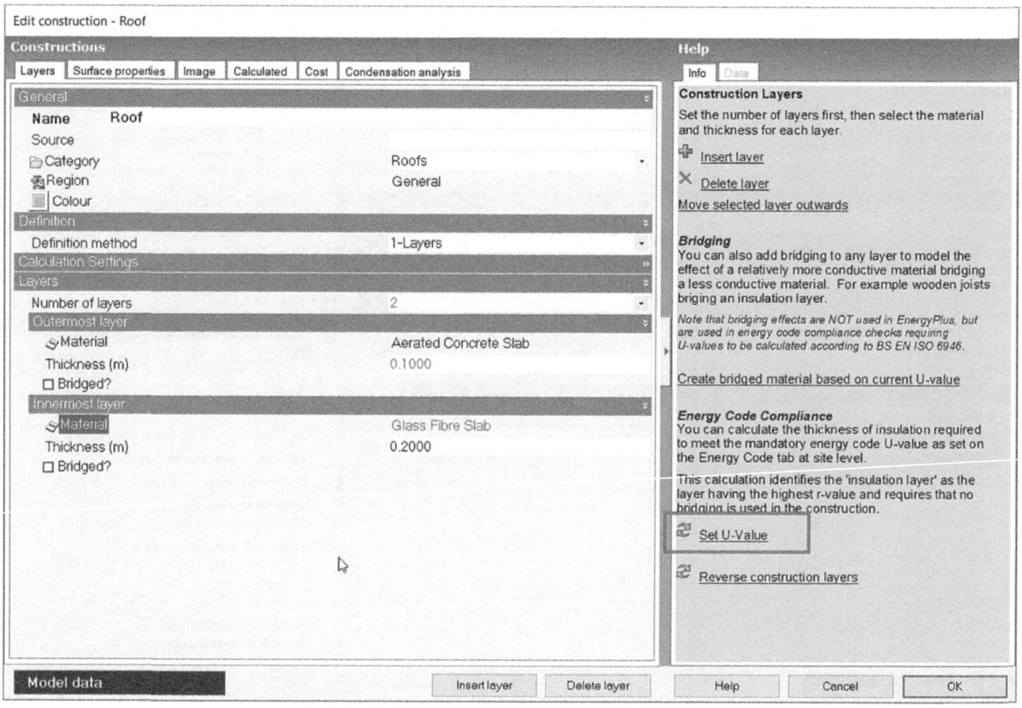

Step 8: Enter the **U-value (W/m^2-K)** as **1.0**. Click **OK**. A confirmation message appears with the updated insulation thickness.

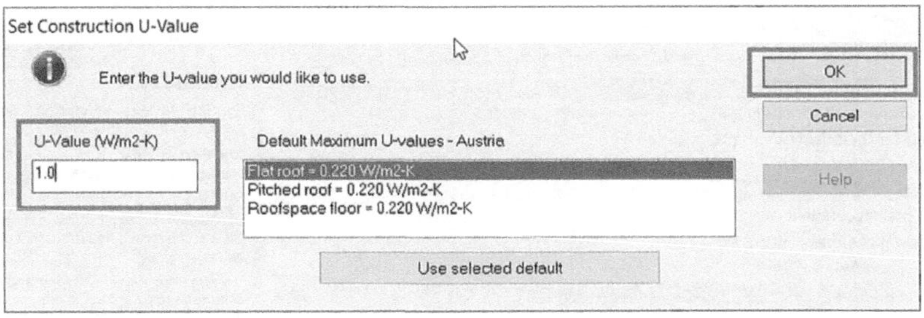

Step 9: Click **OK**. The insulation thickness is updated.

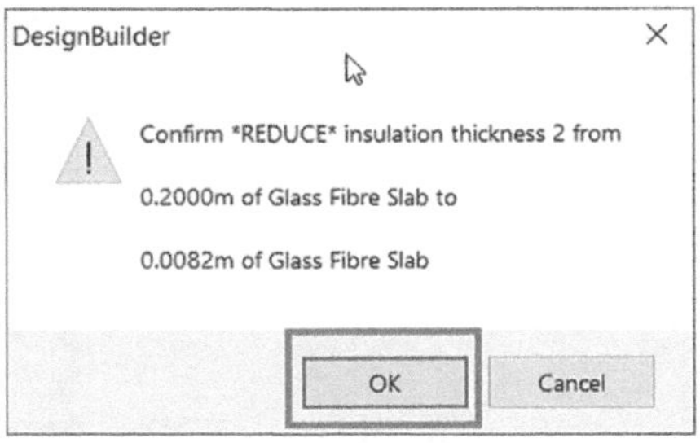

Step 10: Select the **Calculated** tab. The updated **U-value** of the construction appears.

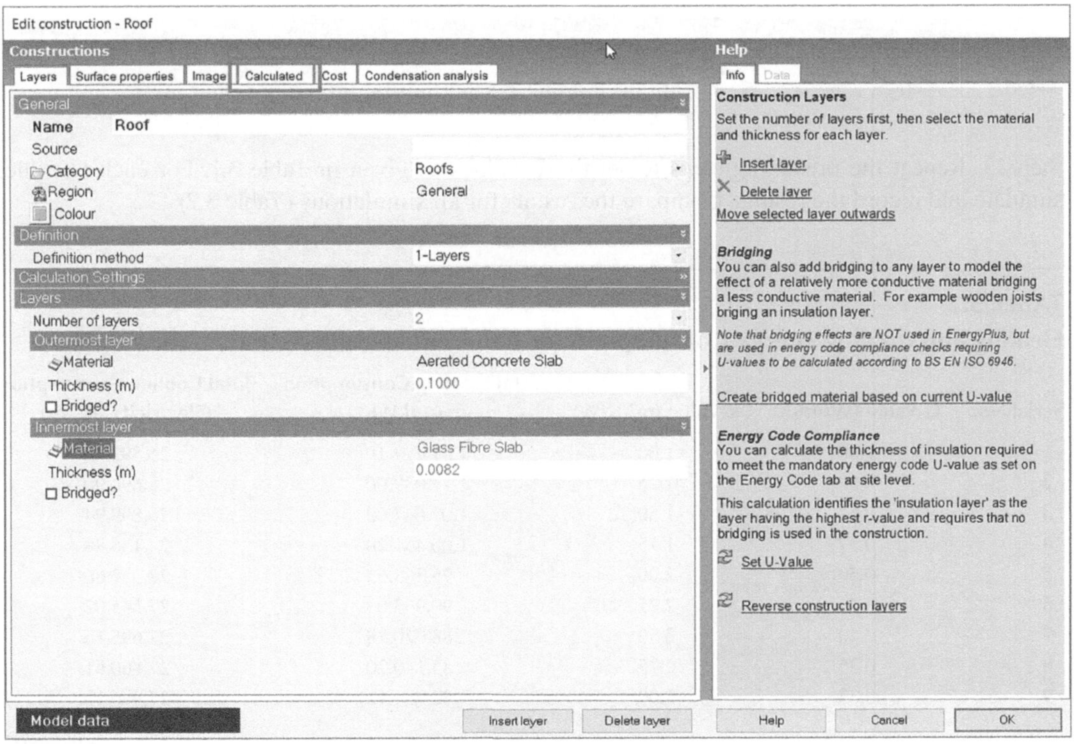

Step 11: Click **OK**.

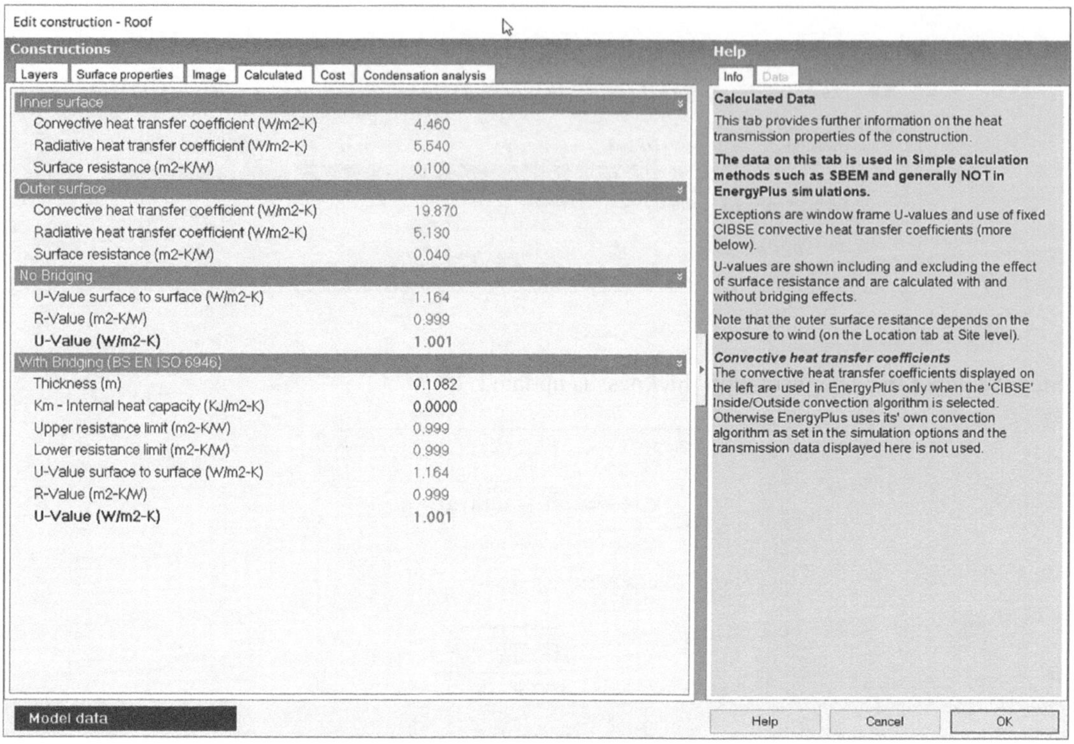

Step 12: Perform an annual energy simulation and record the results.

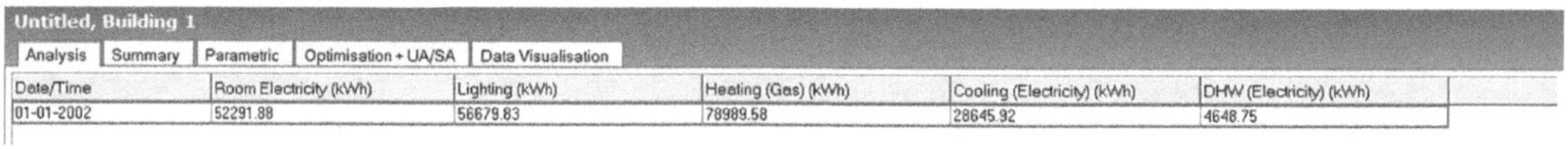

Step 13: Repeat the preceding steps to set the U-values as given in Table 3.1. For each U-value, simulate and record the results. Compare the results for all simulations (Table 3.2).

TABLE 3.2
Heating and Cooling Energy Consumption for Different U-Values

Serial No.	U-Value (W/m²-K)	R-Value (m²-K/W)	Total Heating Consumption (gas) (kWh)	Total Cooling Consumption (Electricity) (kWh)
1	1.00	1.00	1,31,497.10	26,592.62
2	0.80	1.25	1,17,039.00	26,694.90
3	0.67	1.50	1,07,647.00	26,899.94
4	0.57	1.75	1,00,451.20	27,136.88
5	0.50	2.00	95,343.65	27,337.66
6	0.44	2.25	90,963.03	27,545.02
7	0.40	2.50	88,070.78	27,695.16
8	0.36	2.75	85,140.20	27,860.61
9	0.33	3.00	82,909.13	27,996.10

(Continued)

TABLE 3.2 (*Continued*)
Heating and Cooling Energy Consumption for Different U-Values

Serial No.	U-Value (W/m²-K)	R-Value (m²-K/W)	Total Heating Consumption (gas) (kWh)	Total Cooling Consumption (Electricity) (kWh)
10	0.31	3.25	81,453.70	28,091.26
11	0.29	3.50	79,980.94	28,183.73
12	0.27	3.75	78,480.49	28,278.39
13	0.25	4.00	77,018.59	28,376.21
14	0.24	4.25	76,274.17	28,427.13
15	0.22	4.50	74,802.59	28,526.73
16	0.21	4.75	74,072.95	28,577.19
17	0.20	5.00	73,340.41	28,626.81

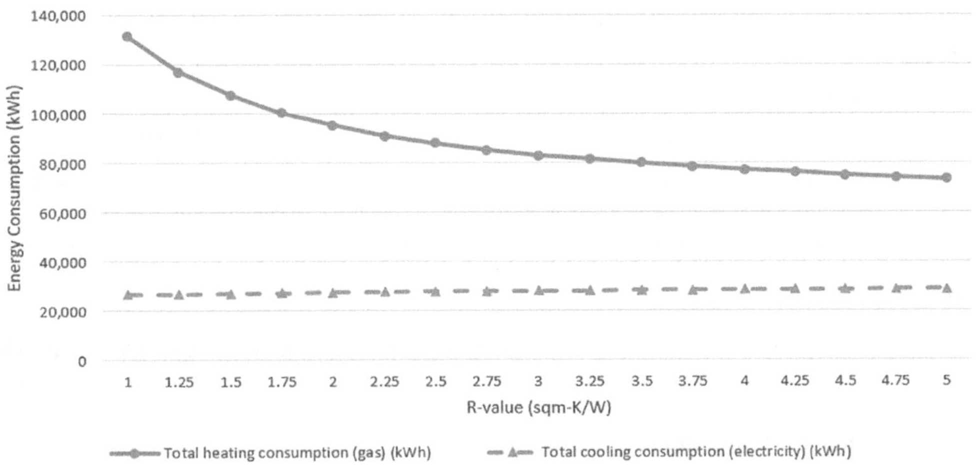

The results presented here are based on simulations conducted for the **WIEN/SCHWECHAT-FLUG, AUSTRIA** weather location, which experiences a temperate climate with cold winters. The impact of insulation is noticeable for this location, as it influences heating energy consumption by reducing heat loss through the roof.

In colder climates, insulation generally has a more pronounced effect on decreasing heating energy consumption. Conversely, in warmer or tropical climates, its impact on cooling energy consumption may be more significant, depending on factors such as roof exposure to solar radiation and overall cooling loads.

The findings emphasize the importance of carefully selecting insulation thickness to achieve optimal energy performance and cost-effectiveness. Effective insulation design involves striking a balance between energy savings and the associated costs of adding more insulation, while also considering the specific climate conditions of the location.

Additionally, the law of diminishing returns is evident in this study. As insulation thickness increases, the reduction in heating energy consumption becomes less pronounced. While initial increases in insulation thickness lead to substantial energy savings, the incremental benefit of adding more insulation decreases as the thickness continues to grow. This highlights the importance of evaluating insulation thickness with respect to cost-effectiveness and energy efficiency.

EXERCISE 3.2

Repeat the preceding steps with the Miami, Florida, location.

TUTORIAL 3.3: EVALUATING THE IMPACT OF THE POSITION OF ROOF INSULATION

Goal

To evaluate the impact of the position of roof insulation (overdeck and underdeck)

What Are You Going to Learn?

- How to edit roof construction

Problem Statement

In this tutorial, you are going to use a 50 m × 25 m model. Use the construction layers as given in Table 3.3. Find out the energy consumption for each variation. Use the weather data for **FRANKFURT MAIN ARPT, Germany** and **DUBAI INTERNATIONAL, United Arab Emirates**.

TABLE 3.3
Construction Layers

Layer	Case I	Case II
Outermost	Concrete, medium density, 0.15 m	XPS extruded polystyrene—HFC blowing, 0.05 m
Innermost	XPS extruded polystyrene—HFC blowing, 0.05 m	Concrete, medium density, 0.15 m

Solution

Step 1: Open a new project. Create a **50 m × 25 m** building. Change the weather location to Frankfurt, Germany.

Step 2: Select the **Construction** tab.

Step 3: Go to the Construction section, click **Flat roof**, and then click the **Add new item** icon. The **Edit construction—Roof** screen appears. Add a roof with underdeck insulation: **0.15 m** of **Concrete, Medium density** (from **Concrete materials**) and **0.05 m** of **XPS Extruded Polystyrene—HFC Blowing** (from **Insulating materials**).

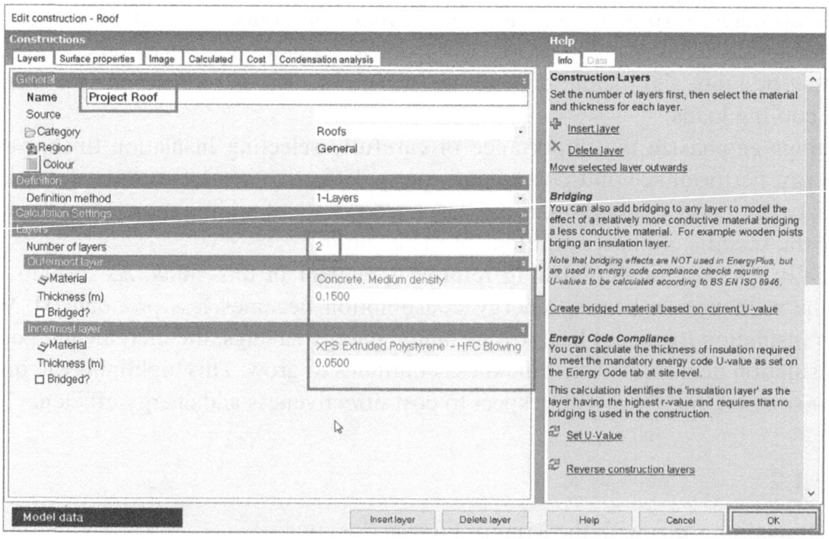

Step 4: Perform an annual simulation and record the results.

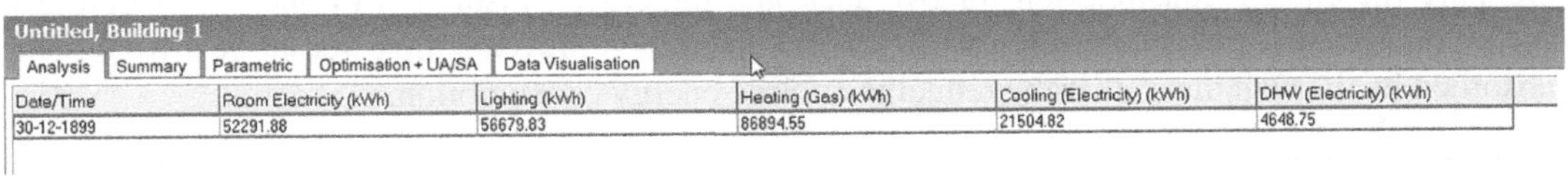

Step 5: Repeat the tutorial for overdeck insulation.

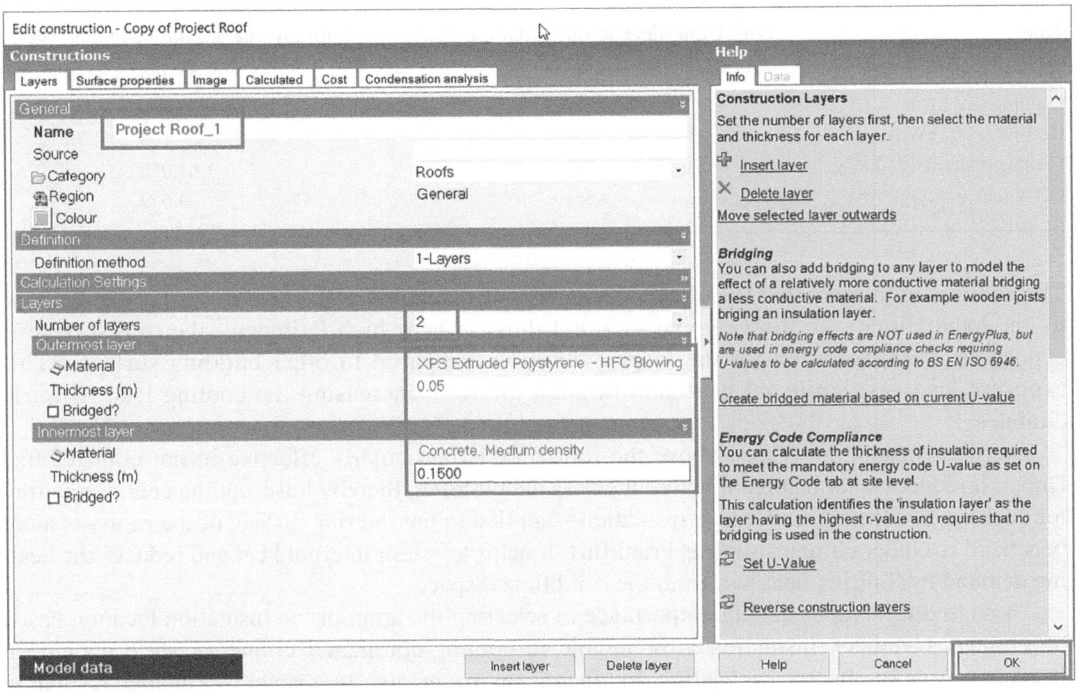

Step 6: Simulate the model and compare the results for the two cases (Table 3.4). Frankfurt requires predominantly heating and underdeck insulation, preventing the internal heat from being absorbed by the slab mass and thereby reducing the heating energy consumption.

TABLE 3.4
Annual Fuel Breakdown for Underdeck and Overdeck Roof Insulation for Frankfurt Location

Type	With Underdeck Roof Insulation	With Overdeck Roof Insulation
Room electricity (kWh)	52,292	52,292
Lighting (kWh)	56,680	56,680
Heating (gas) (kWh)	86,895	92,509
Cooling (electricity) (kWh)	21,505	18,064
DHW (electricity) (kWh)	4,649	4,649

Step 7: Change the weather location to **DUBAI INTERNATIONAL, United Arab Emirates** and compare the energy consumption for overdeck and underdeck insulation (Table 3.5). Cooling is the predominant requirement in Dubai. Overdeck insulation prevents the external heat from being absorbed by the slab mass, thereby reducing cooling energy consumption.

TABLE 3.5
Annual Fuel Breakdown for Underdeck and Overdeck Roof Insulation for Dubai Location

Type	With Underdeck Roof Insulation (kWh)	With Overdeck Roof Insulation (kWh)
Room electricity (kWh)	52,292	52,292
Lighting (kWh)	56,680	56,680
Heating (gas) (kWh)	725	103
Cooling (electricity) (kWh)	1,62,718	1,61,030
DHW (electricity) (kWh)	4,659	4,649

The analysis highlights that the effectiveness of roof insulation placement is strongly influenced by the local climate. In most regions—except those at very high latitudes—the roof surface is exposed to solar radiation for the longest duration compared to other building surfaces. This exposure leads to significant heat gain through the roof, increasing the cooling load in warm climates.

Overdeck insulation, placed above the roof slab, is particularly effective in hot climates like Dubai. It reduces solar heat gain before it enters the building, thereby lowering the energy required for cooling. In contrast, underdeck insulation—applied to the interior surface of the roof—is more beneficial in colder climates such as Frankfurt. It helps to retain internal heat and reduces the heating demand by limiting heat loss from the conditioned space.

These findings emphasize the importance of selecting the appropriate insulation location based on climate. Overdeck insulation is preferable in cooling-dominated climates, while underdeck insulation is more effective in heating-dominated environments. As a general principle, insulation should be positioned as early as possible along the path of heat flow—whether from outside to inside or vice versa—to maximize energy efficiency.

Further exploration could involve assessing the impact of insulation thickness and material properties under various climatic conditions.

EXERCISE 3.3

Repeat this tutorial for the external wall. Use the weather data for **FRANKFURT MAIN ARPT, Germany** and **DUBAI INTERNATIONAL, United Arab Emirates** (Table 3.6).

TABLE 3.6
Construction Layers for External Walls

Layer	Case I	Case II
Outermost	Brickwork 230 mm	XPS extruded polystyrene—HFC blowing, 0.05 m
Innermost	XPS extruded polystyrene—HFC blowing, 0.05 m	Brickwork 230 mm

TUTORIAL 3.4: EVALUATING THE IMPACT OF THE AIR GAP BETWEEN ROOF LAYERS

Goal

To evaluate the effect of the air-gap thickness in roof construction on energy consumption

What Are You Going to Learn?

- How to add the air gap between the roof layers
- How to change the air-gap thickness

Problem Statement

In this tutorial, you are going to use a 50 m × 25 m model with a roof consisting of the following layers (starting with the outermost layer).

1. 0.01 m of cement/plaster/mortar-plaster
2. 0.15 m of concrete, medium density
3. Air gaps with varying thicknesses (as given in what follows)
4. 0.01 m of cement/plaster/mortar-plaster

Thickness of the air gap:

1. No air gap
2. Air gap of 15 mm
3. Air gap of 17 mm
4. Air gap of 25 mm
5. Air gap of 50 mm
6. Air gap of 100 mm
7. Air gap of 300 mm

Find out the energy consumption for **LONDON/GATWICK ARPT, United Kingdom**.

An air gap is different from the airspace layer. An *air gap* is enclosed on either side, whereas an *airspace* is a gap left between exterior finish layers and interior insulation layers—as commonly practiced in lightweight construction. The main role of the airspace layer is to act as a vapor and water drain/barrier compared with an air gap, which improves the overall insulation properties of the wall/roof section.

Solution

Step 1: Open a new project. Create a **50 m × 25 m** building.

Step 2: Select the **Construction** tab.

Step 3: Add a new roof with three layers. Select **Cement/plaster/mortar—cement plaster** from **Plaster** materials with thickness (m) as **0.0100, Concrete, Medium density** from **Concretes** materials with thickness (m) as **0.1500**, and select **Cement/plaster/mortar—cement plaster** from **Plaster** materials with thickness (m) as **0.0100**.

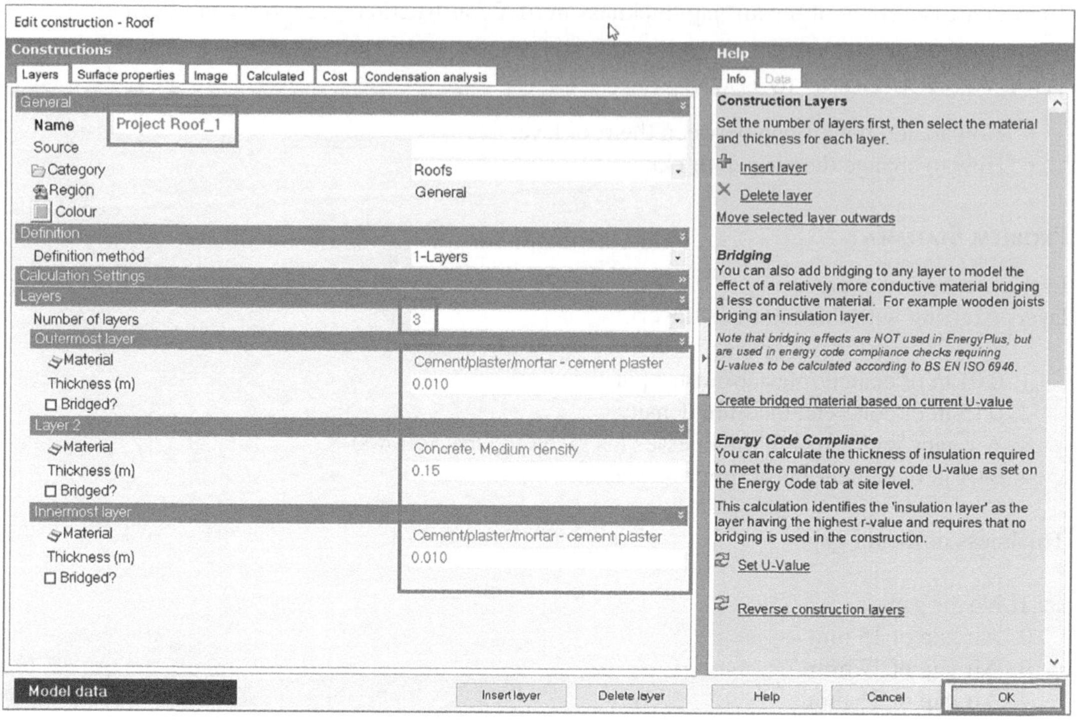

Step 4: Simulate the model and record the results.

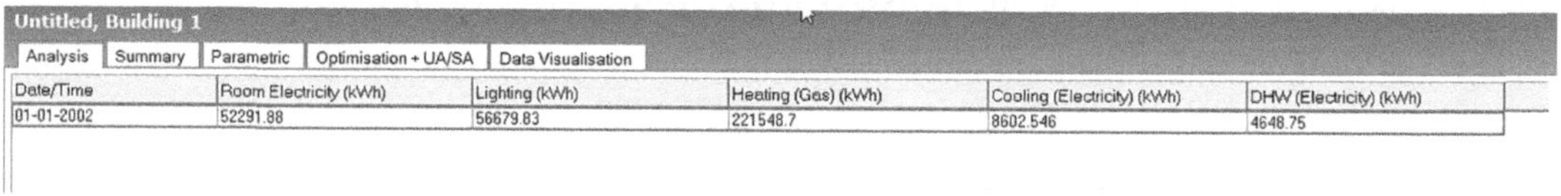

Step 5: Again, select the **Construction** tab. Edit the roof construction.

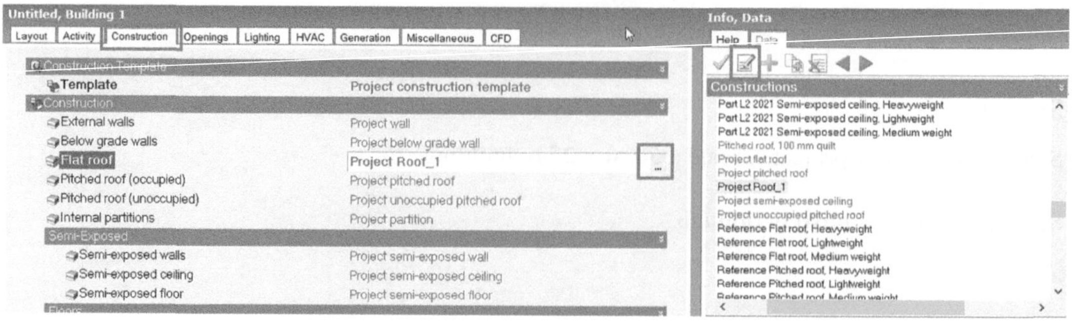

Step 6: Select the **Thickness** of the **Innermost layer** (we want to insert a new layer above this layer; hence, we have to select this layer). Click **Insert layer**. The **Edit Insert Material Layer** screen appears. Click **OK** to insert a new layer.

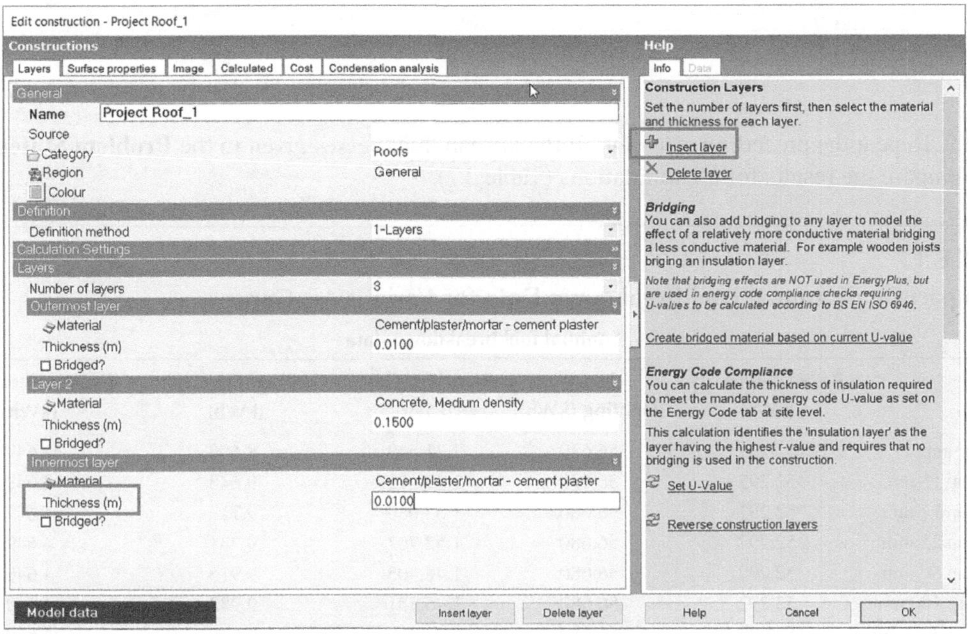

Step 7: Select **Air gap 15mm (downwards)** from **Gases** materials for the newly inserted layer (**Layer 3**) and enter **0.001** as thickness (m).

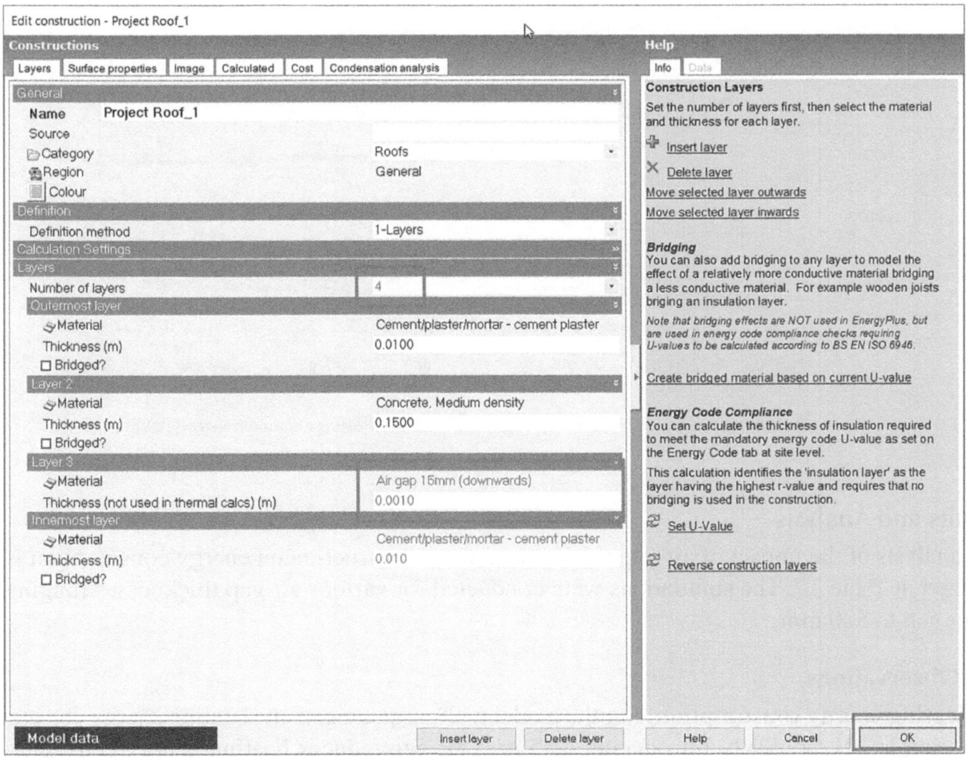

Step 8: Simulate the model and record the results.

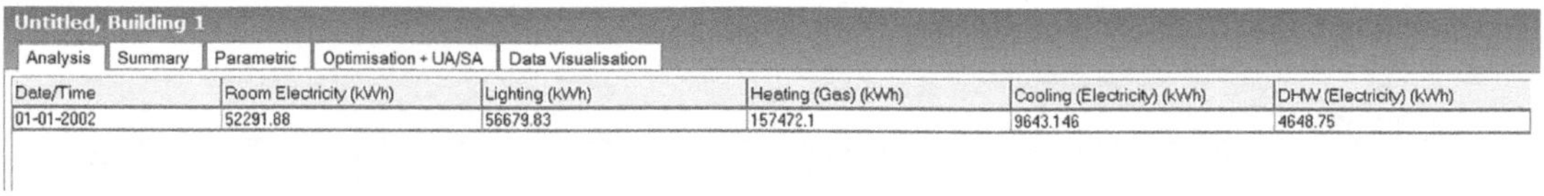

Date/Time	Room Electricity (kWh)	Lighting (kWh)	Heating (Gas) (kWh)	Cooling (Electricity) (kWh)	DHW (Electricity) (kWh)
01-01-2002	52291.88	56679.83	157472.1	9643.146	4648.75

Step 9: Repeat the preceding steps for all the air-gap thicknesses given in the **Problem Statement**. Compare the results for all simulations (Table 3.7).

TABLE 3.7
Comparison of Annual Fuel Breakdown Data for Various Air Gaps

	Annual fuel breakdown data				
Gap	Room Electricity (kWh)	Lighting (kWh)	Heating (Gas) (kWh)	Cooling (Electricity) (kWh)	DHW (Electricity) (kWh)
No air gap	52,292	56,680	2,21,549	8,603	4,649
Air gap 15 mm	52,292	56,680	1,57,472	9,643	4,649
Air gap 17 mm	52,292	56,680	1,55,080	9,712	4,649
Air gap 25 mm	52,292	56,680	1,52,762	9,781	4,649
Air gap 50 mm	52,292	56,680	1,48,405	9,915	4,649
Air gap 100 mm	52,292	56,680	1,46,351	9,982	4,649
Air gap 300 mm	52,292	56,680	1,44,369	10,049	4,649

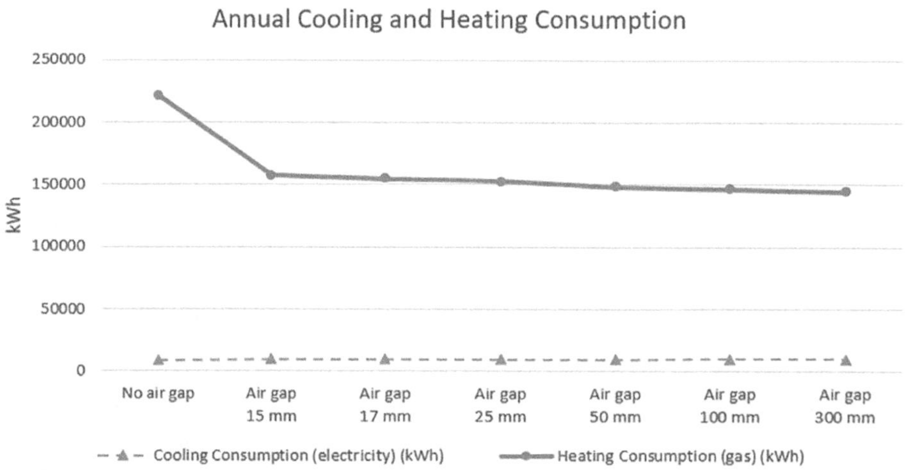

Results and Analysis

The analysis of the impact of air-gap thickness in roof construction on energy consumption is summarized in Table 3.7. The simulations were conducted for various air-gap thicknesses ranging from no air gap to 300 mm.

Key Observations

Heating energy consumption: For the model without an air gap, the heating energy consumption is 221,549 kWh. Introducing a 15 mm air gap reduces heating energy consumption to 157,472 kWh, indicating the positive impact of air-gap insulation. However, beyond 15 mm, the further reduction in heating energy becomes marginal, indicating diminishing returns with further increases in gap thickness. A still layer of air slows down heat transfer by conduction and convection. When you introduce a 15 mm air gap, you already achieve a significant portion of the insulation benefit air can provide. Beyond a certain thickness, convection kicks in and tiny circulating air loops can carry heat across the gap, reducing the insulating effect.

Overall analysis: Introducing an air gap in the roof structure is beneficial for reducing heating energy consumption, particularly with smaller air gaps (e.g., 15–50 mm). The marginal benefit of increasing air-gap thickness beyond 50 mm is minimal due to the occurrence of convection currents within the gap.

TUTORIAL 3.5: EVALUATING THE IMPACT OF ROOF SURFACE REFLECTANCE

Goal

To evaluate the effect of roof surface reflectance on the energy performance of a building

What Are You Going to Learn?

- How to change surface reflectivity

Problem Statement

In this tutorial, you are going to use a 50 m × 25 m model with a roof consisting of the following layers (starting with the outermost layer).

- 0.015 m of cement/plaster/mortar-plaster
- 0.150 m of concrete, medium density
- 0.015 m of cement/plaster/mortar-plaster

Vary the roof solar reflectance of the outermost material from 0.9 to 0.1 in steps of 0.1. Determine the impact of roof solar reflectance on the energy consumption for **SINGAPORE/PAYA LEBA** location.

Surface absorptance is the propensity of the surface material to absorb radiation, and it is the opposite of surface reflectance, which is the ability to reflect radiation. For opaque surfaces, surface absorptance and surface reflectance values are therefore ratios whose sum is always equal to 1.

Solar reflectivity = 1−solar absorptance

Solution

Step 1: Open a new project. Create a **50 m × 25 m** building.

Step 2: Select the **Construction** tab.

Step 3: Add a new roof with three layers. Select **Cement/plaster/mortar—cement plaster** from **Plaster** materials with thickness (m) as **0.01**, **Concrete, Medium density** from **Concretes** materials with thickness (m) as **0.15** and **Cement/plaster/mortar-cement plaster** from **Plaster** materials with thickness (m) as **0.01**.

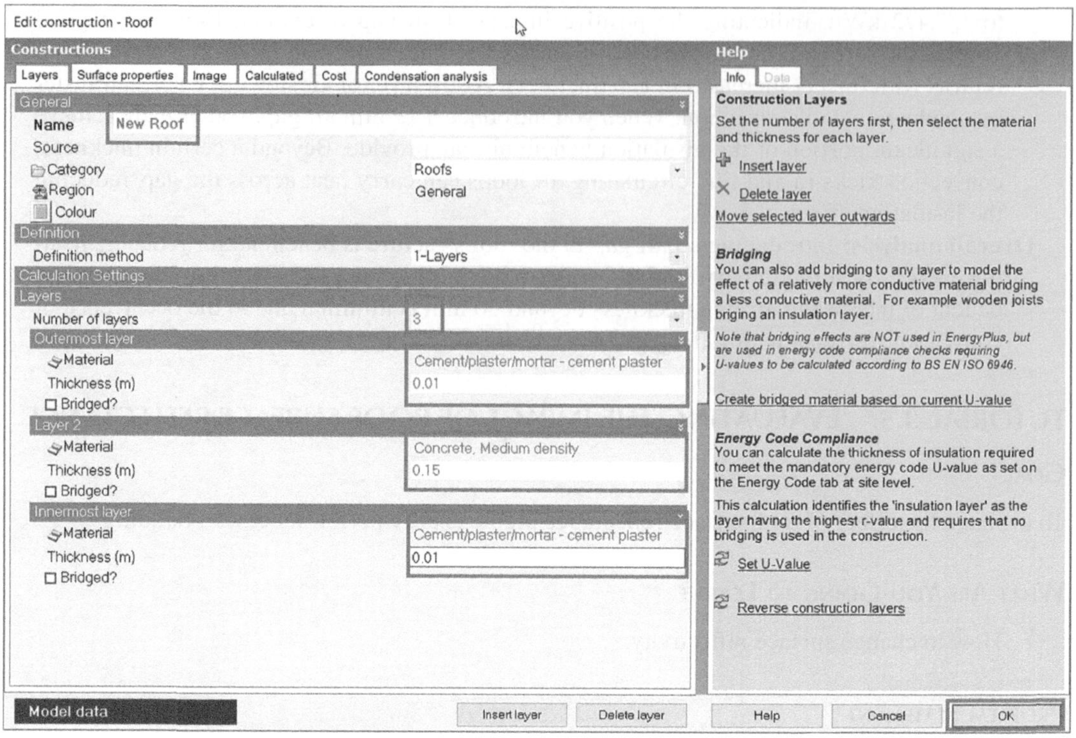

Step 4: Select **Cement/plaster/mortar—cement plaster** in the outermost layer, and then click the **Create copy of highlighted item** icon.

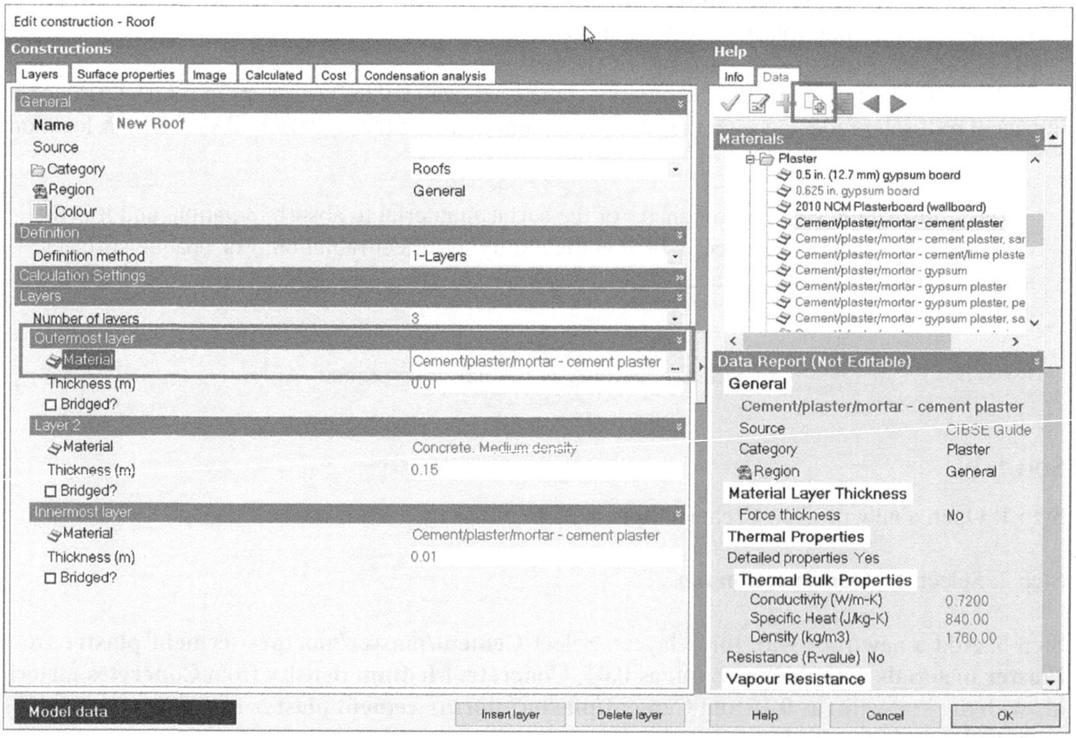

Step 5: Select the **Copy of Cement/plaster/mortar—cement plaster,** and then click the **Edit highlighted item** icon.

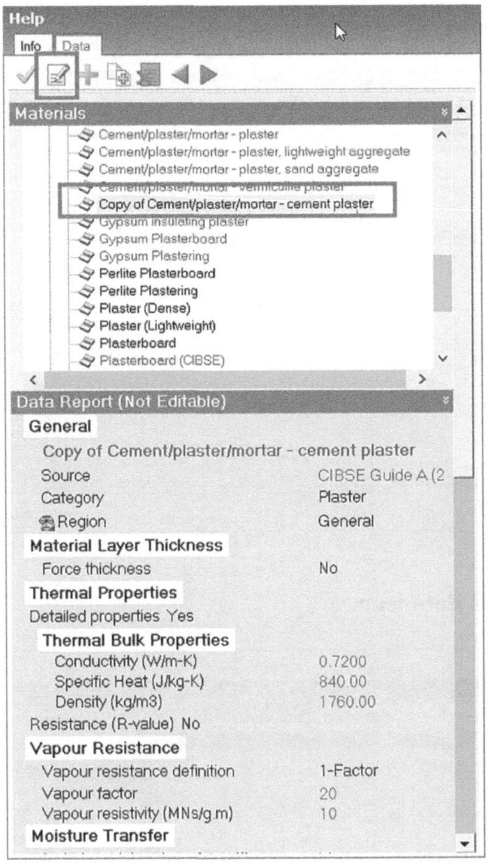

Step 6: Enter **Plaster—Surface Absorptance 0.9** in the **Name** box.

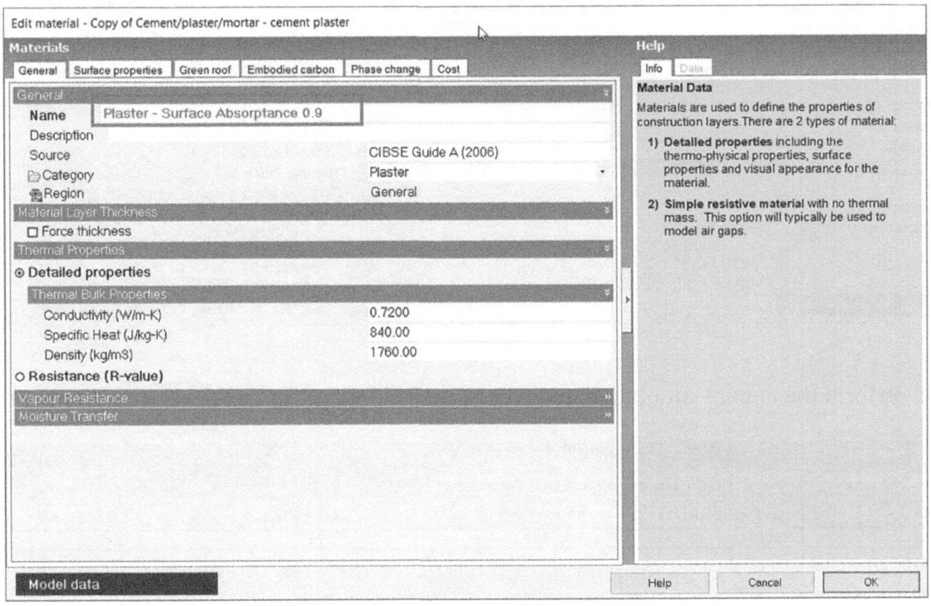

Step 7: Select the **Surface Properties** tab and enter **Solar absorptance** as **0.900**.

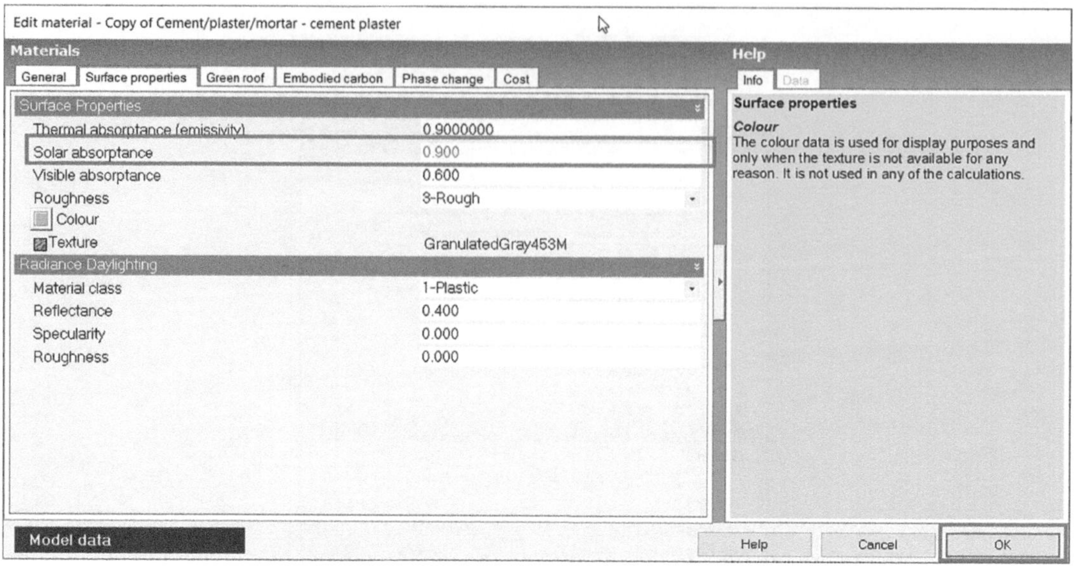

Step 8: Click the **Select this data** icon.

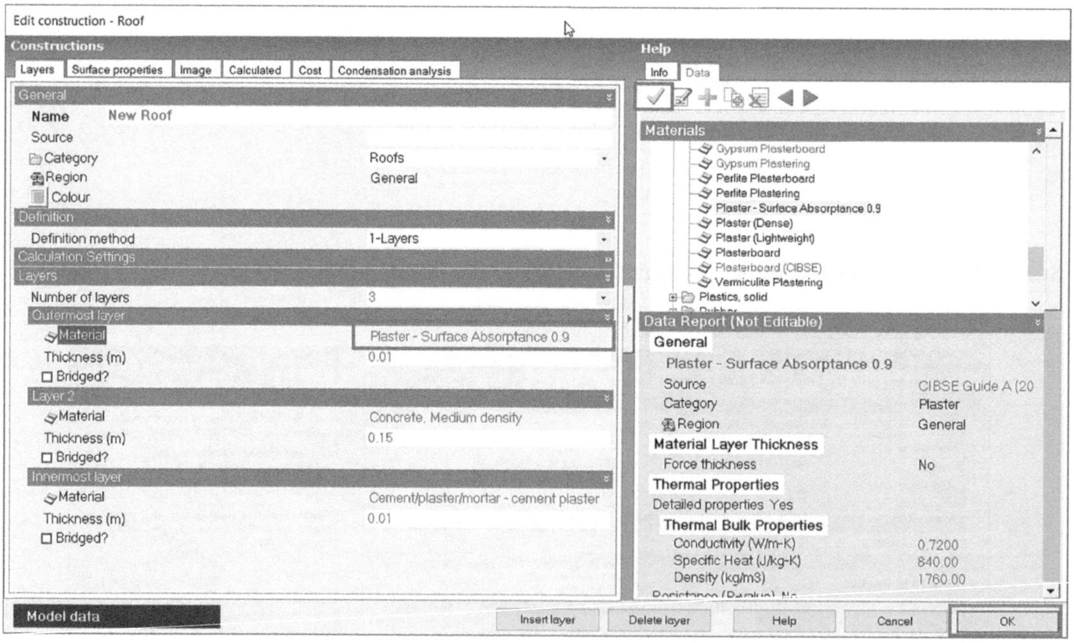

Step 9: Perform the annual simulation and record the results.

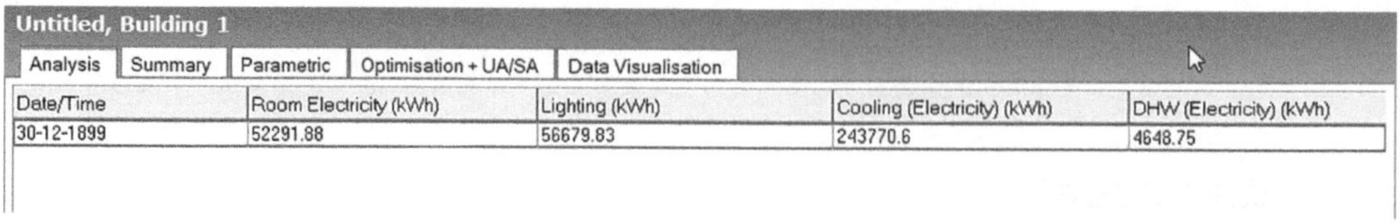

Step 10: Repeat the preceding steps for all the values of solar absorptance given in the problem statement. Compare the results for all simulations (Table 3.8).

TABLE 3.8
Annual Fuel Breakdown Data for Different Surface Absorptances

Surface Absorptance	Surface Reflectance	Room Electricity (kWh)	Lighting (kWh)	Cooling (Electricity) (kWh)	DHW (Electricity) (kWh)
0.9	0.1	52,292	56,680	2,43,771	4,649
0.8	0.2	52,292	56,680	2,34,666	4,649
0.7	0.3	52,292	56,680	2,25,413	4,649
0.6	0.4	52,292	56,680	2,16,073	4,649
0.5	0.5	52,292	56,680	2,06,570	4,649
0.4	0.6	52,292	56,680	1,97,000	4,649
0.3	0.7	52,292	56,680	1,87,287	4,649
0.2	0.8	52,292	56,680	1,77,565	4,649
0.1	0.9	52,292	56,680	1,67,984	4,649

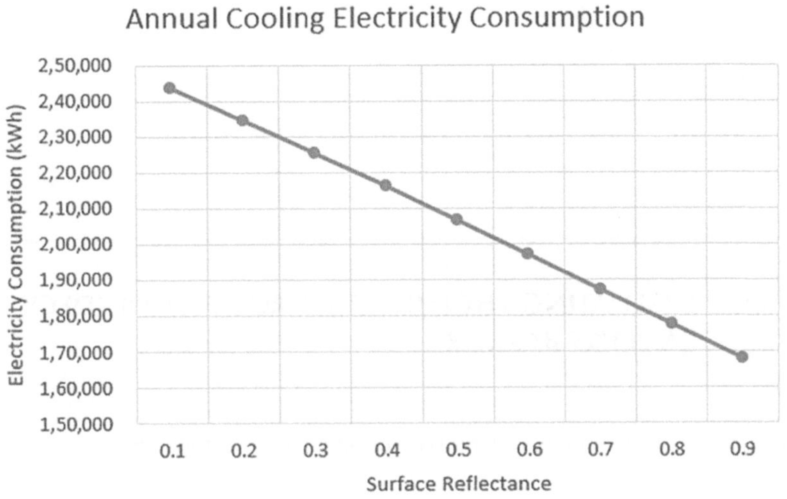

Results and Analysis

The results of the simulations conducted for different surface reflectance are presented in Table 3.8. The impact of varying the roof's surface reflectance from 0.1–0.9 on building energy consumption is evaluated.

Observations

Cooling energy consumption: As the surface reflectance increases from 0.1 to 0.9, the cooling energy consumption decreases for Singapore. This reduction occurs because higher surface reflectance values cause more incident solar radiation to be reflected away from the roof's surface, resulting in less heat being absorbed and transmitted into the building.

Lighting and room electricity: The values for room electricity and lighting remain constant across all simulations. This indicates that surface reflectance of the roof does not affect the lighting energy consumption or room electricity usage.

Implications

Increasing roof surface reflectance proves to be an effective strategy for reducing cooling energy consumption in buildings located in hot climates such as Singapore. The use of high-reflectance roofing materials could significantly lower operational energy requirements, thereby enhancing overall energy efficiency. While increasing roof reflectance influences cooling energy consumption, it does not affect lighting and room electricity consumption.

EXERCISE 3.5

Repeat this tutorial for the **FRANKFURT MAIN ARPT, Germany** weather location (Table 3.9).

TABLE 3.9
Annual Fuel Breakdown Data for Different Surface Absorptance

Surface Absorptance	Surface Reflectance	Room Electricity (kWh)	Lighting (kWh)	Heating (Gas) (kWh)	Cooling (Electricity) (kWh)	DHW (Electricity) (kWh)
0.9	0.1					
0.8	0.2					
0.7	0.3					
0.6	0.4					
0.5	0.5					
0.4	0.6					
0.3	0.7					
0.2	0.8					
0.1	0.9					

TUTORIAL 3.6: EVALUATING THE IMPACT OF ROOF UNDERDECK RADIANT BARRIER

Goal

To evaluate the impact of underdeck radiant barrier on energy performance

What Are You Going to Learn?

- How to change the thermal emittance of a material

Problem Statement

In this tutorial, you are going to use a 50 m × 25 m model with a roof consisting of the following layers (starting with the outermost layer).

- 0.01 m of cement/plaster/mortar-plaster
- 0.15 m of concrete, medium density
- 0.01 m of cement/plaster/mortar-plaster

Vary the thermal emittance on the innermost material of the roof (ceiling) from 0.1–0.9 in steps of 0.1. Determine the impact of roof thermal emittance on the energy consumption for **CA-SAN FRANCISCO INTL, USA**.

THERMAL ABSORPTANCE (EMISSIVITY)

Thermal absorptance represents the fraction of incident long-wavelength radiation that is absorbed by the material. This parameter is used when calculating the long-wavelength radiant exchange between various surfaces and affects the surface heat balances (both inside and outside as appropriate). Values for this field must be between 0.0 and 1.0 (with 1.0 representing "black body" conditions). *Radiant barrier* is the coating used under the roof deck to limit the heat transfer from the inside to the outside and the outside to the inside.

Source: www.designbuilder.co.uk/helpv7.0/Content/SurfaceProperties.htm

SOLUTION

Step 1: Open a new project. Create a **50 m × 25 m** building.

Step 2: Select the **Construction** tab.

Step 3: Add a new roof with three layers. Select **Cement/plaster/mortar—cement plaster** from **Plaster** materials with thickness (m) as **0.01**, **Concrete, Medium density** from **Concretes** materials with thickness (m) as **0.1500**, and **Cement/plaster/mortar—cement plaster** from **Plaster** materials with thickness (m) as **0.0100**.

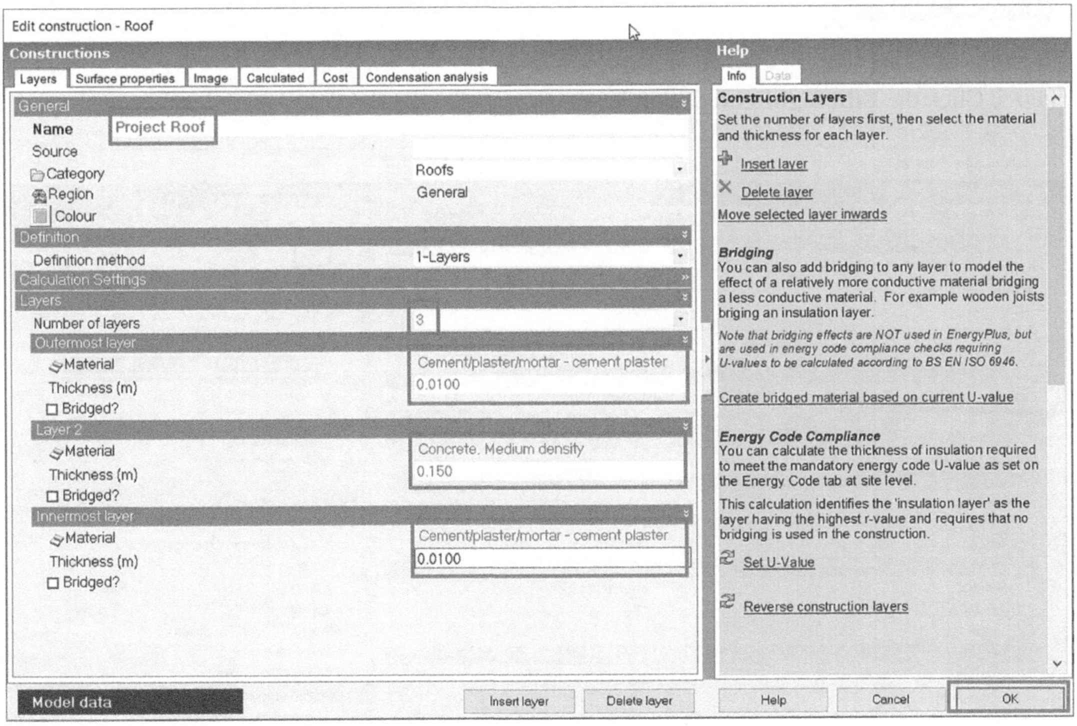

Step 4: Click **Cement/plaster/mortar—cement plaster** in the innermost layer. Create a copy and rename it as **New Cement/plaster/mortar—cement plaster**.

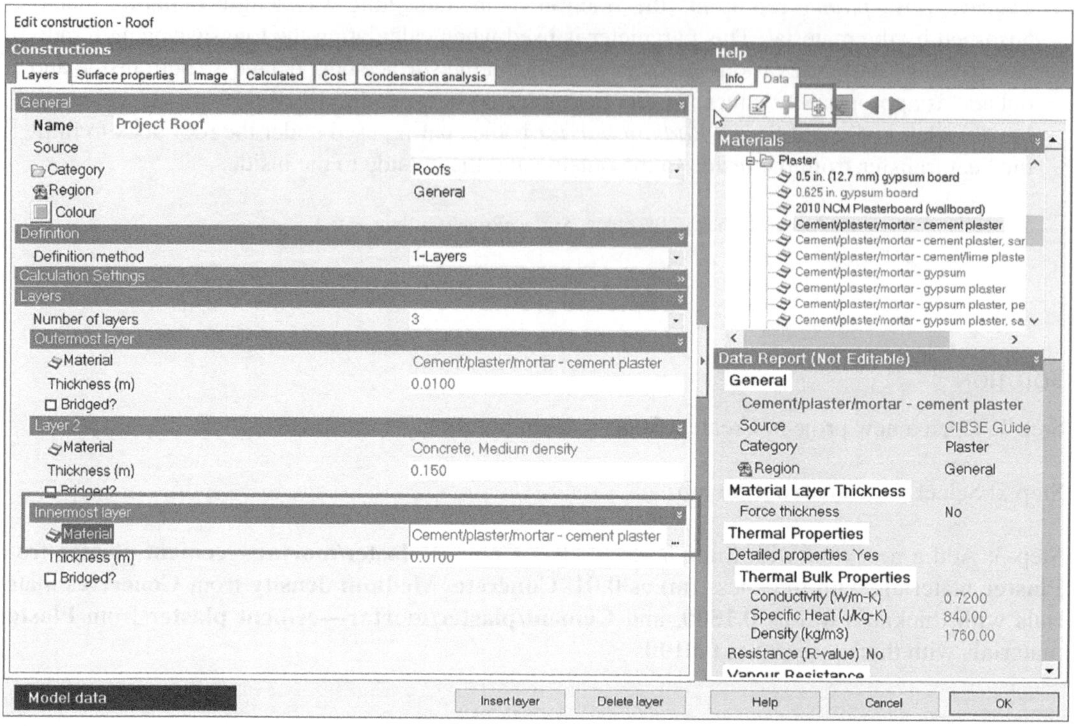

Step 5: Click the **Edit highlighted item** icon.

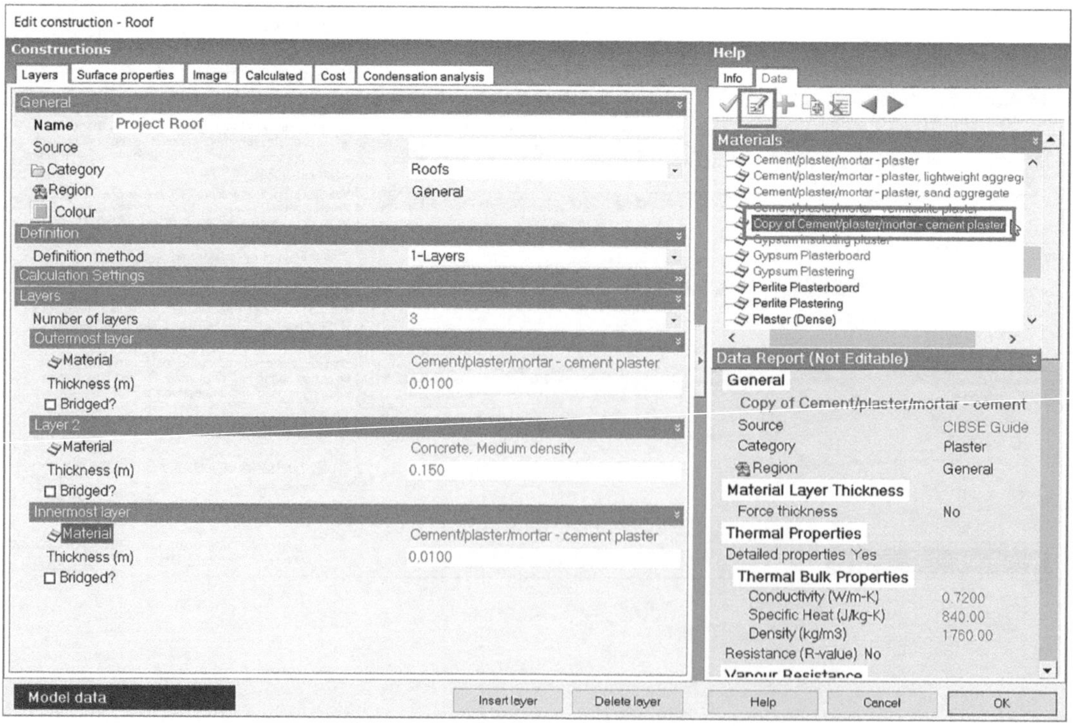

Step 6: Select the **Surface properties** tab and enter **0.900** as the **Thermal absorptance (emissivity)**.

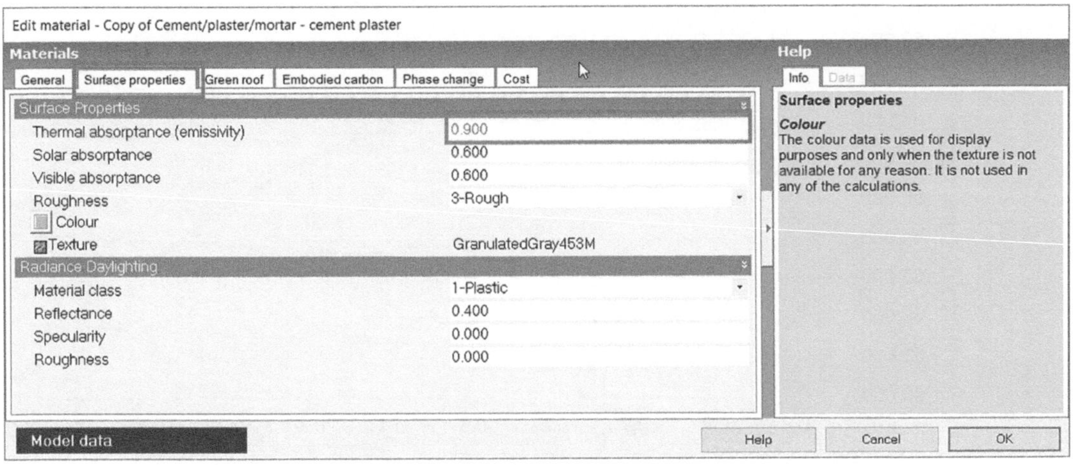

Step 7: Simulate the model and record the results.

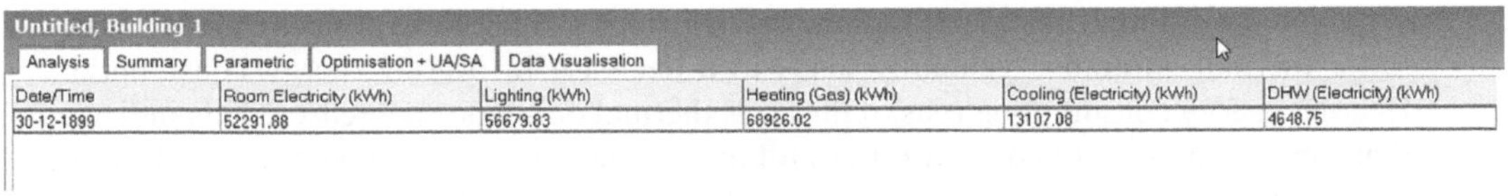

Step 8: Repeat the preceding steps for all the values of thermal emissivity given in the problem statement. Compare the results for all simulations (Table 3.10).

TABLE 3.10
Comparison of Annual Fuel Breakdown Data for Thermal Absorptance

Thermal absorptance (Emissivity)	Room Electricity (kWh)	Lighting (kWh)	Heating (Gas) (kWh)	Cooling (Electricity) (kWh)	DHW (Electricity)
0.9	52,292	56,680	68,926	13,107	4,649
0.8	52,292	56,680	68,659	13,112	4,649
0.7	52,292	56,680	68,338	13,131	4,649
0.6	52,292	56,680	67,946	13,168	4,649
0.5	52,292	56,680	67,471	13,232	4,649
0.4	52,292	56,680	66,860	13,342	4,649
0.3	52,292	56,680	66,044	13,518	4,649
0.2	52,292	56,680	64,942	13,798	4,649
0.1	52,292	56,680	63,398	14,265	4,649

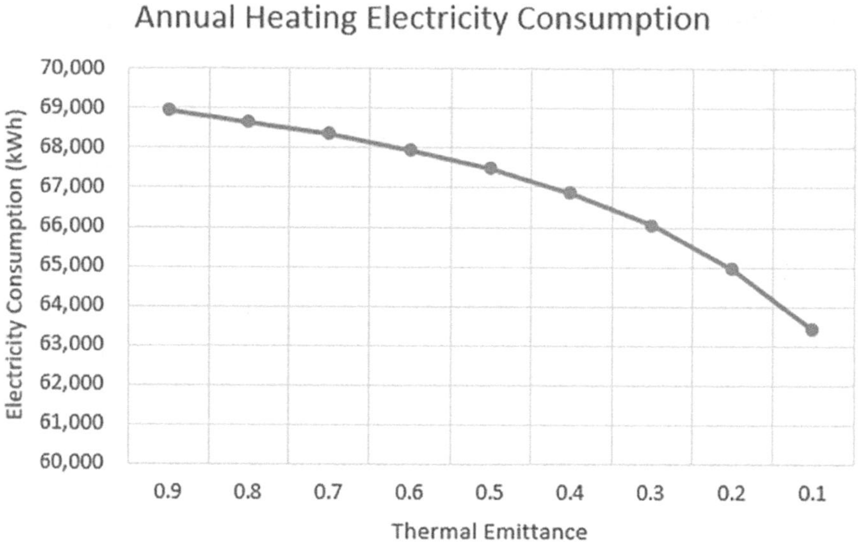

Observations

Energy consumption: Underdeck radiant barrier helps reduce radiant heat transfer, and a lower thermal emittance generally leads to a decrease in overall energy consumption. Heating energy consumption (gas) reduces as thermal emittance decreases. This indicates that more heat is retained within the building, reducing heating requirements. Cooling energy consumption (electricity) increases slightly as thermal emittance decreases, indicating reduced heat transfer to the outside but slightly higher indoor temperatures that require cooling. Room electricity, lighting, and DHW consumption remain unaffected by changes in thermal emittance.

Effectiveness of radiant barrier: The implementation of radiant barriers with low thermal emittance (0.1–0.3) proves to be beneficial for reducing heating loads, especially during colder periods. Lower emittance comes with lower absorptance for the same wavelength, which actually helps in reducing the heat loss through the surface over which the radiant barrier is applied.

Conclusion

The results indicate that lowering the thermal emittance of the innermost roof layer effectively reduces the building's heating energy consumption. Utilizing a radiant barrier with a low-emittance coating (e.g., 0.1–0.3) proves to be a beneficial strategy for improving energy efficiency.

EXERCISE 3.6

Repeat this tutorial for the **DUBAI INTERNATIONAL, United Arab Emirates** weather location (Table 3.11).

TABLE 3.11
Annual Fuel Breakdown Data

Thermal Absorptance (Emissivity)	Room Electricity (kWh)	Lighting (kWh)	Heating (Gas) (kWh)	Cooling (Electricity) (kWh)	DHW (Electricity)
0.9					
0.8					
0.7					
0.6					
0.5					
0.4					
0.3					
0.2					
0.1					

TUTORIAL 3.7: EVALUATING THE IMPACT OF A GREEN ROOF

Goal

To evaluate the impact of a green roof on energy consumption

What Are You Going to Learn?

- How to create a green roof

Problem Statement

In this tutorial, you are going to use a 50 m × 25 m model with a roof consisting of the following layers (starting with the outermost layer).

- 0.01 m of cement/plaster/mortar-plaster
- 0.15 m of concrete, medium density
- 0.01 m of cement/plaster/mortar-plaster

Add a green roof to the outermost layer of the roof. Determine the change in energy consumption with and without a green roof for **DUBAI INTERNATIONAL, United Arab Emirates** location.

> A *green roof*, or *living roof*, is a roof of a building that is partially or completely covered with vegetation and a growing medium planted over a waterproofing membrane.
>
> *Source:* https://designbuilder.co.uk/helpv7.3/Content/GreenRoof.htm

Solution

Step 1: Open a new project. Create a **50 m × 25 m** building.

Step 2: Select the **Construction** tab.

Step 3: Add a **New roof** with three layers. Select **Cement/plaster/mortar—cement plaster** from **Plaster** materials with a thickness (m) as **0.0100**, **Concrete, Medium density** from **Concretes** materials with a thickness (m) as **0.1500,** and **Cement/plaster/mortar—cement plaster** from **Plaster** materials with thickness (m) as **0.0100**.

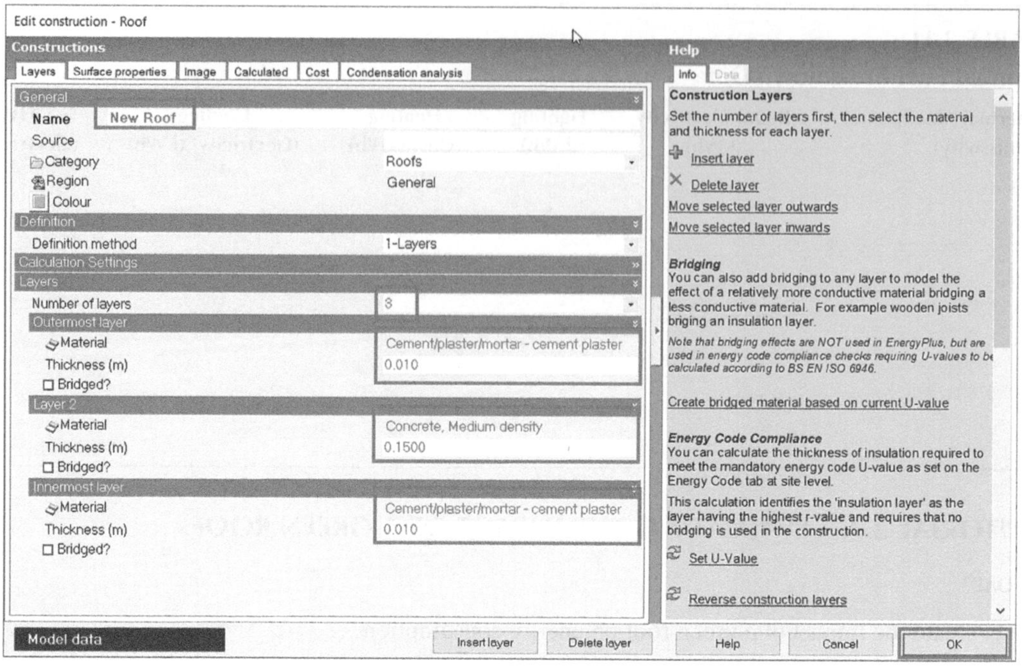

Step 4: Simulate the model and record the results without a green roof. Now you are going to add a green roof.

Step 5: Add a **New roof** with four layers. Select **12 in. Soil at R–0.104/in.** from **Sands, Stones and Soil** materials with thickness (m) as **0.305**, **Ethylene propylene diene monomer (EPDM)** from **Rubber** materials with thickness (m) as **0.015**, **Concrete, Medium density** from **Concretes** materials with thickness (m) as **0.1500**, and **Cement/plaster/mortar—cement plaster** from **Plaster** materials with thickness (m) as **0.0100**.

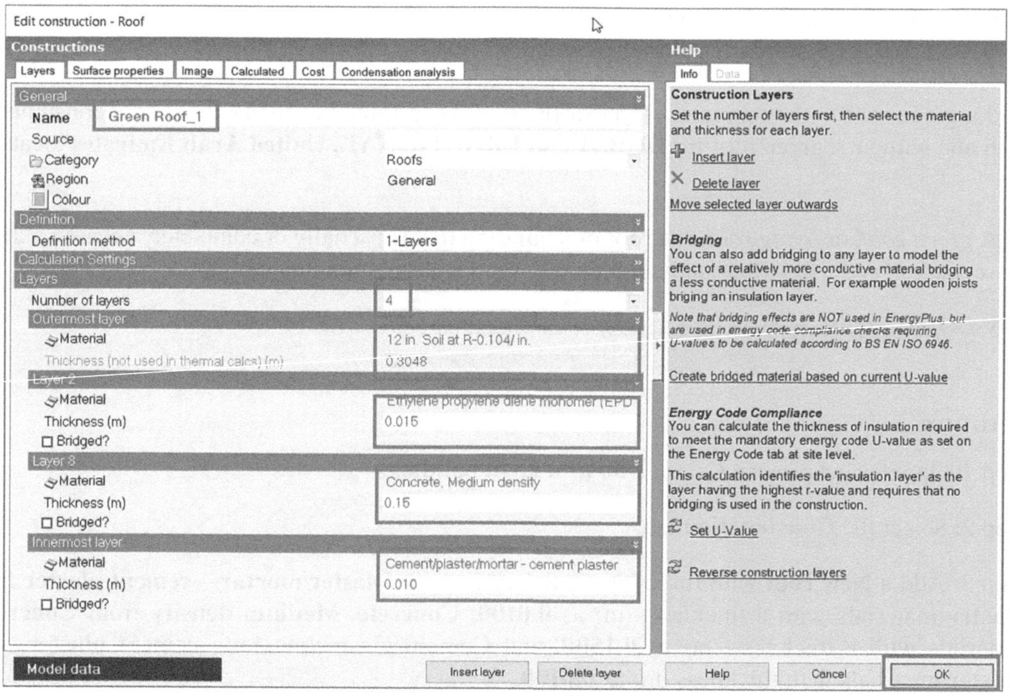

Step 6: Click **12 in. Soil at R–0.104/in**. In the outermost layer, it gets highlighted. Now create a copy.

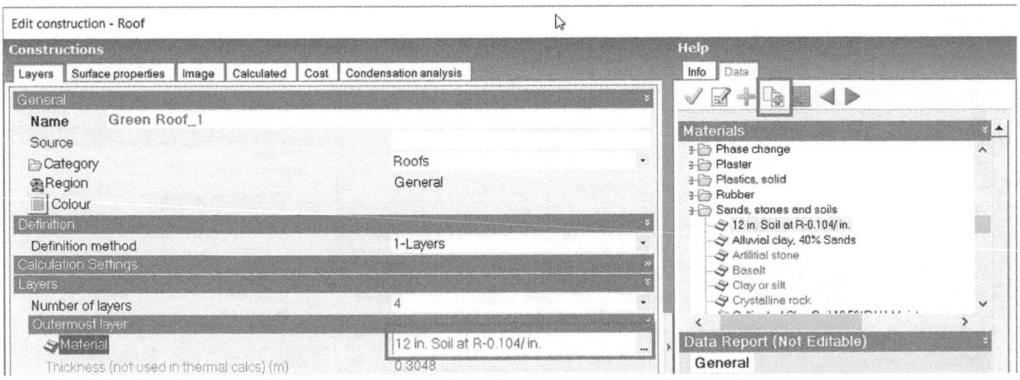

Step 7: Select the **Copy of 12 in. Soil at R–0.104/in**. Rename it **Green Roof Layer**.

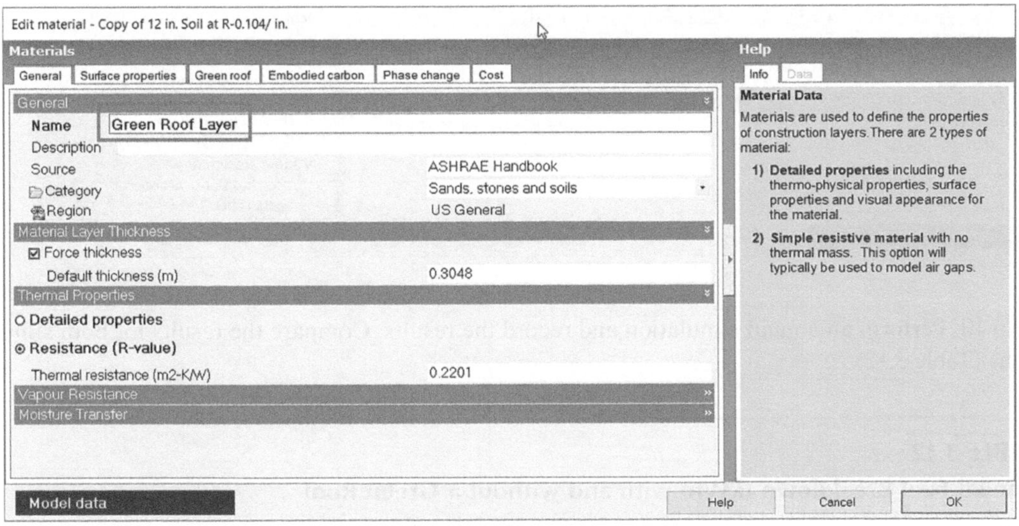

Step 8: Select the **Green roof** tab. Select the **Green roof** checkbox. Select **Simple** from **Moisture diffusion calculation method** drop-down list.

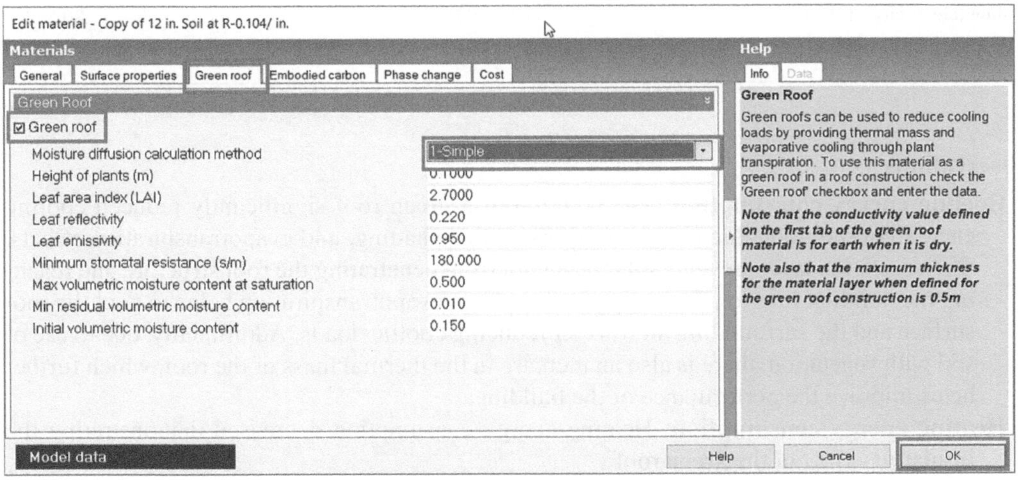

Step 9: Click **OK. Green Roof Layer** appears as the outermost layer.

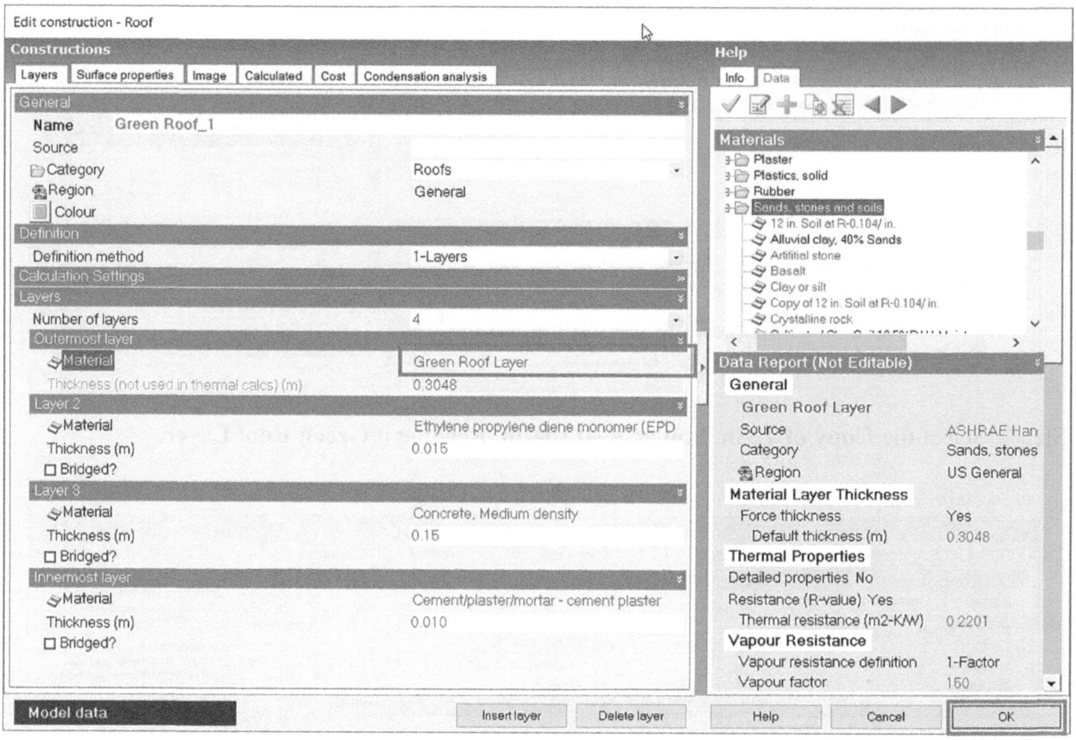

Step 10: Perform an annual simulation and record the results. Compare the results for both simulations (Table 3.12).

TABLE 3.12
Annual Fuel Breakdown (kWh) with and without a Green Roof

Type	Without a Green Roof	With a Green Roof
Room electricity (kWh)	52,292	52,292
Lighting (kWh)	56,680	56,680
Heating (gas) (kWh)	2,732	526
Cooling (electricity) (kWh)	1,89,076	1,80,590
DHW (electricity) (kWh)	4,649	4,649

Observations

Cooling energy consumption: The addition of a green roof significantly reduced cooling energy consumption due to enhanced insulation, shading, and evapotranspiration effects. The vegetation layer prevents solar heat gain from penetrating the roof structure and reaching the building interior. The cooling effect from evapotranspiration helps to cool the roof surface and the surrounding air, further reducing cooling loads. Additionally, due to use of soil with vegetation, there is also an increase in the thermal mass of the roof, which further helps improve the performance of the building.

Heating energy consumption: Heating energy consumption decreased demonstrating the insulating effect of the green roof.

Energy efficiency improvement: The presence of the green roof improved overall energy efficiency by lowering cooling energy demands and enhancing thermal comfort during peak solar gain periods.

Conclusion

The results indicate that incorporating a green roof is an effective strategy for reducing cooling energy consumption in hot climates such as Dubai. The green roof acts as an insulating layer, limits heat transfer, and provides cooling benefits through evapotranspiration, thereby reducing the building's overall energy consumption. Implementing green roofs is recommended for climates where cooling loads dominate the energy consumption profile.

EXERCISE 3.7

Repeat this tutorial for the **CA—San Francisco, USA** weather file (Table 3.13).

TABLE 3.13
Annual Fuel Breakdown Consumption with and without a Green Roof

	Annual Fuel Breakdown Consumption (kWh)	
Type	Without a Green Roof	With a Green Roof
Room electricity		
Lighting		
Heating (gas)		
Cooling (electricity)		
DHW (electricity)		

TUTORIAL 3.8: EVALUATING THE IMPACT OF PHASE-CHANGE MATERIAL IN AN EXTERNAL WALL

Goal

To evaluate the effect of phase-change material (PCM) in external wall construction on cooling energy consumption and wall temperatures

What Are You Going to Learn?

- How to assign PCM construction
- How to determine external and internal surface temperatures

Problem Statement

In this tutorial, you will use a five-zone 50 m × 25 m building model to evaluate the effect of incorporating a PCM layer in the external wall. Starting with a standard 100 mm thick brick wall, you will modify the construction by adding a 25 mm PCM layer.

The objective is to compare the inside and outside surface temperatures of the south-facing wall for both constructions. The simulations are to be conducted using the weather data for **JAIPUR/ SANGANER, India**.

Phase-change material (PCM) works in two phases. By completing a transition between solid and liquid phases, PCMs can absorb large quantities of latent heat. The high latent heat storage capacity of these materials works to effectively increase thermal mass, which tends to moderate interior temperatures and improve comfort conditions. PCMs are frequently used to reduce the need for mechanical cooling and peak load shifting, and to improve solar energy use.

Source: https://designbuilder.co.uk/helpv7.3/Content/PhaseChange.htm

Solution

Step 1: Open a new project file. Create **50 m × 25 m** five-zone building with a 5 m perimeter depth.

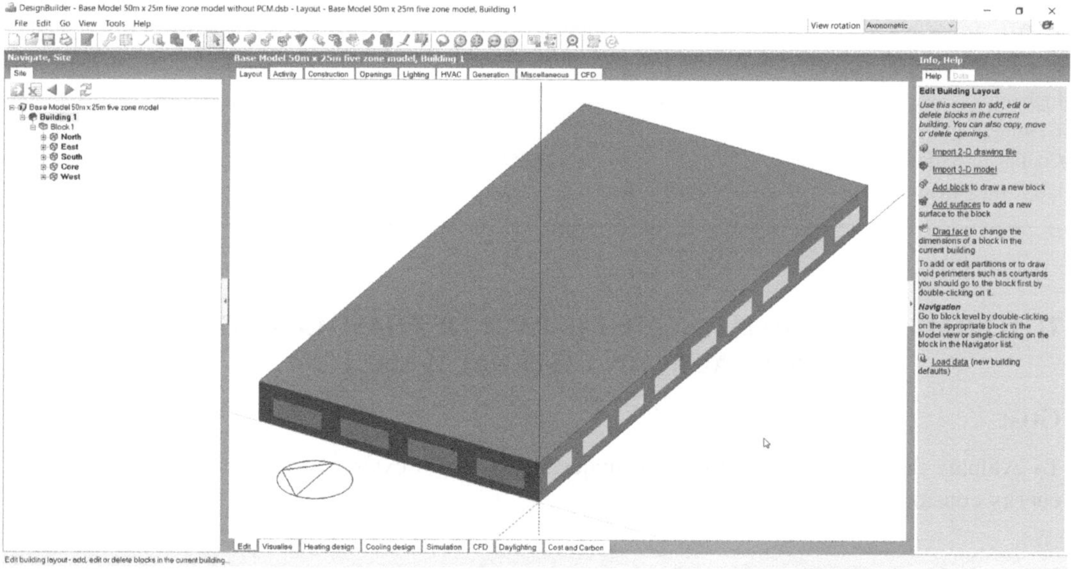

Step 2: Click the **Construction** tab.

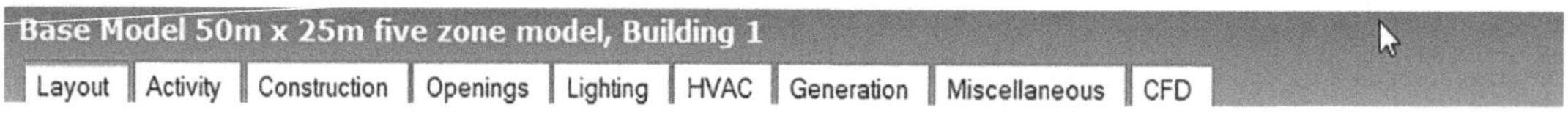

Step 3: In the **Construction** section, click **Project wall, External walls**. Click the **Add new item** icon. The **Edit construction-Wall** screen appears.

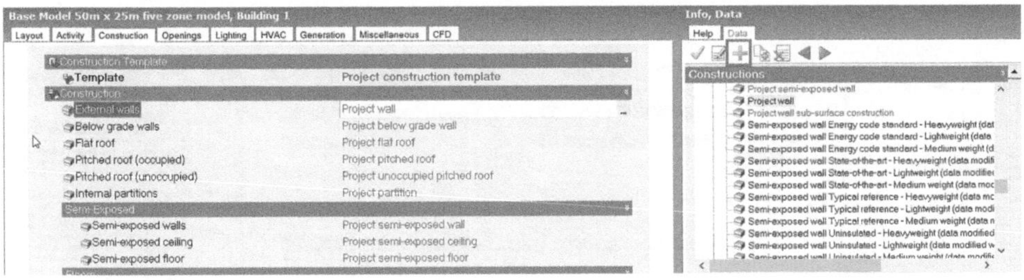

Step 4: Enter **Brick Wall** as the name. Go to the **Layers** section and select **1** from the **Number of layers** drop-down list. Select **Brick** as material.

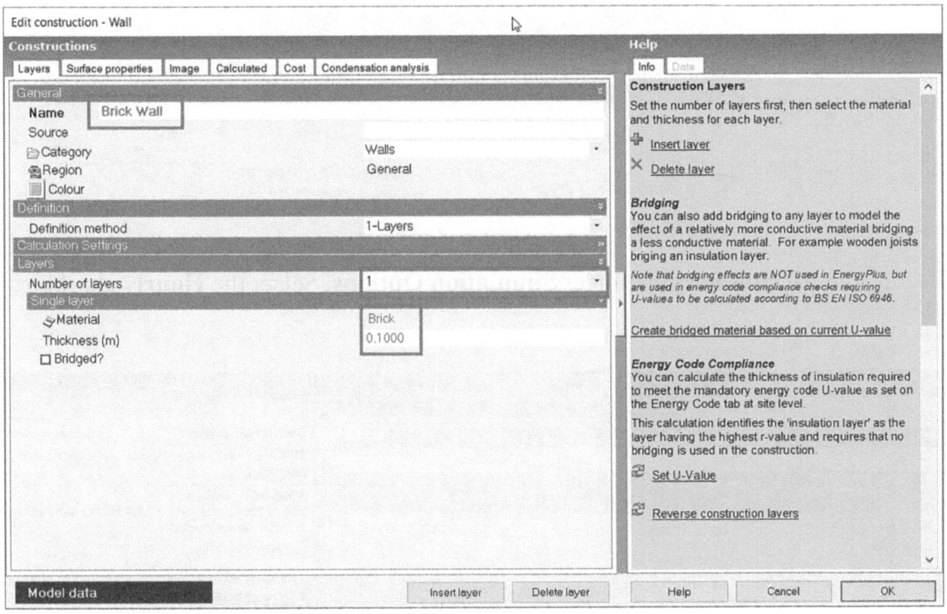

Step 5: Select **Brick Wall** as external wall.

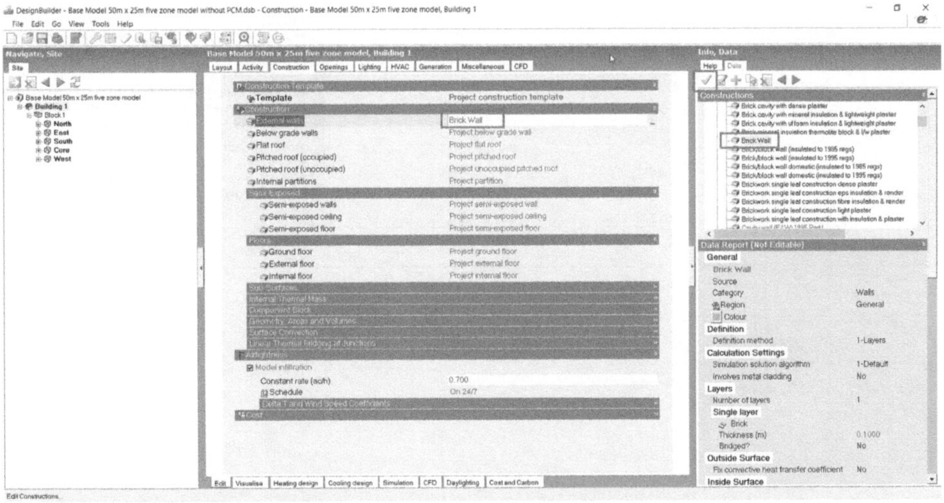

Step 6: Click the **Simulation** screen tab. Click **Update data**. The **Edit Calculation Options** screen appears.

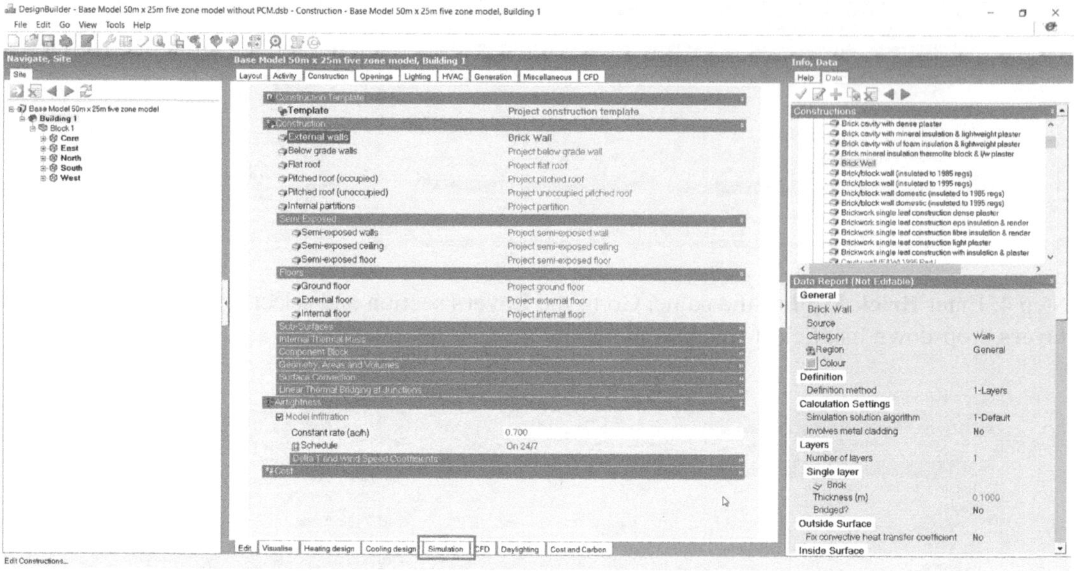

Step 7: Click the **All summer** link under **Simulation Options**. Select the **Hourly** checkbox.

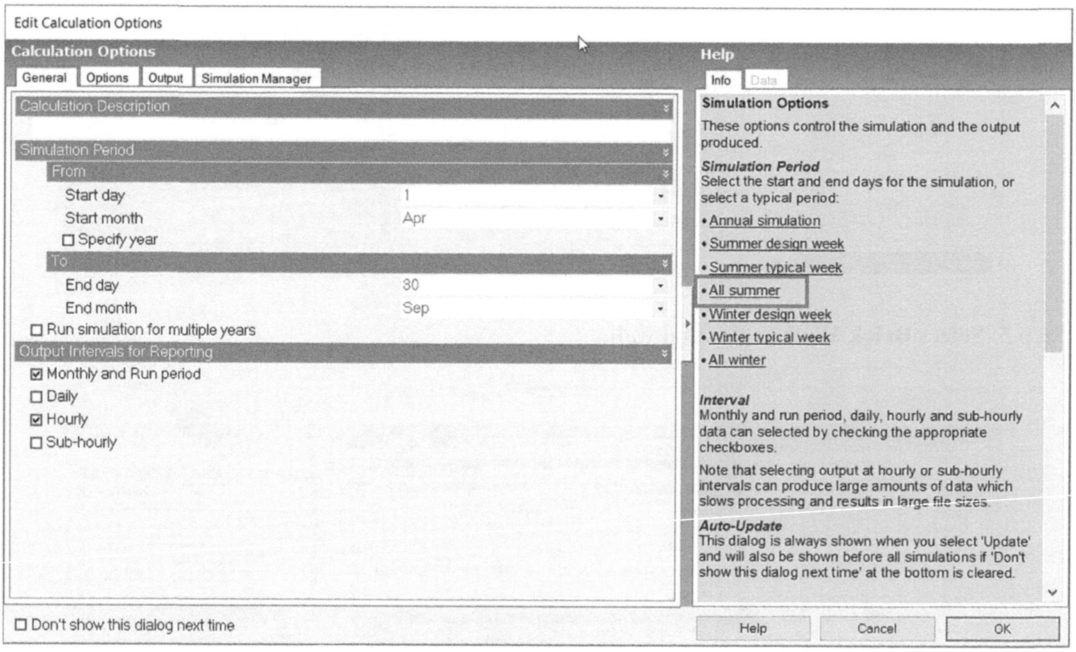

Step 8: Click the **Output** tab. Select the **Store surface output** checkbox. Click **OK**.

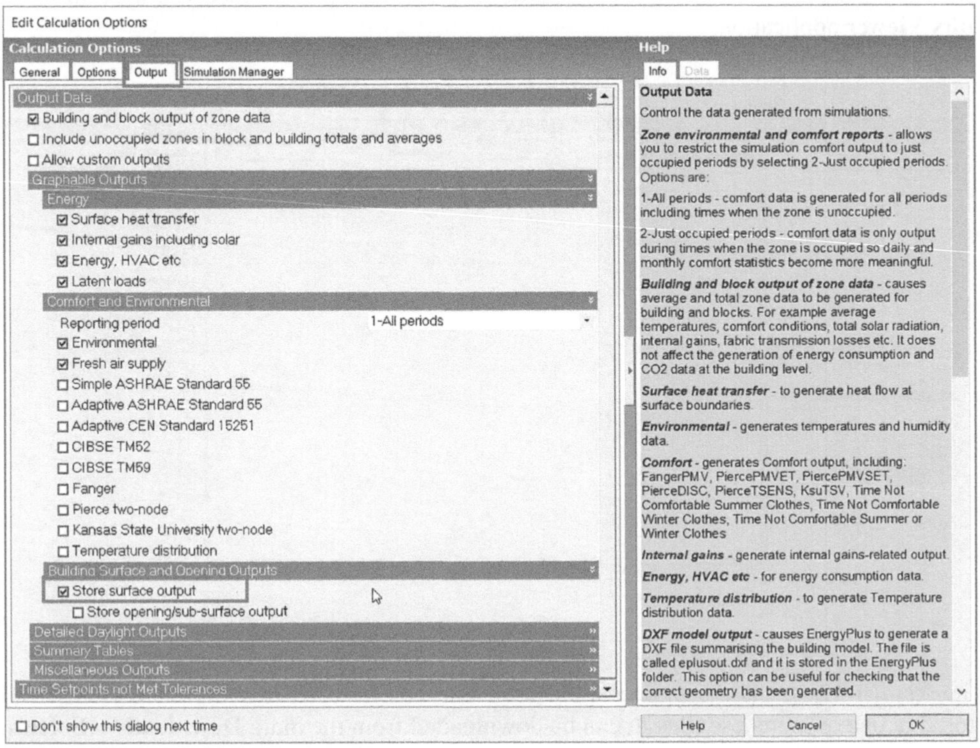

Step 9: Once the simulation is finished, record the energy consumption.

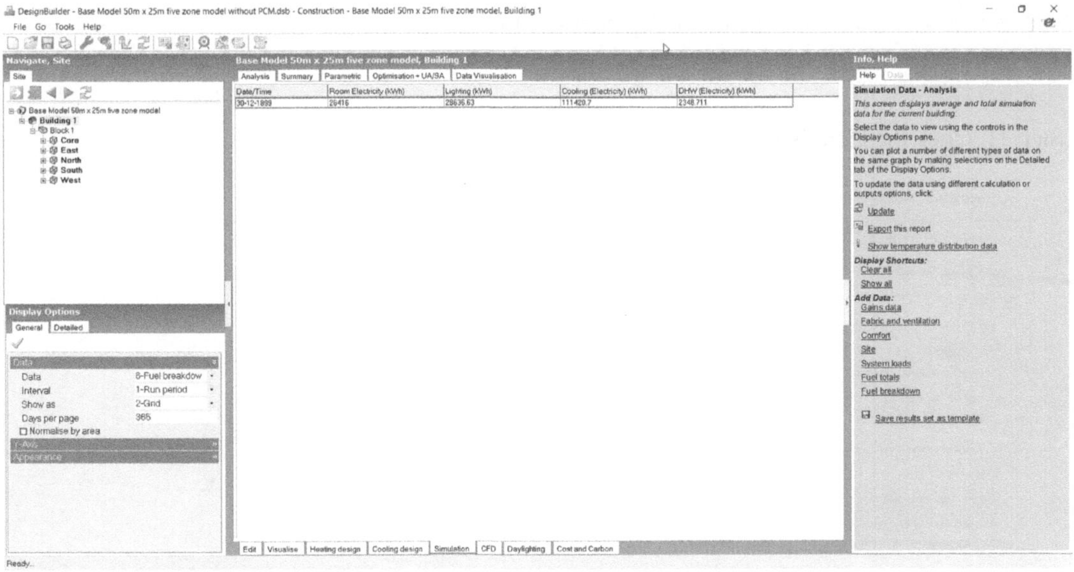

Step 10: Click the **View EnergyPlus results** icon. The **DesignBuilder Results Viewer** appears on the screen. If the **View EnergyPlus results** icon is not displayed, then you need to install the **Results Viewer** application.

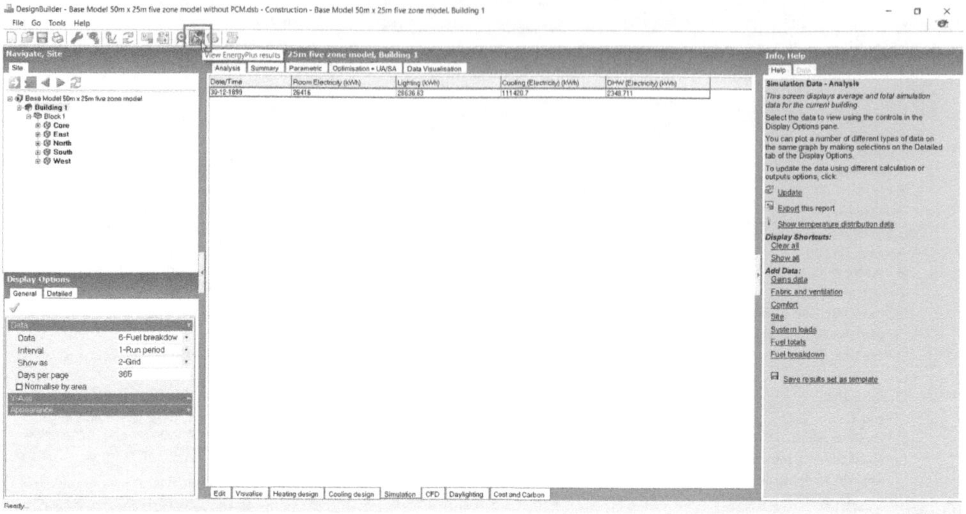

The *Results Viewer* is a separate application that can be used to view EnergyPlus results stored in one or more .eso files. It can be downloaded from the main **Downloads > Software** area of the DesignBuilder website (https://designbuilder.co.uk/download/release-software). When installed, the application allows you to view any results contained within EnergyPlus .eso and .htm files.

Source: https://designbuilder.co.uk/helpv7.3/Content/ResultsViewer.htm

Step 11: Click the **Hourly** tab.

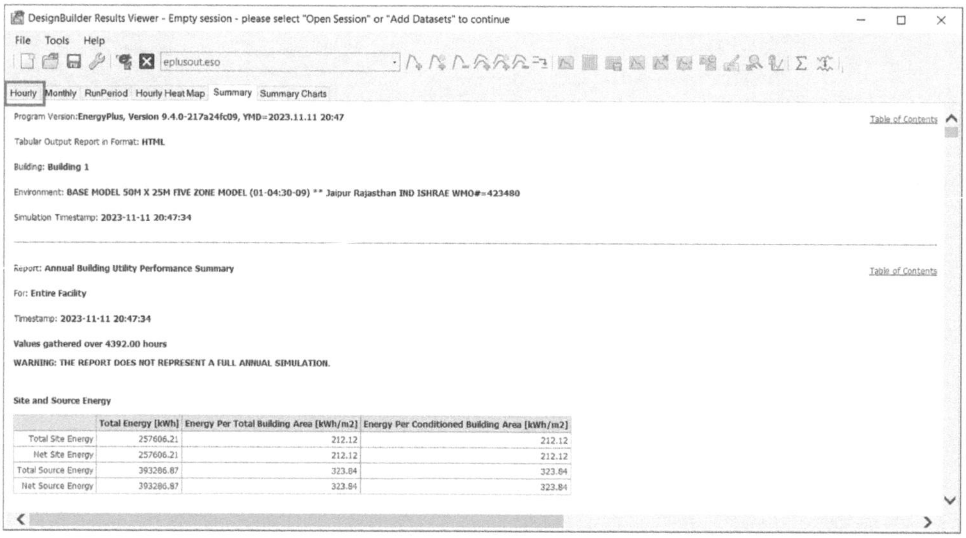

	Total Energy [kWh]	Energy Per Total Building Area [kWh/m2]	Energy Per Conditioned Building Area [kWh/m2]
Total Site Energy	257606.21	212.12	212.12
Net Site Energy	257606.21	212.12	212.12
Total Source Energy	393286.87	323.84	323.84
Net Source Energy	393286.87	323.84	323.84

Step 12: Scroll down and click **Surface Inside Face Temperature** under **Report Type** and **Block1:SOUTH_WALL_5_0_0_0** under **Area**.

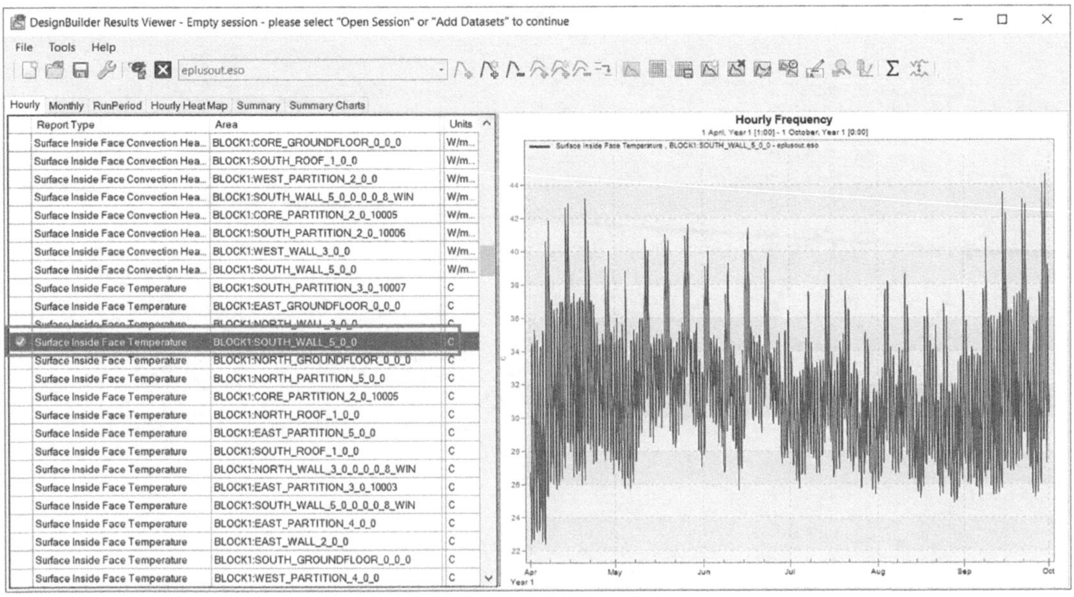

Step 13: Further scroll down and click **Surface Outside Face Temperature** under **Report Type** and **Block1:SOUTH_WALL_5_0_0_0** under **Area**.

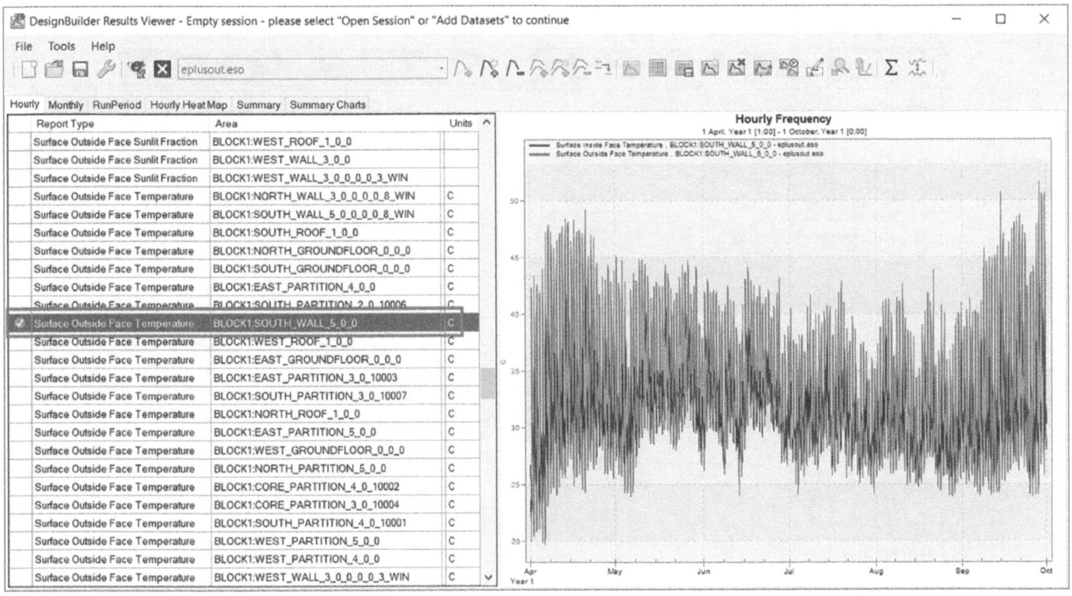

Step 14: Zoom the graph to see the data (from 27 May to 31 May) for the following four days.

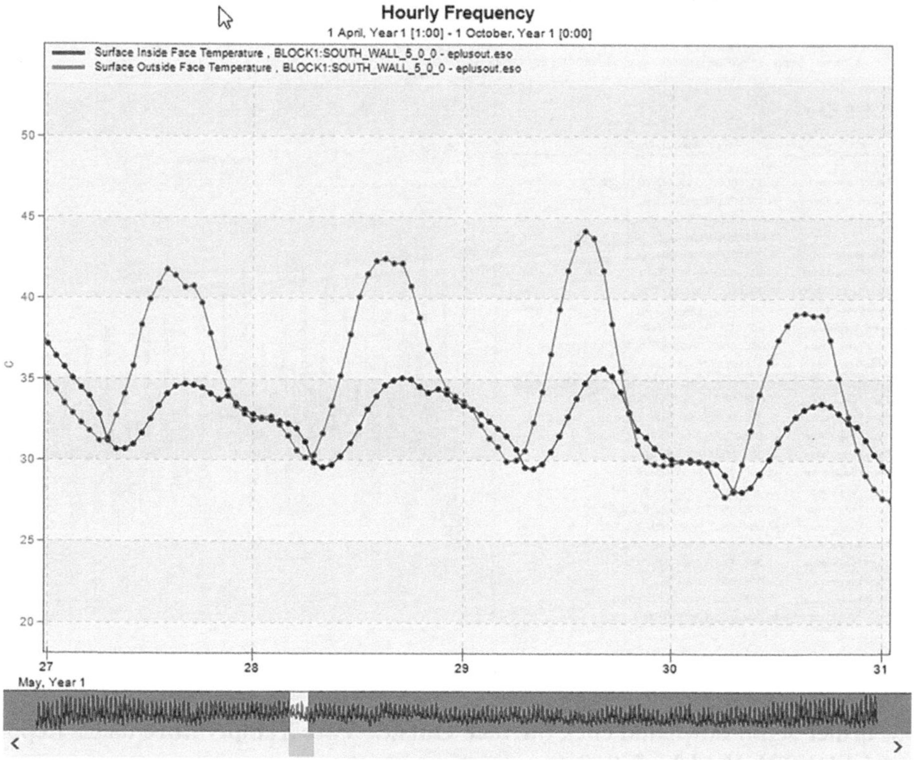

Step 15: Save as the model. Click on the **Edit** screen tab.

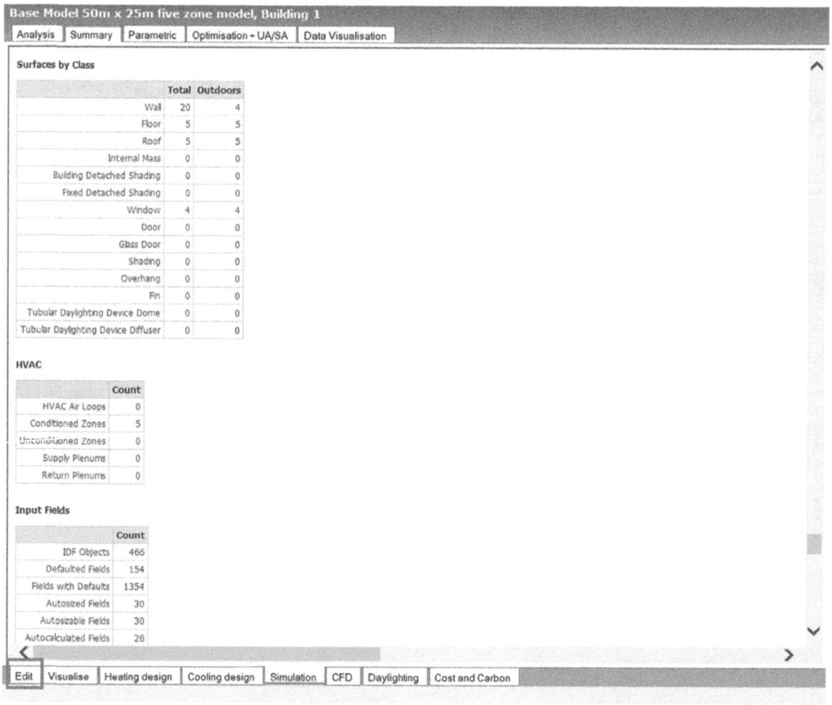

Step 16: Click the **Construction** tab. The **Edit construction** screen appears.

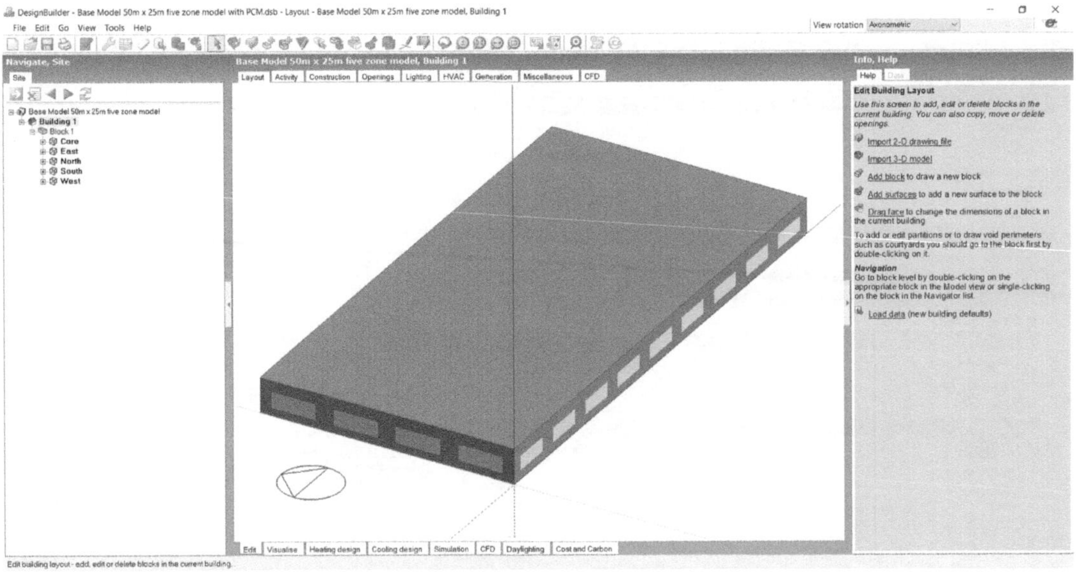

Step 17: Edit the **External wall construction—Brick Wall** (created in the preceding steps of this tutorial). Enter **PCM External Wall** in the **Name** text box. Select **2** from the **Number of layers** drop-down list.

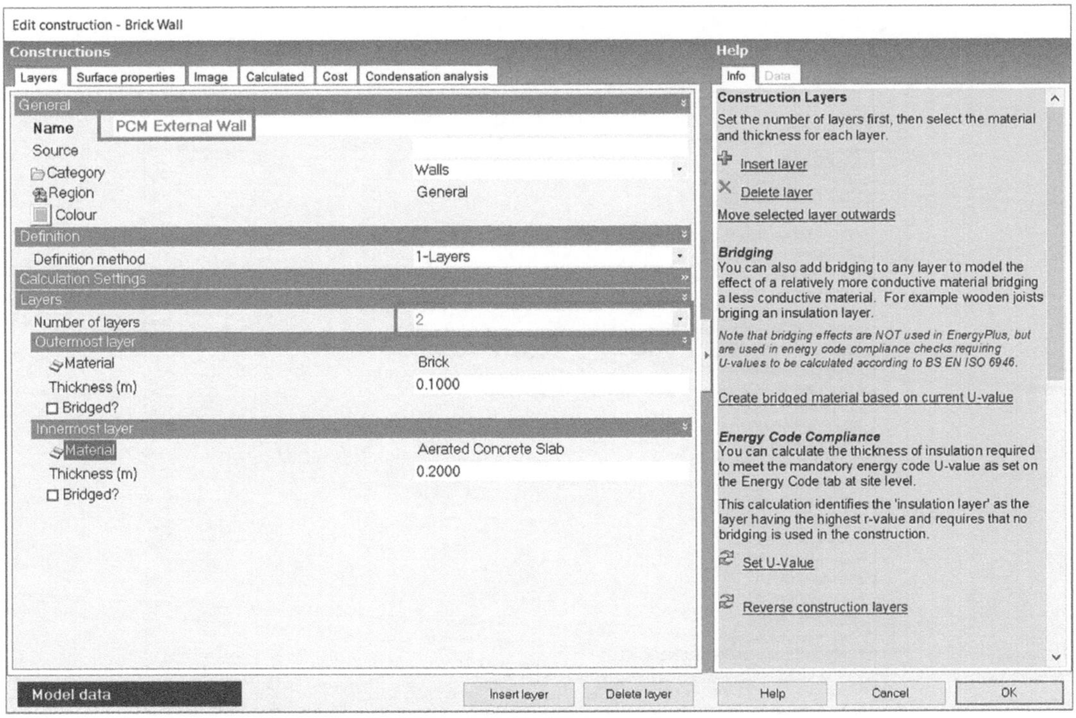

Step 18: Click **Material** under **Innermost layer**. Click on the three dots. Expand **Phase Change**.

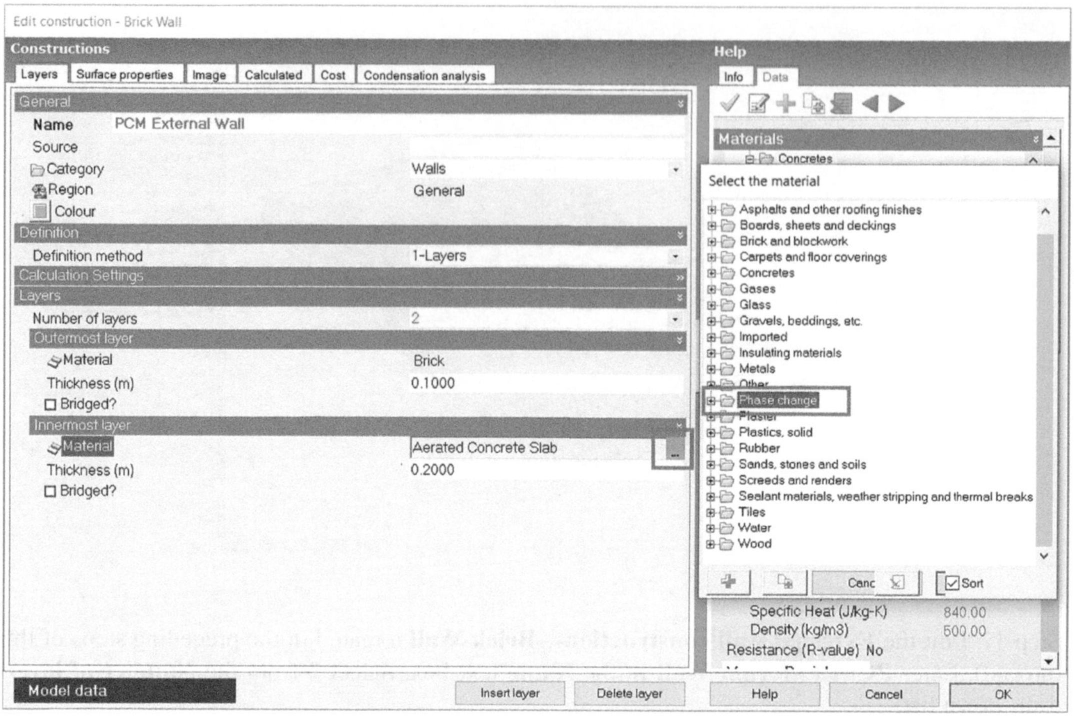

Step 19: Click **InfiniteRPCM29C**.

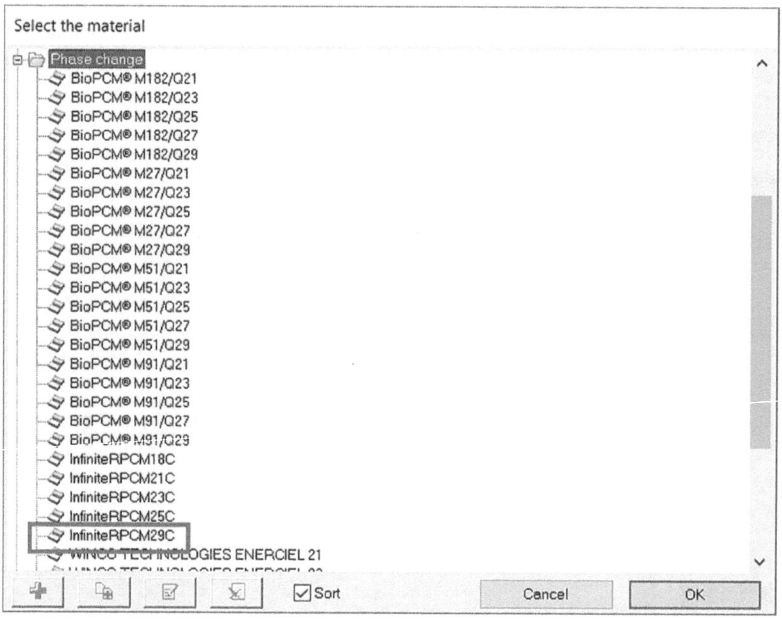

Step 20: Click the **Phase Change** tab to view properties of the PCM.

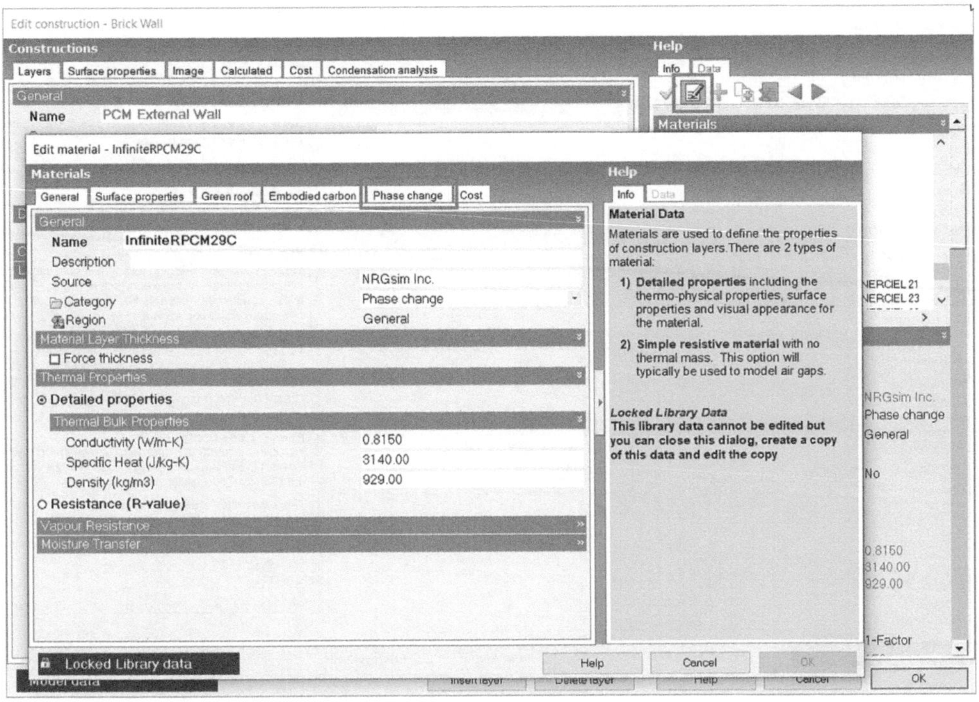

Step 21: Note the peak freezing and melting temperatures.

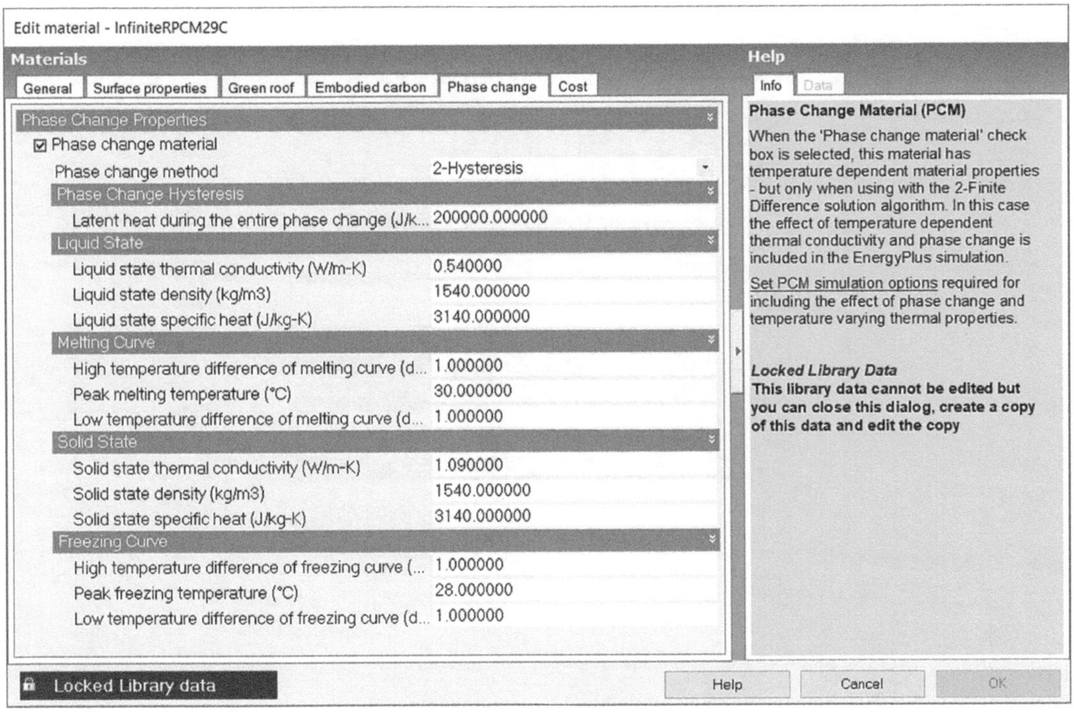

Step 22: Move the **PCM** layer to **Outermost** layer.

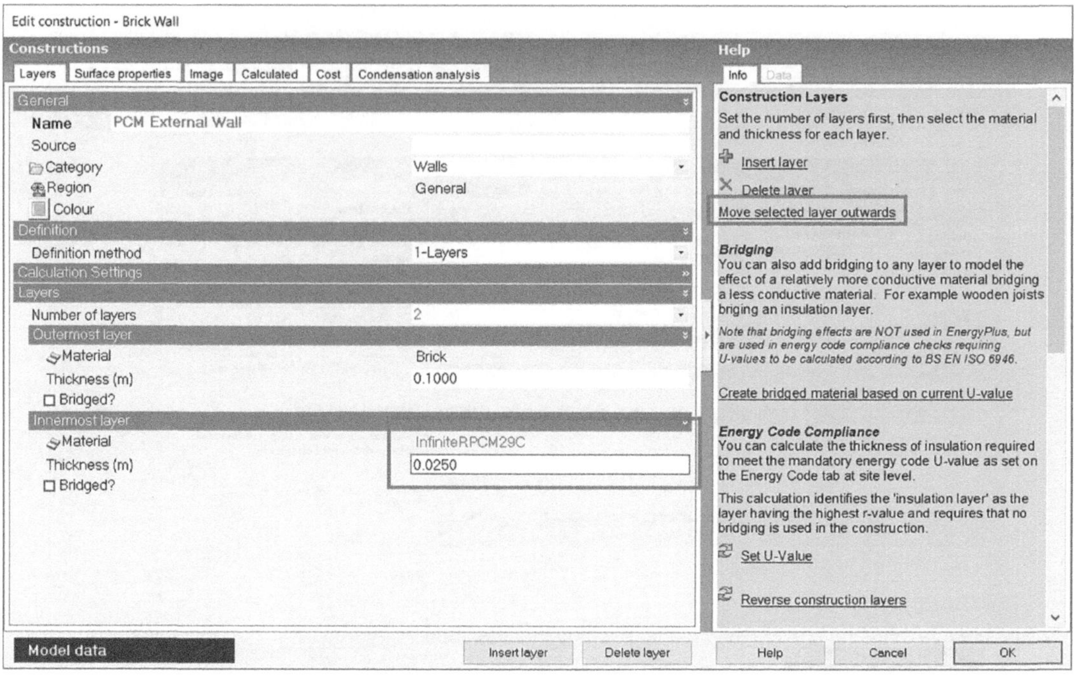

Step 23: Click the **Select item** button. Click the **Simulation** screen tab.

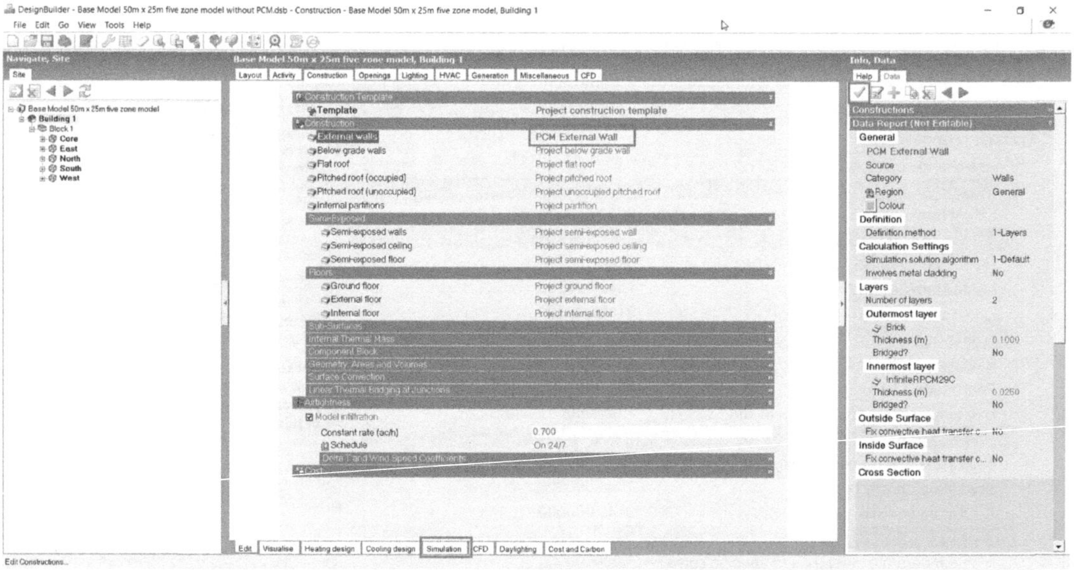

Step 24: Click **Update Calculation**. Click the **Options** tab. Select **12** from **Time steps per hour** drop-down list. Select **2-Finite Difference** from the **Solution algorithm** drop-down list.

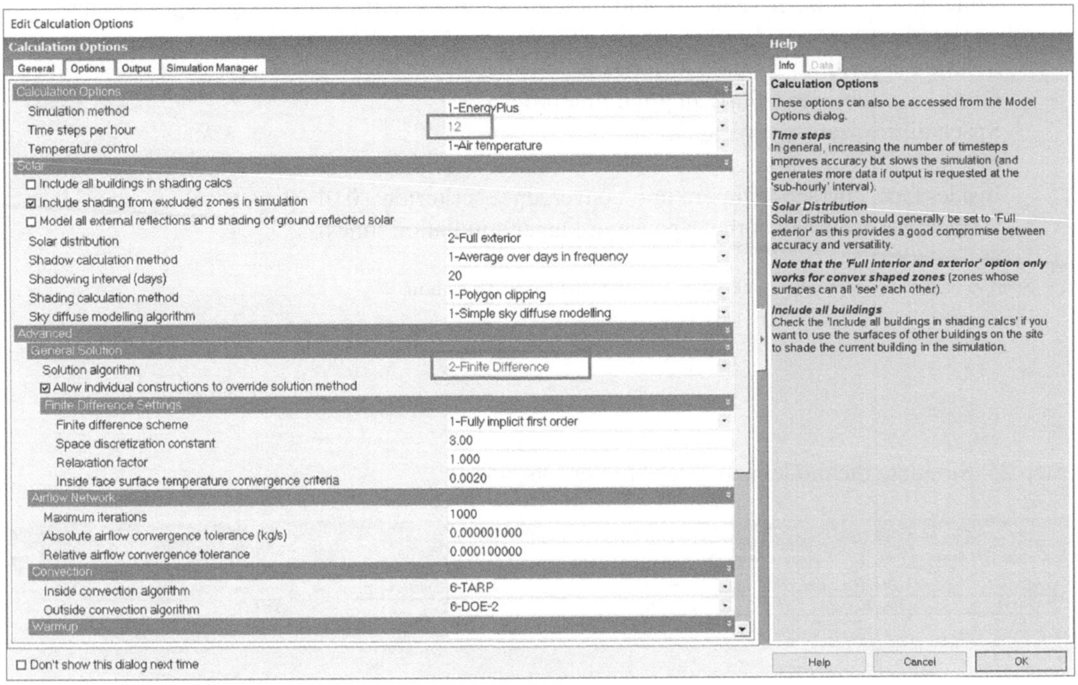

SOLUTION ALGORITHM

The *finite difference solution algorithm* must be used to include the effect of material phase-change properties in simulations. If you use the conduction transfer function (CTF) algorithm, the material will behave as if its PCM option were not selected. Note also that the CTF solution algorithm is the only option available for heating and cooling design, so the phase-change effect of the material will not be seen in these calculations. An error message to this effect will be generated for PCM simulations when the CTF solution method is selected.

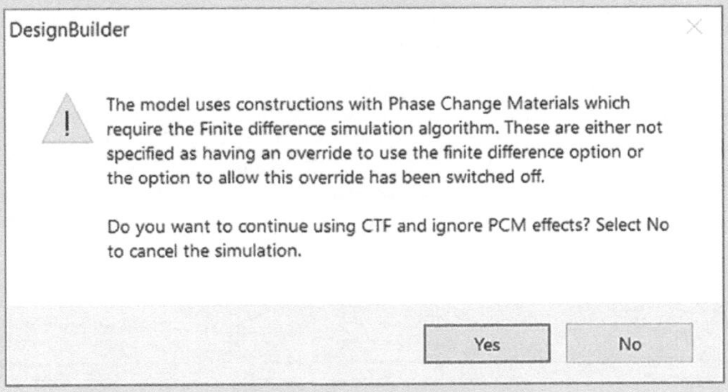

RECOMMENDED SIMULATION PARAMETERS FOR PCM

Time step: 12 (higher values should increase accuracy at the expense of longer simulation times)
Solution algorithm: 2-Finite difference
Difference scheme: 1-Fully implicit first order
Space discretization constant: 3
Relaxation factor: 1
Inside face surface temperature convergence criterion: 0.01 (lower values should increase accuracy at the expense of longer simulation times)

Source: https://designbuilder.co.uk/helpv7.3/Content/PhaseChange.htm

Step 25: Simulate the model and record the results.

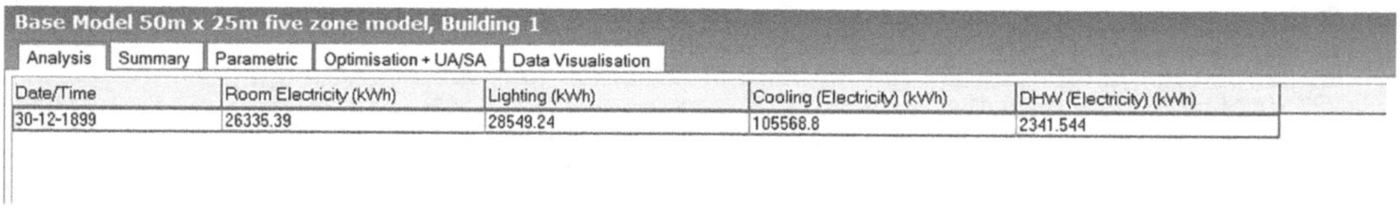
Base Model 50m x 25m five zone model, Building 1

Analysis | Summary | Parametric | Optimisation + UA/SA | Data Visualisation

Date/Time	Room Electricity (kWh)	Lighting (kWh)	Cooling (Electricity) (kWh)	DHW (Electricity) (kWh)
30-12-1899	26335.39	28549.24	105568.8	2341.544

Step 26: Note the inside and outside surface temperatures for the south wall.

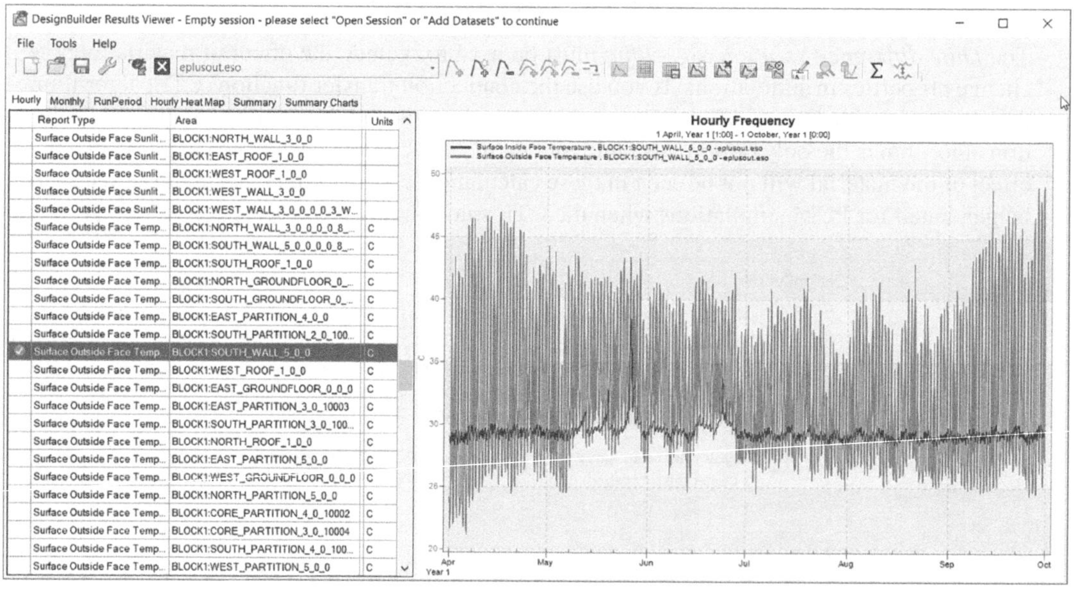

Step 27: Zoom the graph to see the data (from 27 May to 31 May) for the following four days.

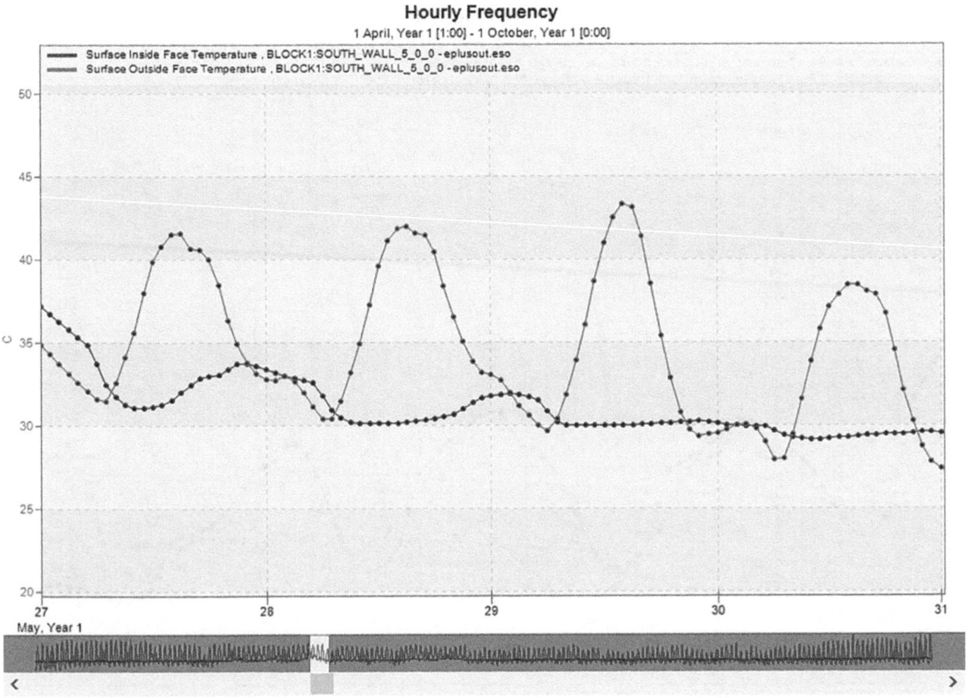

Step 28: Compare the results.

Brick wall construction

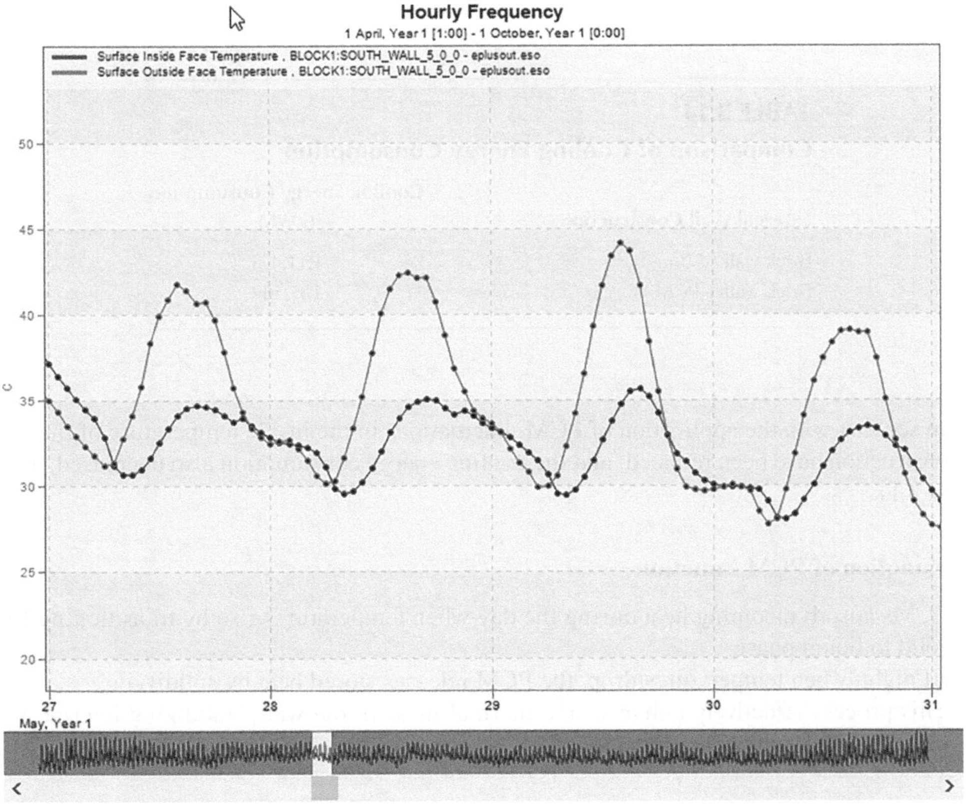

Brick with PCM wall construction

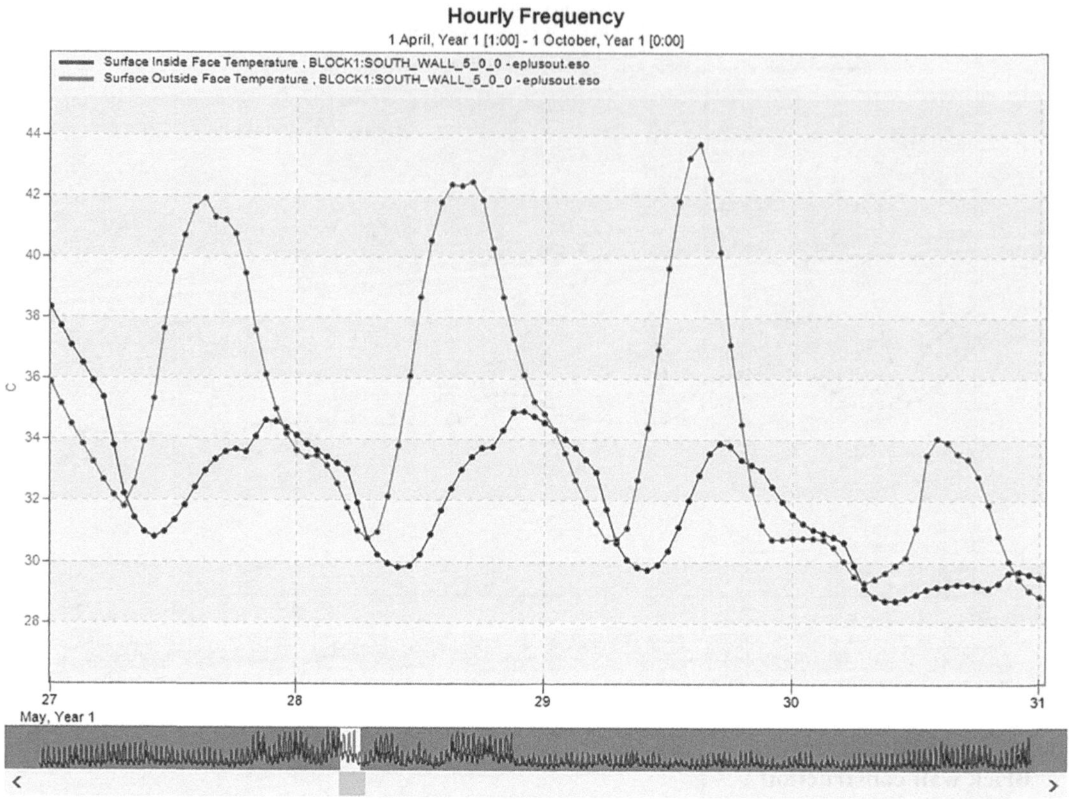

TABLE 3.14
Comparison of Cooling Energy Consumption

External Wall Construction	Cooling Energy Consumption (kWh)
Brick wall	1,11,421
Brick wall + PCM	105,569

You can see that with the application of PCM, fluctuations in the inside temperature of the external wall construction have been reduced, and the cooling energy consumption also decreased, as shown in Table 3.14.

Explanation of PCM Function

- PCMs absorb incoming heat during the day when temperatures rise by transitioning from solid to liquid phase
- At night, when temperatures drop, the PCM releases stored heat by solidifying
- This process effectively enhances the thermal mass of the wall, stabilizing internal temperatures and reducing cooling loads

Observations

- The application of PCM in external walls reduced cooling energy consumption; the PCM moderates indoor temperatures by absorbing excess heat during the day and releasing it during cooler periods
- The fluctuations in the inside temperature of the external wall were reduced, enhancing thermal comfort

Conclusion

Implementing PCM in external wall construction is an effective strategy to reduce cooling energy consumption and improve thermal comfort by stabilizing internal temperatures. The results demonstrate the potential of PCM to enhance energy efficiency in hot climates like Jaipur.

4 Openings and Shading

Openings are essential elements in building envelopes, serving multiple purposes such as providing access to outdoor views, enabling natural ventilation, and allowing daylight into interior spaces. While windows enhance occupant comfort and building aesthetics, they also become critical pathways for heat transfer. To minimize thermal loads while still fulfilling functional and design requirements, optimizing the window-to-wall ratio (WWR) is important.

The type of glazing used further influences building energy use. Key performance indicators for glass include the U-value (thermal transmittance), solar heat gain coefficient (SHGC), and visible light transmittance (VLT). In cooling-dominated climates, the ratio of VLT to SHGC is often used to evaluate glass performance—a higher ratio suggests that the glazing allows more visible light relative to solar heat, which is desirable when trying to reduce cooling loads.

Shading devices provide an effective means of reducing direct solar radiation before it reaches the glazing. By blocking high-angle or low-angle sun, they help moderate interior temperatures and reduce reliance on mechanical cooling systems. Fixed shading devices such as overhangs (horizontal) and fins (vertical) are integrated into the building façade and work passively throughout the day. Operable shades offer additional control, allowing occupants to adjust shading based on time of day or seasonal conditions.

This chapter includes four tutorials designed to help you evaluate the impact of glazing properties, fixed shading elements, and operable shading systems on building energy performance. These exercises will support your ability to analyze façade design strategies that mitigate solar heat gain, especially in hot-climate scenarios.

TUTORIAL 4.1: EVALUATING THE IMPACT OF WINDOW-TO-WALL RATIO AND GLAZING TYPE

Goal

To evaluate the impact of the window-to-wall ratio and glazing type on energy consumption
4.1.2 What Are You Going to Learn?

- How to set the WWR
- How to select glazing type

Problem Statement

In this tutorial, you are going to use a 50 m × 25 m five-zone model with a 5 m perimeter depth.

Using the glass types and configurations specified in Table 4.1, you will simulate and compare energy consumption for each case for **SINGAPORE/PAYA LEBA** location.

> The *light-to–solar gain (L/S) ratio* is the ratio between the VLT and SHGC. It provides a gauge of the relative efficiency of different glass or glazing types in transmitting daylight while blocking heat gain. The higher the number, the more the light is transmitted without adding excessive amounts of heat.

DOI: 10.1201/9781003563013-4

TABLE 4.1
Glass Types and Their Properties

Tutorial Part	Glass	Properties	Light-to-Solar Gain ratio (*L/S*)	WWR	Daylight Controls	Shade
I	Double-glazed	Green 6mm/6mm Air, SHGC-0.49, VLT-0.66	1.35	0–90% in steps of 10%	With and without	None
II	Single-glazed	Sgl Clr 6mm. SHGC-0.81 and VLT-0.88	1.09	0–90% in steps of 10%	With and without	None
III	ASHRAE 90.1 equivalent glass	U-6.81 W/m² K, SHGC-0.25 and VLT 0.53	2.12	0–90% in steps of 10%	With and without	None and with a 0.5 m overhang on all windows

Solution

Step 1: Create a **50 m × 25 m** five-zone model with a 5 m perimeter depth.

Part I: With Dbl Green 6mm/6mm Air Glass

Step 2: Click on the **Openings** tab and select **Glazing type** as **Dbl Green 6mm/6mm Air**. Set WWR to 0.00%.

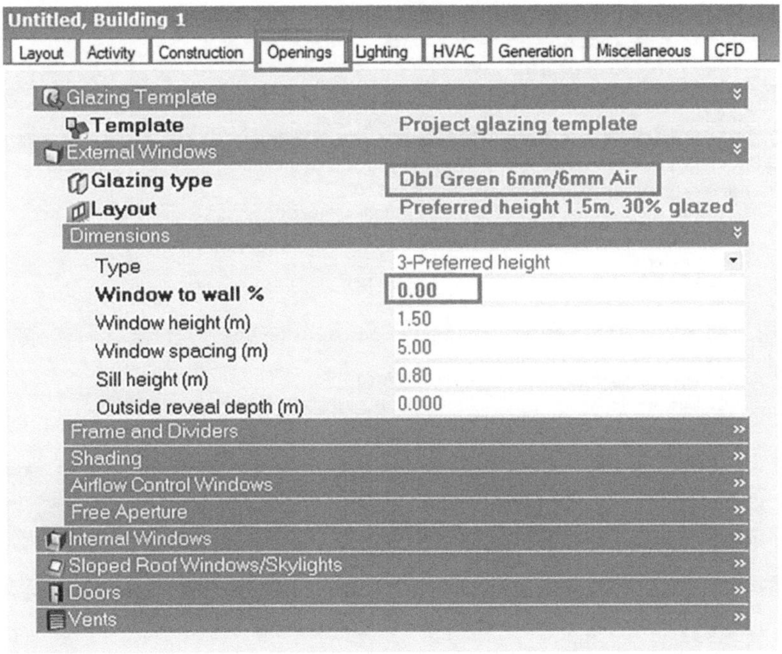

Step 3: Simulate the model and record the results.

Step 4: Repeat the preceding steps to set the WWR (from 10–90% in steps of 10%), as given in the problem statement. Record the results for all WWRs without lighting control (Table 4.2).

TABLE 4.2

Annual Energy Consumption for a Double-Glazed Window without Daylight Controls

Double Glazing without Daylight Controls (Dbl Green 6mm/6mm Air)					
WWR (%)	Room Electricity (kWh)	Lighting (kWh)	Cooling (Electricity) (kWh)	Domestic Hot Water (DHW, Electricity) (kWh)	Annual Consumption (Electricity) (kWh)
0	51,406	55,720	1,77,764	4,570	2,89,460
10	51,406	55,720	1,84,740	4,570	2,96,437
20	51,406	55,720	1,91,721	4,570	3,03,417
30	51,406	55,720	1,98,232	4,570	3,09,928
40	51,406	55,720	2,04,289	4,570	3,15,986
50	51,406	55,720	2,10,025	4,570	3,21,721
60	51,406	55,720	2,15,390	4,570	3,27,087
70	51,406	55,720	2,20,377	4,570	3,32,073
80	51,406	55,720	2,25,220	4,570	3,36,916
90	51,406	55,720	2,29,606	4,570	3,41,303

Step 5: Select the **Lighting** tab. In the **Lighting Control** section, select the **ON** checkbox.

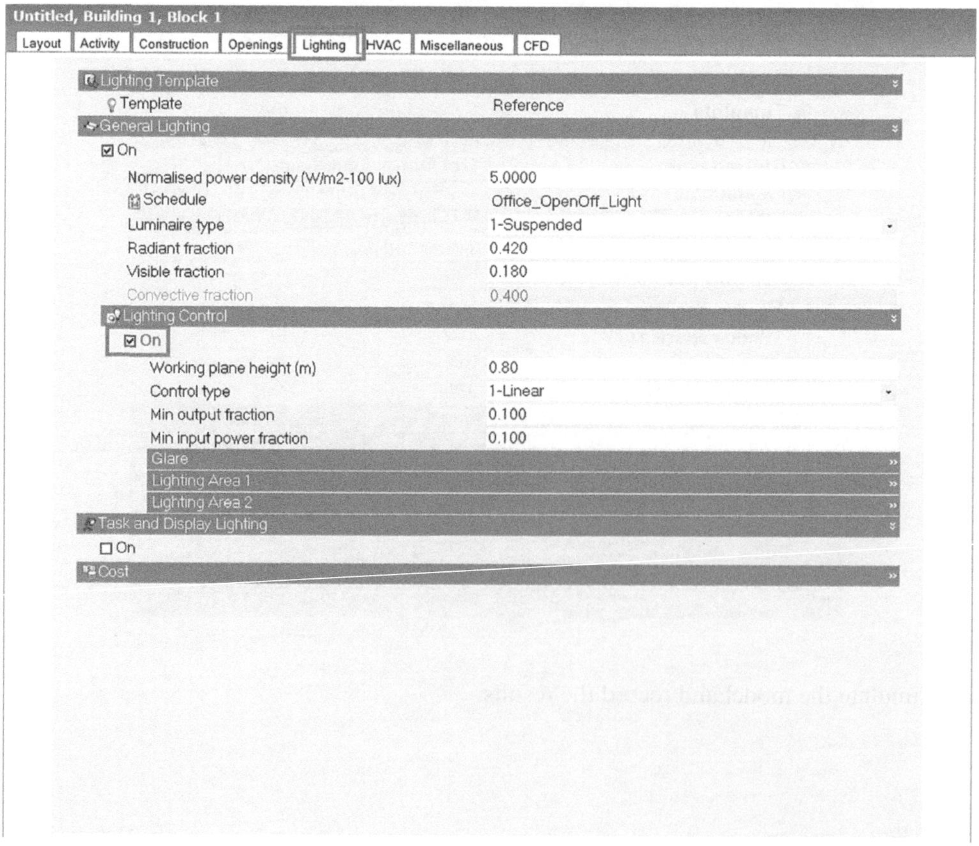

When you select the ON checkbox, you get the daylight sensor placed in all zones. You can see that the lighting energy consumption for the daylit perimeter zones will decrease.

Step 6: Simulate the model and record the results. Record the results for all WWRs with lighting control (Table 4.3).

TABLE 4.3
Annual Energy Consumption for a Double-Glazed Window with Daylight Controls

Double Glazing with Daylight Controls (Dbl Green 6mm/6mm Air)					
WWR (%)	Room Electricity (kWh)	Lighting (kWh)	Cooling (Electricity) (kWh)	DHW (Electricity) (kWh)	Annual Consumption (Electricity) (kWh)
0	51,406	55,720	1,77,764	4,570	2,89,460
10	51,406	35,465	1,72,265	4,570	2,63,706
20	51,406	33,338	1,78,511	4,570	2,67,825
30	51,406	32,694	1,85,066	4,570	2,73,737
40	51,406	32,337	1,91,304	4,570	2,79,617
50	51,406	31,998	1,97,158	4,570	2,85,133
60	51,406	31,763	2,02,692	4,570	2,90,432
70	51,406	31,608	2,07,858	4,570	2,95,442
80	51,406	31,748	2,12,963	4,570	3,00,688
90	51,406	31,598	2,17,514	4,570	3,05,088

Step 7: Compare the results with and without daylight controls (Table 4.4).

TABLE 4.4
Comparison of the Total Annual Energy Consumption for a Double-Glazed Window with and without Daylight Controls

WWR (%)	Annual Energy Consumption without Daylight Controls (kWh)	Annual Energy Consumption with Daylight Controls (kWh)
0	2,89,460	2,89,460
10	2,96,437	2,63,706
20	3,03,417	2,67,825
30	3,09,928	2,73,737
40	3,15,986	2,79,617
50	3,21,721	2,85,133
60	3,27,087	2,90,432
70	3,32,073	2,95,442
80	3,36,916	3,00,688
90	3,41,303	3,05,088

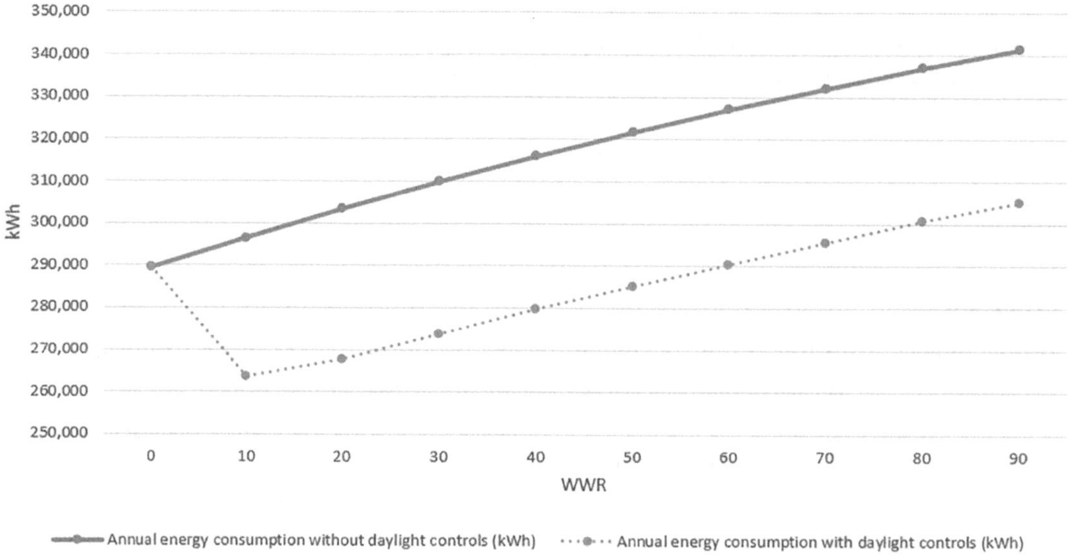

The results show that with double-glazed windows, the building consumes minimum energy at 20% WWR when daylight controls are installed in the building in all its daylit perimeter spaces.

Part II: With Sgl Clr 6mm Glass

Step 8: Repeat the preceding steps to get simulation results with **Sgl Clr 6mm** glass (Table 4.5, Table 4.6, and Table 4.7).

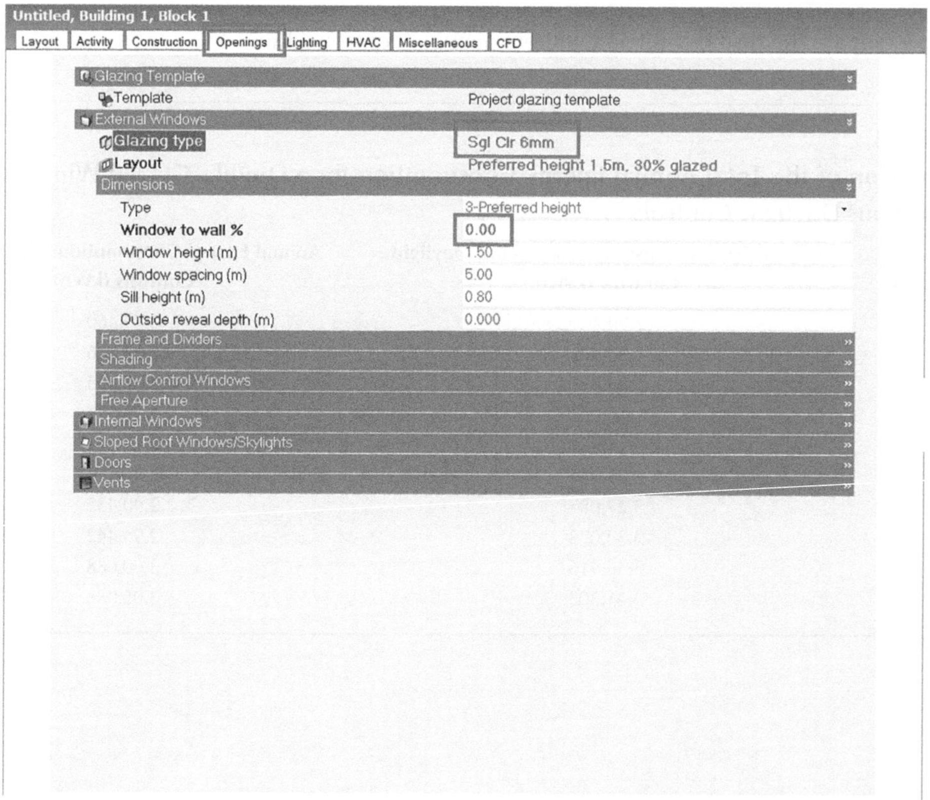

TABLE 4.5
Annual Energy Consumption for a Single-Glazed Window without Daylight Controls

Single Glazing without Daylight Controls (Sgl Clr 6mm)					
WWR (%)	Room Electricity (kWh)	Lighting (kWh)	Cooling (Electricity) (kWh)	DHW (Electricity) (kWh)	Annual Consumption (Electricity) (kWh)
0	51,406	55,720	1,77,764	4,570	2,89,460
10	51,406	55,720	1,88,962	4,570	3,00,658
20	51,406	55,720	1,99,695	4,570	3,11,391
30	51,406	55,720	2,09,219	4,570	3,20,916
40	51,406	55,720	2,17,789	4,570	3,29,485
50	51,406	55,720	2,25,760	4,570	3,37,456
60	51,406	55,720	2,33,091	4,570	3,44,787
70	51,406	55,720	2,39,740	4,570	3,51,437
80	51,406	55,720	2,45,847	4,570	3,57,544
90	51,406	55,720	2,51,458	4,570	3,63,154

TABLE 4.6
Annual Energy Consumption for a Single-Glazed Window with Daylight Controls

Single Glazing with Daylight Controls (Sgl Clr 6mm)					
WWR (%)	Room Electricity (kWh)	Lighting (kWh)	Cooling (Electricity) (kWh)	DHW (Electricity) (kWh)	Annual Consumption (Electricity) (kWh)
0	51,406	55,720	1,77,764	4,570	2,89,460
10	51,406	34,241	1,76,041	4,570	2,66,259
20	51,406	32,804	1,86,641	4,570	2,75,421
30	51,406	32,283	1,96,454	4,570	2,84,713
40	51,406	31,988	2,05,295	4,570	2,93,259
50	51,406	31,708	2,13,485	4,570	3,01,169
60	51,406	31,519	2,21,020	4,570	3,08,515
70	51,406	31,384	2,27,908	4,570	3,15,268
80	51,406	31,507	2,34,296	4,570	3,21,779
90	51,406	31,376	2,40,095	4,570	3,27,448

TABLE 4.7
Comparison of the Total Annual Energy Consumption for a Single-Glazed Window with and without Daylight Controls

WWR (%)	Annual Energy Consumption without Daylight Controls (kWh)	Annual Energy Consumption with Daylight Controls (kWh)
0	2,89,460	2,89,460
10	3,00,658	2,66,259
20	3,11,391	2,75,421
30	3,20,916	2,84,713
40	3,29,485	2,93,259
50	3,37,456	3,01,169
60	3,44,787	3,08,515
70	3,51,437	3,15,268
80	3,57,544	3,21,779
90	3,63,154	3,27,448

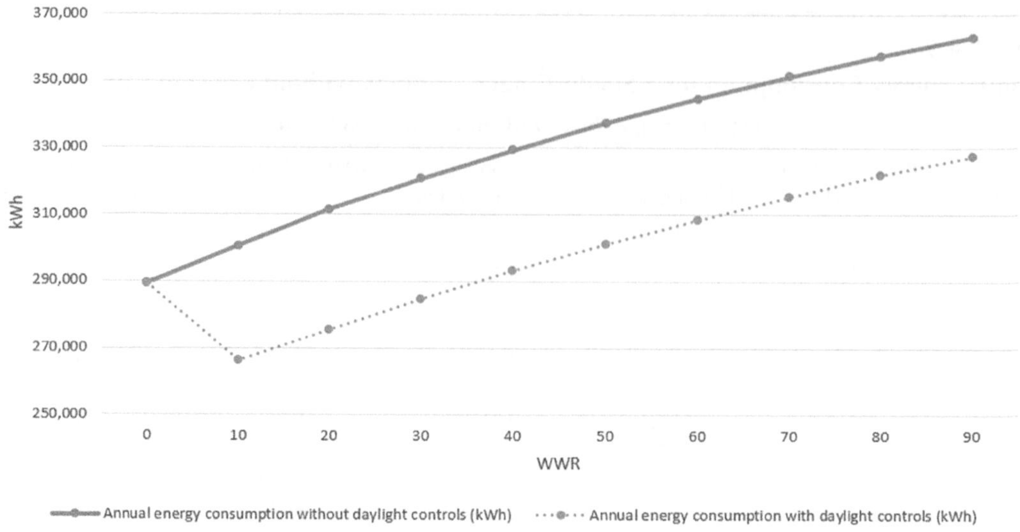

When there are no daylight controls, annual energy consumption increases linearly with increases in the WWR. With the use of daylight controls, first, energy decreases and then increases; minimum energy consumption is at 10% WWR.

Part III(a): With ASHRAE 90.1-2007 Equivalent Glass (U-1.20 [6.81], SHGC-0.25, and VLT 53%)

Step 9: Now refer to the preceding steps to get simulation results with **ASHRAE 90.1** equivalent glass, **Vertical glazing, 0%–40% of wall, U-1.20 (6.81) and SHGC-0.25** (Table 4.8 and Table 4.9).

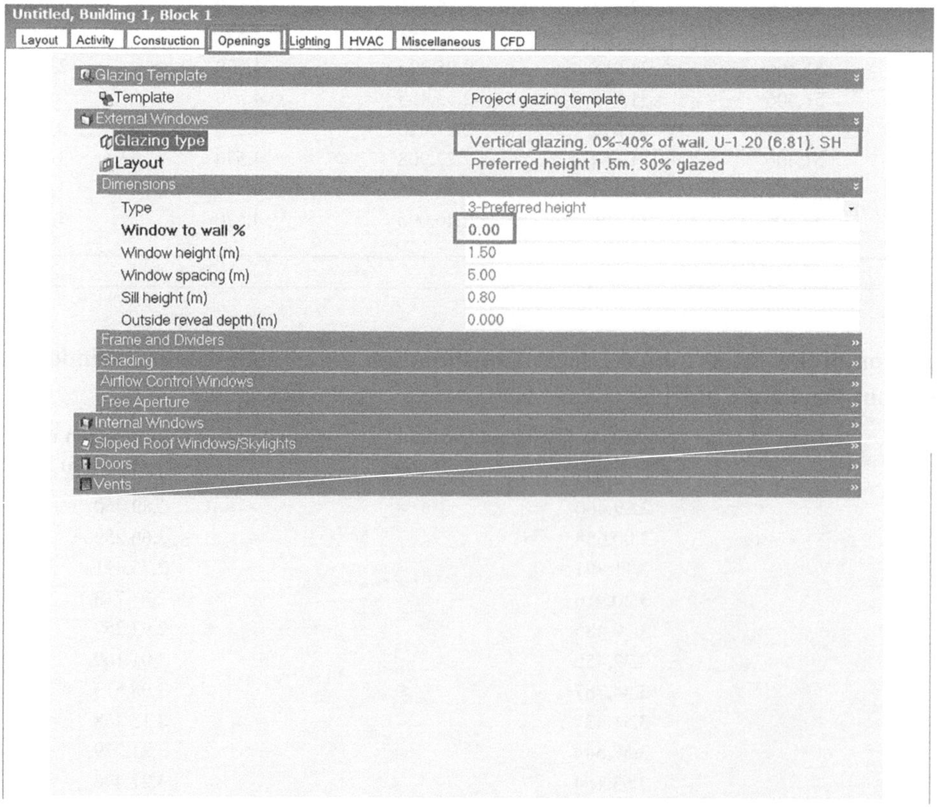

TABLE 4.8
Annual Energy Consumption with ASHRAE 90.1 Equivalent Glass without Daylight Controls

ASHRAE 90.1 Glass without Daylight Controls (Vertical Glazing, 0–40% of wall, U-1.20 [6.81] and SHGC-0.25)

WWR (%)	Room Electricity (kWh)	Lighting (kWh)	Cooling (Electricity) (kWh)	DHW (Electricity) (kWh)	Annual Consumption (Electricity) (kWh)
0	51,406	55,720	1,77,764	4,570	2,89,460
10	51,406	55,720	1,81,425	4,570	2,93,121
20	51,406	55,720	1,85,137	4,570	2,96,833
30	51,406	55,720	1,88,583	4,570	3,00,280
40	51,406	55,720	1,91,800	4,570	3,03,496
50	51,406	55,720	1,94,758	4,570	3,06,454
60	51,406	55,720	1,97,470	4,570	3,09,166
70	51,406	55,720	2,00,028	4,570	3,11,724
80	51,406	55,720	2,02,563	4,570	3,14,259
90	51,406	55,720	2,04,773	4,570	3,16,470

TABLE 4.9
Annual Energy Consumption with ASHRAE 90.1 Equivalent Glass with Daylight Controls

Good Glazing with Daylight Controls (Vertical Glazing, 0–40% of wall, U-1.20 [6.81] and SHGC-0.25)

WWR (%)	Room Electricity (kWh)	Lighting (kWh)	Cooling (Electricity) (kWh)	DHW (Electricity) (kWh)	Annual Consumption (Electricity) (kWh)
0	51,406	55,720	1,77,764	4,570	2,89,460
10	51,406	36,154	1,69,290	4,570	2,61,420
20	51,406	33,617	1,71,933	4,570	2,61,526
30	51,406	32,893	1,75,398	4,570	2,64,266
40	51,406	32,511	1,78,819	4,570	2,67,306
50	51,406	32,146	1,81,935	4,570	2,70,057
60	51,406	31,889	1,84,836	4,570	2,72,701
70	51,406	31,715	1,87,544	4,570	2,75,235
80	51,406	31,873	1,90,415	4,570	2,78,264
90	51,406	31,704	1,92,798	4,570	2,80,478

Step 10: Compare the results with and without daylight controls (Table 4.10).

TABLE 4.10
Comparison of Simulation Results of Glass with and without Daylight Controls

WWR (%)	Annual Energy Consumption without Daylight Controls (kWh)	Annual Energy Consumption with Daylight Controls (kWh)
0	2,89,460	2,89,460
10	2,93,121	2,61,420
20	2,96,833	2,61,526
30	3,00,280	2,64,266
40	3,03,496	2,67,306
50	3,06,454	2,70,057

(Continued)

TABLE 4.10 (*Continued*)
Comparison of Simulation Results of Glass with and without Daylight Controls

WWR (%)	Annual Energy Consumption without Daylight Controls (kWh)	Annual Energy Consumption with Daylight Controls (kWh)
60	3,09,166	2,72,701
70	3,11,724	2,75,235
80	3,14,259	2,78,264
90	3,16,470	2,80,478

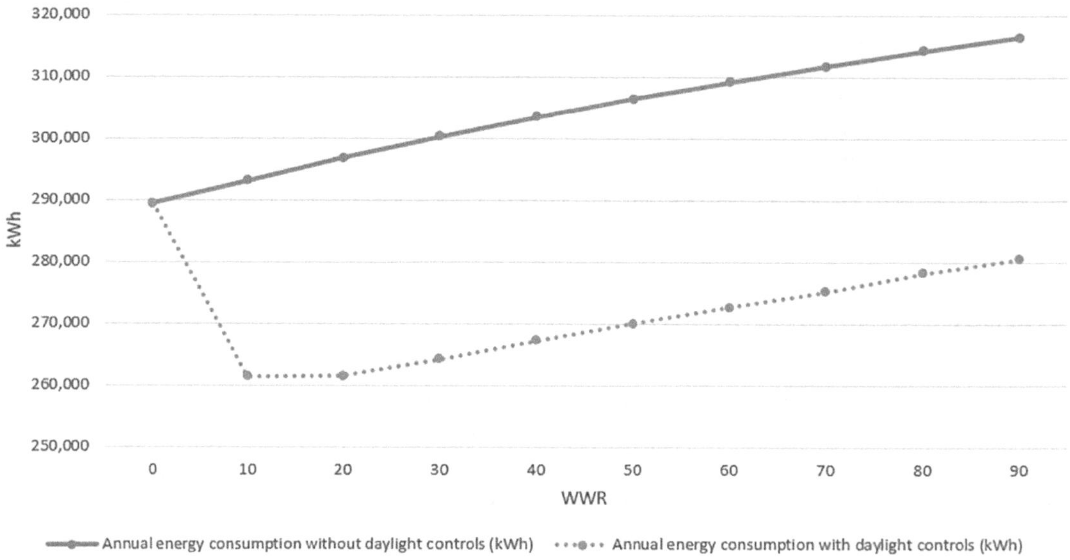

Step 11: Now compare the results for all simulations of double-glazed, single-glazed, and ASHRAE 90.1 equivalent glass with lighting controls (Table 4.11).

TABLE 4.11
Comparison of the Total Annual Energy Consumption for All Glass Types with Daylight Controls

WWR (%)	Double Glazing (*L/S* = 1.35) (kWh)	Single Glazing (*L/S* = 1.09) (kWh)	ASHRAE 90.1 Equivalent Glazing (*L/S* = 2.12) (kWh)
0	2,89,460	2,89,460	2,89,460
10	2,63,706	2,66,259	2,61,420
20	2,67,825	2,75,421	2,61,526
30	2,73,737	2,84,713	2,64,266
40	2,79,617	2,93,259	2,67,306
50	2,85,133	3,01,169	2,70,057
60	2,90,432	3,08,515	2,72,701
70	2,95,442	3,15,268	2,75,235
80	3,00,688	3,21,779	2,78,264
90	3,05,088	3,27,448	2,80,478

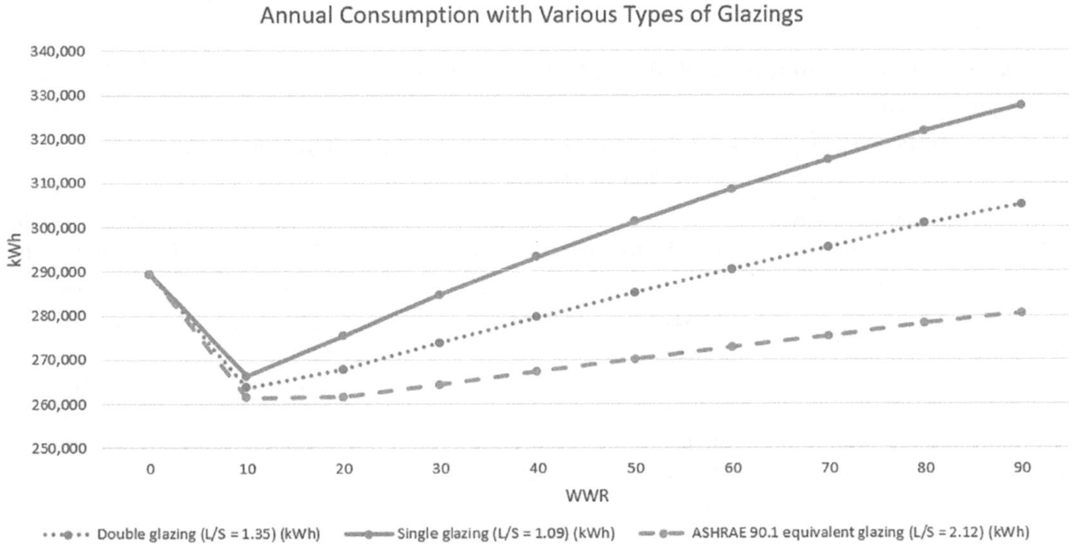

Part III(b): With ASHRAE 90.1-2007 Equivalent Glass (U-1.20 [6.81], SHGC-0.25, and VLT 53%) with the Shading of a 0.5 m Overhang

Step 12: Select the **Openings** tab and check the **Local shading** checkbox under the **Shading** section. Select **0.5m Overhang** from the **Type** drop-down list (Table 4.12).

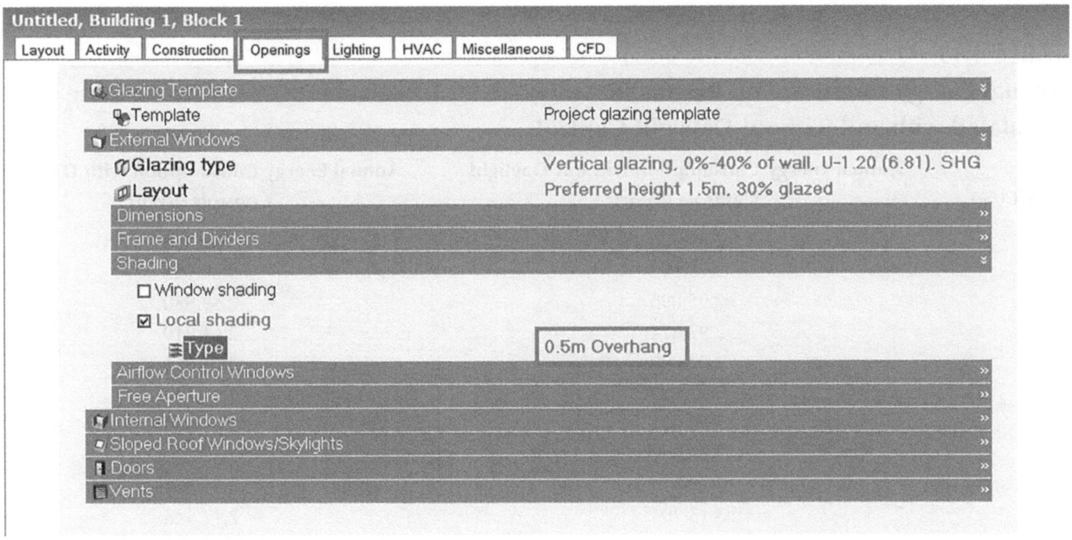

TABLE 4.12
Annual Energy Consumption with ASHRAE 90.1 Equivalent Glass with Fixed Shade and Daylight Controls

ASHRAE 90.1 Equivalent Glass with Daylight Controls (Vertical Glazing, 0–40% of Wall, U-1.20 (6.81), and SHGC-0.25)

WWR (%)	Room Electricity (kWh)	Lighting (kWh)	Cooling (Electricity) (kWh)	DHW (Electricity) (kWh)	Annual Consumption (Electricity) (kWh)
0	51,406	55,720	1,77,764	4,570	2,89,460
10	51,406	36,341	1,68,611	4,570	2,60,928
20	51,406	33,683	1,70,220	4,570	2,59,880
30	51,406	32,941	1,72,802	4,570	2,61,719
40	51,406	32,551	1,75,353	4,570	2,63,881
50	51,406	32,169	1,78,280	4,570	2,66,425
60	51,406	32,184	1,81,378	4,570	2,69,538
70	51,406	31,953	1,84,150	4,570	2,72,079
80	51,406	32,177	1,87,108	4,570	2,75,262
90	51,406	31,913	1,89,539	4,570	2,77,428

Compare the results with and without shades for ASHRAE 90.1 equivalent glass and daylight controls (Table 4.13).

TABLE 4.13
Annual Energy Consumption Results for ASHRAE 90.1 Equivalent Glass with Shading Simulated with and without Daylight Controls

WWR (%)	Annual Energy Consumption without Daylight Controls (kWh)	Annual Energy Consumption with Daylight Controls (kWh)
0	2,89,460	2,89,460
10	2,92,323	2,60,928
20	2,95,096	2,59,880
30	2,97,674	2,61,719
40	3,00,098	2,63,881
50	3,02,873	2,66,425
60	3,05,641	2,69,538
70	3,08,268	2,72,079
80	3,10,879	2,75,262
90	3,13,204	2,77,428

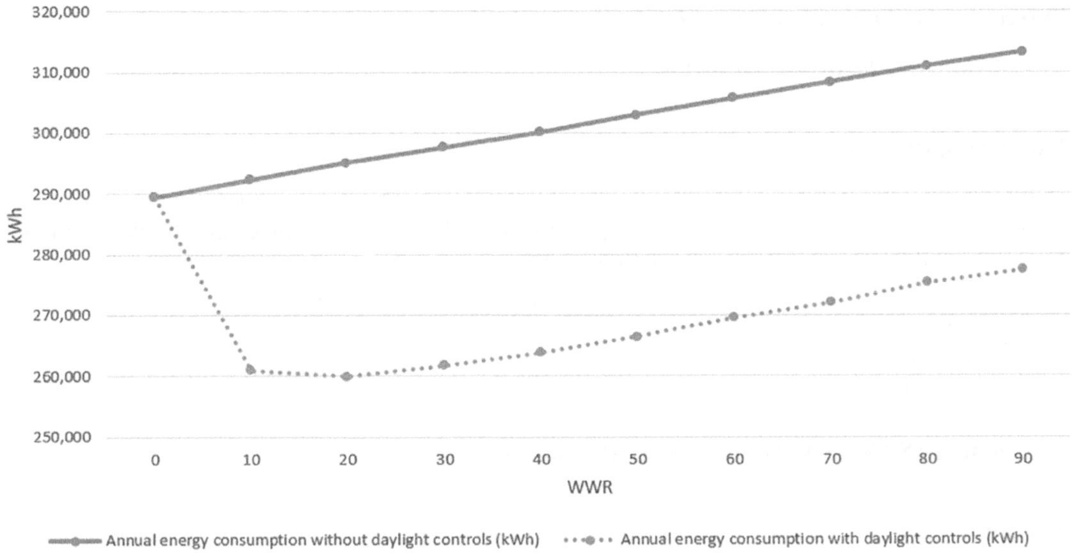

Analysis

Impact of glass type: The ASHRAE 90.1 equivalent glass has the highest L/S ratio of 2.12, which allows for better daylight transmission while limiting heat gain. This results in the lowest annual energy consumption across all WWRs when compared to other glass types used in this tutorial.

Impact of daylight controls: Incorporating daylight controls generally reduces energy consumption, particularly for larger WWRs. The daylight controls are most effective when WWR is moderate (10–30%), where the daylight contribution to perimeter zones is maximized without significantly increasing cooling loads.

Impact of shading devices: The use of a 0.5 m overhang with ASHRAE 90.1 glass demonstrates a further reduction in energy consumption by limiting direct solar heat gain. This shading is particularly effective at high WWRs when cooling loads are prominent.

General observations: Without daylight controls, energy consumption increases linearly with increasing WWR for all glass types. The energy-saving potential of daylight controls is more pronounced for glass types with higher L/S ratios.

Recommendations

- For warm/hot climates such as Singapore, glass types with higher L/S ratios should be preferred to reduce cooling energy demand while maximizing daylight use
- Implementing shading devices like overhangs can further enhance energy performance, particularly at higher WWRs
- The use of daylight controls is beneficial for optimizing energy savings

EXERCISE 4.1

Repeat this tutorial for **WIEN/HOHE VARTE, Austria** location.

TUTORIAL 4.2: EVALUATING THE IMPACT OF OVERHANGS AND FINS

GOAL

To evaluate the impact of window overhangs and fins on energy performance

What Are You Going to Learn?

- How to model overhangs and fins

Problem Statement

In this tutorial, you are going to use a 30 m × 30 m model with a 5 m perimeter depth. Use square windows of 1.5 m. You need to select the **Split no fresh air** HVAC system for the simulation model. You also need to simulate the model with the following options.

1. No shades
2. Overhangs
3. Vertical fins

Use **DUBAI INTERNATIONAL, United Arab Emirates** location.

Solution

Step 1: Open a new project and create a **30 m × 30 m** building with a 5 m perimeter depth.

Step 2: Select the **Openings** tab. Select the **Single glazing, clear, no shading** template. Select **Sgl Grey 6mm** in **Glazing** type.

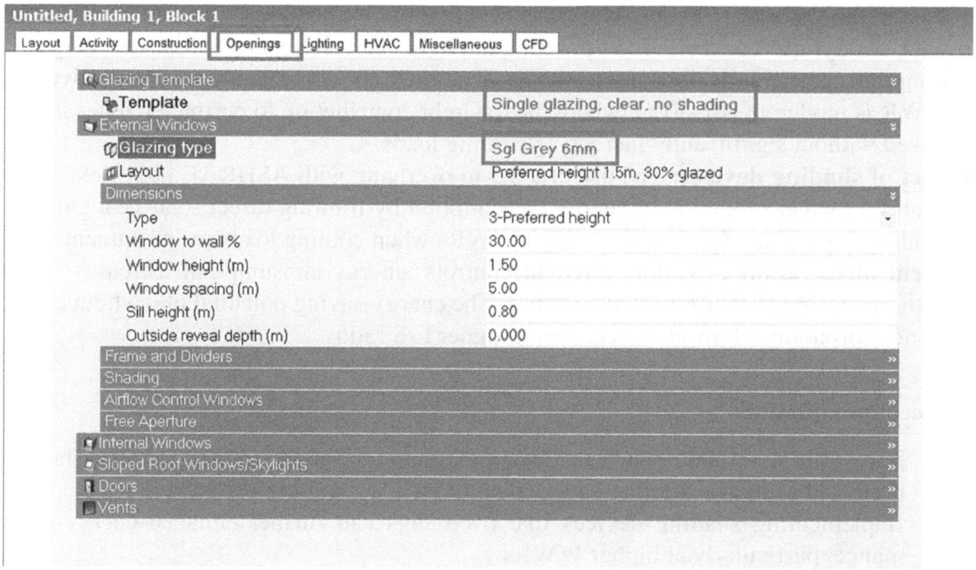

Step 3: Select **4-Fixed width** and **Height** from the **Type** drop-down list. Enter **1.5** in the **Window width (m)** and **Window height (m)** textbox. Enter **2.00** in the **Window spacing (m)** textbox.

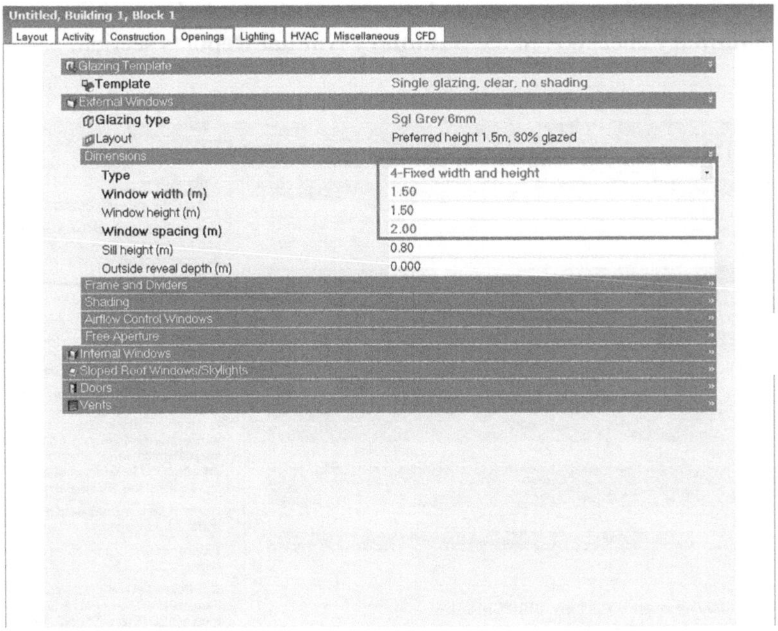

Step 4: Select the **HVAC** tab. Select the **Split no fresh air** template. Clear the **Heated** checkbox under the **Heating** section and clear the **On** checkbox under **DHW**. Select the checkbox under **Cooling system** and enter **3.0** as **CoP**.

When you use the **Split no fresh air** template, you can get the cooling energy consumption for all zones separately.

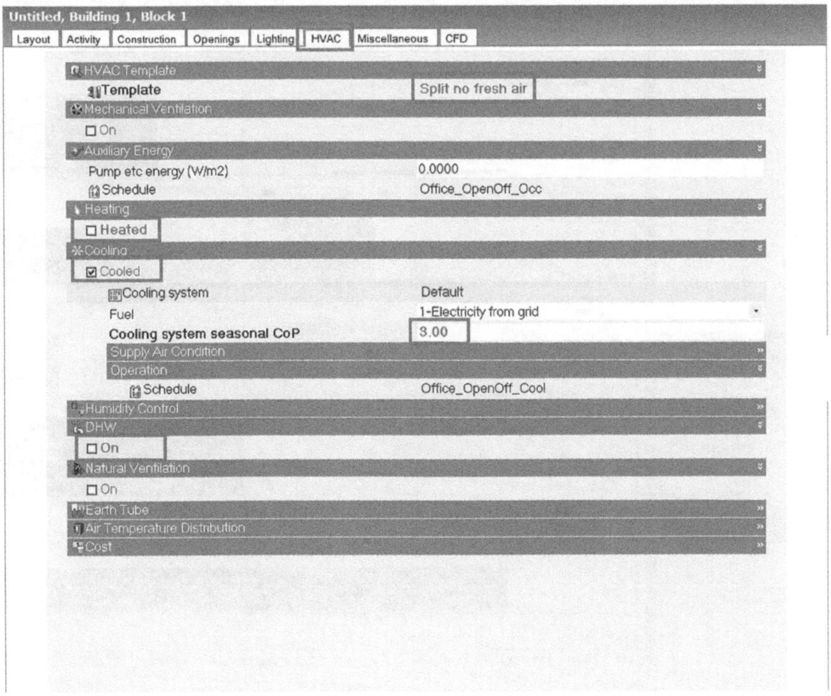

Step 5: Select the **Simulation** screen tab and then the **Output** tab. Expand **Summary Tables** and select the **All Summary** checkbox in the **Summary Annual Reports** section.

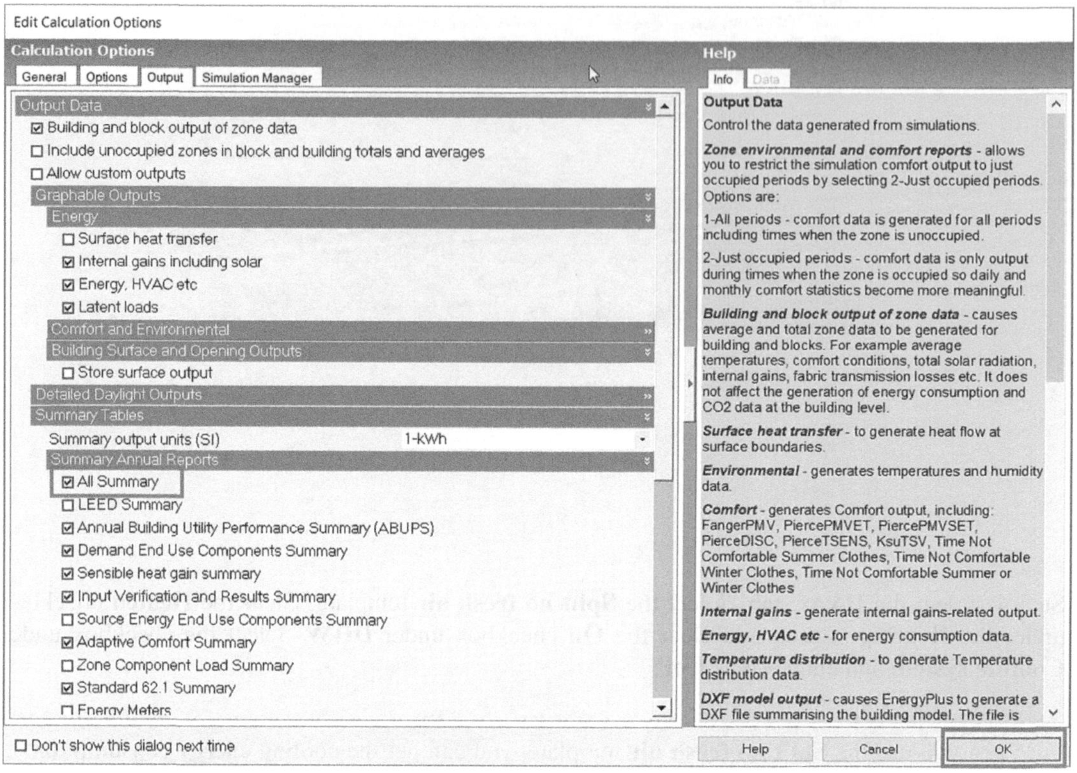

Step 6: Perform an annual simulation and view the data in graphical format.

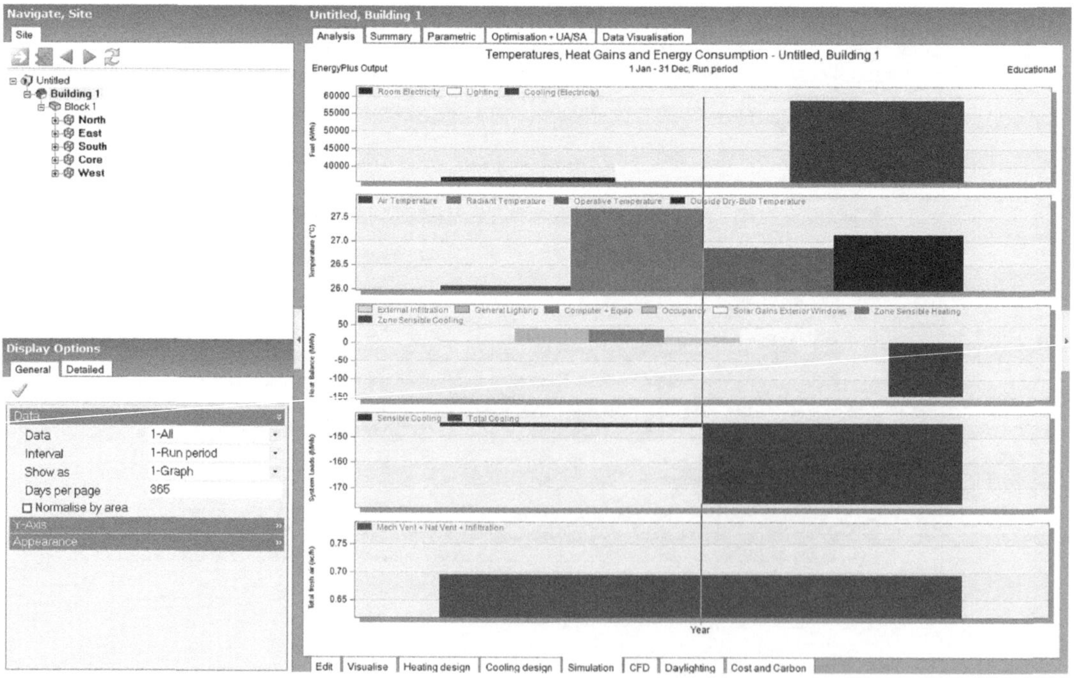

Step 7: Select the **Summary** tab.

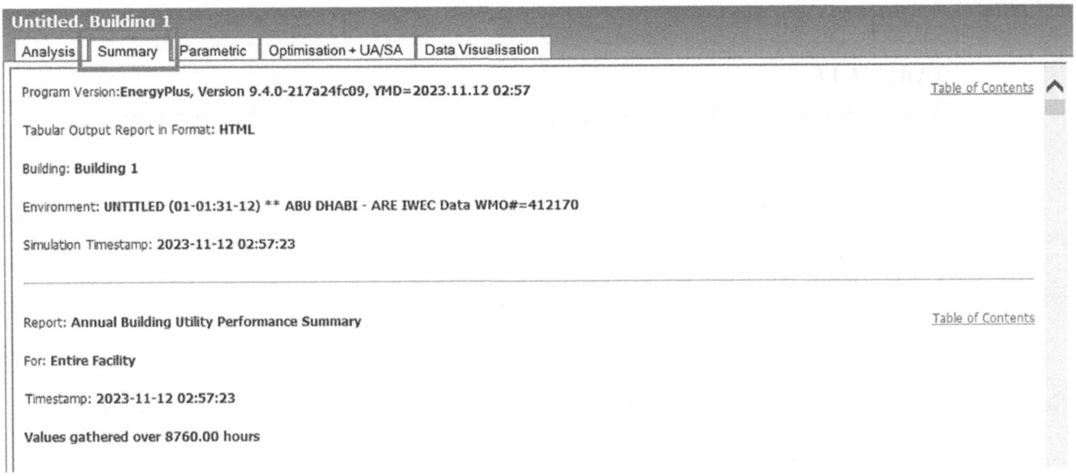

Step 8: Scroll down to read **Annual and Peak Values—Other**.

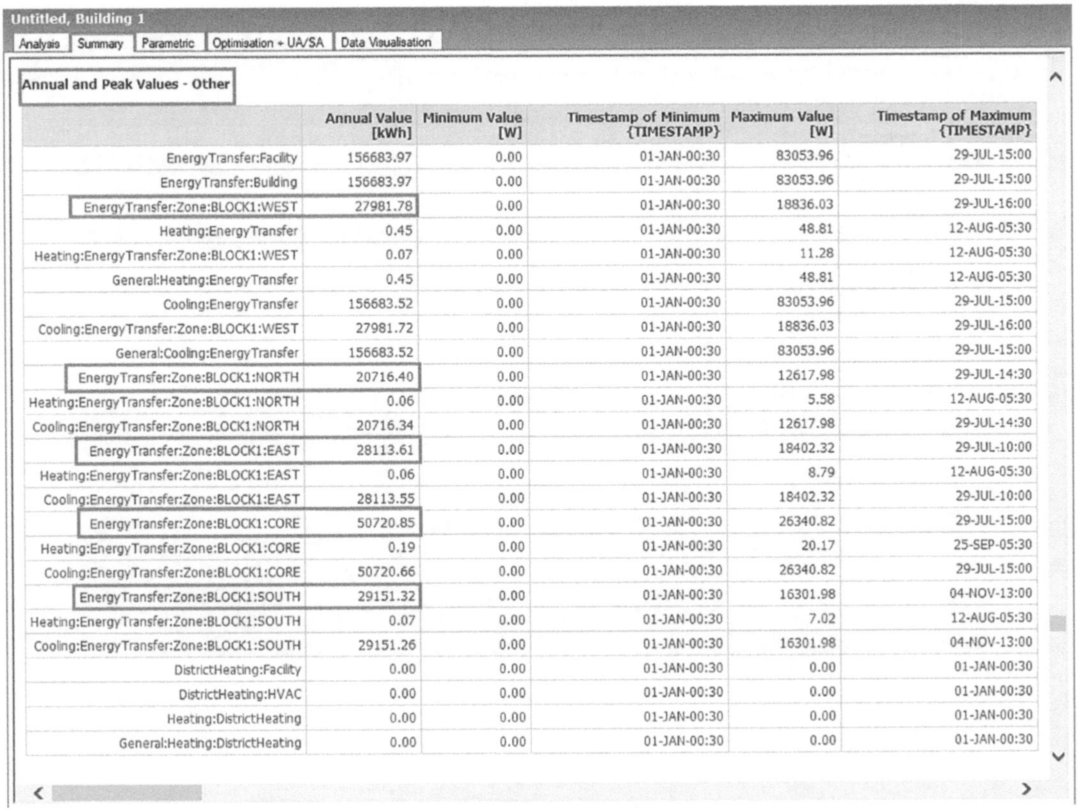

	Annual Value [kWh]	Minimum Value [W]	Timestamp of Minimum {TIMESTAMP}	Maximum Value [W]	Timestamp of Maximum {TIMESTAMP}
EnergyTransfer:Facility	156683.97	0.00	01-JAN-00:30	83053.96	29-JUL-15:00
EnergyTransfer:Building	156683.97	0.00	01-JAN-00:30	83053.96	29-JUL-15:00
EnergyTransfer:Zone:BLOCK1:WEST	27981.78	0.00	01-JAN-00:30	18836.03	29-JUL-16:00
Heating:EnergyTransfer	0.45	0.00	01-JAN-00:30	48.81	12-AUG-05:30
Heating:EnergyTransfer:Zone:BLOCK1:WEST	0.07	0.00	01-JAN-00:30	11.28	12-AUG-05:30
General:Heating:EnergyTransfer	0.45	0.00	01-JAN-00:30	48.81	12-AUG-05:30
Cooling:EnergyTransfer	156683.52	0.00	01-JAN-00:30	83053.96	29-JUL-15:00
Cooling:EnergyTransfer:Zone:BLOCK1:WEST	27981.72	0.00	01-JAN-00:30	18836.03	29-JUL-16:00
General:Cooling:EnergyTransfer	156683.52	0.00	01-JAN-00:30	83053.96	29-JUL-15:00
EnergyTransfer:Zone:BLOCK1:NORTH	20716.40	0.00	01-JAN-00:30	12617.98	29-JUL-14:30
Heating:EnergyTransfer:Zone:BLOCK1:NORTH	0.06	0.00	01-JAN-00:30	5.58	12-AUG-05:30
Cooling:EnergyTransfer:Zone:BLOCK1:NORTH	20716.34	0.00	01-JAN-00:30	12617.98	29-JUL-14:30
EnergyTransfer:Zone:BLOCK1:EAST	28113.61	0.00	01-JAN-00:30	18402.32	29-JUL-10:00
Heating:EnergyTransfer:Zone:BLOCK1:EAST	0.06	0.00	01-JAN-00:30	8.79	12-AUG-05:30
Cooling:EnergyTransfer:Zone:BLOCK1:EAST	28113.55	0.00	01-JAN-00:30	18402.32	29-JUL-10:00
EnergyTransfer:Zone:BLOCK1:CORE	50720.85	0.00	01-JAN-00:30	26340.82	29-JUL-15:00
Heating:EnergyTransfer:Zone:BLOCK1:CORE	0.19	0.00	01-JAN-00:30	20.17	25-SEP-05:30
Cooling:EnergyTransfer:Zone:BLOCK1:CORE	50720.66	0.00	01-JAN-00:30	26340.82	29-JUL-15:00
EnergyTransfer:Zone:BLOCK1:SOUTH	29151.32	0.00	01-JAN-00:30	16301.98	04-NOV-13:00
Heating:EnergyTransfer:Zone:BLOCK1:SOUTH	0.07	0.00	01-JAN-00:30	7.02	12-AUG-05:30
Cooling:EnergyTransfer:Zone:BLOCK1:SOUTH	29151.26	0.00	01-JAN-00:30	16301.98	04-NOV-13:00
DistrictHeating:Facility	0.00	0.00	01-JAN-00:30	0.00	01-JAN-00:30
DistrictHeating:HVAC	0.00	0.00	01-JAN-00:30	0.00	01-JAN-00:30
Heating:DistrictHeating	0.00	0.00	01-JAN-00:30	0.00	01-JAN-00:30
General:Heating:DistrictHeating	0.00	0.00	01-JAN-00:30	0.00	01-JAN-00:30

Step 9: Record the annual value for the variable energy transfer zone (kWh) for all perimeter zones (Table 4.14).

TABLE 4.14
Annual Values for the Energy Transfer for Each Zone with No Shading

Zone	Annual Value for Energy Transfer (kWh)
North	20,716
East	28,114
West	27,982
South	29,151

Save the simulation model for the next step. Now you are going to model window overhangs.

Step 10: Select the **Openings** tab.

Step 11: Click the **Shading** section. **Shading options** appears.

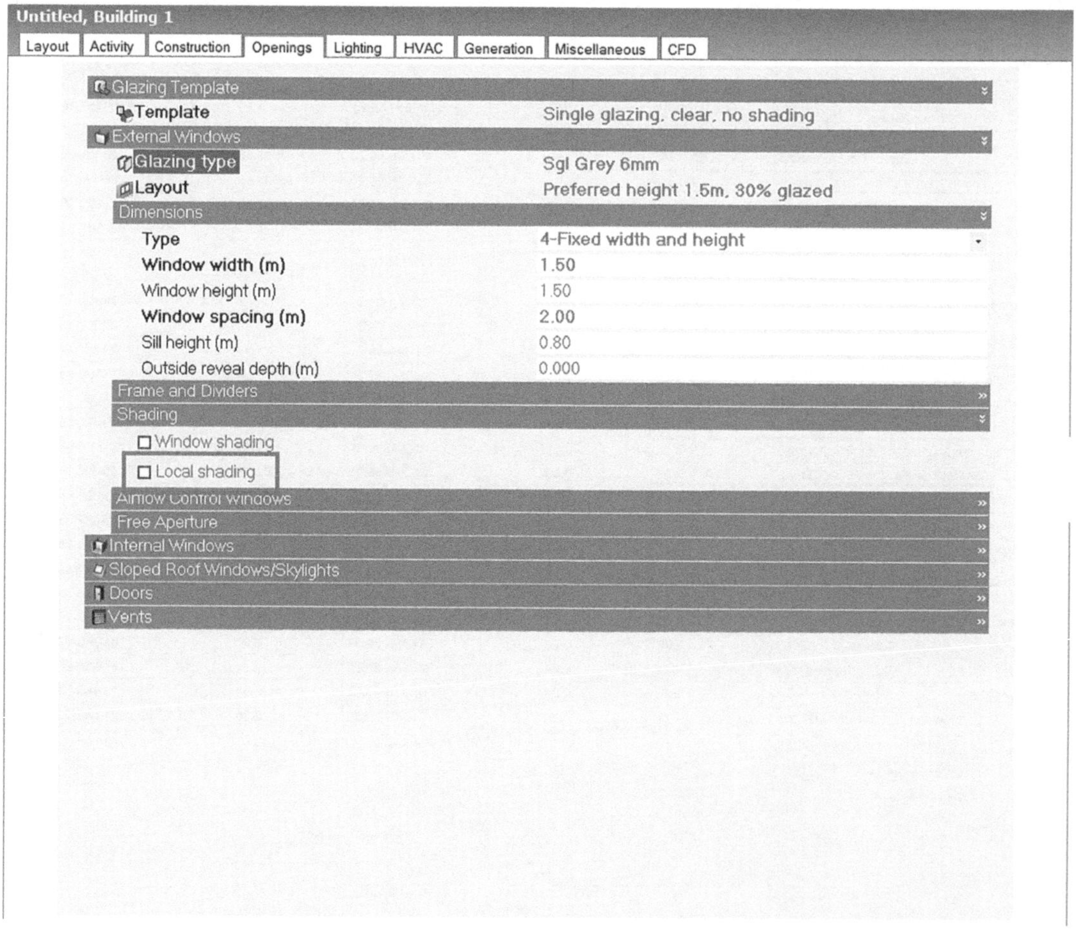

Step 12: Select the **Local shading** checkbox. It displays shading type. Select **0.5 m Overhang** in **Type**.

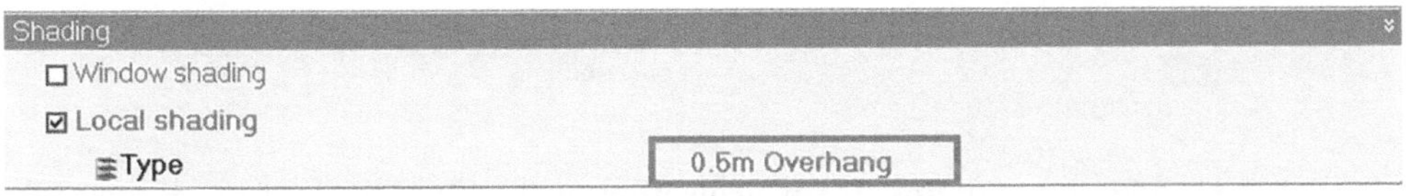

Step 13: Select the **Visualize** screen tab. Select the **Rendered View** tab. It displays the rendering of the building. Make sure that all windows are modeled with an overhang.

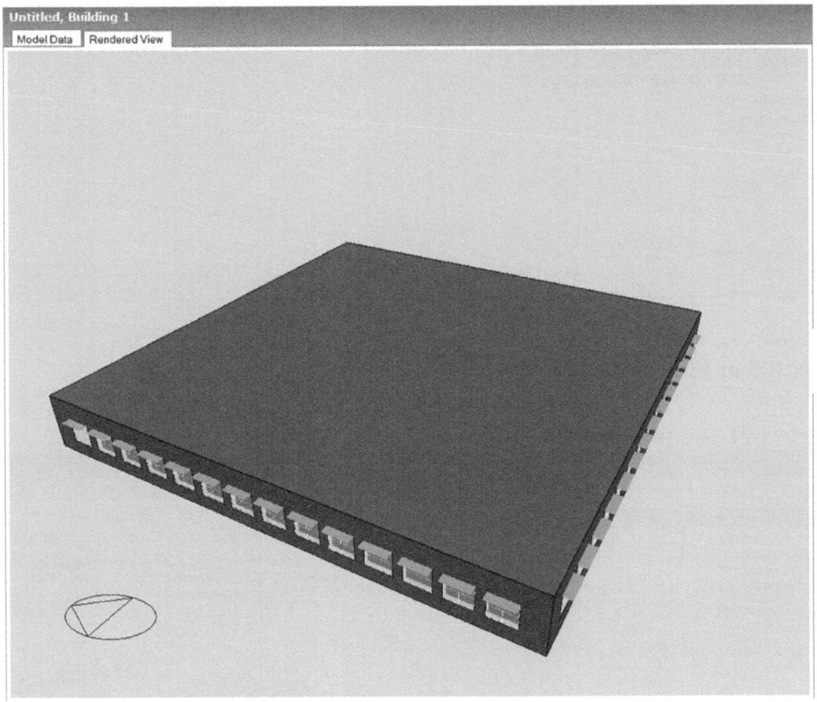

Step 14: Repeat the preceding steps to get simulation results (Table 4.15).

TABLE 4.15
Annual Value for the Energy Transfer with Overhangs

Zone	Annual Value for Energy Transfer (kWh)
North	20,176
East	25,237
West	25,203
South	25,408

Save the simulation model with the name **DB_overhang**. In the following steps, you are going to model vertical fins.

Step 15: Open the saved simulation model.

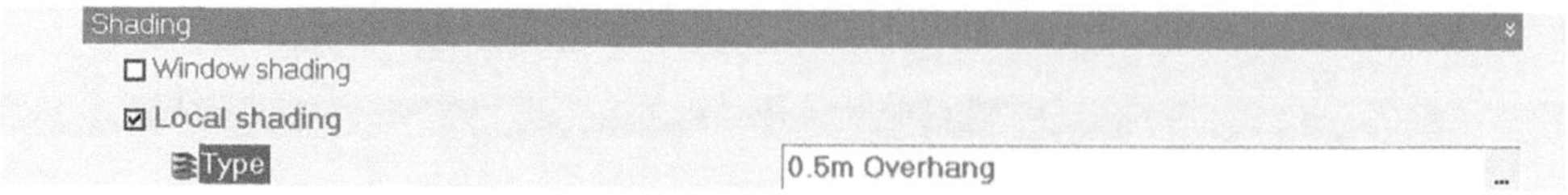

Step 16: Select **Overhang + sidefins (0.5m projection)** and make a copy of the current selection and edit for changes.

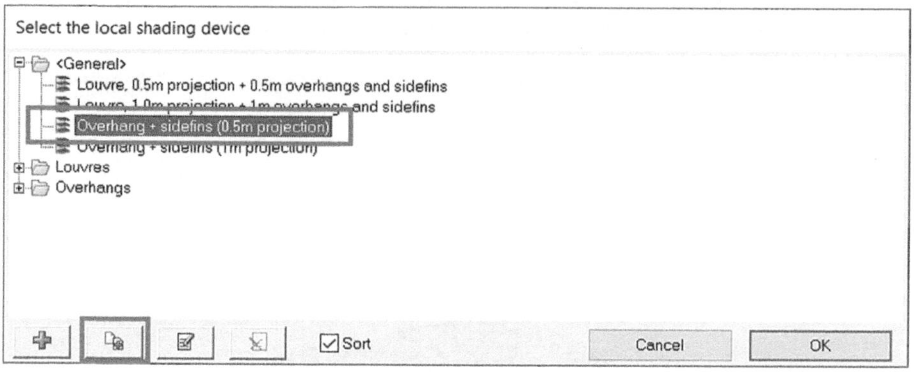

Step 17: Enter **0.5 m Fins** as the name.

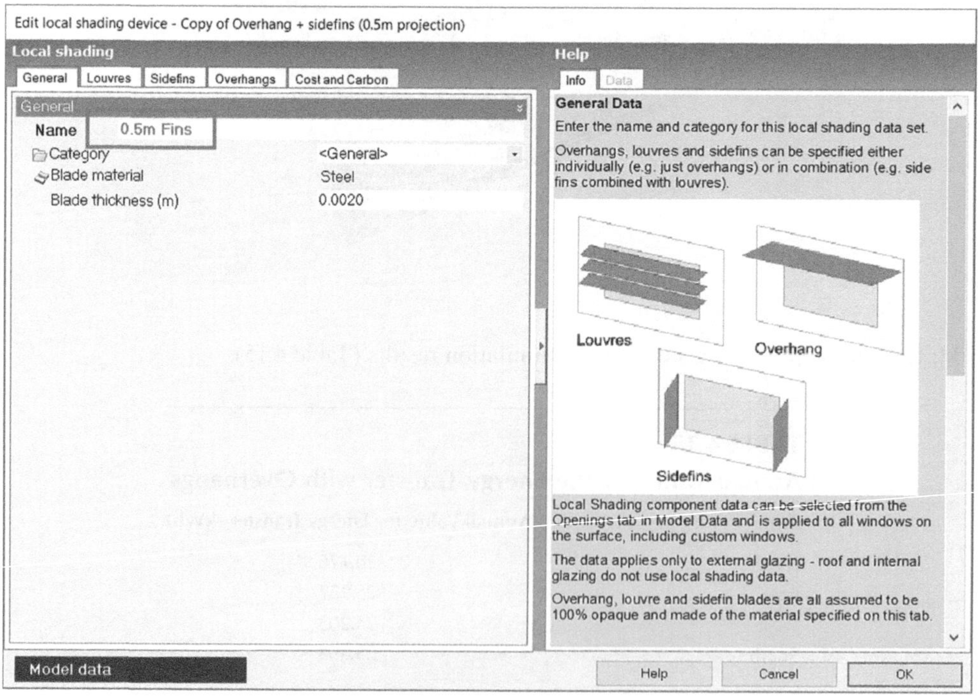

Step 18: Select the **Overhangs** tab.

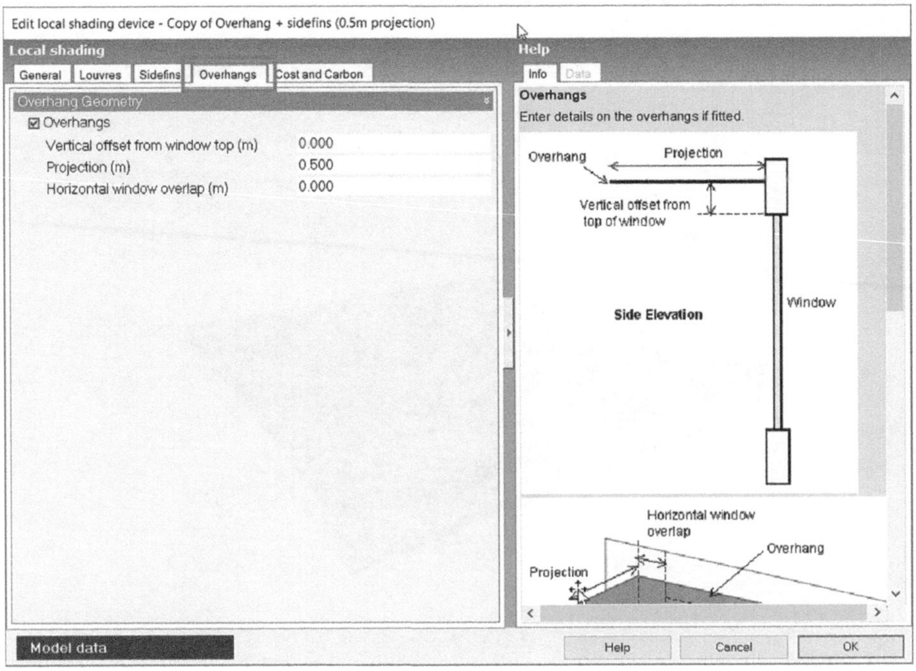

Step 19: Clear the **Overhangs** checkbox.

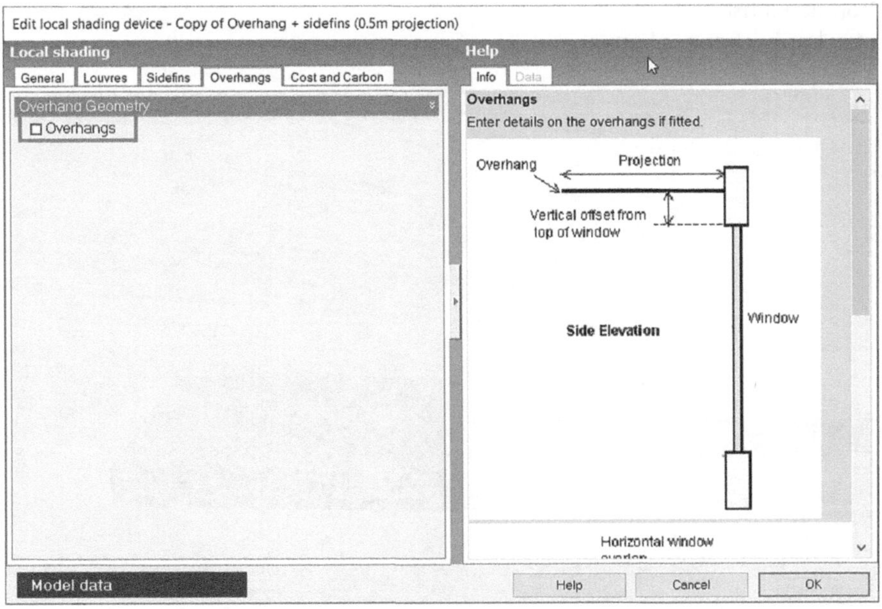

Step 20: Select the **Visualize screen** tab. Select the **Rendered View** tab. It displays the rendering of the building. Make sure that all windows are modeled with side fins.

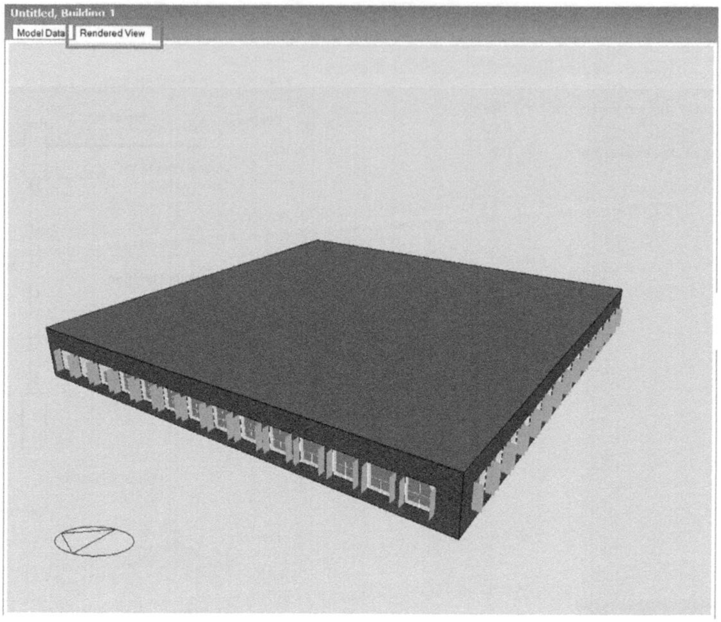

Step 21: Under the **Display options**, select the **Show sunpath** checkbox. Select the **Show shadows** checkbox. Under **Period**, enter **17:00 Time (dec hrs)** and **21** in the **Day** textbox. Select **Jun** in **Month** drop-down list.

Click the **Apply changes** button.

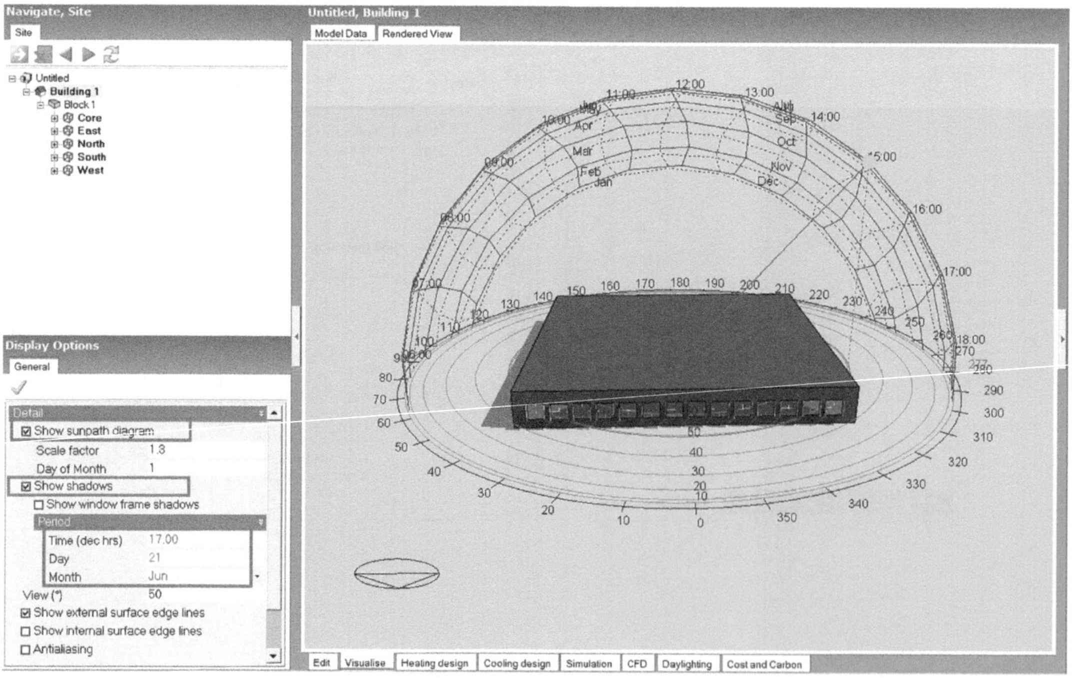

Step 22: Perform an annual simulation and record the results (Table 4.16).

TABLE 4.16
Annual Value of Energy Transfer for Vertical Fins

Zone	Annual Value for Energy Transfer (kWh)
North	19,995
East	26,741
West	26,624
South	26,661

Compare the results for all cases (Table 4.17).

TABLE 4.17
Comparison of the Annual Cooling Energy for All Types of Shades

	Annual Cooling Energy (kWh)		
Zone	No Shade	With Overhangs	With Fins
North	20,716	20,176	19,995
East	28,114	25,237	26,741
West	27,982	25,203	26,624
South	29,151	25,408	26,661

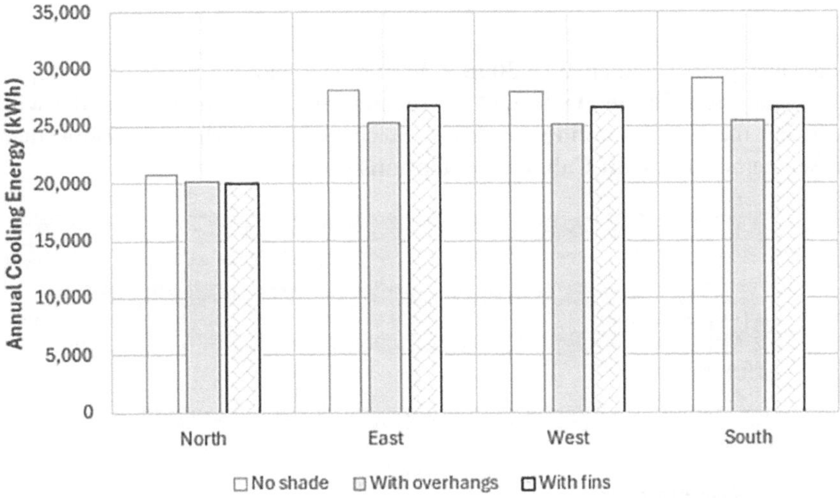

Analysis

Overall performance: Incorporating shading devices (overhangs or fins) generally reduces the annual cooling energy consumption across all zones compared to the case without shades.

Effectiveness of overhangs: Overhangs are particularly effective in reducing cooling loads on the south-facing façade. This reduction highlights the effectiveness of overhangs in minimizing solar heat gain from high-angle sunlight, which is most prevalent on the south façade.

Effectiveness of fins: Fins are more effective in the north-facing façade, with energy consumption reducing from 20,716 kWh (no shade) to 19,995 kWh (with fins). This suggests that vertical fins are better suited to block low-angle sunlight coming from the north direction. On other façades (facing east, west, and south), fins also provide a reduction in cooling energy consumption but are generally less effective than overhangs, especially on the south façade.

EXERCISE 4.2

Repeat this tutorial for both overhangs and fins to observe the combined effect. Compare the energy consumption for all the perimeter zones.

TUTORIAL 4.3: EVALUATING THE IMPACT OF INTERNAL OPERABLE SHADES

Goal

To evaluate the impact of window internal operable shades on energy consumption

What Are You Going to Learn?

- How to model operable shades

Problem Statement

In this tutorial, you are going to use the **30 m × 30 m** model with a 5 m perimeter depth. Determine the energy consumption and solar gain for all perimeter zones. You need to simulate the model with the following options.

1. Overhang with a 0.5 m depth
2. Overhang and internal operable shades with solar control

Use the **New Delhi/Palam, India** weather location.

Solution

Step 1: Open a new project and create a **30 m × 30 m** model with a 5 m perimeter depth. Select the **Openings** tab and select **Single glazing, clear, no shading** Template. Change the window height to **1.00 m**. Select the **Local shading** checkbox. Select **0.5 m** overhang in **Type**, then select the **Simulation** screen tab. The **Edit Calculation Options** screen appears.

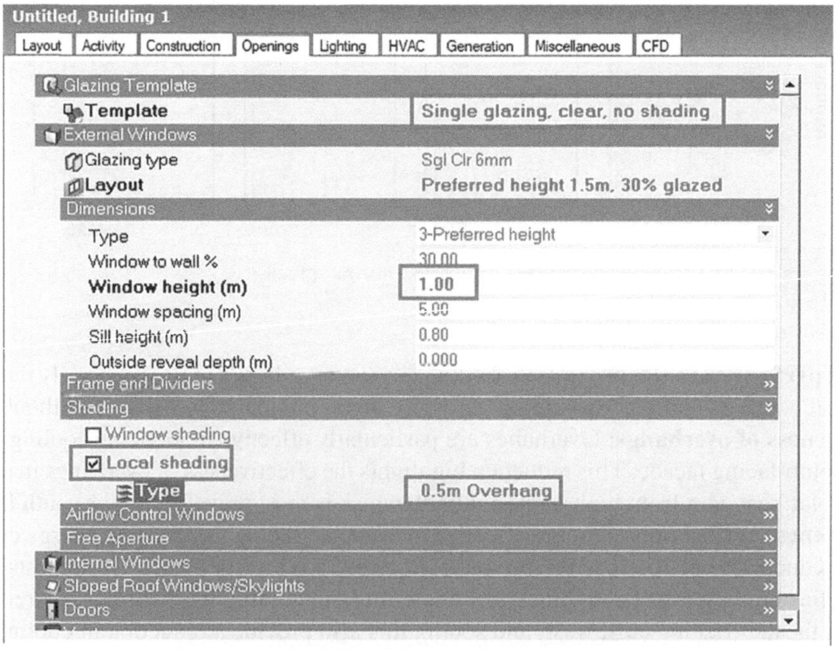

Step 2: Simulate the model and select the **Sub-hourly** checkbox. Click **OK**. The results are displayed.

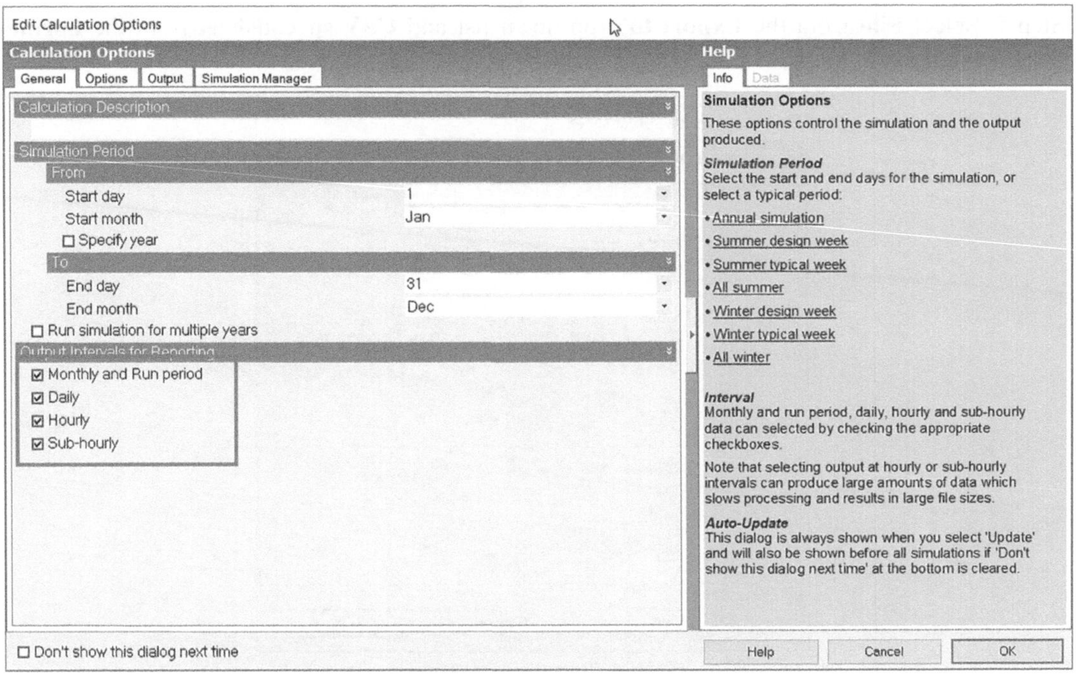

Step 3: Select **Internal gains** from the **Data** drop-down list. Click **East** in the navigation tree. It shows internal gains for the east zone.

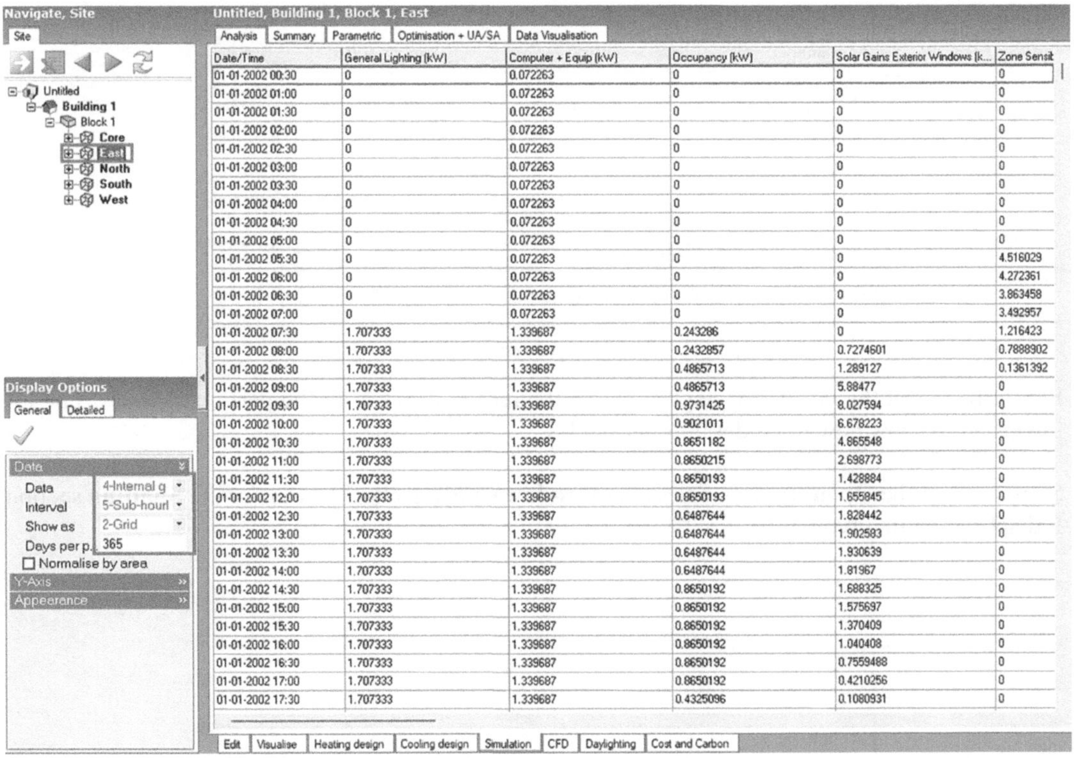

Step 4: Click the **Export data** button. The **Export Results Spreadsheet** screen appears.

Step 5: Select **File** from the **Export to** drop-down list and **CSV spreadsheet** from the **Format** drop-down list. Click **OK**.

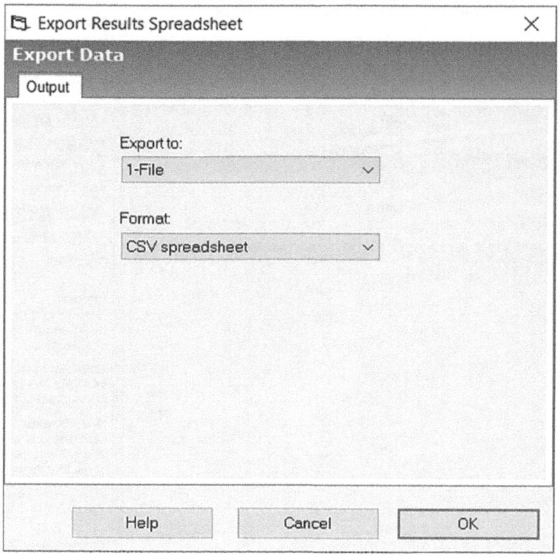

Step 6: Name the file **East Zone Results** and save it for a comparison of results.

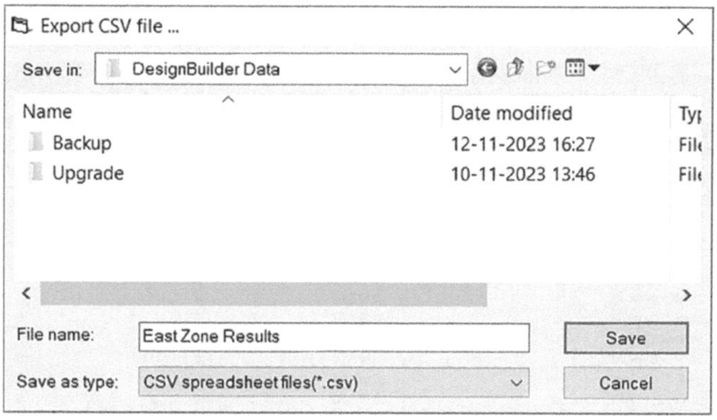

Similarly, repeat the preceding steps and save the results files for all the perimeter zones. The following steps show how to model internal shades.

Step 7: Select the **Openings** tab. Select the **Window shading** checkbox under the **Shading** section. It displays shading type.

Step 8: Select **High reflectance—low transmittance shade** from the **Type** drop-down list. Select **4-Solar** from the **Control type** drop-down list. Set **250** as solar setpoint (W/m^2).

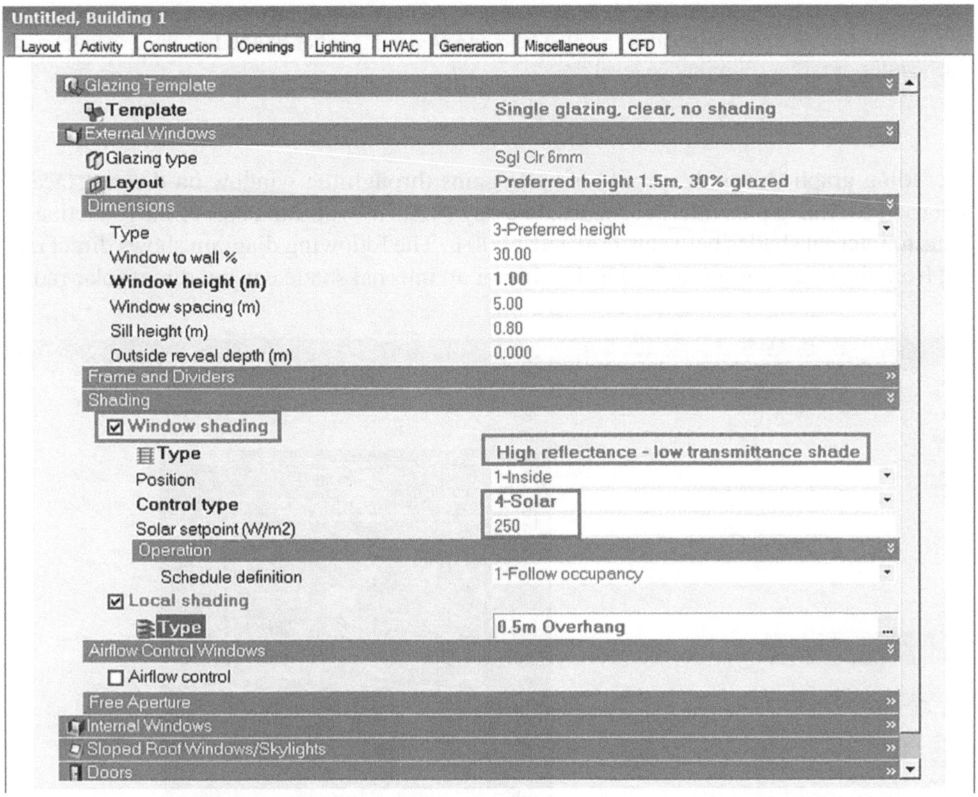

Step 9: Simulate the model and record the results for all four zones. Compare the results for two cases in each orientation: only overhang and overhang with internal shade for solar gains (kW) for 4 April.

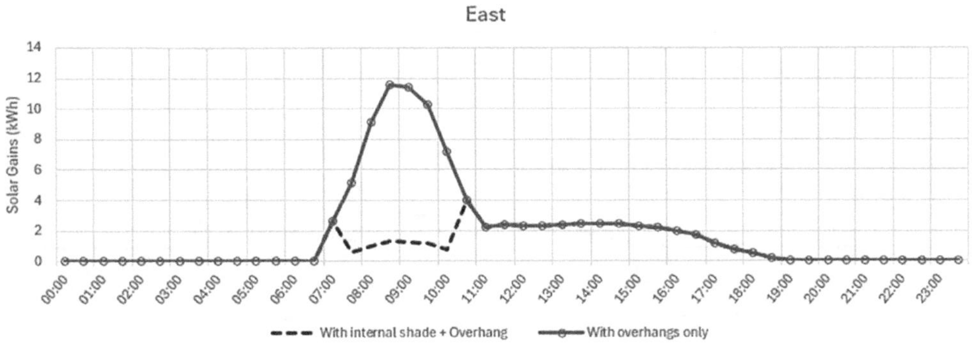

The solar control type operates indoor shading devices based on incident solar radiation levels. In this example, a solar setpoint of 250 W/m² is used. When the solar radiation on the window exceeds this threshold, the shading device is deployed to reduce heat gain through solar radiation. Indoor operable shades can also be controlled using other methods, such as predefined schedules or target illuminance levels.

The preceding graph shows the profile of solar gains through the window on the east façade with overhang and overhang plus internal shades for a day (4 April); you can observe the reduction in solar gains due to internal shades between 7:00 and 10:00 h. The following diagram shows direct radiation coming from the east window at 7:00 h. The use of an internal shade can cut direct solar radiation.

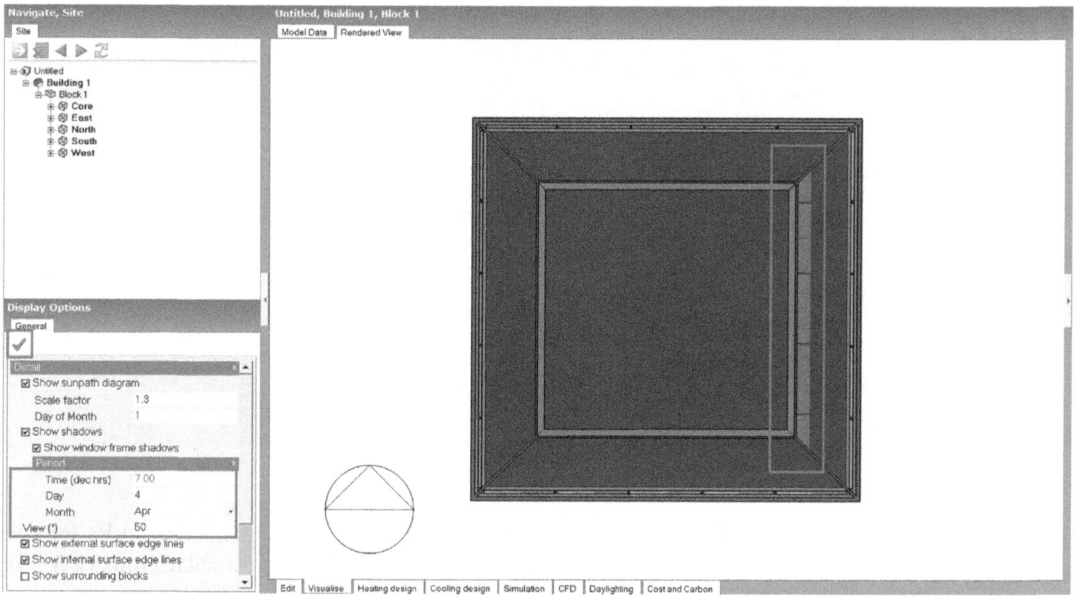

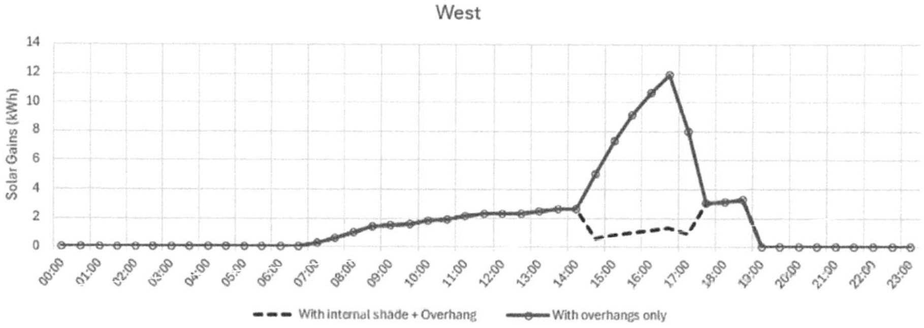

You can observe from the preceding graph that the internal shade is effective on the west window from 14:00–19:00 h on 4 April. It can also be seen from the following sunpath diagram that when there are no internal shades, west-facing windows get direct solar radiation at 16:00 h.

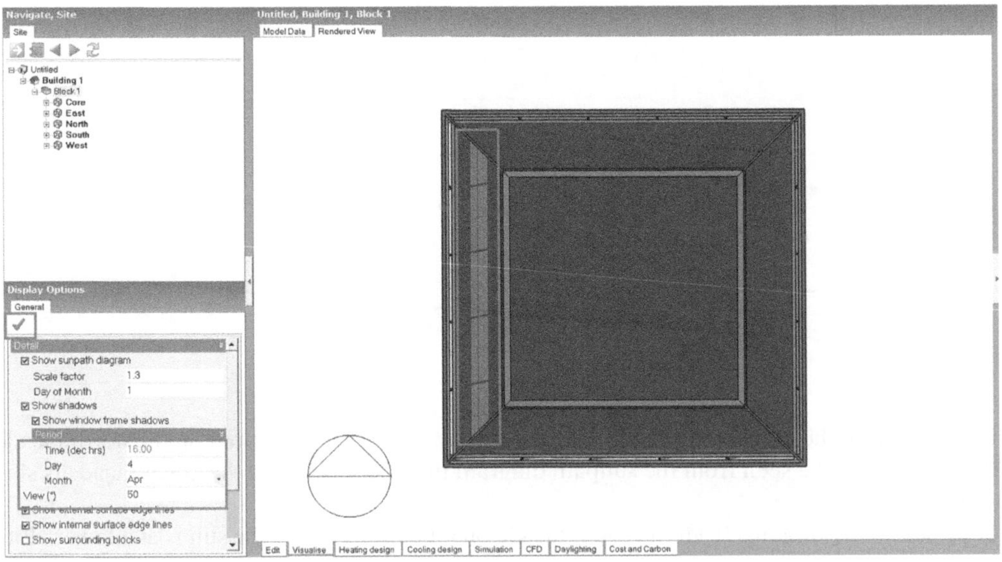

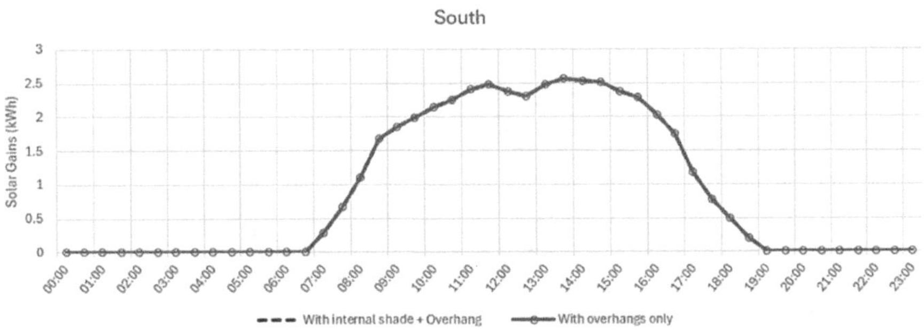

You can see from the preceding graph that internal shades are not required on the south window on 4 April for this building. It can also be seen from the following diagram that the south zone is not getting any direct glare with an overhang of 0.5 m. This is due to the fact that the altitude of the sun in daytime, except for sunrise and sunset hours, is high enough on 4 April to not cause glare through the windows.

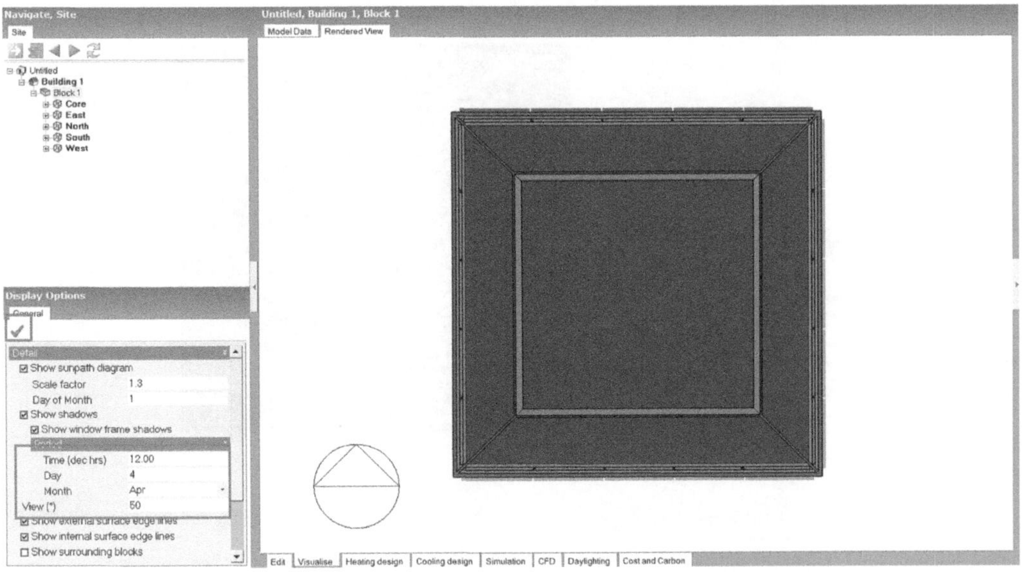

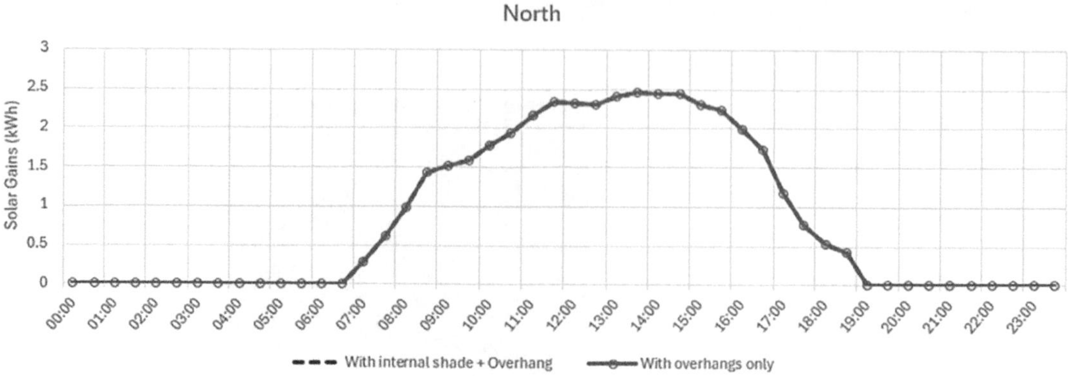

You can see from the preceding graph that the internal shade is not required on the north façade on 4 April. It can also be seen from the sunpath diagram that the window in the north façade is shaded on 4 April.

Now plot the solar gains from the south façade on 6 February when the sun is at a lower altitude.

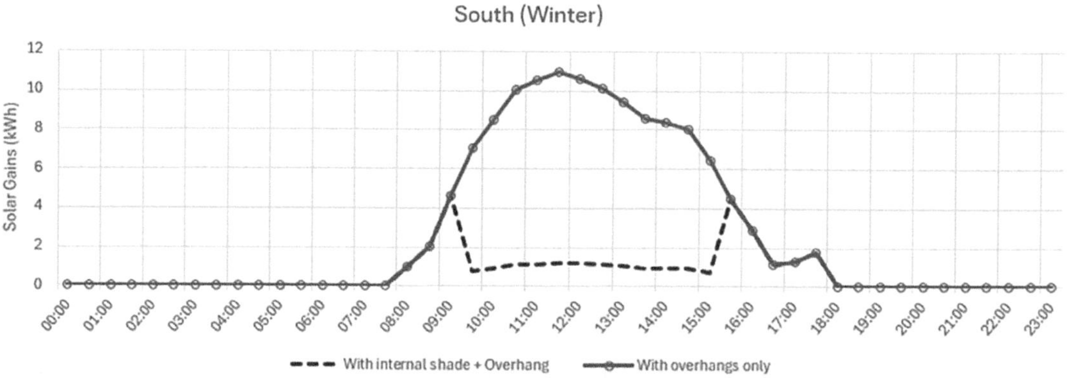

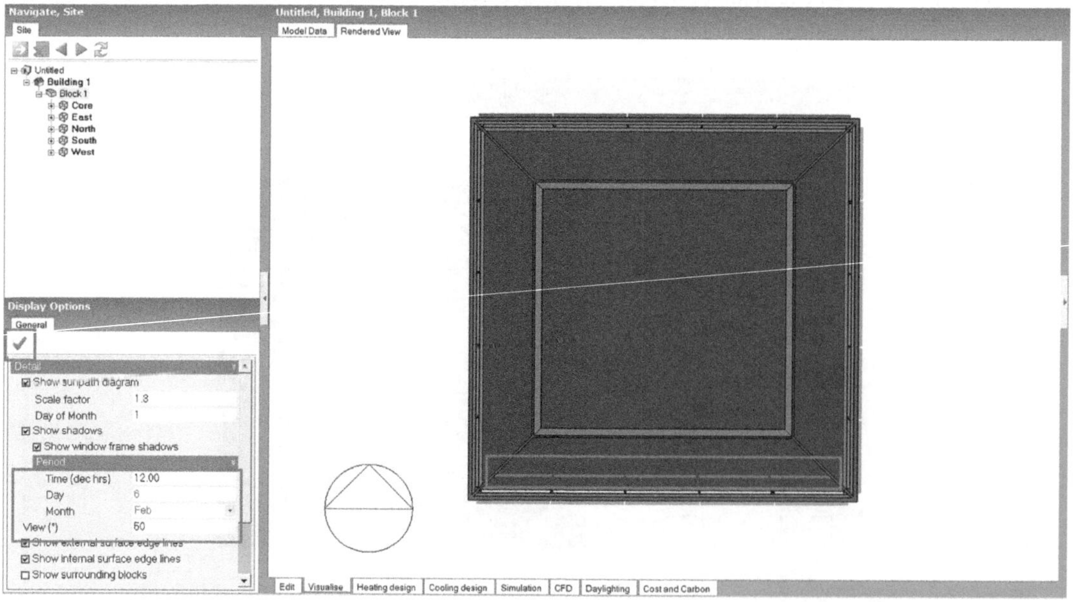

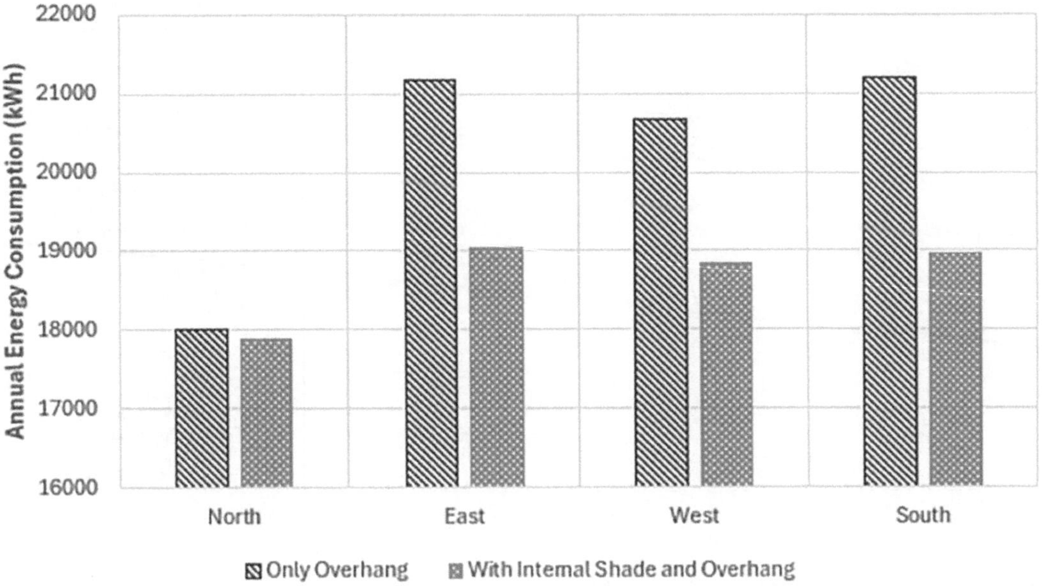

You can see that with internal shades, there is a reduction in solar gains from the south window between 12:00 h and 16:00 h. You can also see from the preceding graph that internal shades are effective for the east, west, and south zones for New Delhi.

The following two graphs present global horizontal radiation and direct normal radiation data for 4 April 4 and 6 February. On both days, solar radiation levels are lower in the afternoon, suggesting that—for this specific location and these particular dates—shading on the east side is more effective than on the west.

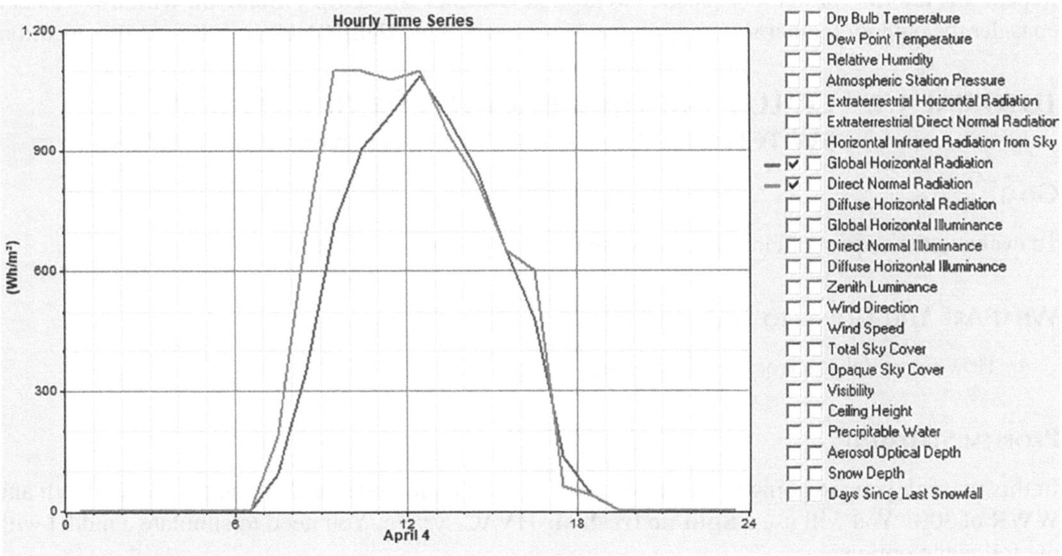

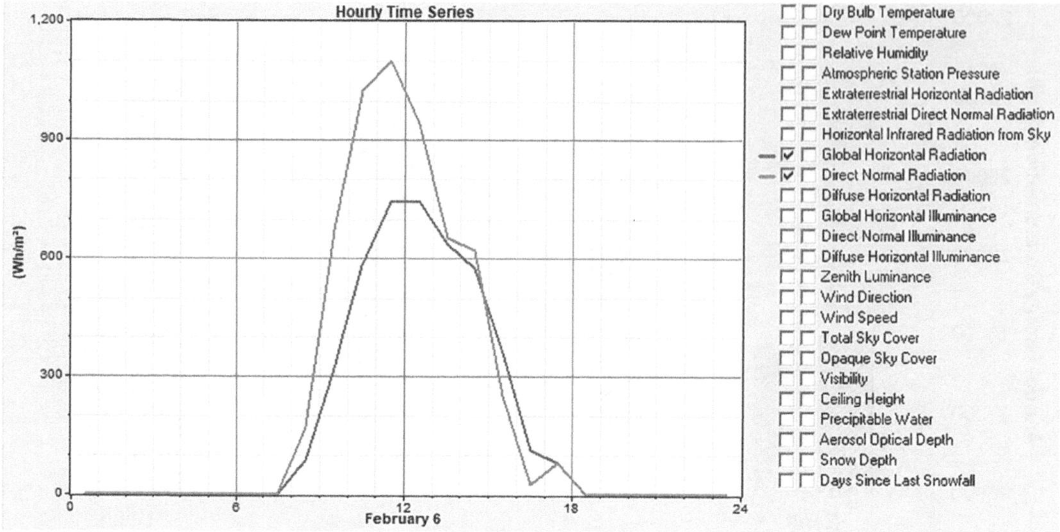

The use of internal operable shades combined with overhangs is effective in reducing solar gains, especially for east- and west-facing windows. Their impact is most pronounced during early morning (east) and late afternoon (west) hours. On 4 April, internal shades are not needed for the south and north orientations. However, on 6 February, they offer substantial benefits for the south façade due to the low sun angle.

The results indicate that using internal shades can help lower cooling energy demand in locations comparable to that of New Delhi.

EXERCISE 4.3

Repeat this tutorial for operable shade–controlled **Outdoor air temp + Solar on window**. You can consider the outdoor temperature setpoint as 35°C for the **New Delhi/Palam, India** weather location.

TUTORIAL 4.4: EVALUATING THE IMPACT OF ELECTROCHROMIC SWITCHABLE GLAZING ON WINDOWS' SOLAR GAINS

Goal

To evaluate the impact of electrochromic switchable glazing on windows' solar gains

What Are You Going to Learn?

- How to model electrochromic glazing

Problem Statement

In this tutorial, you are going to use a five-zone 30 m × 30 m model with a 5 m perimeter depth and WWR of 30%. You will use a **Split no fresh air** HVAC system. You need to simulate a model with the following options.

1 Single-glazed (**Sgl Clear 6mm**)
2 Electrochromic reflective 6 mm (darkened state whenever the solar radiation incident on the outside of the window is greater than 120 W/m^2)

Compare solar gains from exterior windows in the south zone. Use weather data for **Ganzhou, China** for the simulations.

Electrochromic windows are part of a new generation of technologies called *switchable glazing* or *smart windows.* Switchable glazing can change the light transmittance, transparency, or shading of windows in response to an environmental signal such as sunlight, temperature, or an electrical control. Electrochromic windows change from transparent to tinted by applying an electric current. Potential uses for electrochromic technology include daylighting control, glare control, solar heat control, and fading protection in windows and skylights. By automatically controlling the amount of light and solar energy that can pass through the window, electrochromic windows can help save energy in buildings.

Source: https://designbuilder.co.uk/helpv7.3/Content/Electrochromic_Glazing.htm

Solution

Step 1: Open a new project and create a **30 m × 30 m** building with five zones with a 5 m perimeter depth. Rename all zones. Click the **Openings** tab.

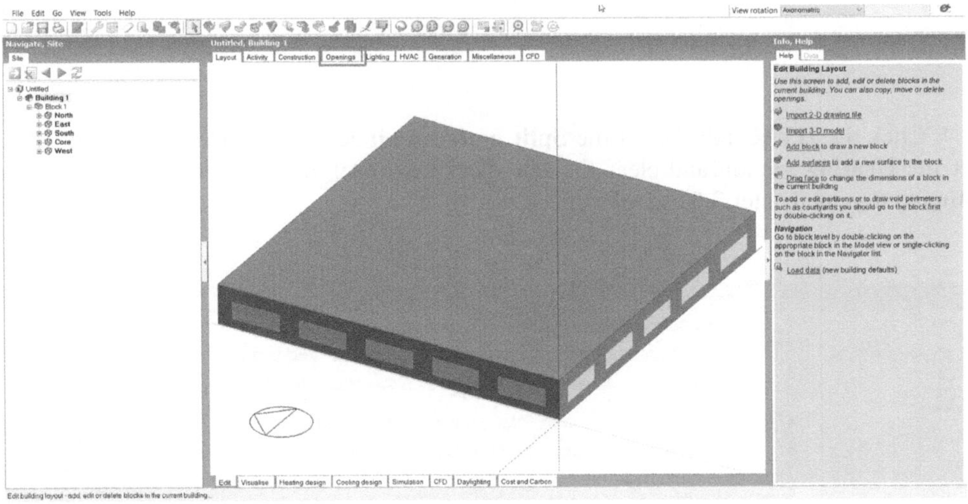

Step 2: Select the **Single glazing, clear, no shading** template. Select **Sgl Clr 6mm** in **Glazing** type.

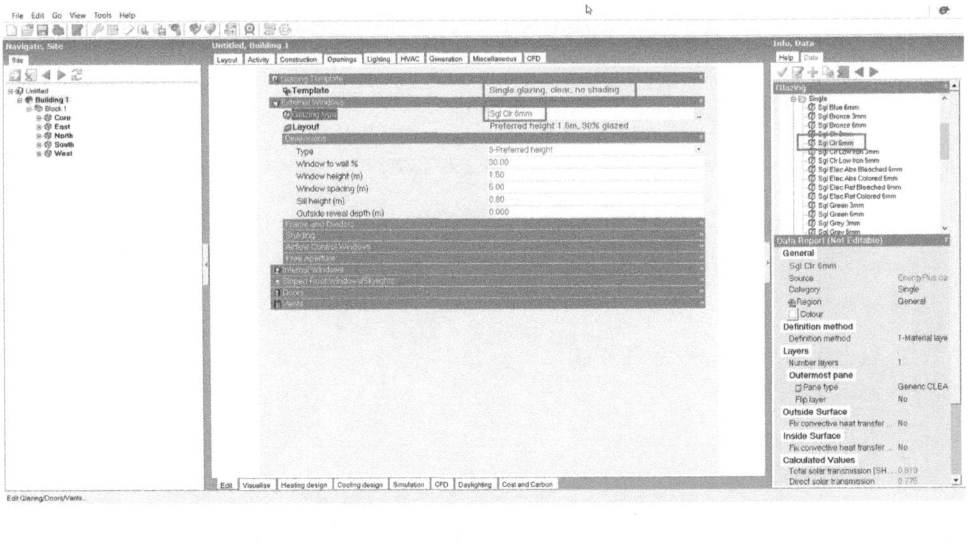

Step 3: Make sure that the **Window shading** checkbox is clear.

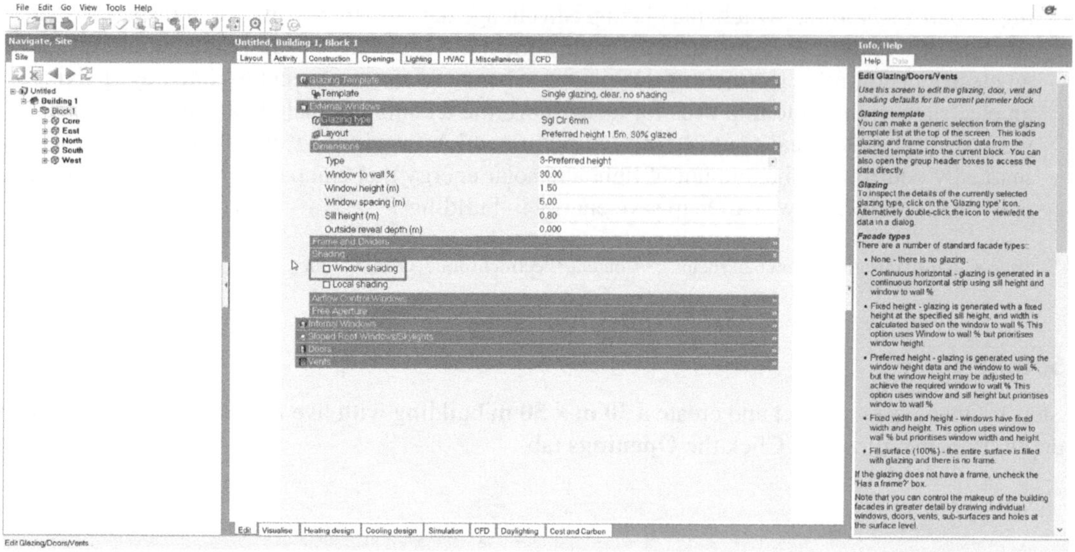

Step 4: Click the **HVAC** tab. Select the **Split no fresh air** template. Clear the **Heated** checkbox under the **Heating** section and clear the **On** checkbox under **DHW**. Select the checkbox under **Cooling system** and enter **3.0** as **CoP**.

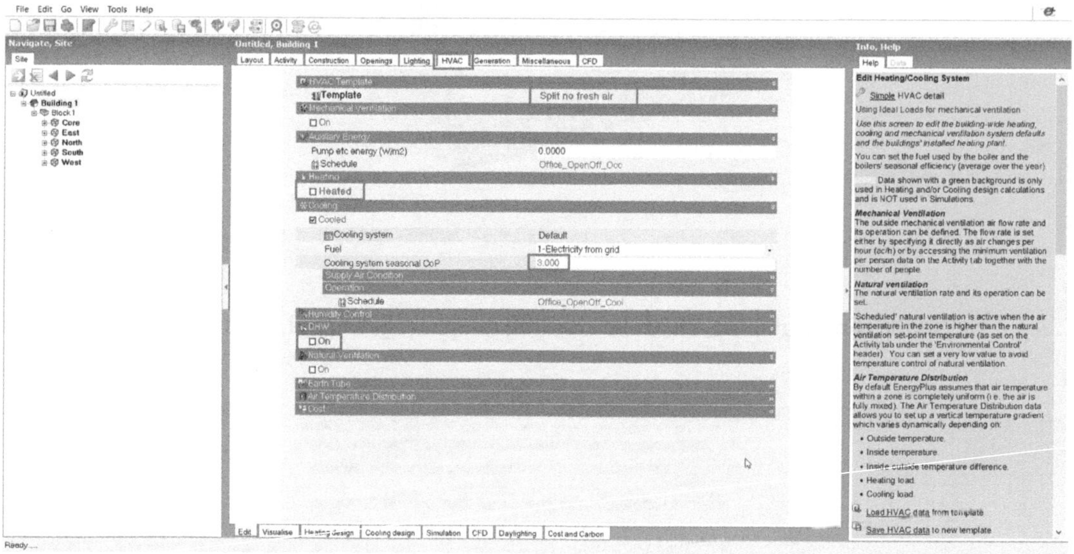

Step 5: Perform the simulation for **Summer design week**.

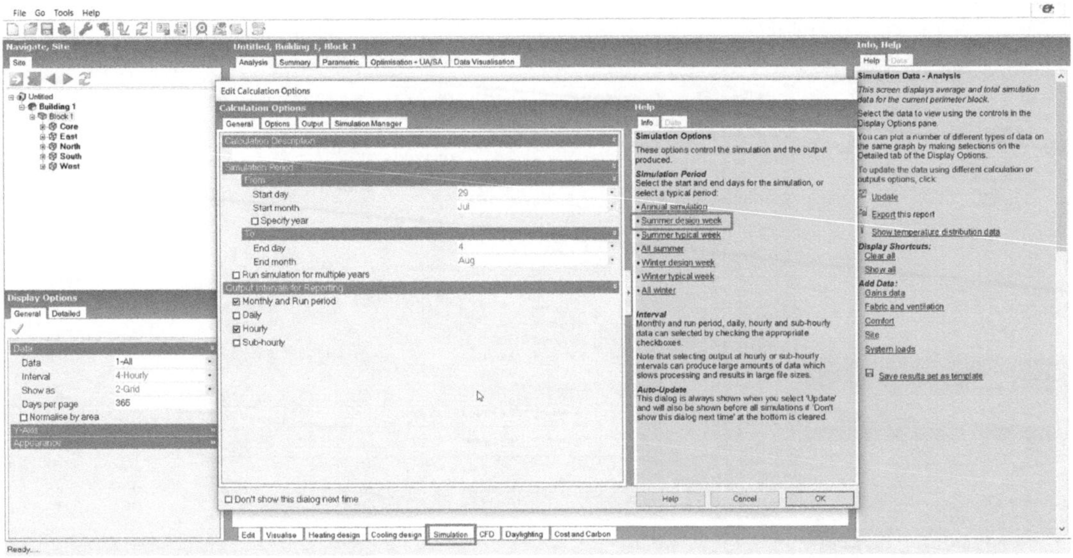

Step 6: Click the **South** zone, and then click the **Export data** icon.

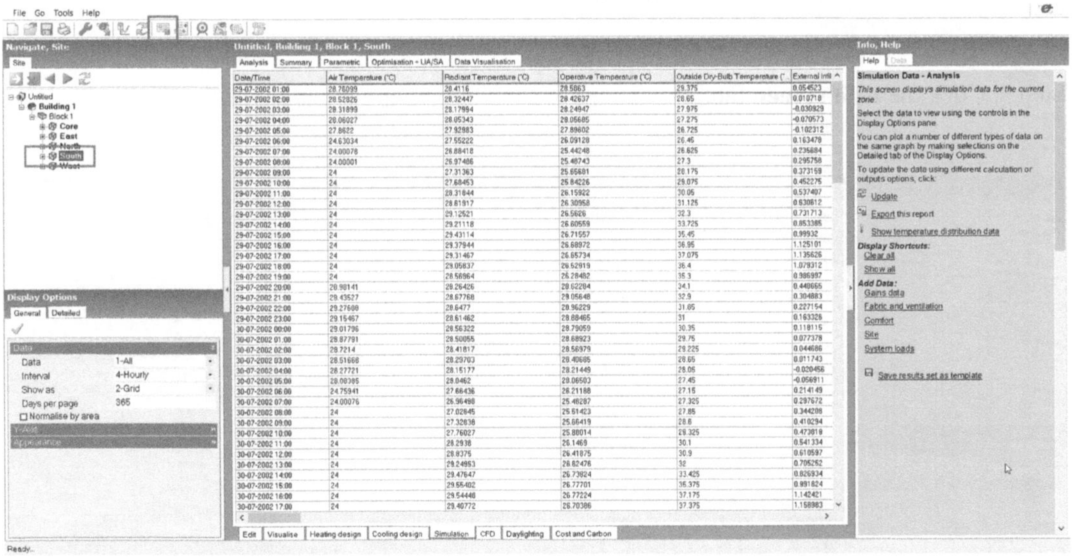

Step 7: Save the results to a spreadsheet program.

Step 8: Click the **File** menu and **Save file As**.

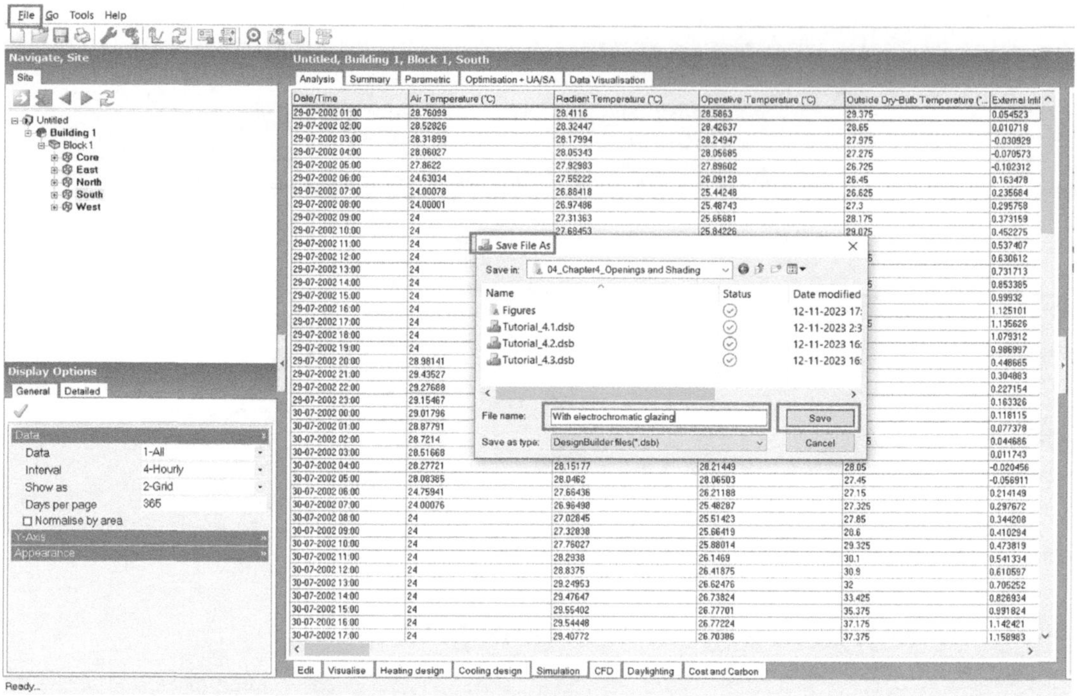

Step 9: Click the **Edit screen** tab.

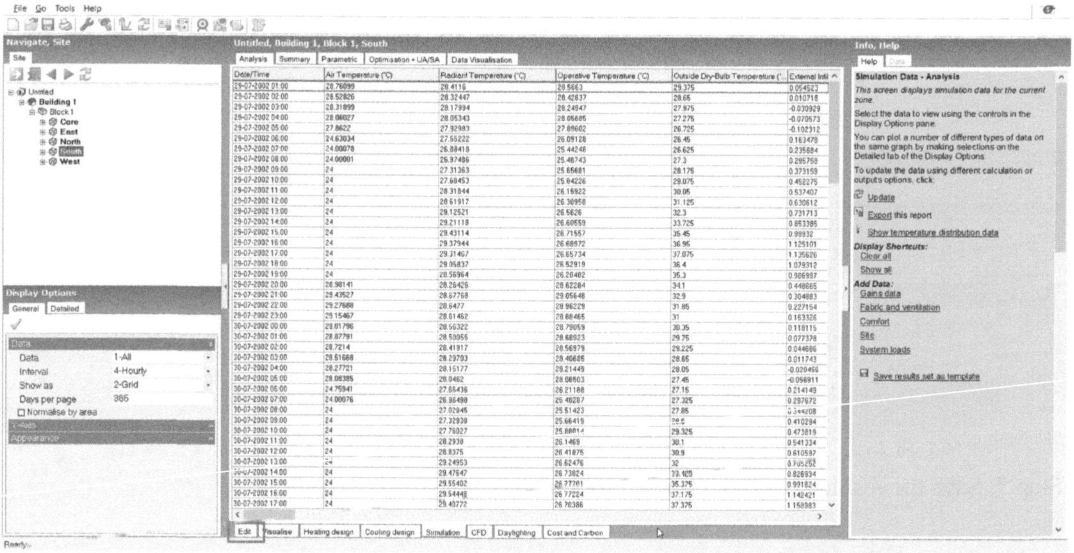

Step 10: Click the **Openings** tab.

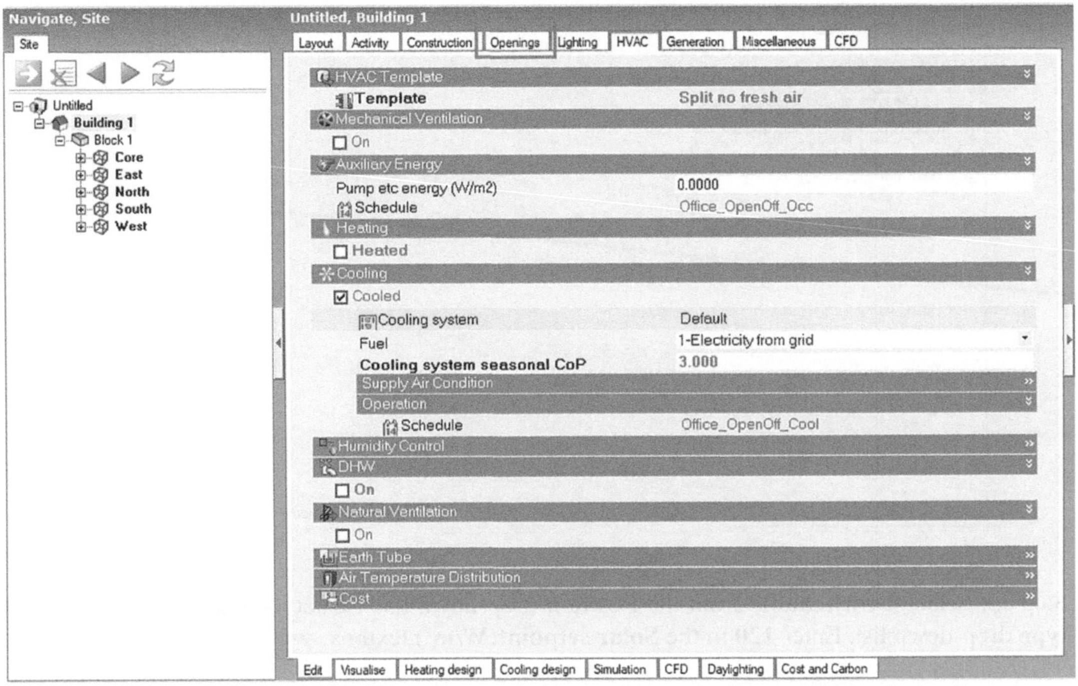

Step 11: Select the **Window shading** checkbox. Click on the three dots to change **Type**. The **Select the Window shade** screen appears.

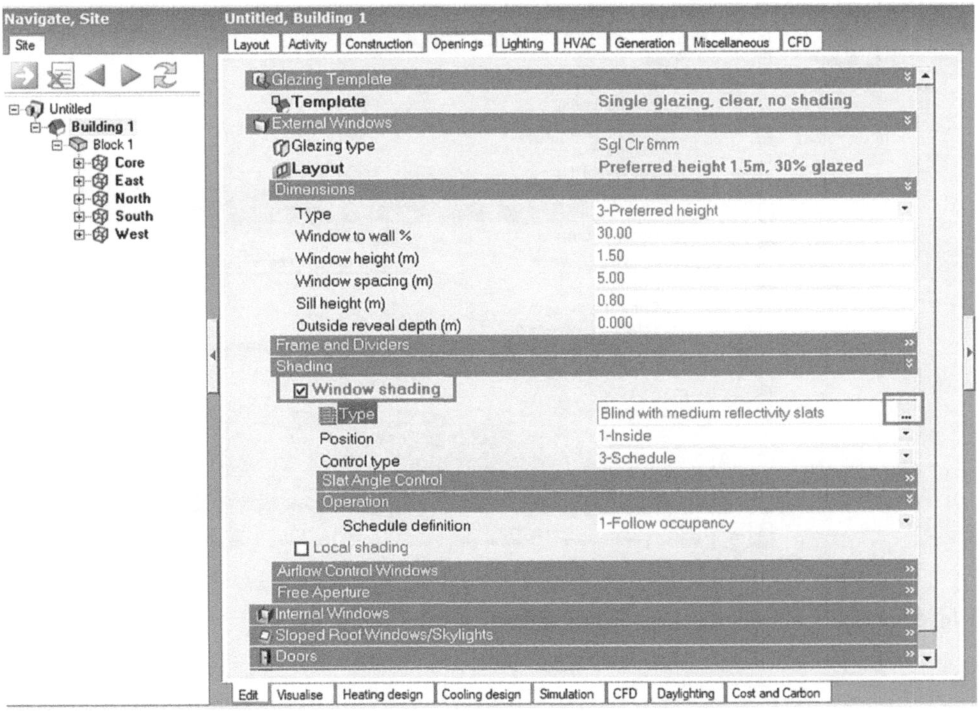

Step 12: Expand the **Electrochromic switchable** tree and select **Electrochromic reflective 6mm**.

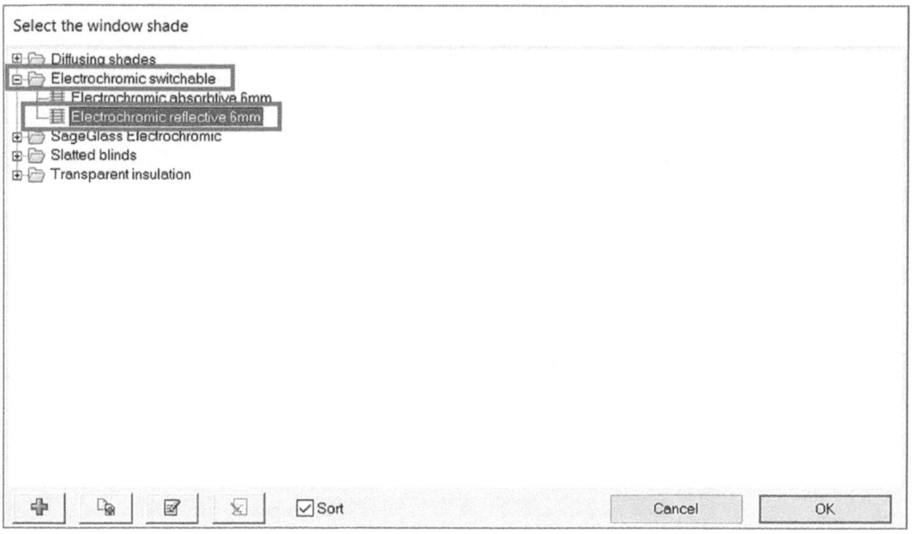

Step 13: Select **4-Switchable** from the **Position** drop-down list. Select **4-Solar** from the **Control type** drop-down list. Enter **120** in the **Solar setpoint(W/m²)** textbox.

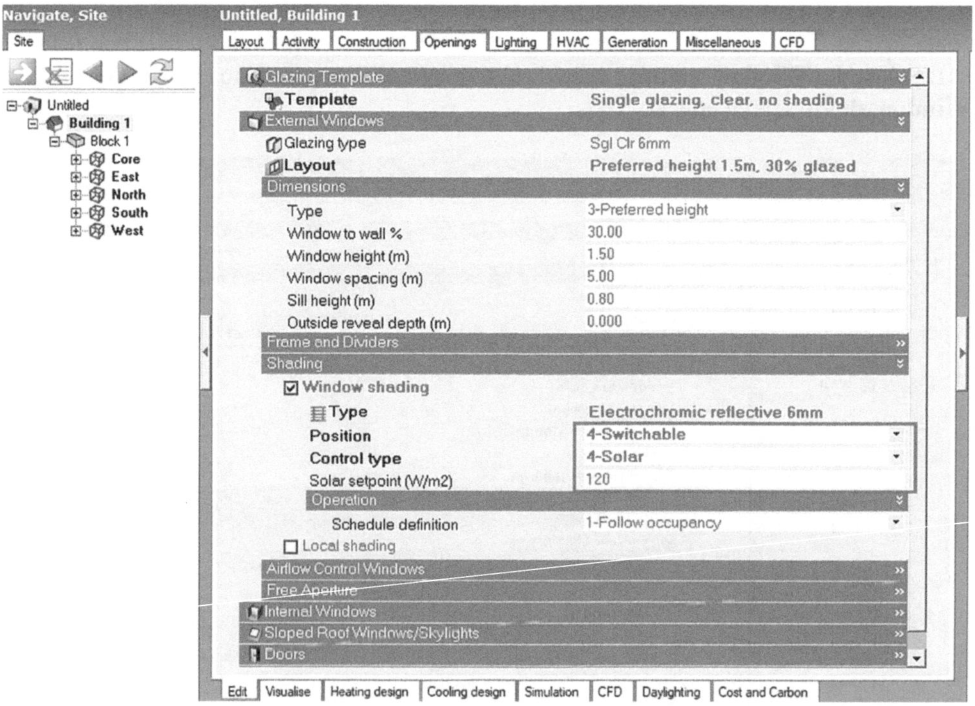

Step 14: Simulate the model and record the results for the south zone.

Step 15: Compare hourly results for **Exterior Windows Solar Gain** (plot for 29 and 30 July).

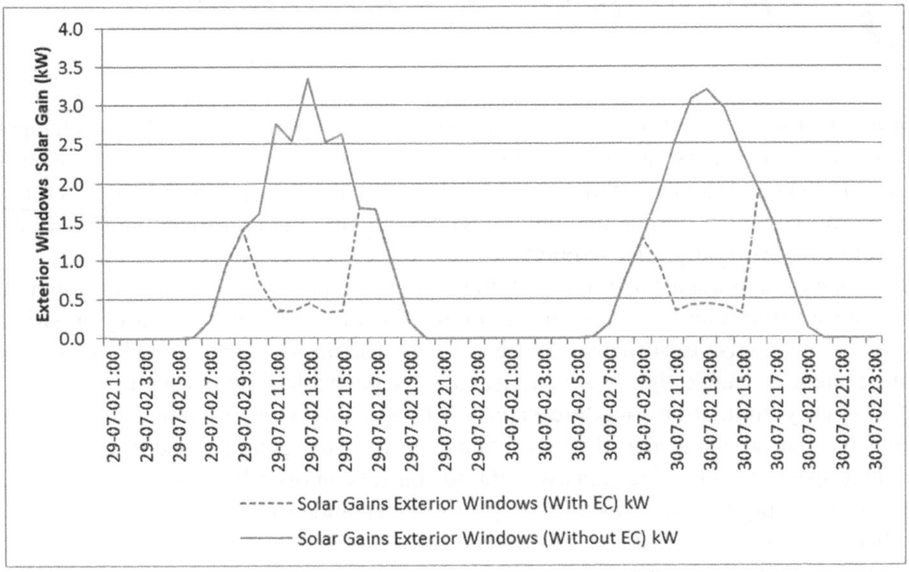

The results demonstrate the effectiveness of electrochromic switchable glazing in reducing solar heat gains. By automatically switching to a reflective state when incident solar radiation exceeded the threshold, the system effectively reduced the peak solar heat gains. This behavior is particularly advantageous during peak summer conditions, when cooling loads can be reduced by controlling excessive heat ingress.

Comparing the two scenarios, it is evident that **Electrochromic Reflective 6mm glazing** offers better performance in terms of solar heat gain reduction. This can contribute to enhanced thermal comfort, reduced cooling energy demand, and improved energy efficiency of the building.

Furthermore, the ability to dynamically adjust the glazing properties based on environmental conditions makes electrochromic glazing a versatile and energy-efficient solution for modern buildings.

The findings from this analysis support the use of electrochromic switchable glazing as an effective strategy for minimizing solar heat gains, particularly for south-facing façades in hot climates.

5 Lighting and Controls

Daylight availability in buildings varies continuously throughout the day due to changes in sun position, weather conditions, and obstructions. As a result, daylight alone is often insufficient to maintain consistent and targeted illuminance levels within interior spaces. To address this variability, photosensors can be integrated into lighting systems to monitor real-time illuminance and detect deficits relative to desired lighting setpoints.

These sensors can modulate artificial lighting output to supplement the available daylight, ensuring that required illumination levels are maintained at all times. By dimming or switching off electric lights in response to sufficient daylight, these systems—commonly referred to as daylight harvesting systems—enable significant energy savings, particularly in perimeter zones.

Modern energy simulation tools are capable of modeling such lighting control strategies by accounting for sensor placement, daylight contribution, and lighting schedules. These tools can evaluate the impact of daylight-responsive lighting controls on overall energy consumption, helping designers optimize lighting systems for both performance and efficiency.

This chapter includes four tutorials. The first two tutorials demonstrate how daylight sensors can be modeled and how their integration with artificial lighting systems can reduce energy use. The third and fourth tutorials introduce the process of conducting daylight simulations, which provide spatial and temporal analysis of daylight distribution within a space. Together, these exercises provide a practical framework for understanding and applying daylight-integrated lighting strategies in building design.

TUTORIAL 5.1: EVALUATING THE IMPACT OF DAYLIGHTING-BASED CONTROLS

Goal

To evaluate the effect of daylighting-based controls on energy consumption

What Are You Going to Learn?

- How to model daylight controls

Problem Statement

In this tutorial, you are going to use a 50 m × 25 m five-zone model with a 5 m perimeter depth. You are going to evaluate the following lighting controls.

1. No lighting control
2. Linear control

Use the **WIEN/HOHE WARTE, Austria** location.

Solution

Step 1: Open a new project and create a **50 m × 25 m** building with a **5 m** perimeter depth.

Step 2: Select the **Openings** tab and select **Dbl Grey 6mm/6mm Air** as the **Glazing type** from the drop-down list.

DOI: 10.1201/9781003563013-5

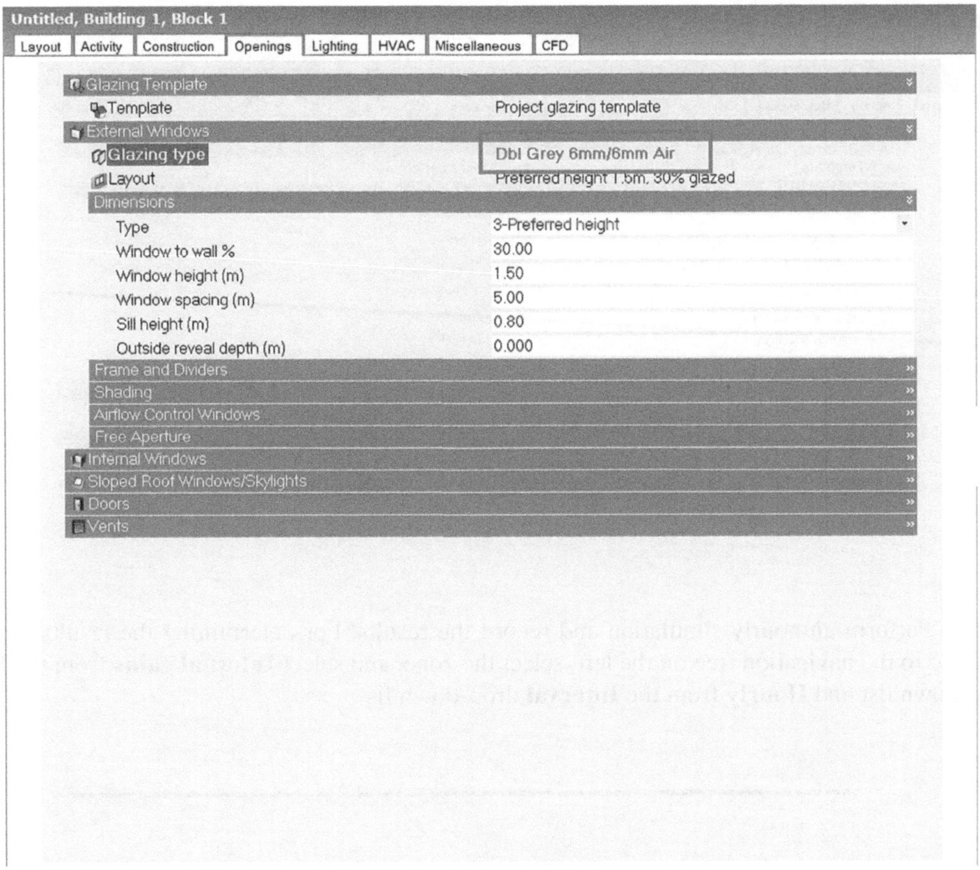

Step 3: Select the **Activity** tab. Select the **24×7 Generic Office Area** template from the **Miscellaneous 24hr activities** folder.

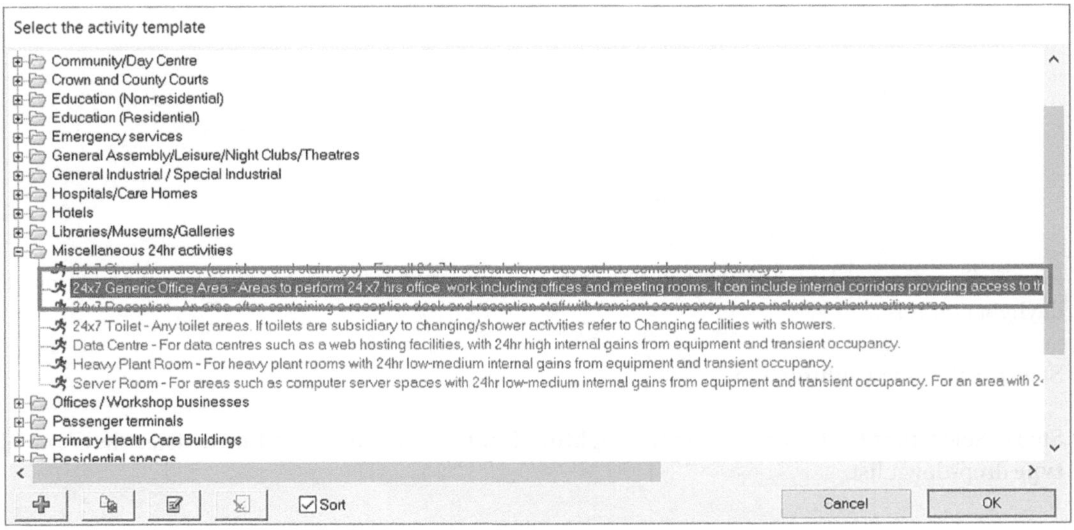

Step 4: Select the **Lighting** tab. Make sure to clear the **On** checkbox under **Lighting Control**.

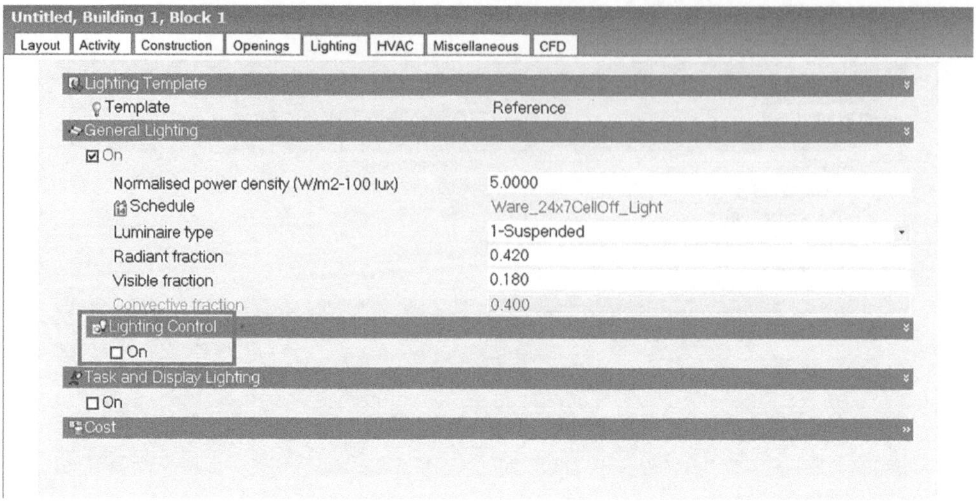

Step 5: Perform an hourly simulation and record the results. For determining the results of each zone, go to the navigation tree on the left, select the zone, and select **Internal gains** from the **Data** drop-down list and **Hourly** from the **Interval** drop-down list.

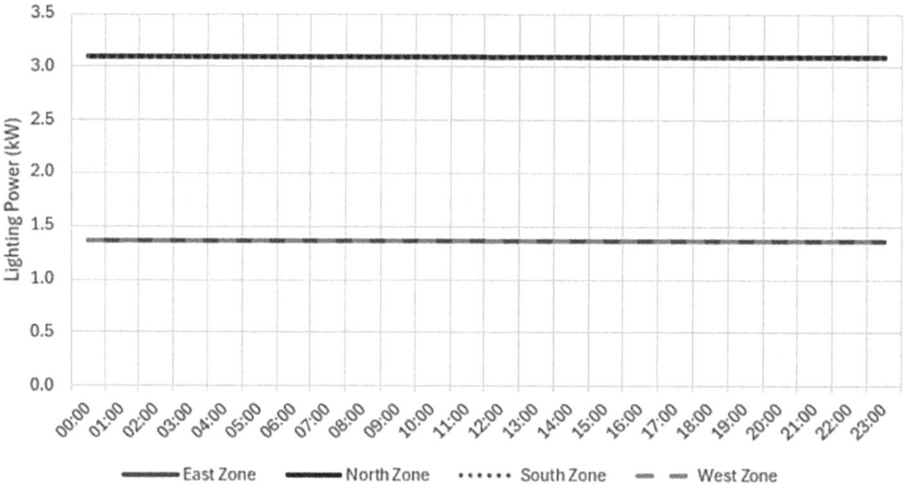

Step 6: Save the model using the **Save as** option. In the following steps, you are going to install daylight controls.

Step 7: Select the **Lighting** tab.

Step 8: Select the **On** checkbox under the **Lighting Control** section. Select **Linear** from the **Control type** drop-down list.

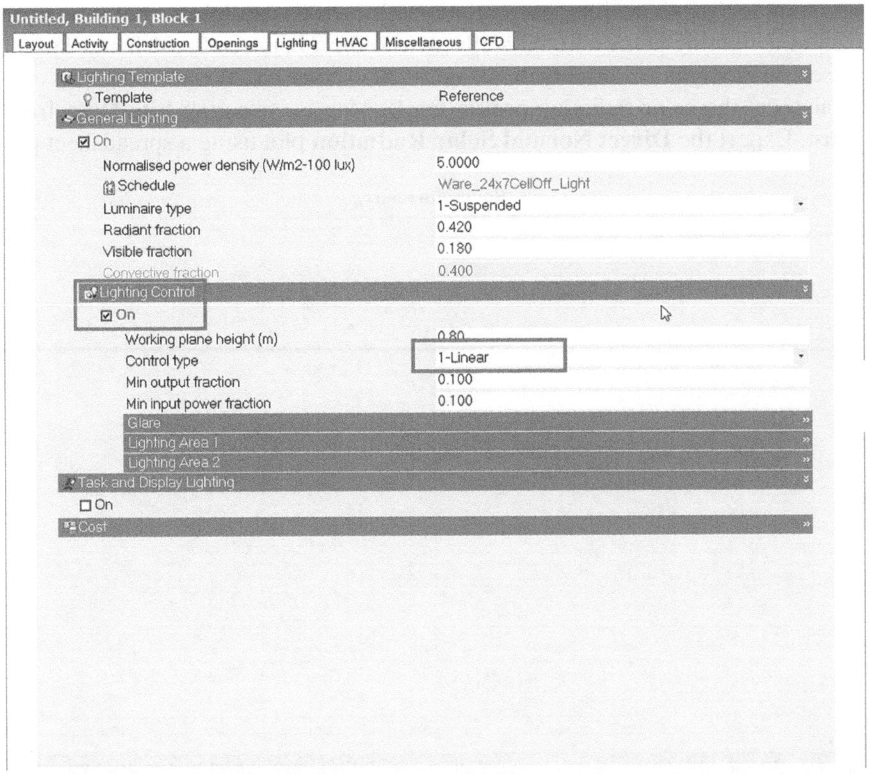

Based on the control type, lighting controls dim or turn off the internal lighting when the assigned illuminance level is met. This reduces the lighting energy consumption, as well as the internal heat gain due to artificial lighting. This reduction in heat gain decreases the cooling load and hence the cooling energy consumption.

Step 9: Select the **Hourly Simulation** tab and record the results. To determine the results of each zone, go to the navigation tree on the left, select the zone, and select **Internal gains** from the **Data** drop-down list and **Hourly** from the **Interval** drop-down list.

Step 10: Compare the hourly lighting energy consumption for all perimeter zones for 18 May. You can use a spreadsheet program to plot the comparative graphs.

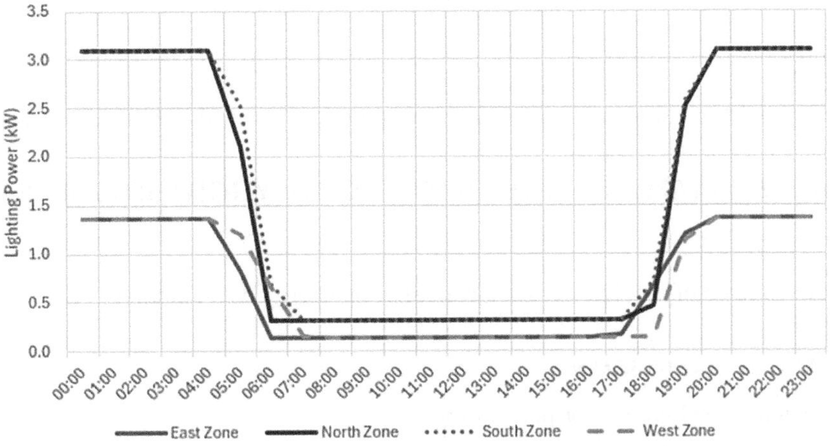

From this graph, it is clear that with the installation of controls, the lighting power consumption has decreased with the lighting being automatically switched off between 06:00 h and 18:00 h in different zones based on daylight availability.

You can also get the solar radiation profile for 18 May by selecting **Site Data** from the **Data** drop-down list. Export the **Direct Normal Solar Radiation** plot using a spreadsheet program.

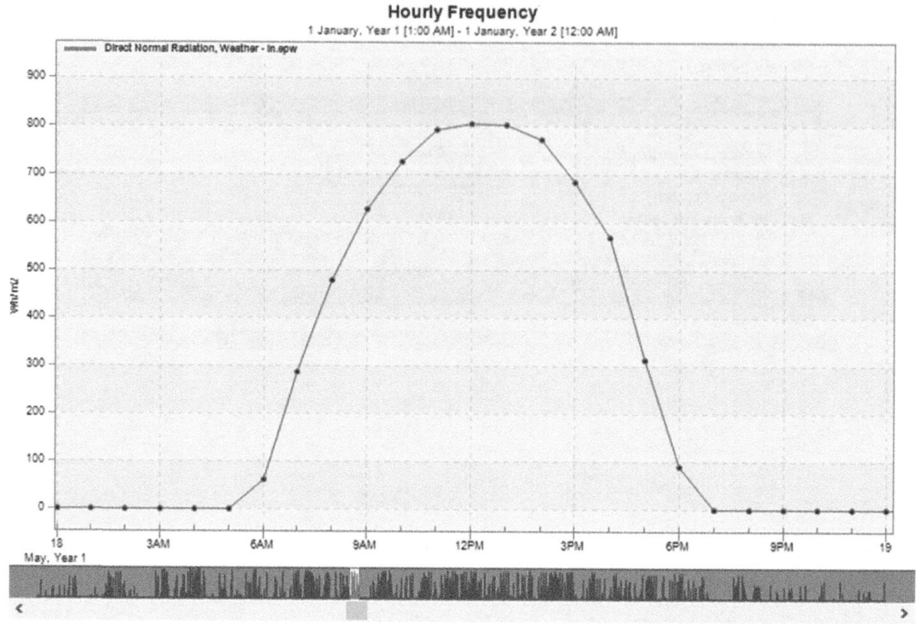

The date of 18 May is selected for the simulation. The solar radiation curve is smooth, with no perturbations.

Results and Analysis

In this study, two lighting control strategies were evaluated: no lighting control and linear daylight-based control. In the no-control scenario, artificial lighting operates continuously during occupied hours, regardless of available daylight. This leads to the highest lighting energy consumption and significantly increases internal heat gains, which in turn raises the cooling load. In contrast, the linear control strategy automatically dims or switches off artificial lighting based on daylight availability. When daylight levels are sufficient, the system reduces or eliminates electric lighting. This results in a marked reduction in lighting energy use and lowers internal heat gains, leading to a corresponding decrease in cooling energy consumption.

EXERCISE 5.1

Repeat this tutorial and compare the energy consumption for stepped lighting controls.

TUTORIAL 5.2: EVALUATING THE IMPACT OF DAYLIGHT SENSOR PLACEMENT

Goal

To evaluate the impact of daylight sensor positioning on energy consumption

What Are You Going to Learn?

- How to define daylight sensor positioning

Problem Statement

In this tutorial, you are going to use a 10 m × 10 m single-zone model with a single window located on the south façade. The window area is 40% of the gross south façade area. Determine the change in internal lighting gains within the zone, with and without the use of a daylight sensor employing **Linear/Off** control. Simulations will be conducted by placing daylight sensor(s) at the following locations:

- One sensor at 2 m from the window
- One sensor at 8 m from the window
- Two sensors at 2 m and 8 m from the window

Use weather file for **New Delhi/Palam, India** location.

Solution

Step 1: Open a new project and create a **10 m × 10 m** single-zone model. Press the **ESC** button to get out of **Edit mode**.

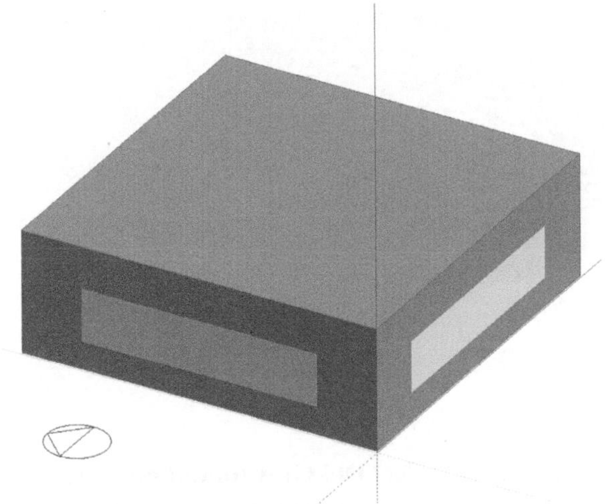

Step 2: Click on the west window; this highlights the window.

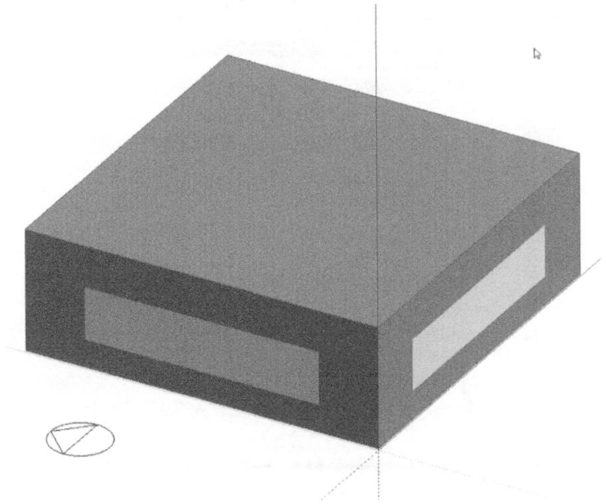

Step 3: Click **Delete selected object(s) (Del)**, then click **Yes** on the message box. The window disappears. Similarly, you can delete windows on the north and east façades with the help of **Dynamic orbit**. The south window remains.

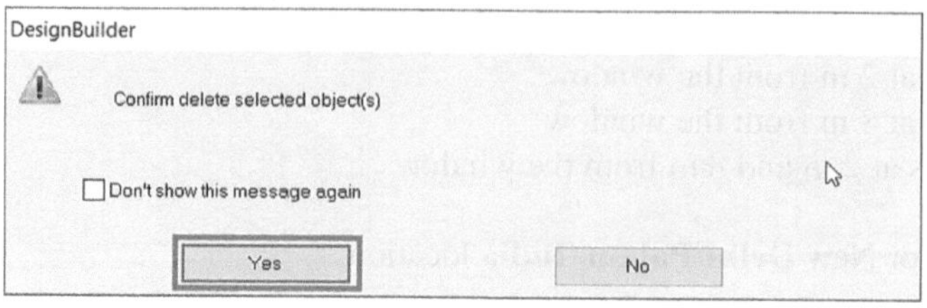

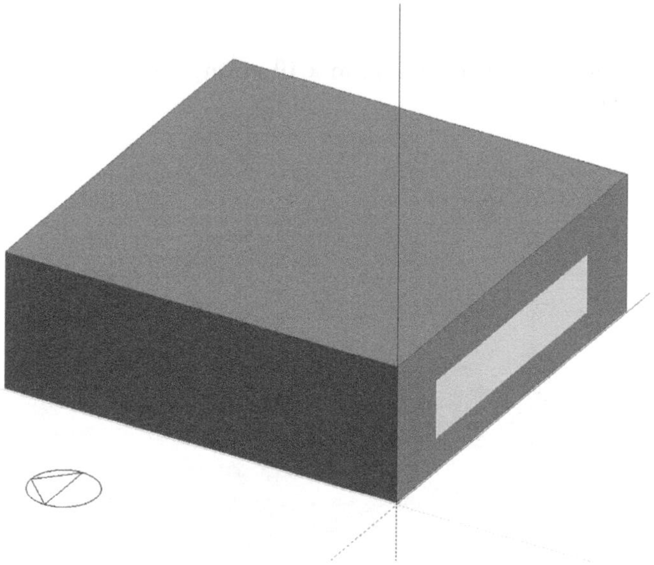

Step 4: Select the **Openings** tab and select **Dbl Grey 6mm/6mm Air** from the glazing type. Enter **40.00** in the **Window to wall %** text box. You can cross-check the window area on the south wall in **Navigation** area.

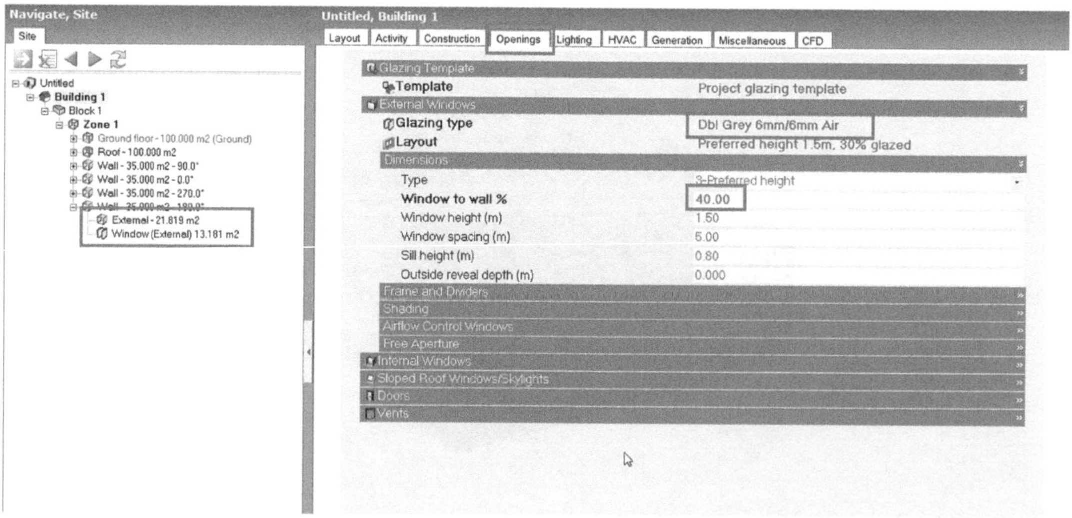

Step 5: Select the **Activity** tab. Select the **24 × 7 Generic Office Area** template from the **Miscellaneous 24hr activities** branch. Select the **Construction** tab. Select **Simple** from the **Geometry convention template**.

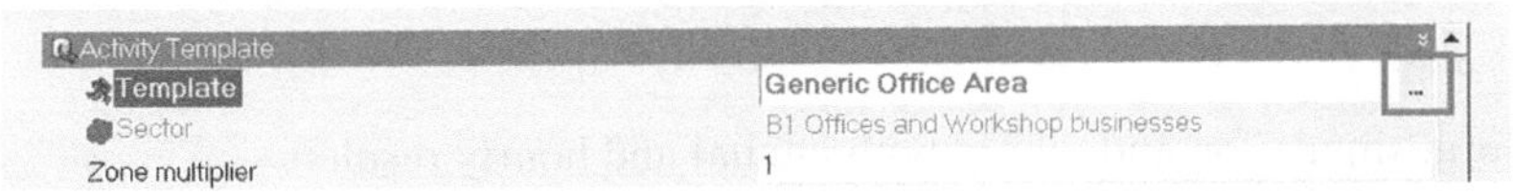

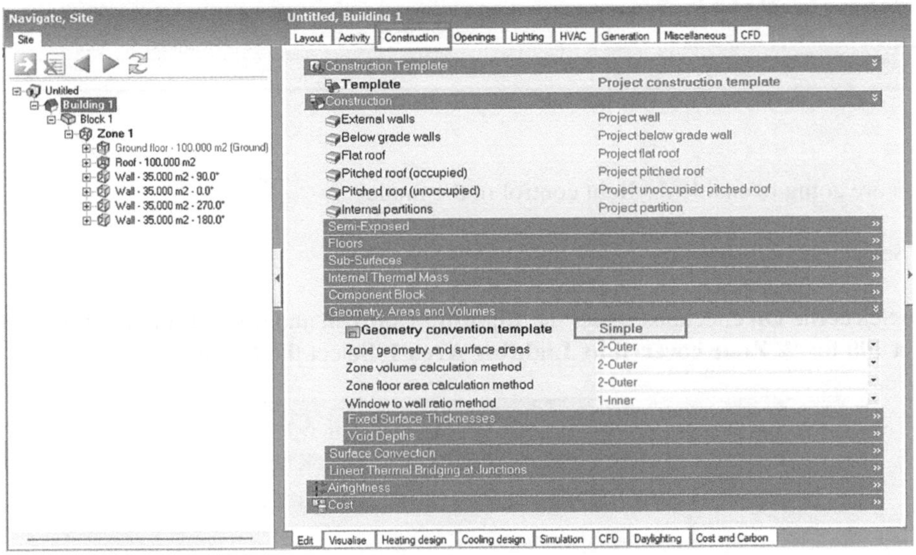

Step 6: Select the **Lighting** tab. Set the **Normalised power density (W/m^2-100 lux)** to **3**.

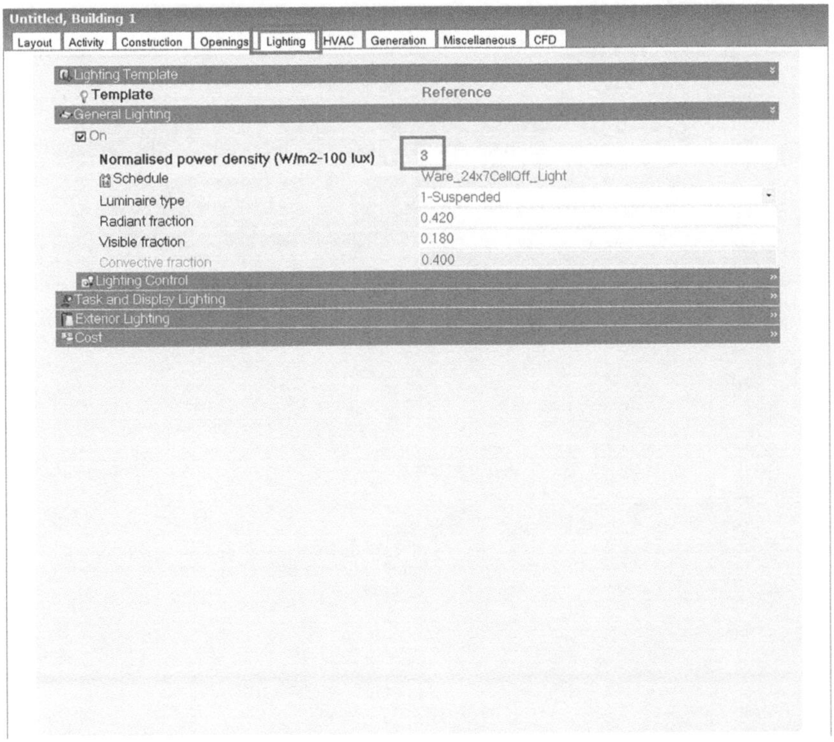

Step 7: Make sure to clear the **On** checkbox under the **Lighting Control** section.

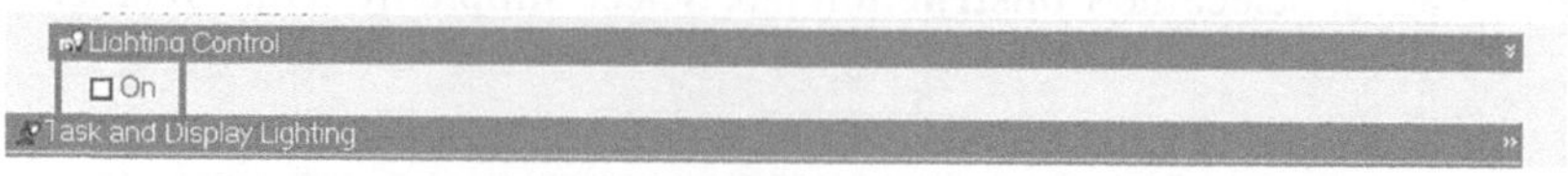

Step 8: Perform simulation and record both annual and hourly results.

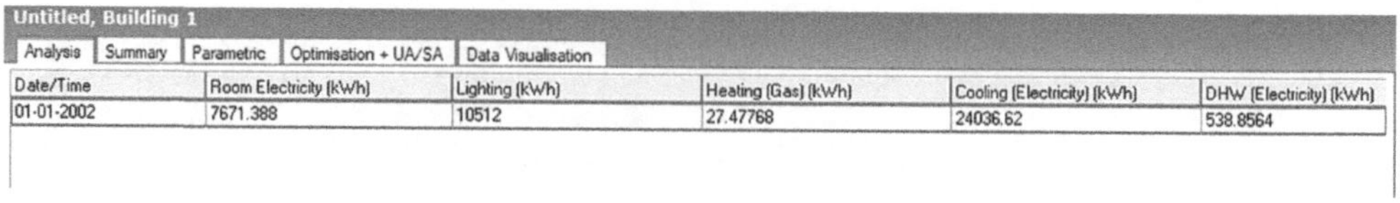

Now you are going to enable daylight control in the model.

Step 9: Select the **Lighting** tab.

Step 10: Select the **On** checkbox under the **Lighting Control**, and select **Linear/off** in the **Control type**. Set **100** for **% Zone covered by Lighting Area 1**. Select the **Layout** tab.

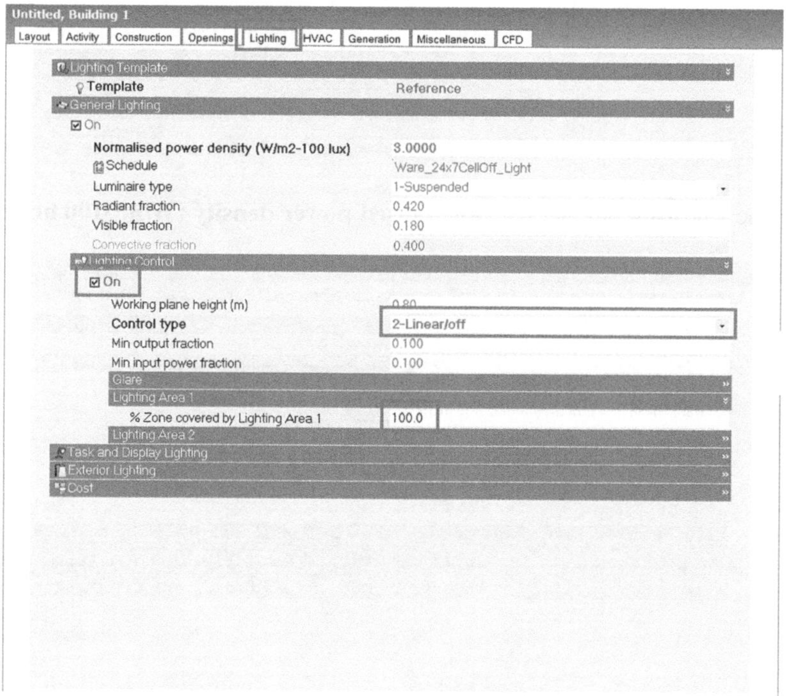

Step 11: Select the **Activity** tab, and make sure that the target illuminance (lux) is **400**.

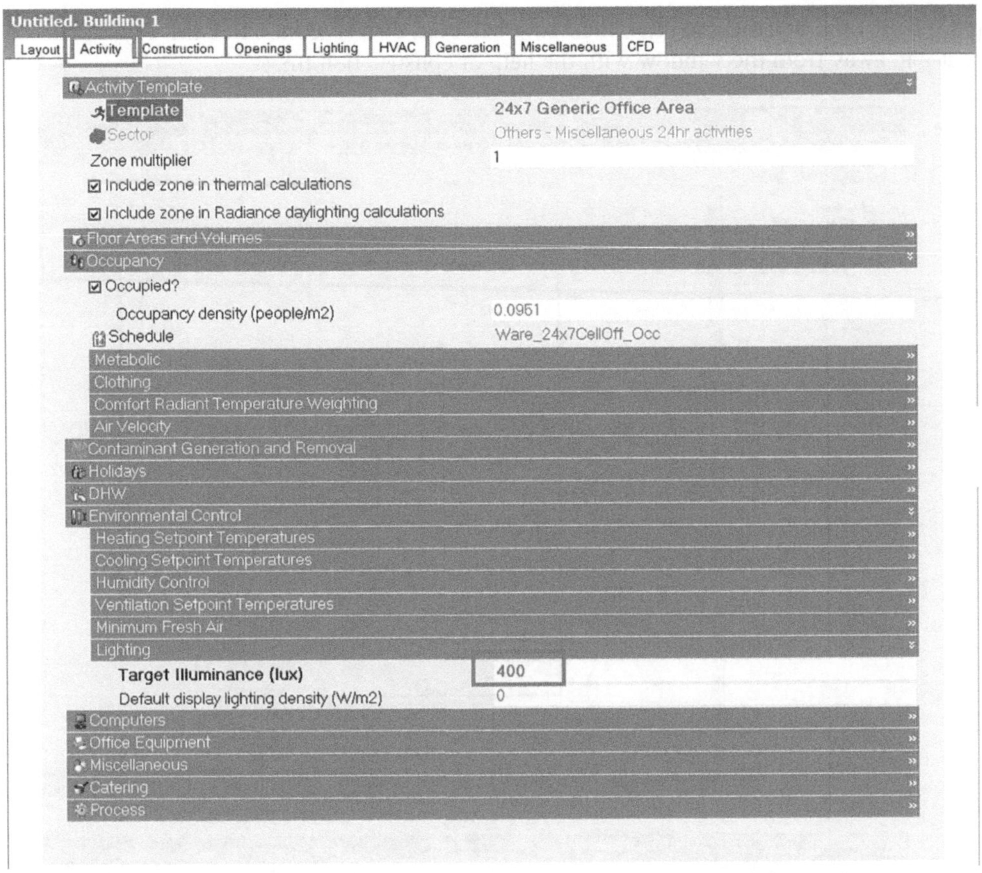

Step 12: Select **Zone 1** from the navigation tree. The daylight sensor can be seen in the layout.

Step 13: Select the **Sensor** and click the **Move selected object** icon. Click the daylight sensor and place it 2 m away from the window with the help of construction lines.

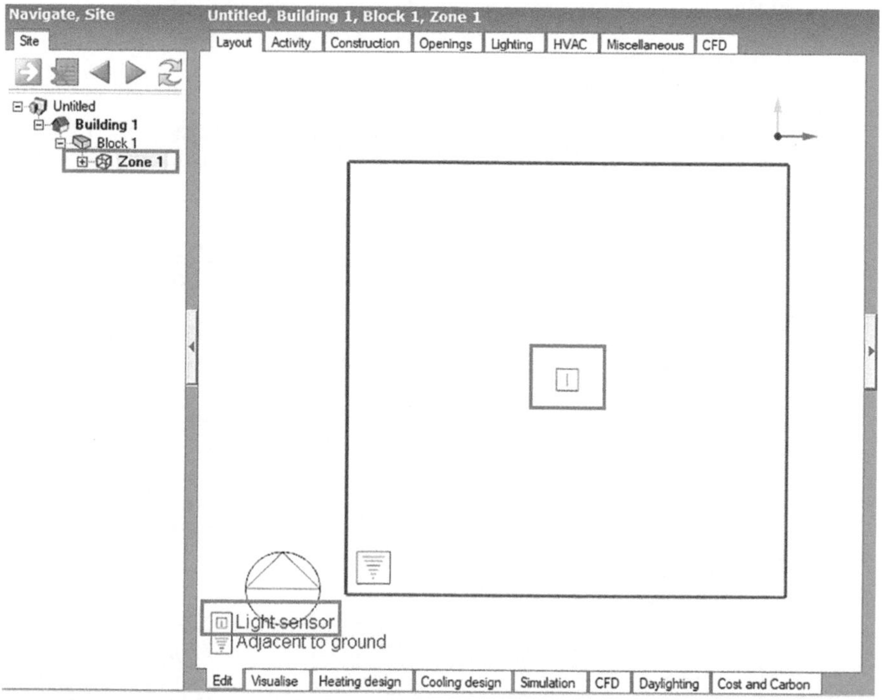

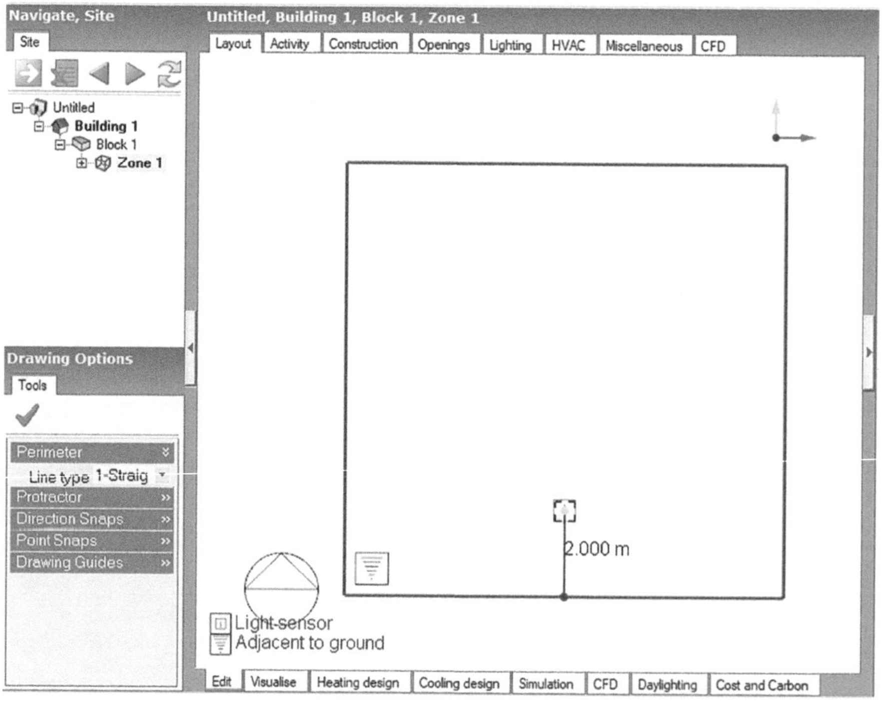

Step 14: Select the **Simulation** screen tab. The **Edit Calculation Options** screen appears. Select the **Output** tab and click the **Detailed Daylight Outputs** section. Select the **Daylight map output** checkbox. Click **OK**. Perform both annual and hourly simulations.

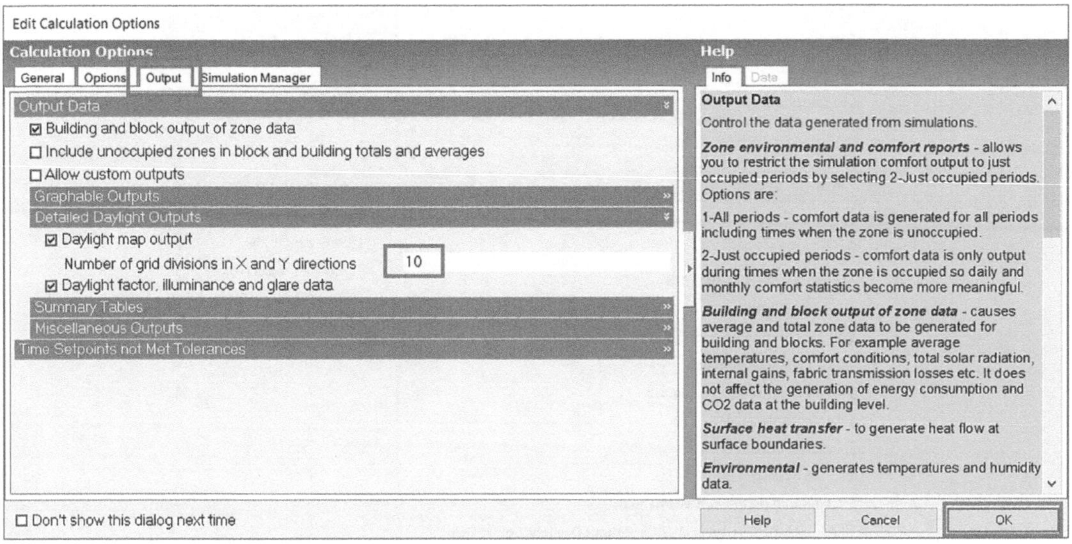

Step 15: Record the results. Plot hourly lighting data for 1 April.

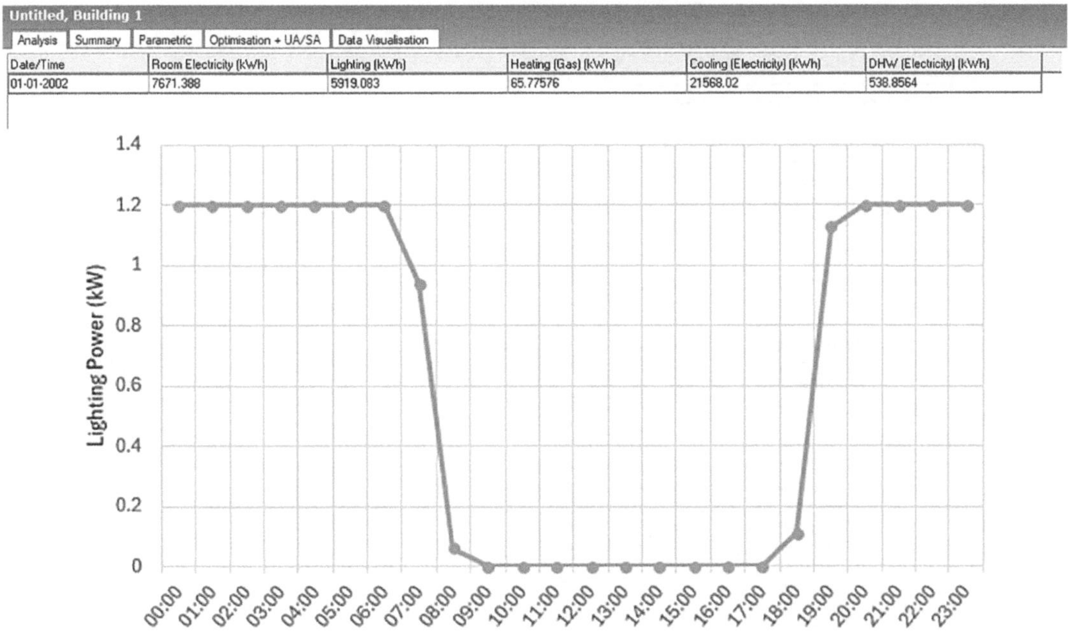

Date/Time	Room Electricity (kWh)	Lighting (kWh)	Heating (Gas) (kWh)	Cooling (Electricity) (kWh)	DHW (Electricity) (kWh)
01-01-2002	7671.388	5919.083	65.77576	21568.02	538.8564

In the preceding graph, you can see that the lighting consumption is zero between 09:00 h and 17:00 h. To get the lux level at this time, you need to get the illuminance map.

Step 16: Open the **eplusmap** file that exists in the **EnergyPlus** folder.

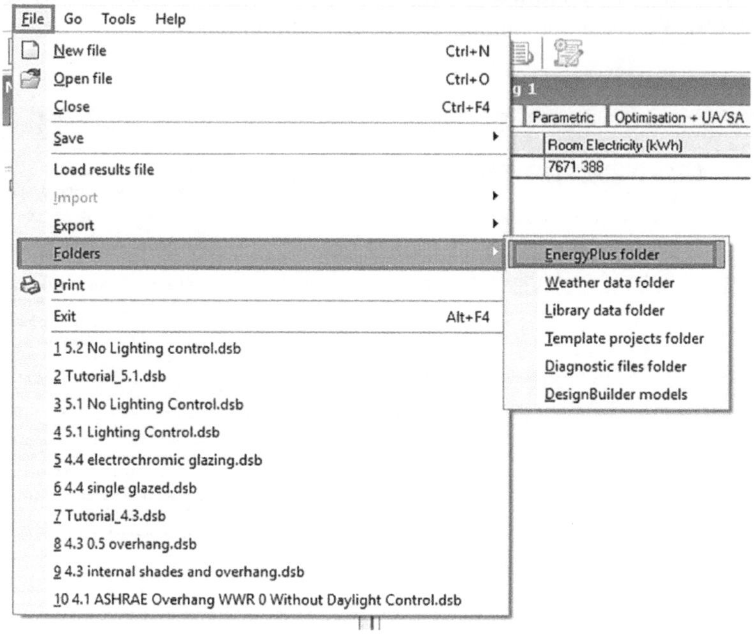

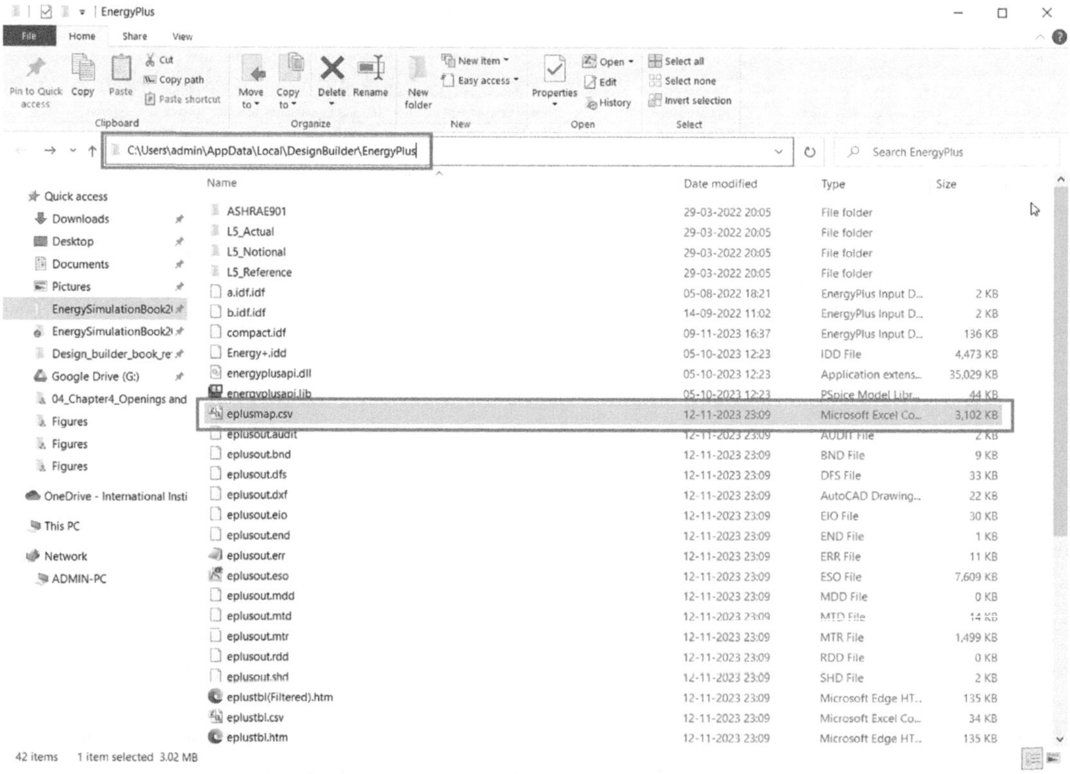

Step 17: Get the data for 11:00 h. Table 5.1 shows the illuminance map when the sensor is placed near the window. In the map, each cell reports the illuminance (in lux) at the location specified by the (X; Y) coordinates in the column and row headers. These are XY pairs separated by a semicolon for ease in importing into the spreadsheet. In the **eplusmap** file, the Z coordinate of the map is shown in the title (the illuminance map is set in a plane), and the date and time are indicated in the upper-left cell of the map. You can see that at 11:00 h, all artificial lights are off because of sufficient illuminance near the sensor. This leads to low illuminance in the interiors of the space.

DAYLIGHT MAP OUTPUT

EnergyPlus generates a file called **eplusmap.csv** when one or more zones have daylight control and the **Daylight map** simulation output option has been selected for those zones. This file contains a series of grids for each zone with daylight controls and has the daylight map output requested and for each hour of the day in the simulation.

Source: https://designbuilder.co.uk/helpv7.3/Content/Daylighting.htm

TABLE 5.1
Illuminance Map for 1 April: Reference Point 1 (RefPt1) = (4.98:2.00:0.80)

01/04 11:00	(0.18; 0.21)=	(1.24; 0.21)=	(2.31; 0.21)=	(3.38; 0.21)=	(4.44; 0.21)=	(5.51; 0.21)=	(6.58; 0.21)=	(7.64; 0.21)=	(8.71; 0.21)=	(9.78; 0.21)=
(0.18; 0.21)=	833	17855	17879	18112	17780	18037	17910	17836	18120	643
(0.18; 1.28)=	17078	17444	17508	17519	17521	17521	17513	17481	13568	799
(0.18; 2.35)=	947	1087	1137	1153	1157	1153	1136	1079	911	699
(0.18; 3.41)=	763	825	856	869	871	865	846	800	721	639
(0.18; 4.48)=	676	707	725	733	734	728	711	682	643	604
(0.18; 5.55)=	630	647	658	662	662	656	645	627	605	584
(0.18; 6.61)=	604	614	621	624	623	619	611	599	586	573
(0.18; 7.68)=	586	592	596	598	597	593	588	580	572	564
(0.18; 8.75)=	574	578	580	581	580	577	574	569	563	558
(0.18; 9.81)=	566	568	569	570	569	567	564	561	557	554

You can also get the solar radiation profile on 1 April by selecting **Site Data** from the **Data** drop-down list. Export the **Direct Normal Solar Radiation** and plot using a spreadsheet program.

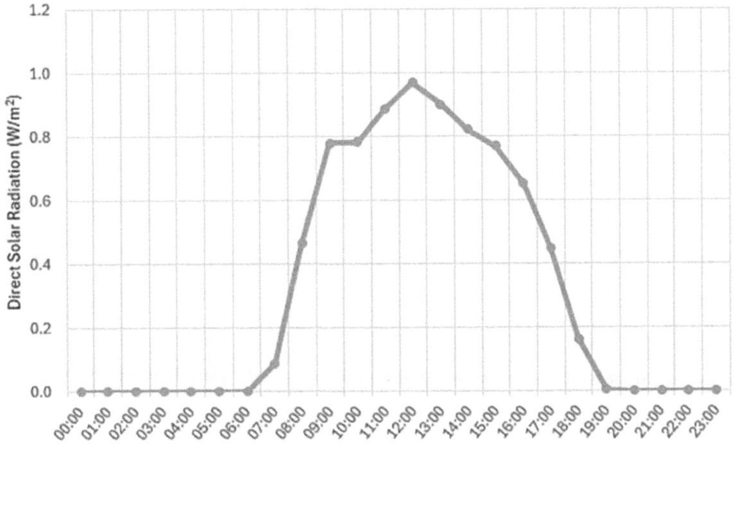

Step 18: Save the model using the **Save as** option. Now you are going to change the location of the daylight sensor to 8 m from the south façade window.

Step 19: Move the daylight sensor 8 m away from the window toward the north (as explained earlier).

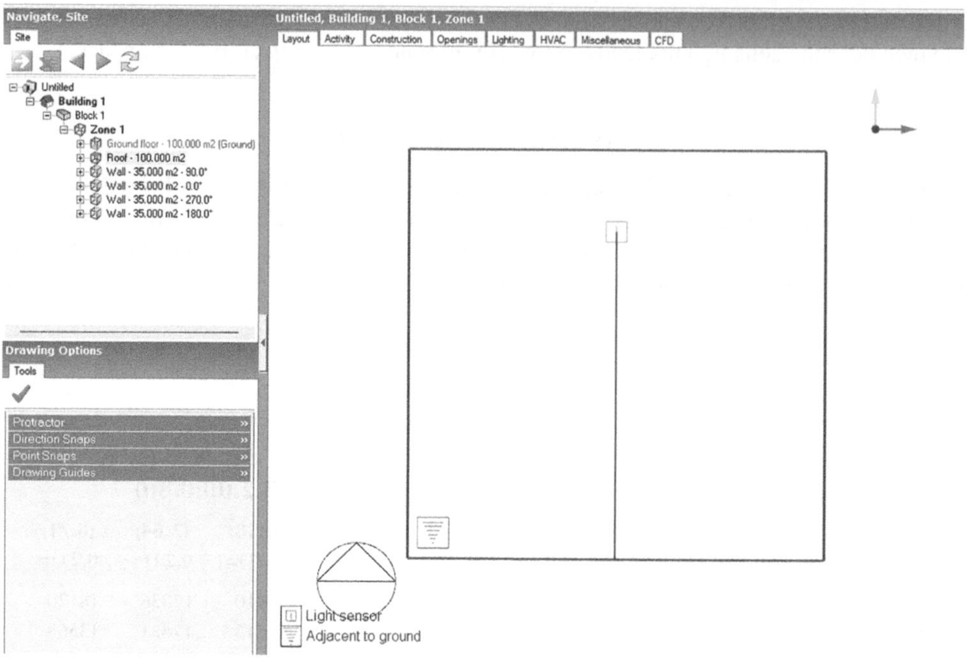

Step 20: Perform an hourly simulation. Record the annual results. Also record hourly lighting data for 1 April.

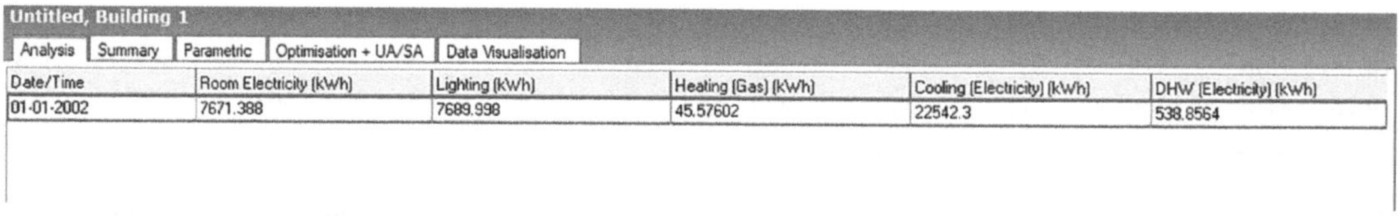

Date/Time	Room Electricity (kWh)	Lighting (kWh)	Heating (Gas) (kWh)	Cooling (Electricity) (kWh)	DHW (Electricity) (kWh)
01-01-2002	7671.388	7689.998	45.57602	22542.3	538.8564

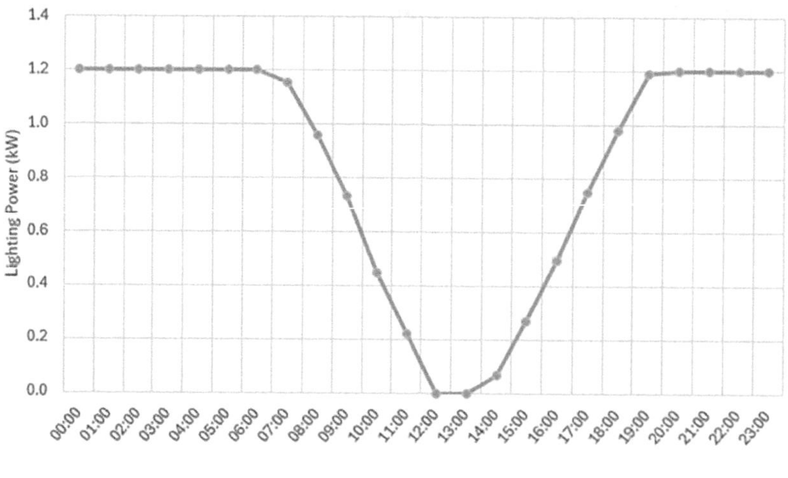

The preceding figure shows the hourly lighting load for 1 April (with the daylight sensor 8 m from the window). In this graph, you can see that lighting energy consumption between 08:00 h and 17:30 h is higher than when the sensor is placed near the window. When the sensor is moved away from the window, it is in a darker portion of the room and is triggered when more daylight enters the room. This increase in daylight might take some time, thereby delaying the time when the lights are switched off. This increases the lighting energy consumption but ensures that sufficient light is available, even in the interior areas of the room when the lights are switched off. However, this might lead to higher illuminance levels near the window, resulting in visual discomfort, so there is a need to place two sensors at different positions to get both energy savings and visual comfort.

Step 21: Save the model using the **Save as** option. Now you are going to install one more daylight sensor.

Step 22: Select the **Lighting** tab and select the **Second lighting area** checkbox. Enter **50.0** as the **% Zone covered by Lighting Area 1** and **% Zone covered by Lighting Area 2**. Make sure that the **Target Illuminance (lux)** is set to **400**.

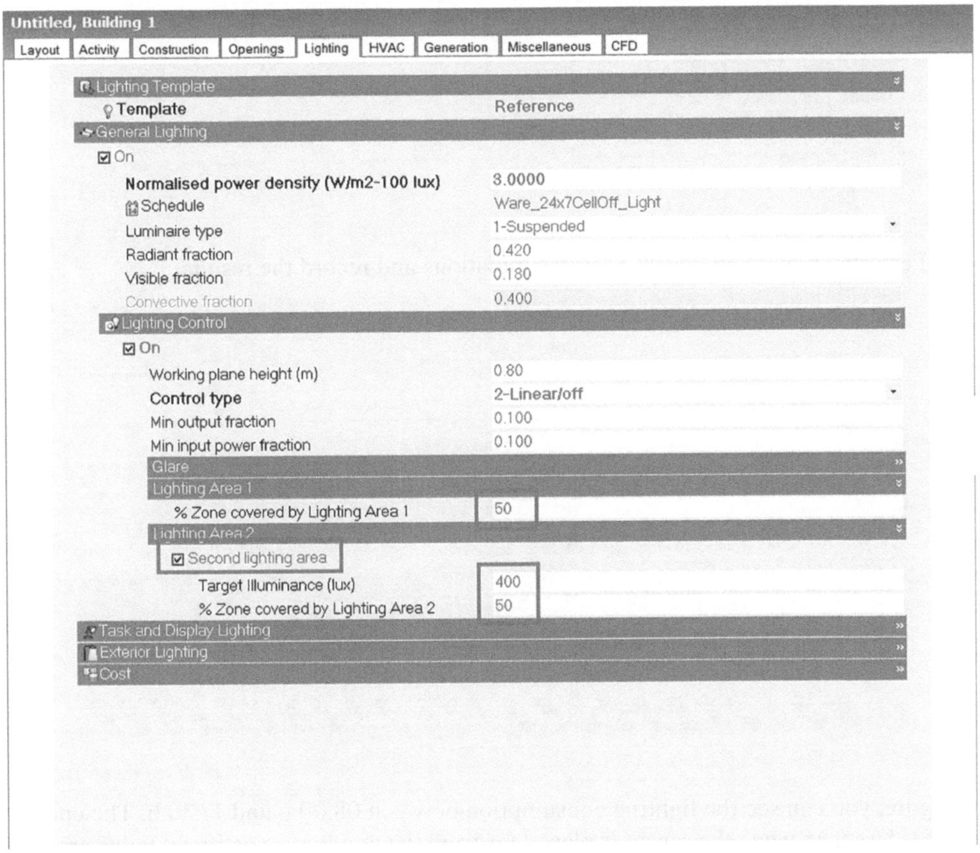

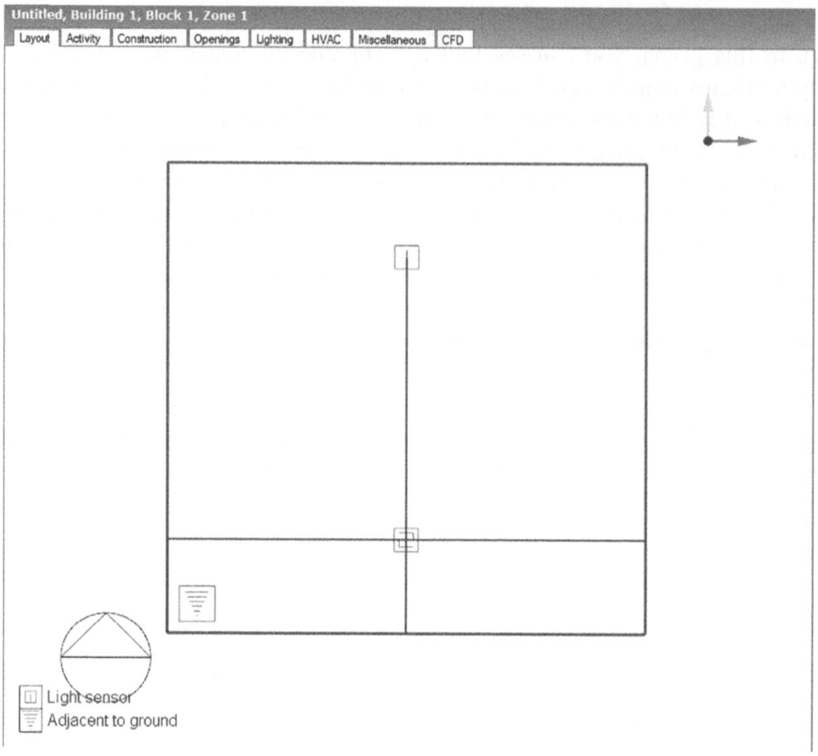

Step 23: Perform annual and hourly energy simulations and record the results.

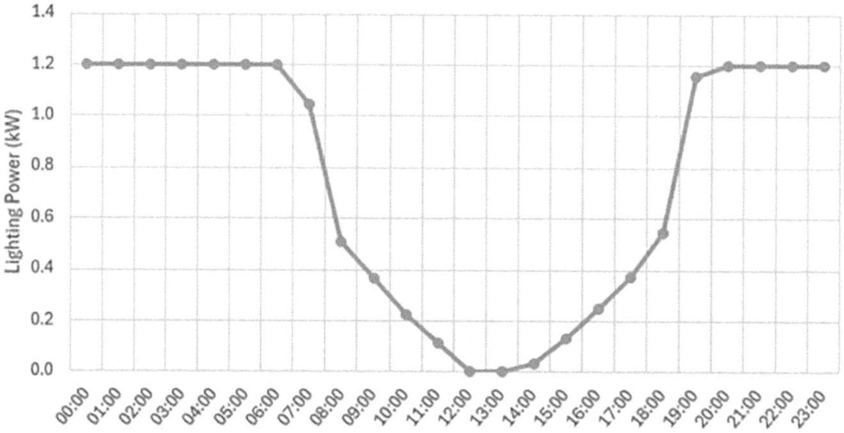

In this figure, you can see the lighting consumption between 08:00 h and 17:30 h. The energy consumption is less than when the sensor is placed far from the window, as artificial lights are switched off near the window when there is sufficient daylight. Get the data for 11:00 h for illuminance levels (Table 5.2). Now compare the annual energy consumption for all three cases (Table 5.3).

TABLE 5.2
Illuminance Map for 1 April: RefPt1 = (4.98:8.06:0.80), RefPt2 = (4.98:2.19:0.80)

01/04 11:00	(0.18; 0.21)=	(1.24; 0.21)=	(2.31; 0.21)=	(3.38; 0.21)=	(4.44; 0.21)=	(5.51; 0.21)=	(6.58; 0.21)=	(7.64; 0.21)=	(8.71; 0.21)=	(9.78; 0.21)=
(0.18;0.21)=	833	17855	17879	18112	17780	18037	17910	17836	18120	643
(0.18;1.28)=	17078	17444	17508	17519	17521	17521	17513	17481	13568	799
(0.18;2.35)=	947	1087	1137	1153	1157	1153	1136	1079	911	699
(0.18;3.41)=	763	825	856	869	871	865	846	800	721	639
(0.18;4.48)=	676	707	725	733	734	728	711	682	643	604
(0.18;5.55)=	630	647	658	662	662	656	645	627	605	584
(0.18;6.61)=	604	614	621	624	623	619	611	599	586	573
(0.18;7.68)=	586	592	596	598	597	593	588	580	572	564
(0.18;8.75)=	574	578	580	581	580	577	574	569	563	558
(0.18;9.81)=	566	568	569	570	569	567	564	561	557	554

TABLE 5.3
Annual Lighting Energy Consumption with Different Sensor Placements

Sensor Placement	Annual Lighting Energy Consumption (kWh)
No sensor	10512
Near the window	5919.09
Far from the window	7690
With two sensors	6804.55

The analysis shows that the placement and number of daylight sensors greatly influence lighting control effectiveness. Without any sensors, lights remain switched on throughout the day, leading to unnecessary energy use. A single sensor near the window quickly responds to daylight and reduces artificial lighting, but it may leave interior zones poorly lit. When the sensor is placed deeper into the space, it maintains better lighting distribution but delays energy-saving actions, leading to higher energy use. Using two sensors—one near the window and one further inside—creates zonal control that adjusts lighting more precisely based on daylight availability. This approach offers a practical balance between reducing energy consumption and maintaining consistent lighting levels across the space.

TUTORIAL 5.3: EVALUATING THE IMPACT OF WINDOW EXTERNAL SHADES AND WWR ON DAYLIGHT PERFORMANCE

Goal

To evaluate the impact of window external shades and WWR on daylight performance

What are You Going to Learn?

- How to perform simulations for daylight assessment

Problem Statement

In this tutorial, you will model a 10 m × 10 m single-zone building using the glazing template **Double glazing; clear no shading (Dbl Clr 6 mm/13 mm Air)**. The occupancy schedule will be set from

TABLE 5.4
Window-to-Wall ratio (WWR) and Shading Variations

Serial No.	Glass	WWR (%)	Overhang and Fins
1	Dbl Clr 6 mm/13 mm Air	30	No shade
2	Dbl Clr 6 mm/13 mm Air	30	Overhang (1 m) and fins (1 m)
3	Dbl Clr 6 mm/13 mm Air	25	Overhang (1 m) and fins (1 m)

09:00–17:00 h. You are required to simulate and analyze three scenarios as outlined in Table 5.4 to evaluate the impact of different daylight shading configurations.

Find the floor areas that meet or exceed the illuminance level between 100 and 2,000 lux for at least 90% of the potential daylit time in a year. Consider a work plane height of 0.75 m above the finished floor. Use the **New Delhi Palam, India** location.

Useful daylight illuminance (UDI) is defined as the annual occurrence of daylight between 100 and 2,000 lux on a work plane. The degree to which UDI is not achieved because illuminances exceed the upper limit is indicative of the potential for occupant discomfort.

Source: Nabil, A., and Mardaljevic, J. (2005). Useful daylight illuminance: a new paradigm for assessing daylight in buildings. *Lighting Research & Technology* 37(1):41–57. Available at https://doi.org/10.1191/1365782805li128oa

SOLUTION

Step 1: Open a new project and create a single-zone **10 m × 10 m** building. Click the **Activity** tab.

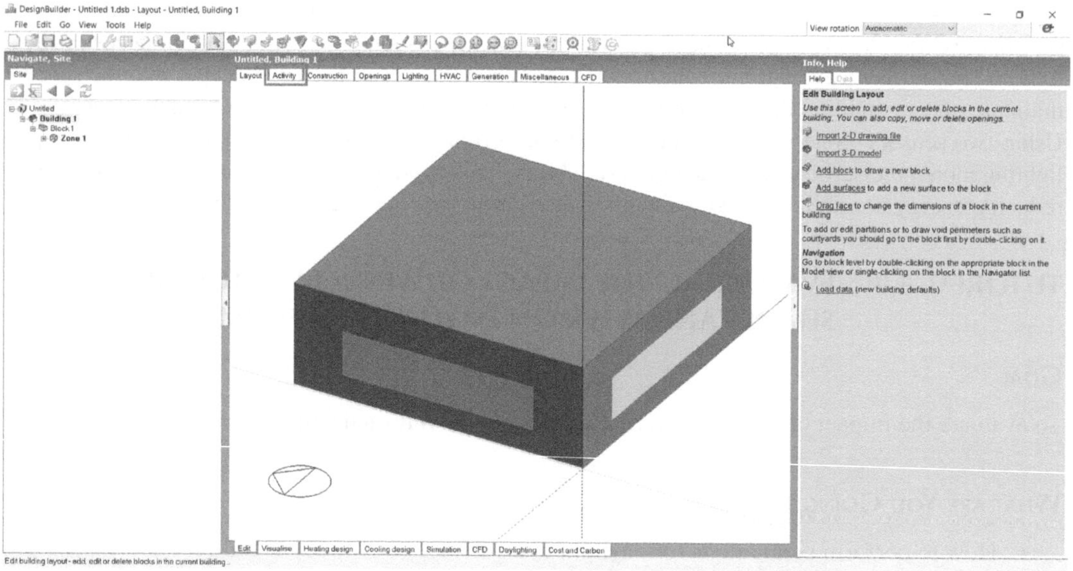

Step 2: Select **Schedule** under **the Occupancy** tab. Copy and edit the existing schedule.

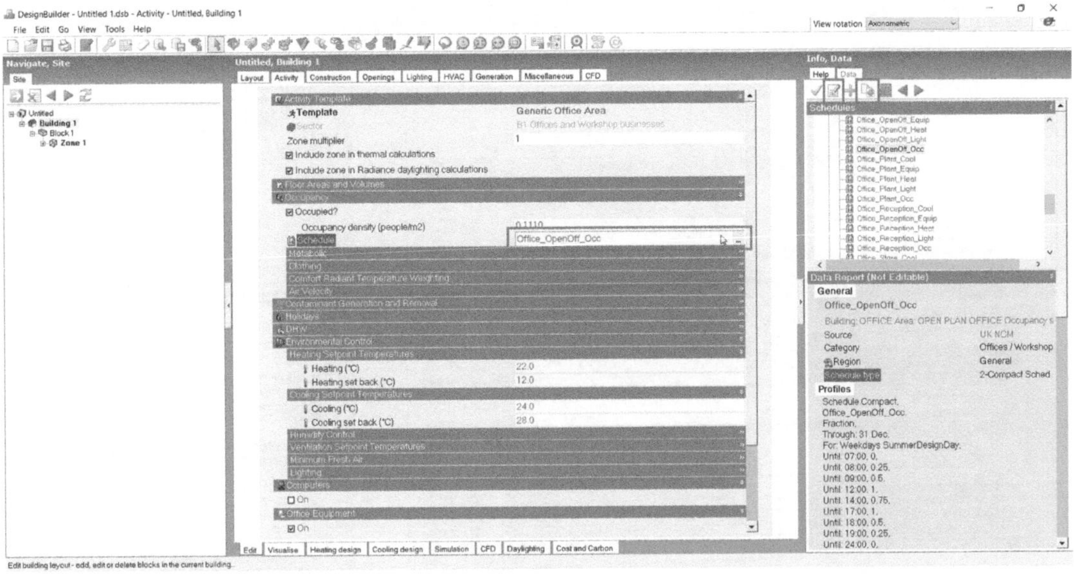

Step 3: Type **Office_OpenOff_Occ 9 AM to 5 PM** in the **Name** textbox. Set **1** from 12:00–17:00 h and the remaining to **0**.

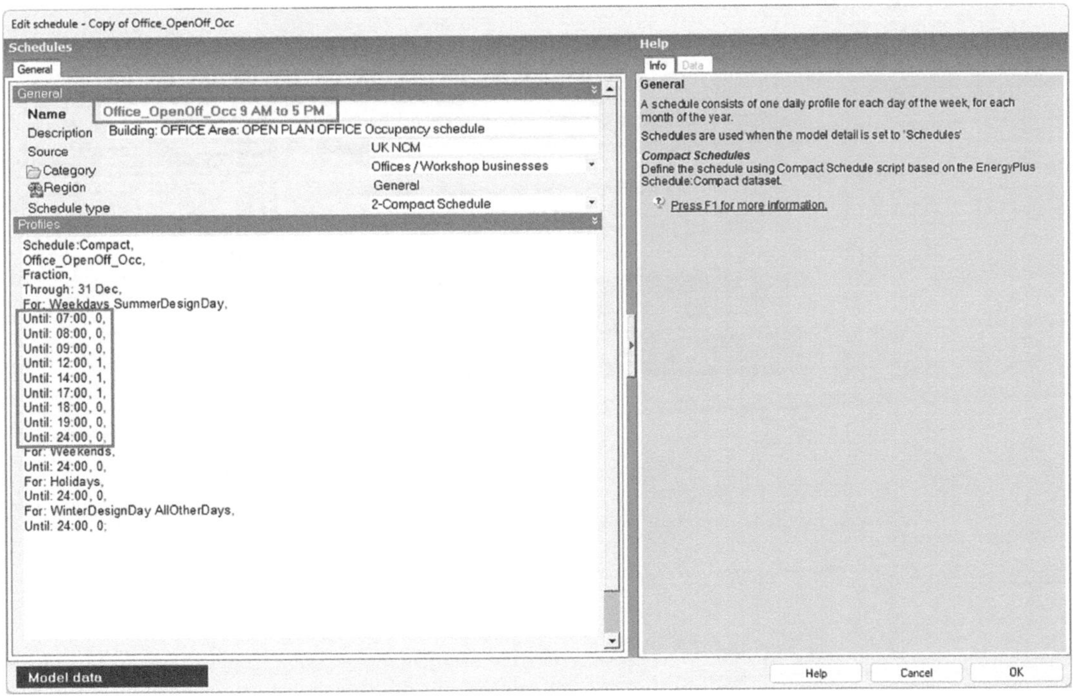

Step 4: Apply the **Office_OpenOff_Occ 9 AM to 5 PM** schedule.

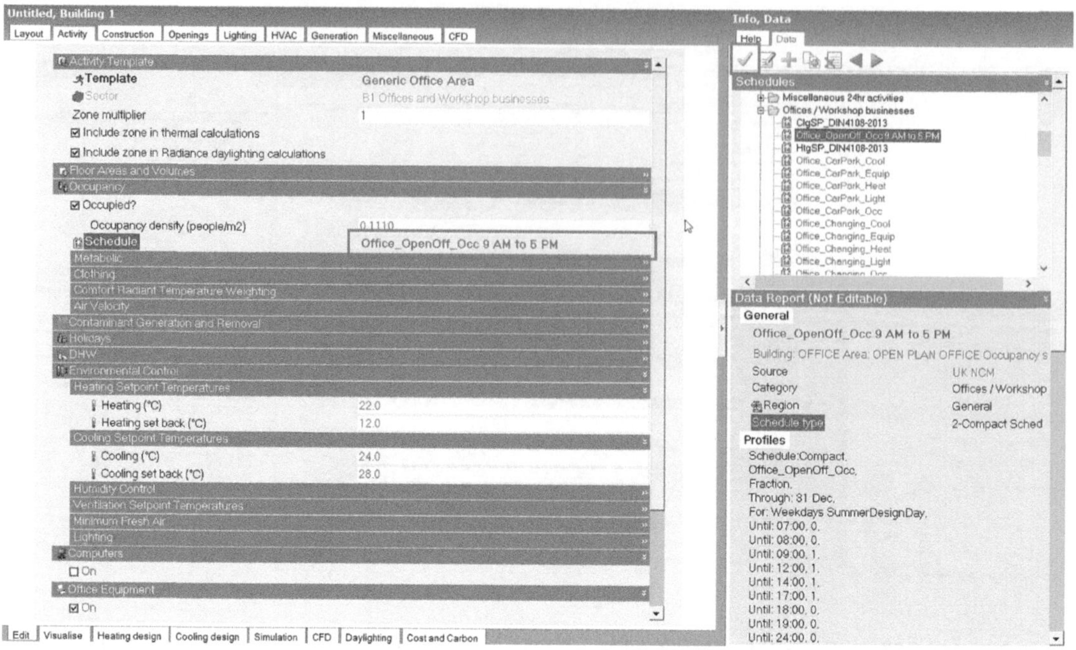

Step 5: Click **Openings** tab.

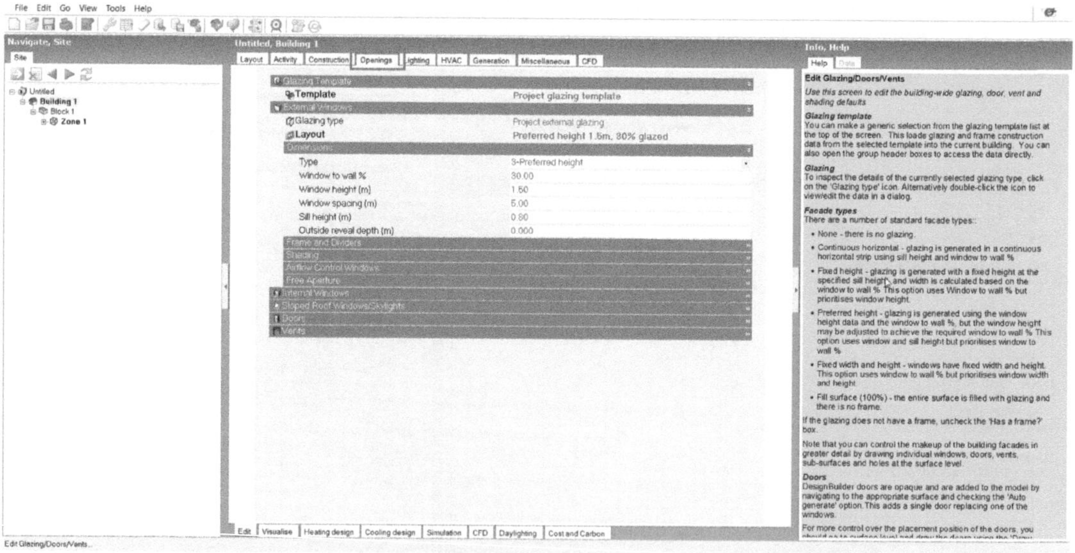

Step 6: Select the **Double glazing, clear, no shading** template.

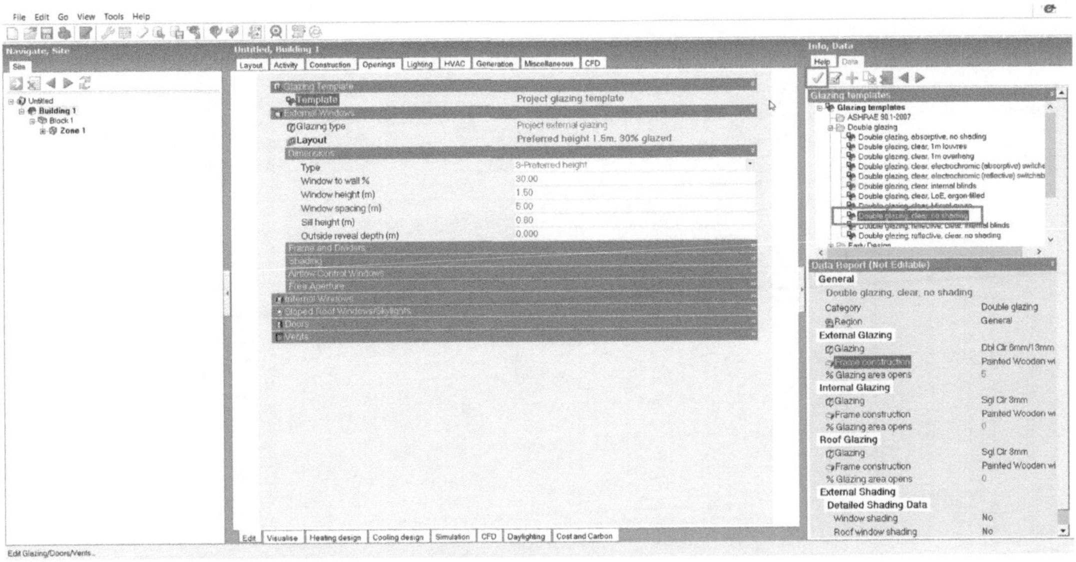

Step 7: Make sure that the **Glazing** type is **Dbl Clr 6mm/13mm Air** and that the WWR is set to **30%**.

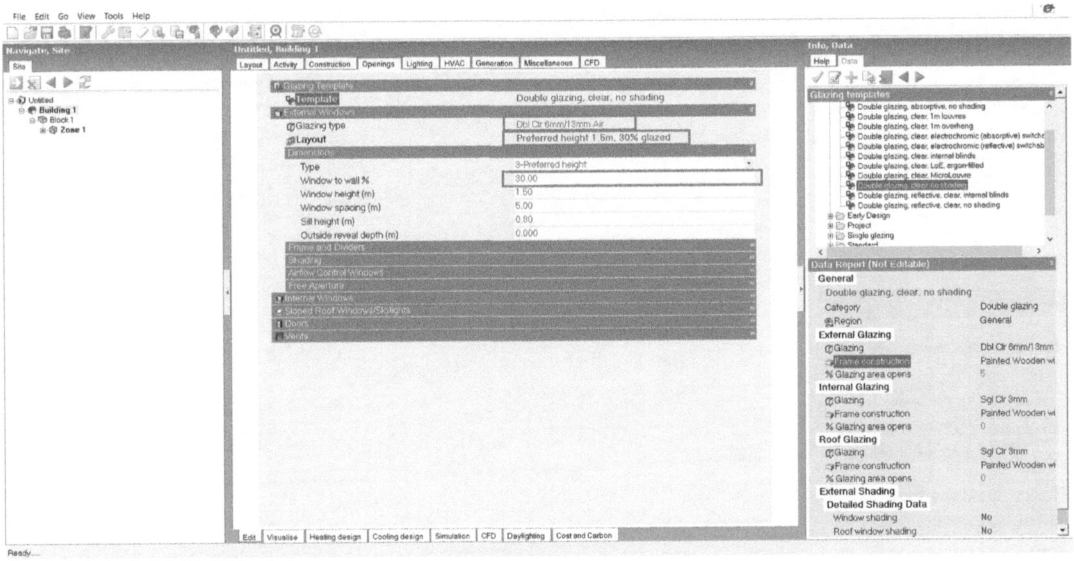

Step 8: Click the **Daylighting** screen tab.

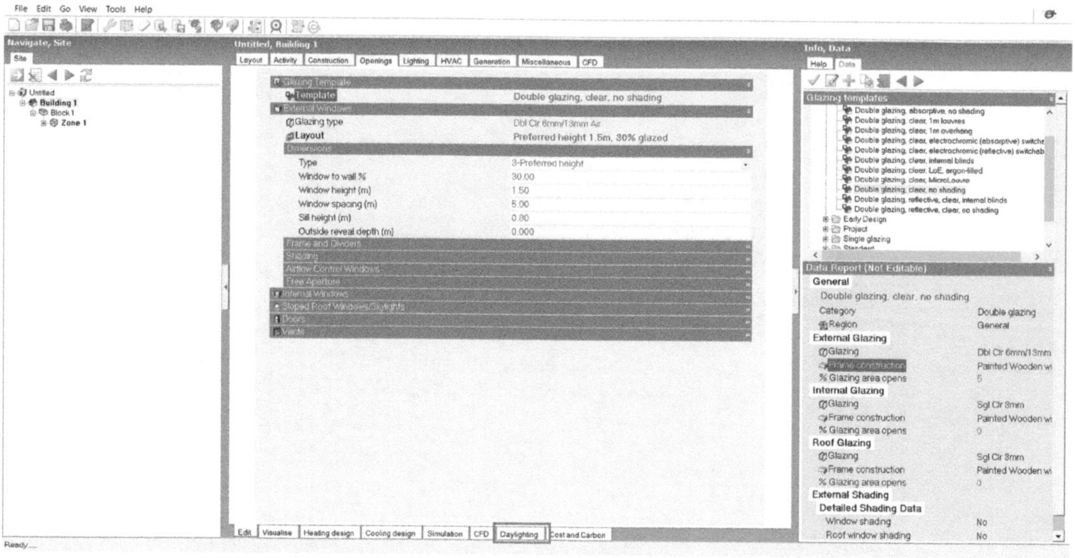

Step 9: Select **2-Annual** from **Option** drop-down.

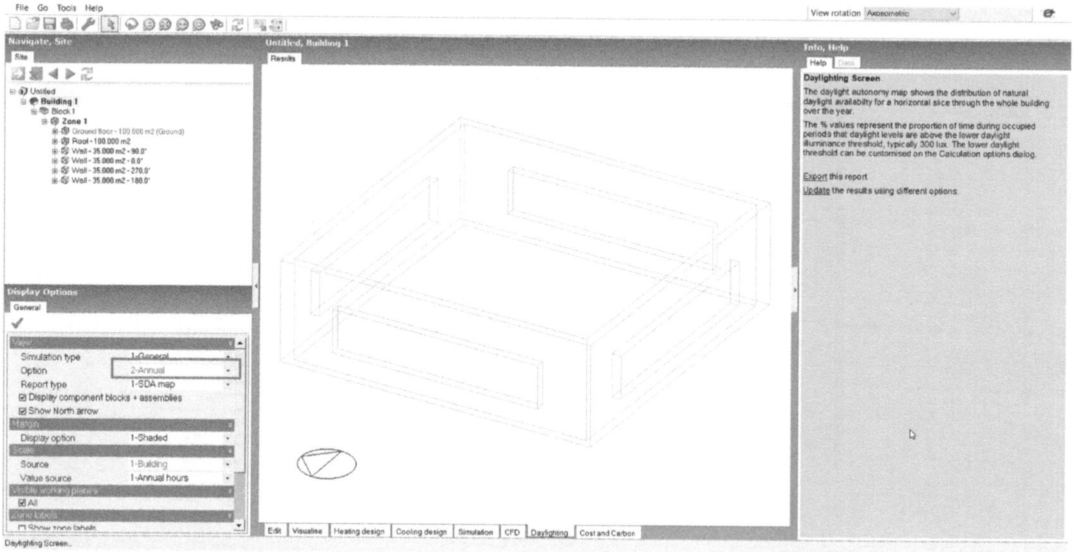

Step 10: Click **Update data**. The **Edit Calculation Option** screen appears.

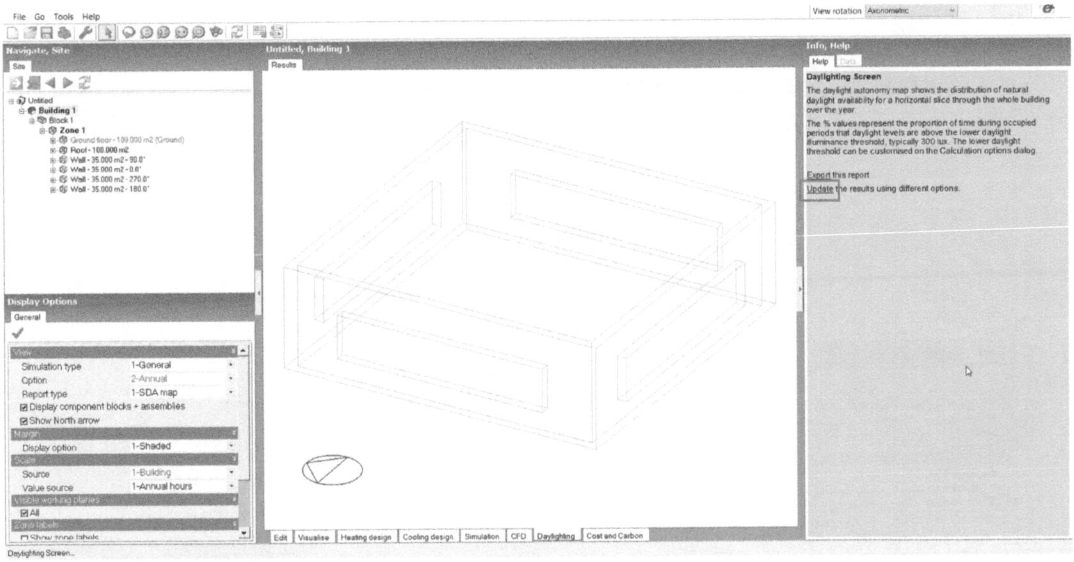

Step 11: Select **Perform annual simulation** check box and clear **Perform point-in-time simulation** check box. Expand the **Annual Options** section. Enter **100** in the **UDI lower illuminance threshold (lux)** text box and **2000** in the **UDI upper illuminance threshold (lux)** text box. Click **OK** to perform the simulations.

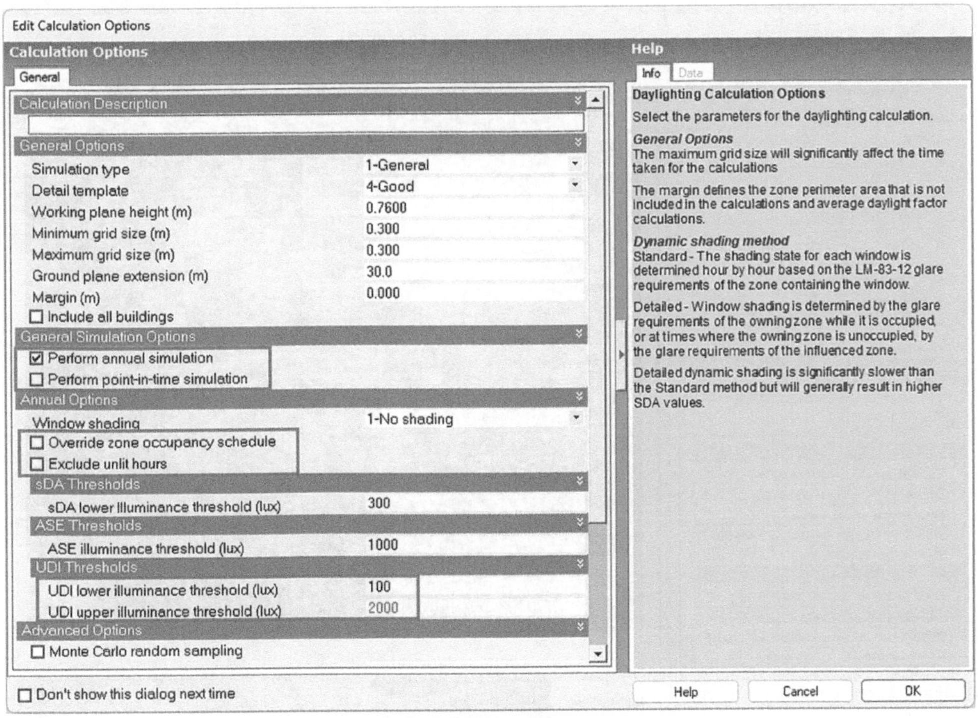

Step 12: Select **4-Grid** from the **Report type** drop-down list. Enter **90** in the **UDI percentage annual hours** text box. Record the results. Note the value of **UDI Area in Range (%)**.

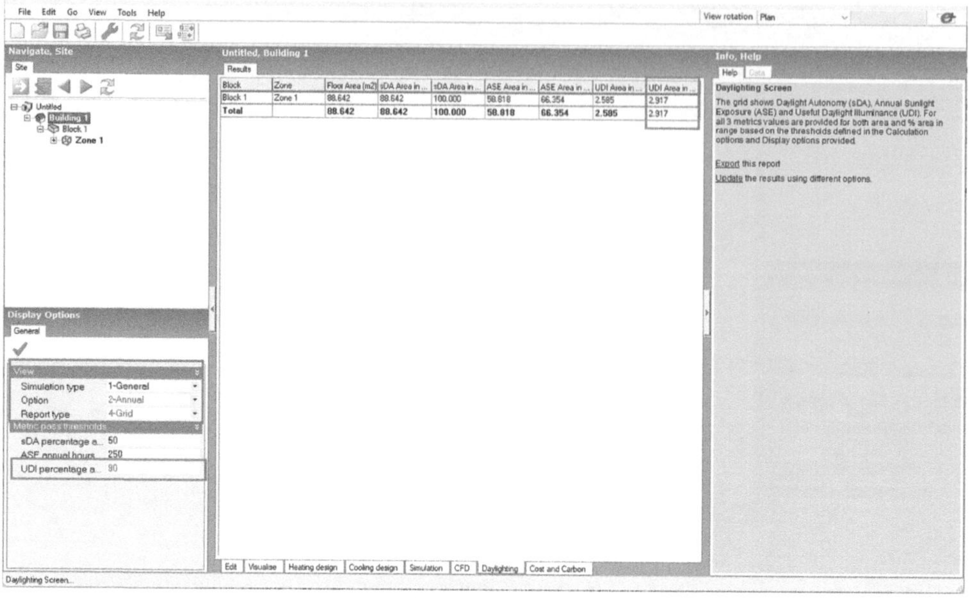

Step 13: Select the **3-UDI map** from the **Report type** drop-down list.

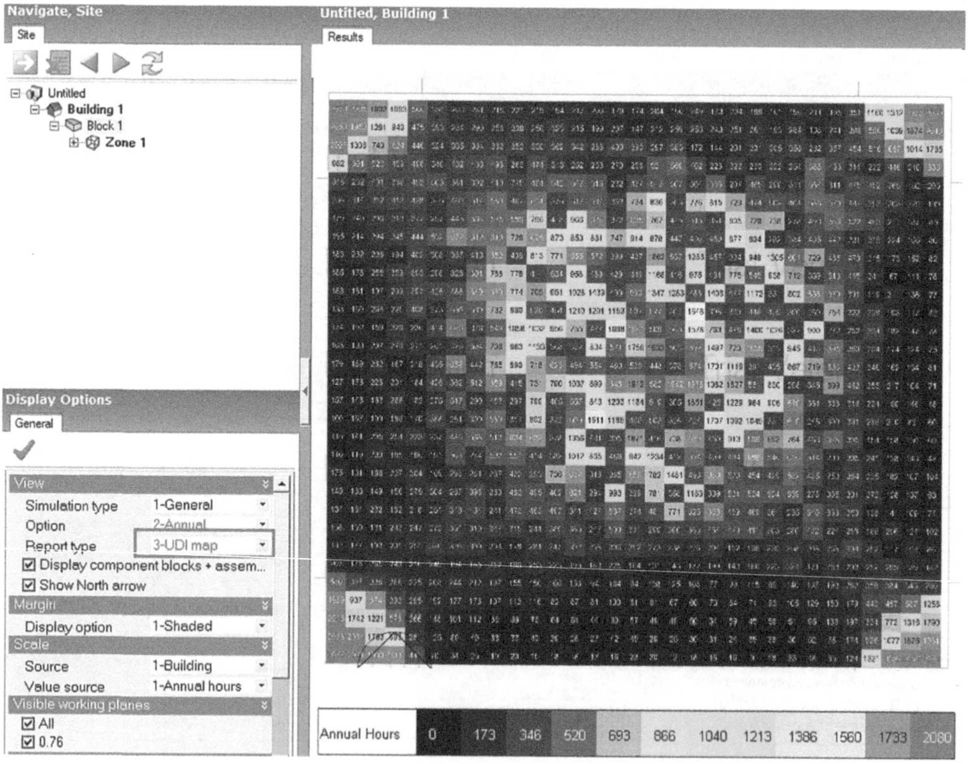

You can see that for most of the area, especially near windows, the UDI percentage is low because of excess daylight available near the windows. To increase the percentage of UDI area in range, you can provide overhangs and fins.

Step 14: Click the **Edit** tab.

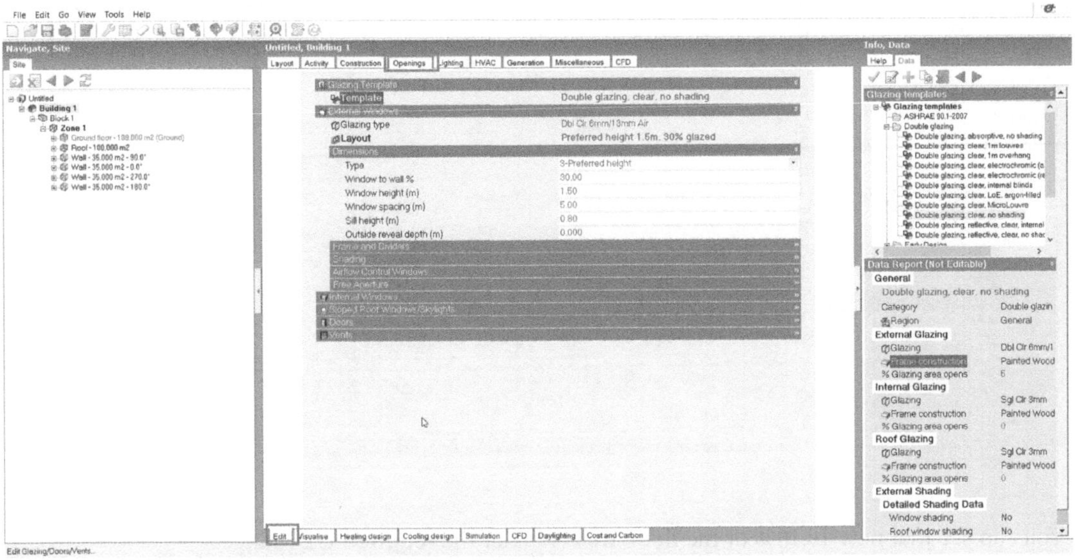

Step 15: Select the **Local shading** checkbox. Select **Overhang + sidefins (1m projection)** for **Type**.

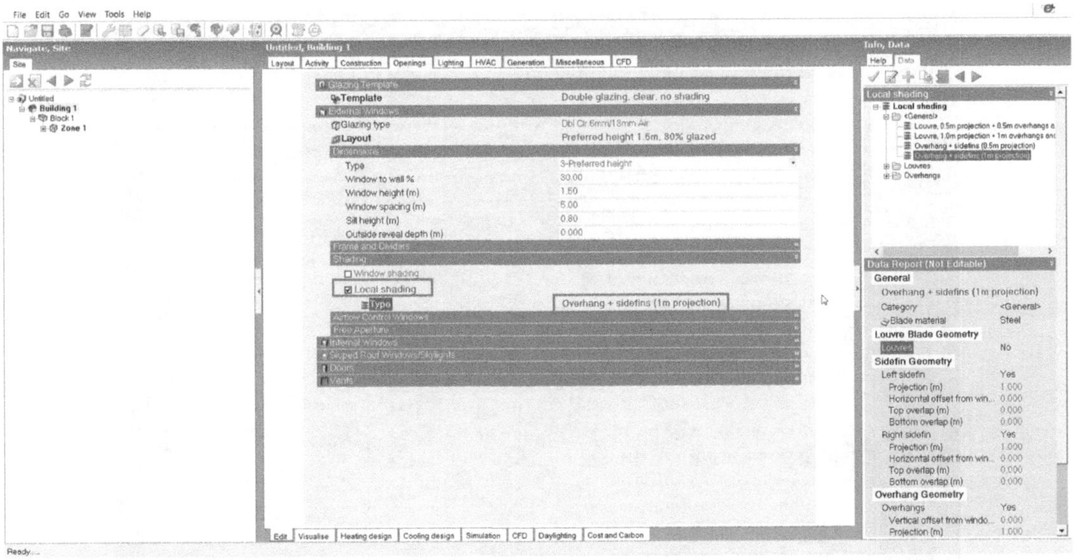

Step 16: Simulate the model for daylighting analysis and record the UDI area.

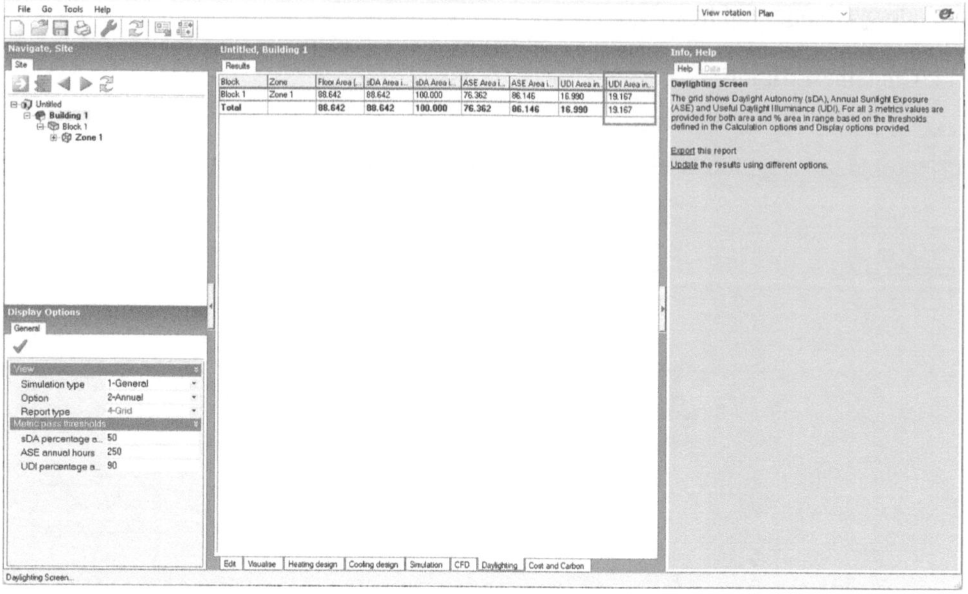

You can see that now 19.16% of the floor area is under the required UDI range.

Step 17: Select **3-UDI map** from the **Report type** drop-down list.

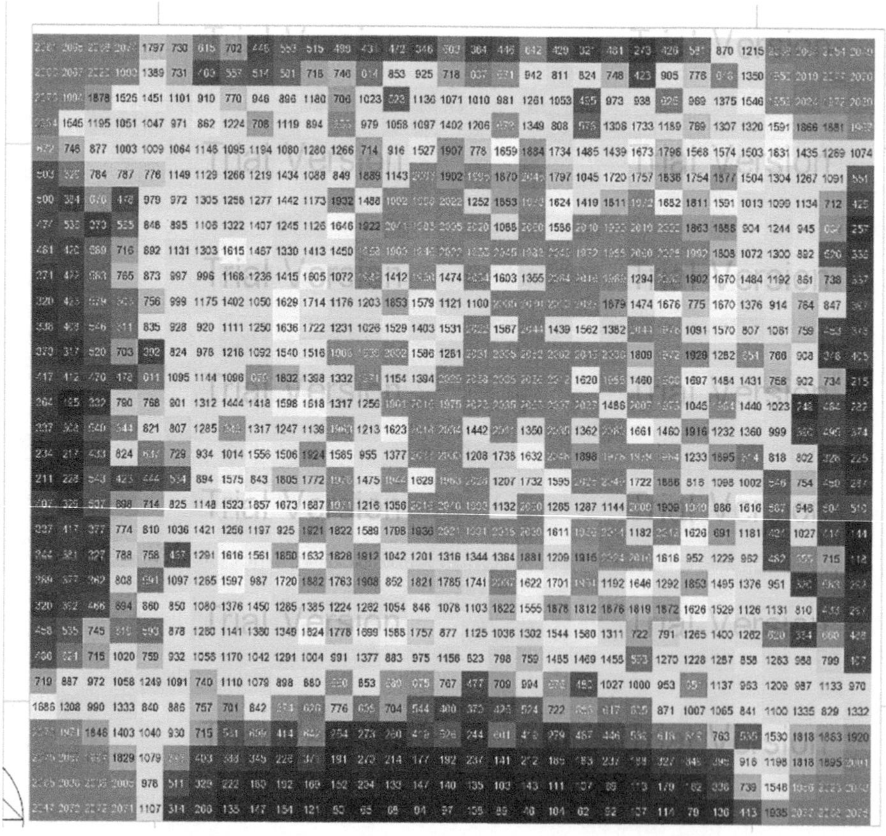

Step 18: Click the **Edit** tab. Click the **Openings** tab. Enter **25.00** in the **Window to wall %** text box.

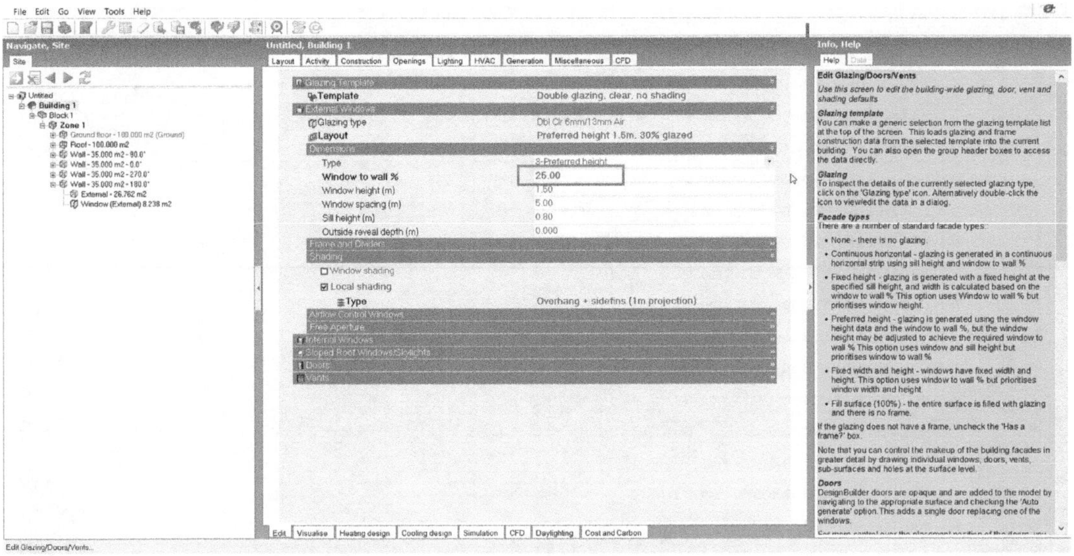

Step 19: Simulate the model for daylighting analysis and record the results.

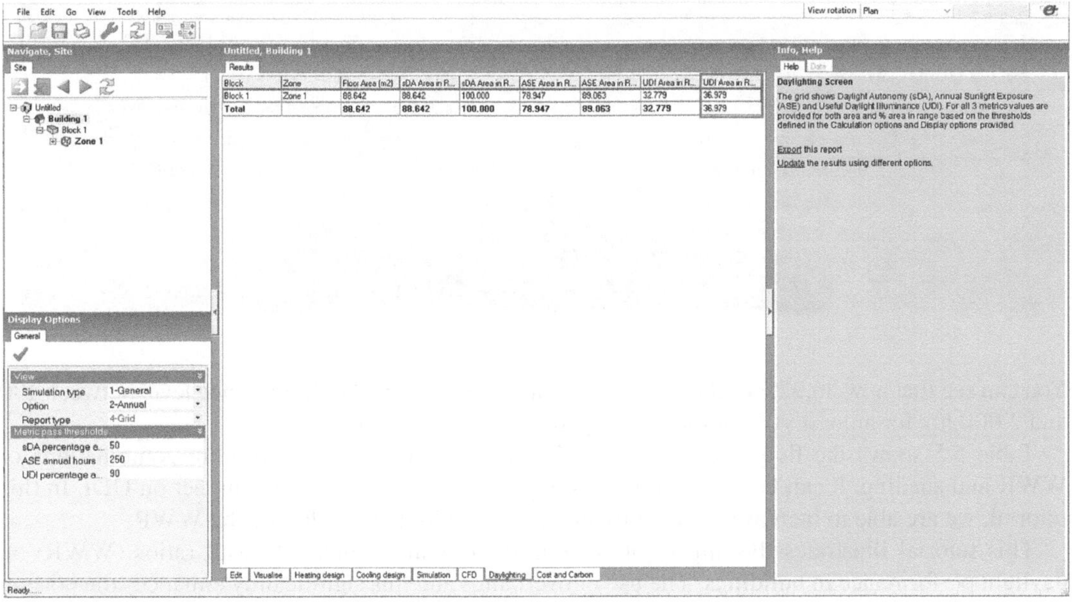

Step 20: Select **3-UDI map** from the **Report type** drop-down list.

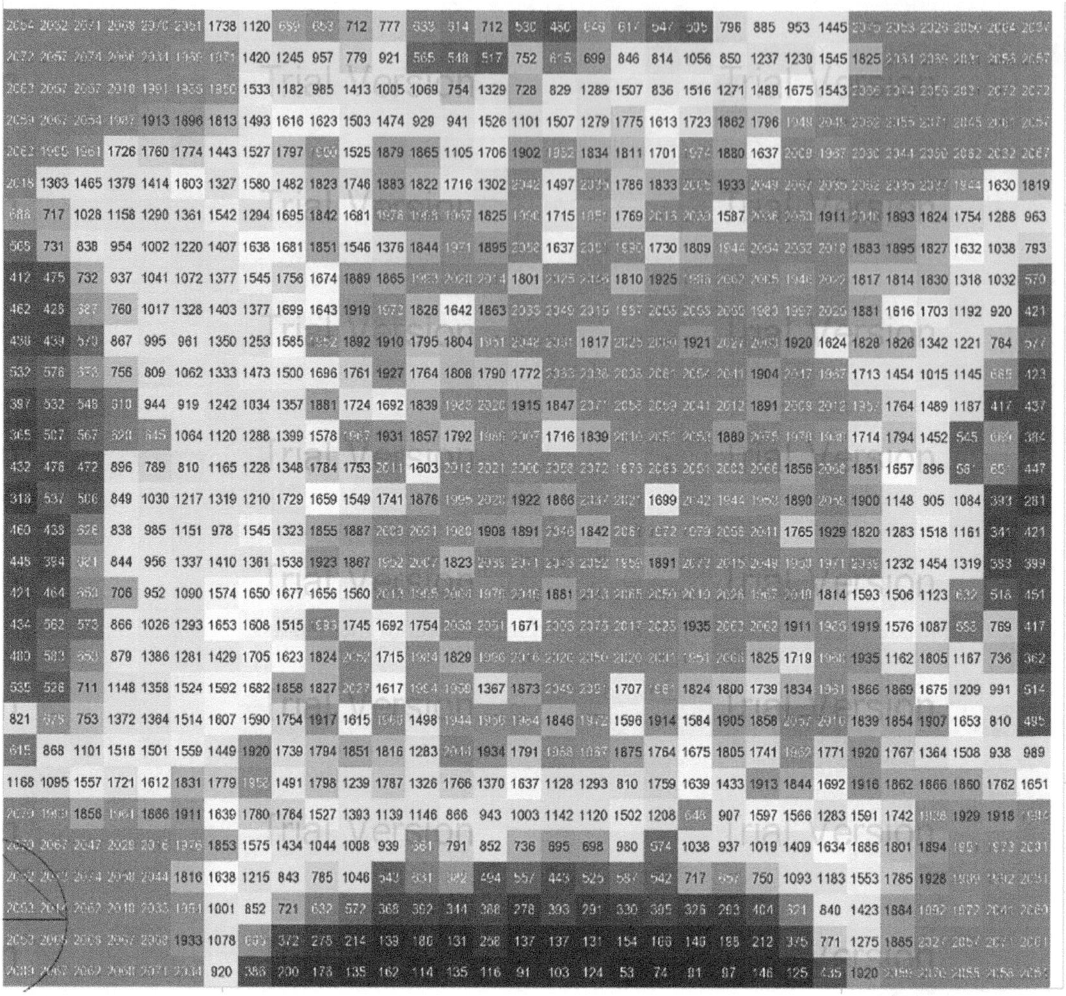

You can see that now 36.97% of the above-grade floor area meets the illuminance level between 100 and 2,000 lux for at least 90% of the potential daylit time in a year.

Table 5.5 shows the floor area meeting UDI requirements (%) for different combinations of WWR and shading. It can be seen that excess daylight can have a negative impact on UDI. In this tutorial, we are able to increase the UDI by providing shading and reducing the WWR.

This tutorial illustrates the impact of shading devices and window-to-wall ratios (WWR) on daylight performance in buildings. The use of overhangs and fins significantly enhances the portion of the floor area that meets acceptable daylight levels by mitigating excessive illumination near windows. These shading elements help control the amount of direct sunlight entering the space, thereby maintaining illuminance within a desirable range.

Additionally, decreasing the WWR from 30% to 20% enhances daylight distribution by limiting excessive direct daylight penetration, thereby making it easier to achieve desirable UDI levels in the building model.

TABLE 5.5
Floor Area Meeting UDI Requirements (%) for Different Combinations of WWR and Shading

Serial No.	Glass	WWR (%)	Overhang and Fins	Floor Area Meeting UDI Requirements (%)
1	Dbl Clr 6 mm/13 mm Air	30	No Shade	2.91
2	Dbl Clr 6 mm/13 mm Air	30	Overhang (1 m) and Fins (1 m)	19.16
3	Dbl Clr 6 mm/13 mm Air	25	Overhang (1 m) and Fins (1 m)	36.97

TABLE 5.6
Window-to-Wall Ratio (WWR) and Shading Variations

Serial No.	Glass
1	No internal windows
2	Internal windows (Sgl Clr 3 mm) (8 m × 2.5 m)
3	With overhang and fins (0.5 m) on south, east and west external windows

TUTORIAL 5.4: EVALUATING THE IMPACT OF INTERNAL WINDOWS ON DAYLIGHT PERFORMANCEGOAL

To evaluate the impact of internal windows on daylight performance

What are You Going to Learn?

- How to add internal windows
- How to perform simulations for daylight assessment

Problem Statement

In this tutorial, you are going to use a 15 m × 15 m five-zone building model with a 3 m perimeter depth. The building will use the glazing template **Double glazing; clear no shading (Dbl Clr 6 mm/13 mm Air)**, and the occupancy schedule will be set from 08:00–18:00 h. The external windows will be fixed, each measuring 1.5 meters in height and 0.5 meters in width, with a spacing of 0.75 meters. You are required to simulate three different scenarios, as outlined in Table 5.6, to evaluate the impact of these configurations on daylight and energy performance.

Find the floor areas that meet or exceed the illuminance level 300 lux for at least 50% of the potential daylit time in a year (*spatial daylight autonomy*) and work plane exceeding 1000 lux for more than 250 occupied hours per year (*annual sunlight exposure*). Consider a work plane height of 0.75 m above the finished floor.

Demonstrate through annual simulations that spatial daylight autonomy (sDA300/50%) of at least 55% is achieved.

Demonstrate through annual simulations that annual sunlight exposure 1000,250 (ASE1000,250) of no more than 10% is achieved

Use weather file for **New Delhi Palam, India** location.

Daylight autonomy (sDA) is a measure of how often a minimum work plane illuminance threshold (often 300 lux) can be maintained by daylight alone. It is expressed as the percentage of occupied time during the year when a minimum work plane illuminance threshold of 300 lux can be maintained by daylight alone. It defines the extent to which a space has enough natural daylight to allow the prescribed activity to be carried out without the need for electric lighting. This is a recommended metric by the Illuminating Engineering Society of North America (IESNA). When default settings are used, the zone summary sDA outputs represent the percentage of the floor area that exceeds 300 lux for at least 50% of the time.

Annual sunlight exposure (ASE) is the number of hours per year a point on the working plane receives direct sunlight greater than a threshold value. This metric is often used alongside sDA to quantify the risk of visual discomfort due to glare.

Source: https://designbuilder.co.uk/helpv7.0/Content/AnnualCBDM.htm

Solution

Step 1: Open a new project and create a five-zone **15 m × 15 m** building with a 3 m perimeter.

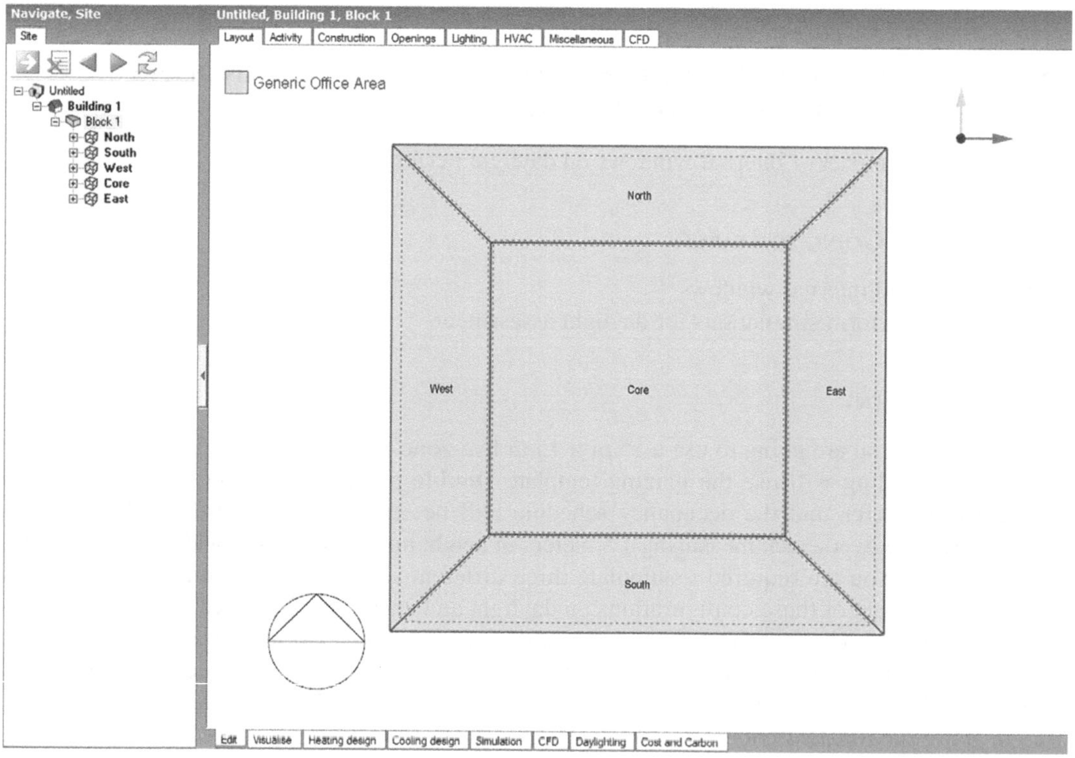

Step 2: Follow steps 2–6 from **Tutorial 5.3** to set the schedule and window template.

Make sure that the **Glazing** type is **Dbl Clr 6mm/13mm Air**. Select **Layout** as **Fixed windows—height:1.0 m, width:0.5**. Click the **Daylighting** screen tab.

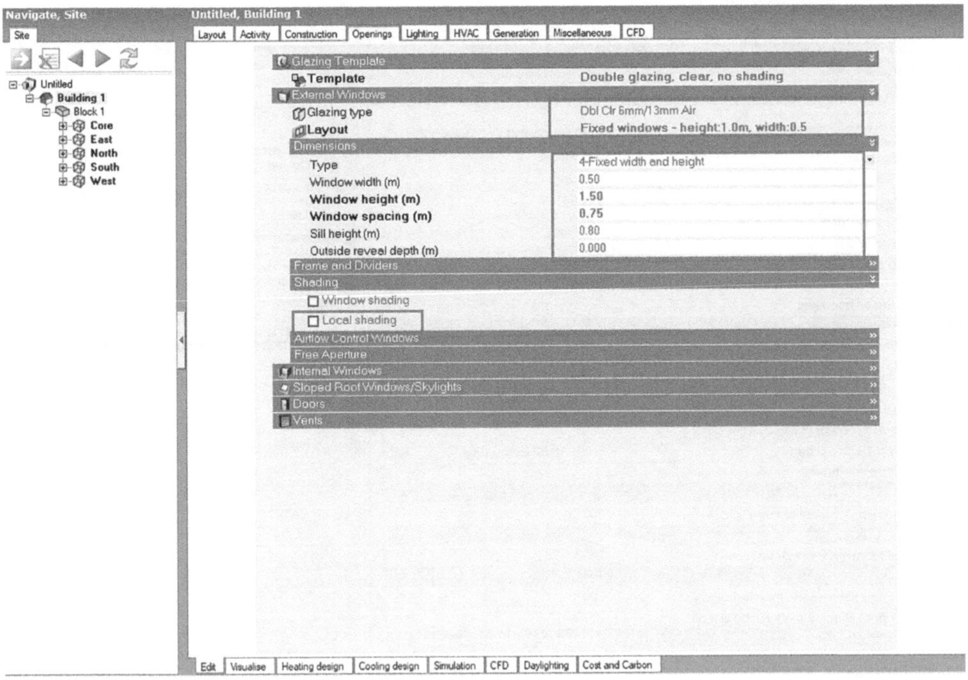

Step 3: Click **Update data**. The **Edit Calculation Option** screen appears.

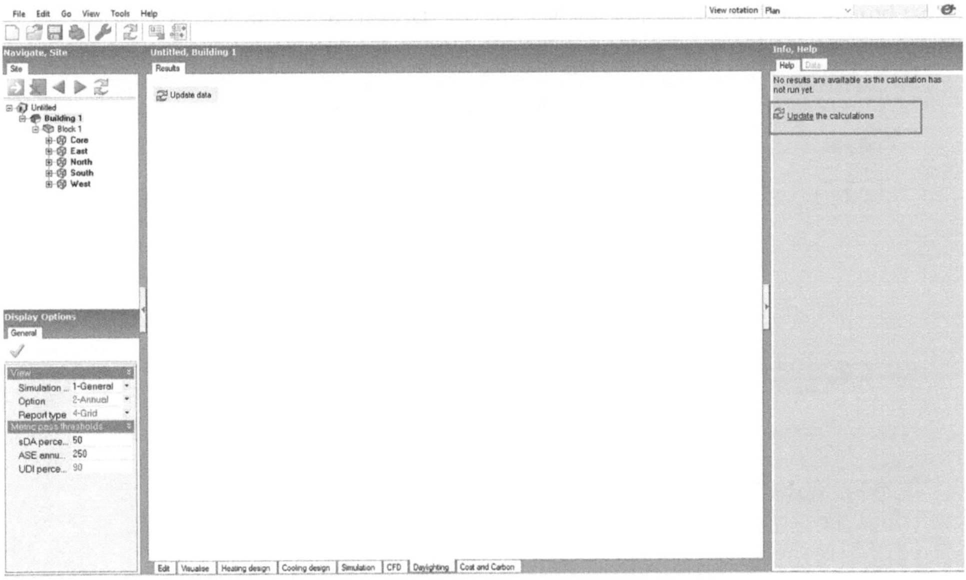

Step 4: Select the **Perform annual simulation** check box and clear the **Perform point-in-time simulation** check box. Expand the **Annual Options** section. Select **1-No shading** from **Window Shading** Drop-down. Make sure to clear the **Exclude unlit hours** check box. Enter **300** in the **sDA lower illuminance threshold (lux)** text box and **1000** in the **ASE illuminance threshold (lux)** text box. Enter **100** in the **UDI lower illuminance threshold (lux)** text box and **2000** in the **UDI upper illuminance threshold (lux)** text box. Click **OK** to perform the simulations.

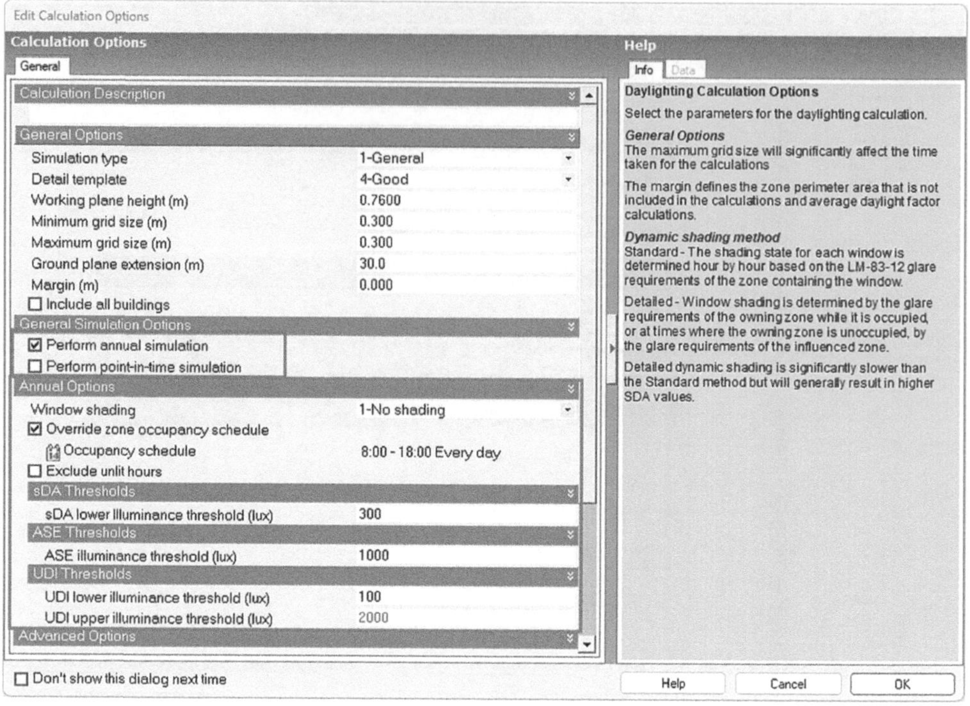

Step 5: Select **4-Grid** from the **Report type** drop-down list. Enter **50** in the **sDA percentage annual hours** text box. Record the results. Note the value of **sDA Area in Range (%)**.

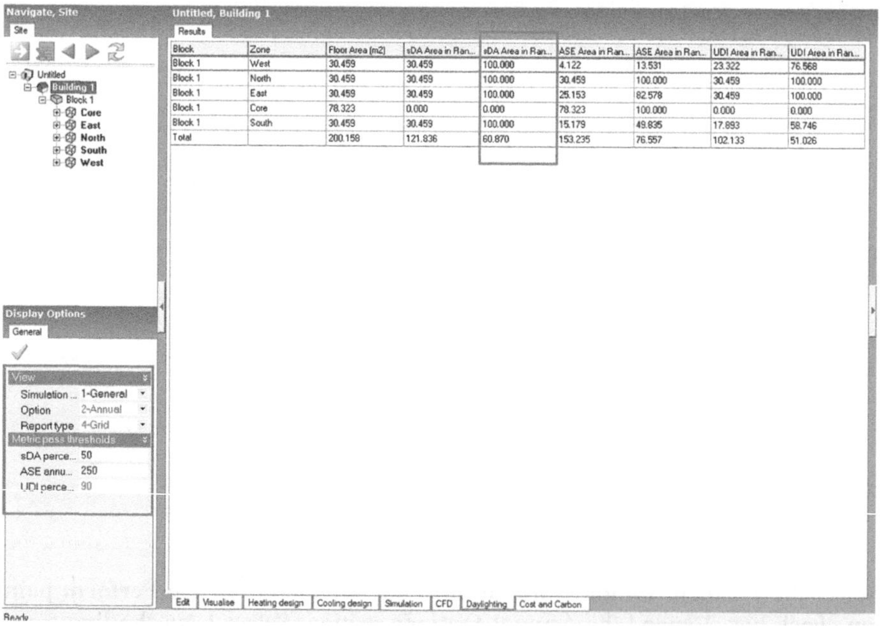

You can see that in the core zone, the sDA percentage is zero because there is no penetration of daylight into the core zone. Internal windows can be provided to get access to the daylight in the core zone and improve the percentage of sDA. Also, 60.87% floor area is in the ASE range.

Adding Internal Windows

Step 6: To add Windows on internal partitions, select **Partition-31.375 m² (Block1, East)**. Click on **add windows** icon. Draw an 8 m × 2.5 m window (sill height 0.5 m, distance from left 0.5 m). Similarly, add windows to other partition walls.

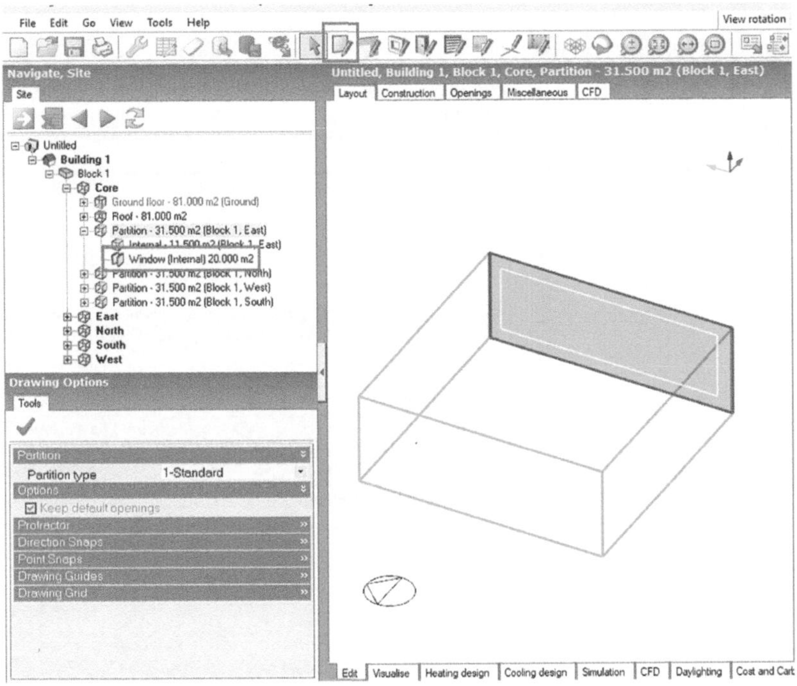

Step 7: Simulate the model for daylighting analysis and record the **sDA Area in Range (%)** and **ASE area in Range (%)**.

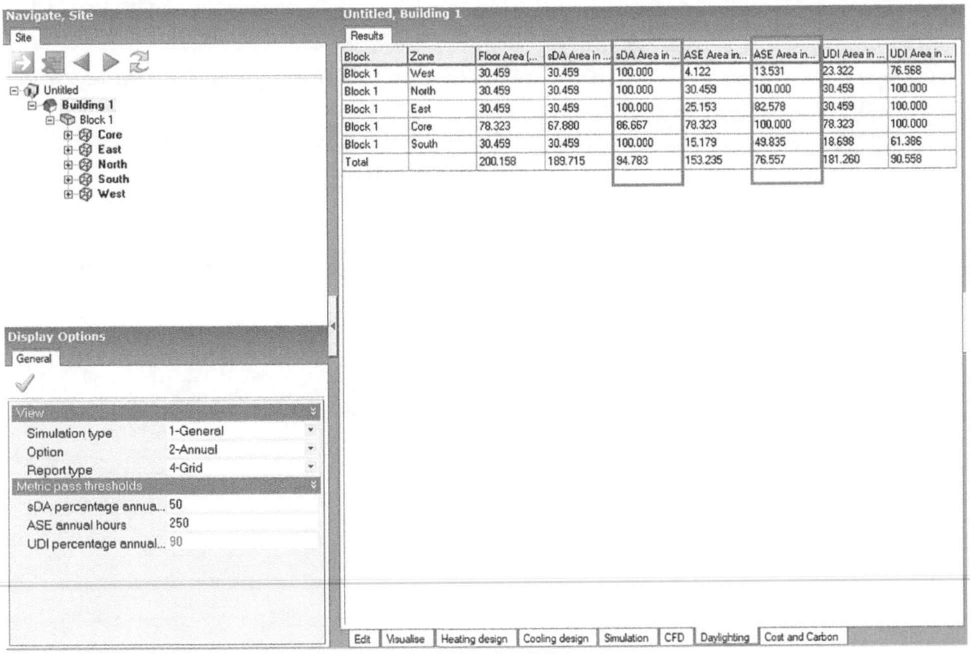

You can observe that after adding internal windows, 94.78% of the floor area is under the required sDA range, but only 76.55% area is in the ASE range.

Adding Overhangs and Side Fins to Control ASE

Step 8: Click **Edit** screen tab. Click **Openings** tab. Select **Local Shading** check box.

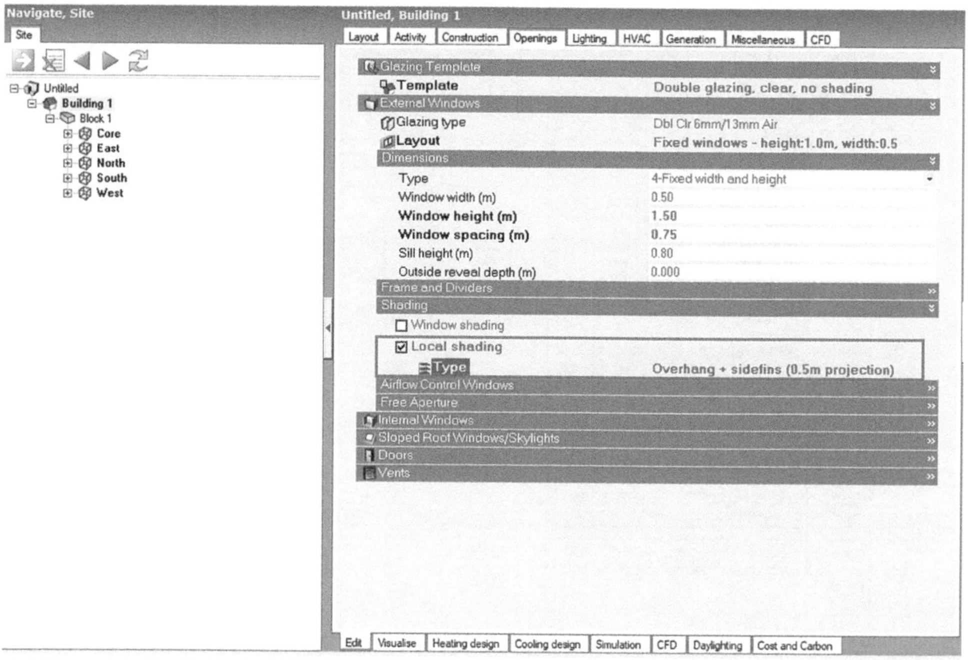

Step 9: Select **North** and uncheck the **Local Shading** check box.

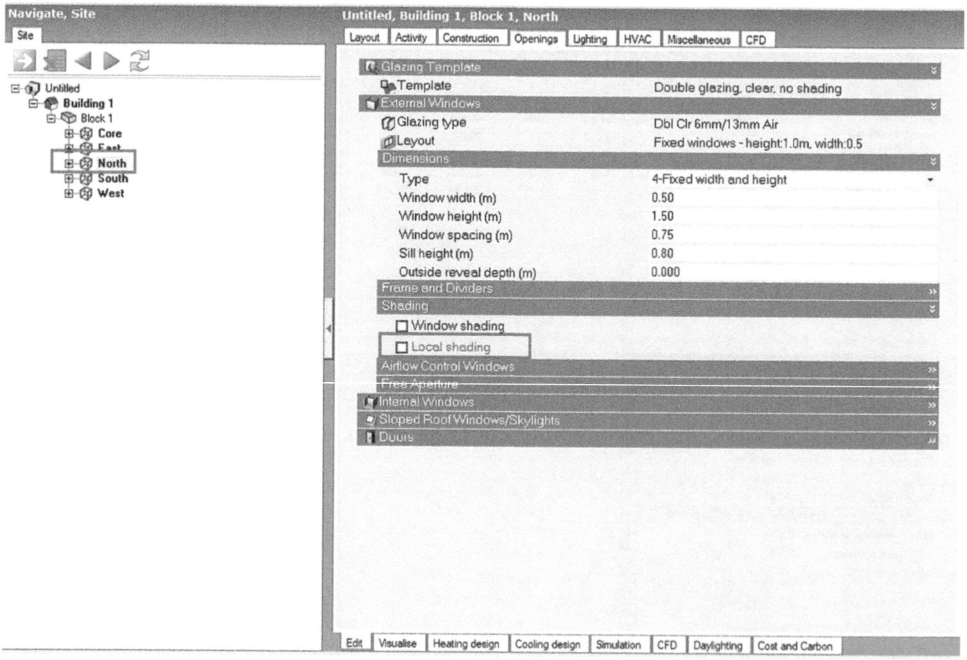

Step 10: Simulate the model daylighting and record the results.

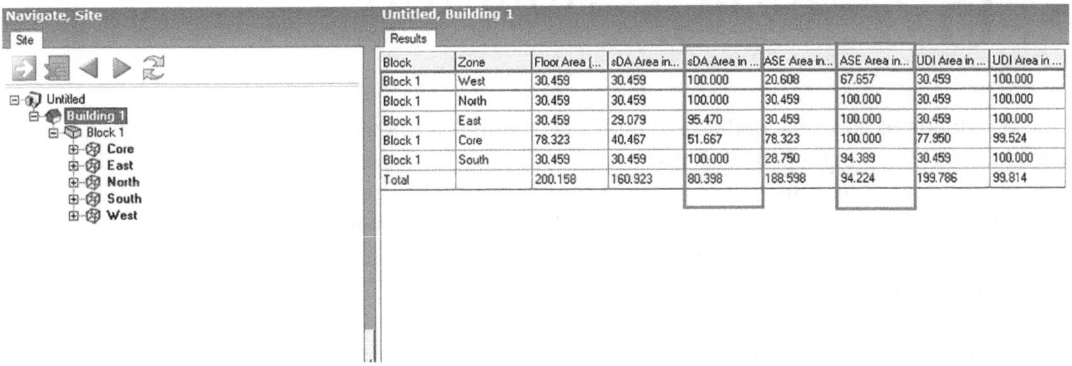

Block	Zone	Floor Area (...	sDA Area in...	sDA Area in ...	ASE Area in...	ASE Area in...	UDI Area in ...	UDI Area in ...
Block 1	West	30.459	30.459	100.000	20.608	67.657	30.459	100.000
Block 1	North	30.459	30.459	100.000	30.459	100.000	30.459	100.000
Block 1	East	30.459	29.079	95.470	30.459	100.000	30.459	100.000
Block 1	Core	78.323	40.467	51.667	78.323	100.000	77.950	99.524
Block 1	South	30.459	30.459	100.000	28.750	94.389	30.459	100.000
Total		200.158	160.923	80.398	188.598	94.224	199.786	99.814

TABLE 5.7
Floor Area Meeting sDA Requirements (%) for Different Combinations of WWR and Shading

Serial No.	Case	Floor Area Meeting sDA Requirements (%)	Floor Area Meeting ASE Requirements (%)
1	No internal windows	60.87	76.55
2	Internal windows (Sgl Clr 3 mm) (8 m × 2.5 m)	94.78	76.55
3	With overhang and fins (0.5 m) on south, east and west	80.39	94.22

Table 5.7 shows the floor area meeting sDA and ASE requirements (%) for different combinations of WWR and shading.

The analysis demonstrates that internal windows significantly enhance daylight penetration, particularly into core zones, leading to a substantial improvement in spatial daylight autonomy (sDA). While a model without internal windows showed limited daylight distribution and moderate potential for visual discomfort, introducing internal windows improved daylight autonomy without affecting the area exposed to excessive sunlight. Adding overhangs and fins further helped control glare and reduce direct solar exposure by lowering the ASE, although this came at the cost of some reduction in daylight levels due to shading and reduced sky view. Overall, the use of internal windows is highly effective for improving daylight access, while external shading devices play a critical role in mitigating glare and controlling overexposure.

6 Natural Ventilation

The simulation of non–air-conditioned buildings requires a fundamentally different approach compared to buildings equipped with conventional heating and cooling systems. In such buildings, indoor temperature is primarily maintained through natural ventilation, typically achieved using operable windows or ventilation fans. Simulation tools are capable of modeling these ventilation methods and predicting indoor temperature based on factors such as building geometry, weather data, and control strategies.

Natural ventilation relies on the interaction between building openings and environmental forces like wind and temperature differences. Simulation tools can model both wind-driven and buoyancy-driven ventilation by calculating airflow through windows, cracks, and other openings. This enables designers to evaluate how outdoor conditions affect indoor environments, identify periods when natural ventilation is effective, and understand its limitations in hot, humid, or still weather conditions.

A mixed-mode ventilation system combines natural ventilation during periods of moderate outdoor conditions with mechanical heating or cooling during periods of thermal discomfort. Simulation of mixed-mode buildings involves defining control strategies that dictate when windows open or close, and when HVAC systems activate. These strategies are typically based on indoor or outdoor temperature thresholds and time-of-day schedules. Simulating such scenarios allows for the assessment of how often the building can operate in passive mode, the energy savings achieved, and the impact on indoor conditions.

This chapter explains the modeling of non-air-conditioned and mixed-mode buildings through six tutorials. The tutorials guide the reader step-by-step through the process of defining window openings, assigning schedules for their operation, and analyzing their impact on indoor thermal conditions. Readers learn how to model constant or variable window openings, and to explore the relationship between wind speed and ventilation rate. In later tutorials, more advanced features such as temperature-controlled automatic window operation and the integration of natural and mechanical ventilation in a mixed-mode setup are introduced.

TUTORIAL 6.1: EVALUATING THE IMPACT OF WIND SPEED ON NATURAL VENTILATION

Goal

To evaluate the impact of a change in wind speed on the ventilation rate

What Are You Going to Learn?

- How to model natural ventilation
- How to define glazing area openings for natural ventilation

There are two general approaches to natural ventilation and infiltration modeling in Design Builder depending on the setting of the natural ventilation model option.

- Scheduled, whereby the natural ventilation change rate is explicitly defined for each zone in terms of a maximum air changes per hour (ACH) value and a schedule, and

DOI: 10.1201/9781003563013-6

infiltration air change rate is defined by a constant ACH value. A range of control options is provided.

- Calculated, whereby natural ventilation and infiltration are calculated based on window openings, cracks, buoyancy and wind-driven pressure differences, crack dimensions and so on. Control options are provided.

Source: www.designbuilder.co.uk/helpv7.3/Content/CalculatedNatVent.htm

Problem Statement

In this tutorial, you are going to use a 10 m × 10 m single-zone model. Find the zone air changes for the model with natural ventilation. Use the **New Delhi/Safdarjung, India** location.

Solution

Step 1: Open a new blank project file and create a **10 m × 10 m** single-zone building. Select the **Activity** tab.

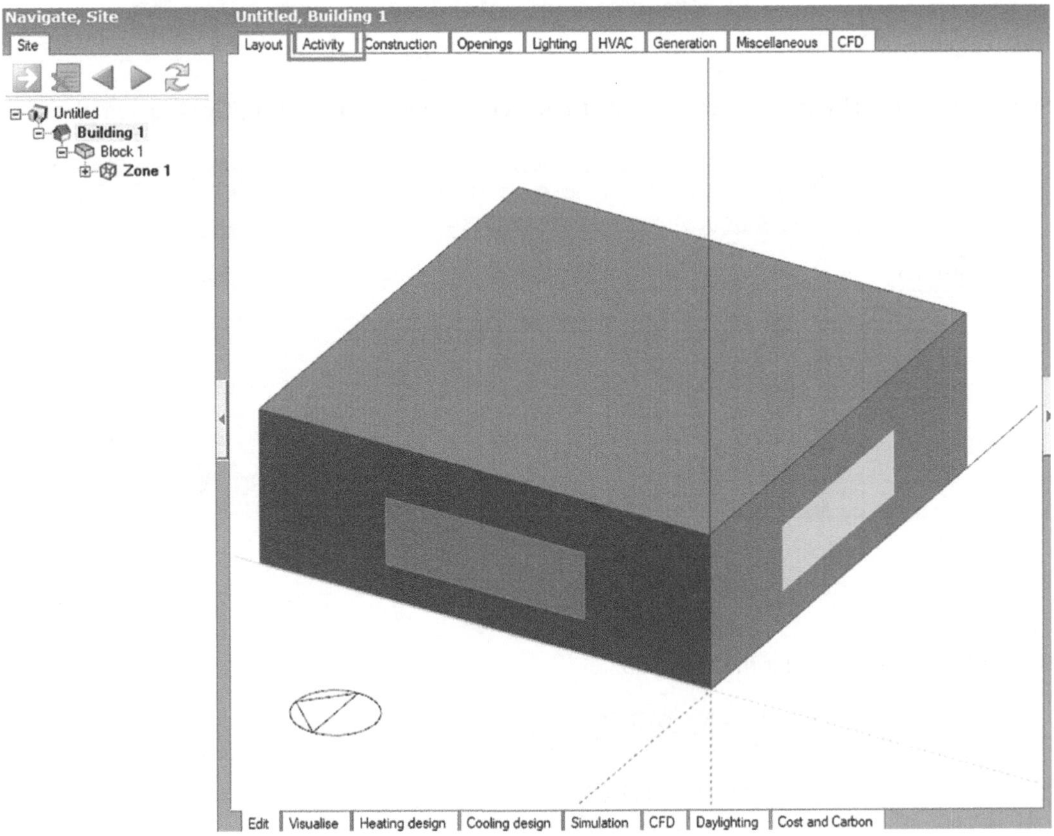

Step 2: Select the **24×7 Generic Office Area** template. Set **Fresh air (l/s-person)** and **Mech vent per area (l/s-m²)** to **0**.

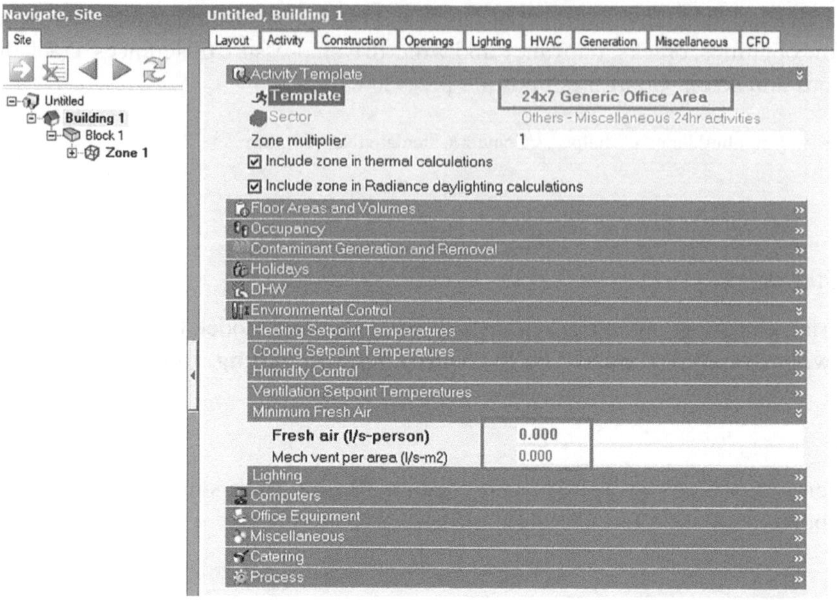

Step 3: Select the **HVAC** tab. Select the **Natural ventilation—No Heating/Cooling** template.

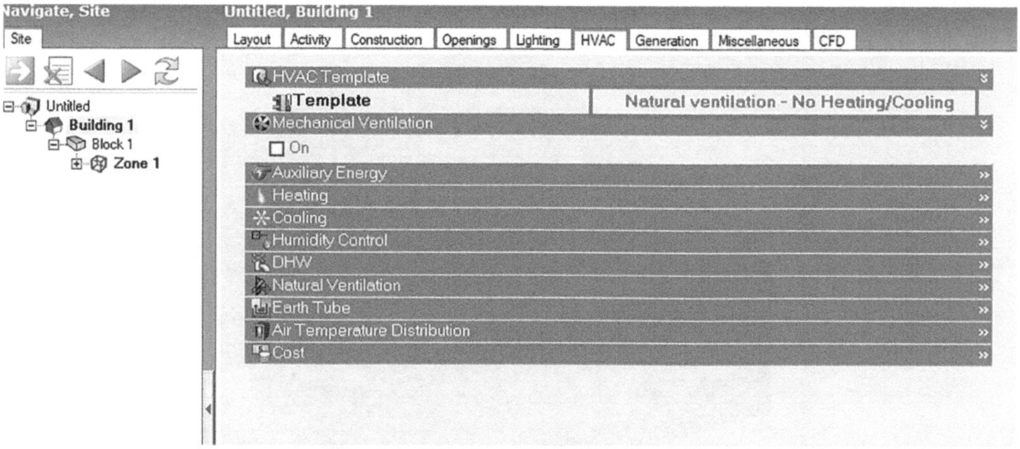

Step 4: Click **Simple** under the **Help** tab.

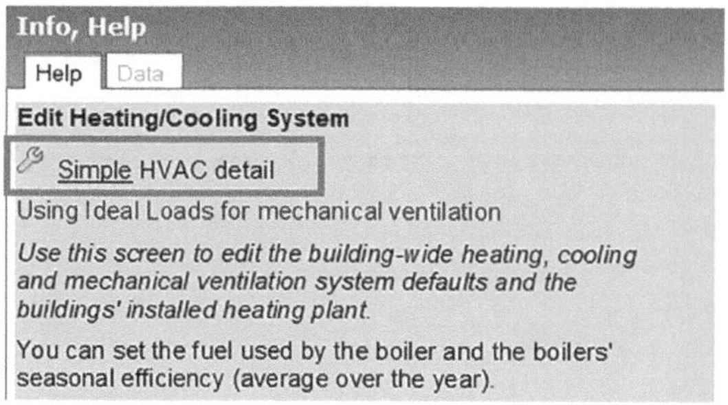

Step 5: Click **Calculated** under **Natural ventilation**. Click **OK**.

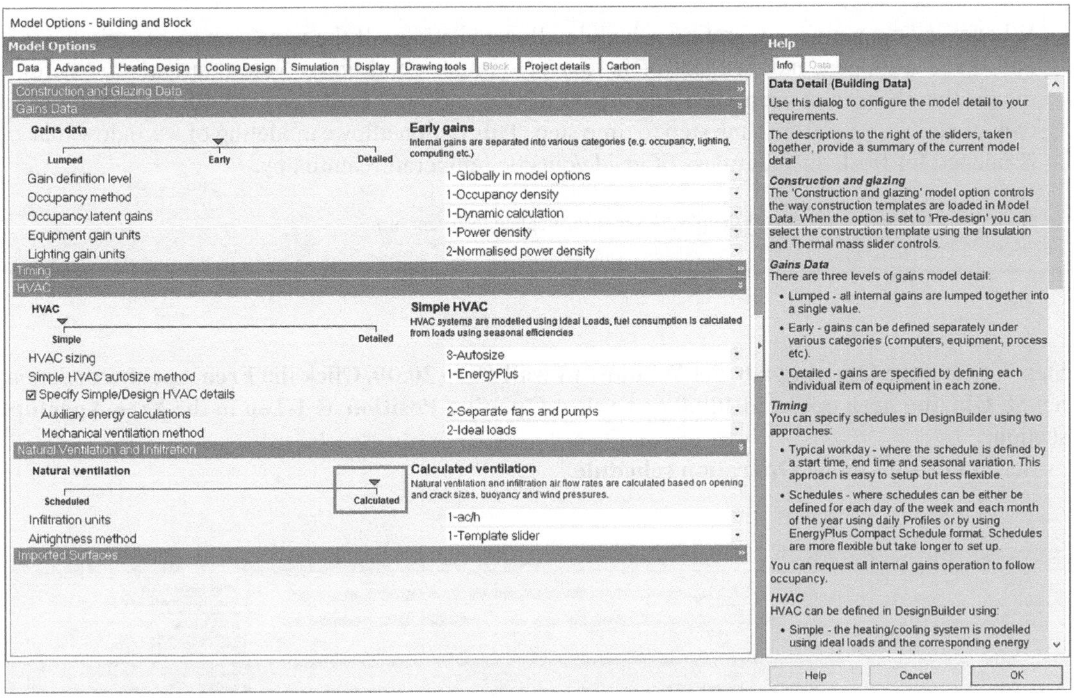

Step 6: Click the **Options** subsection in the **Natural Ventilation** section and select **Constant** for the **External control mode** and **Internal control mode** drop-down lists.

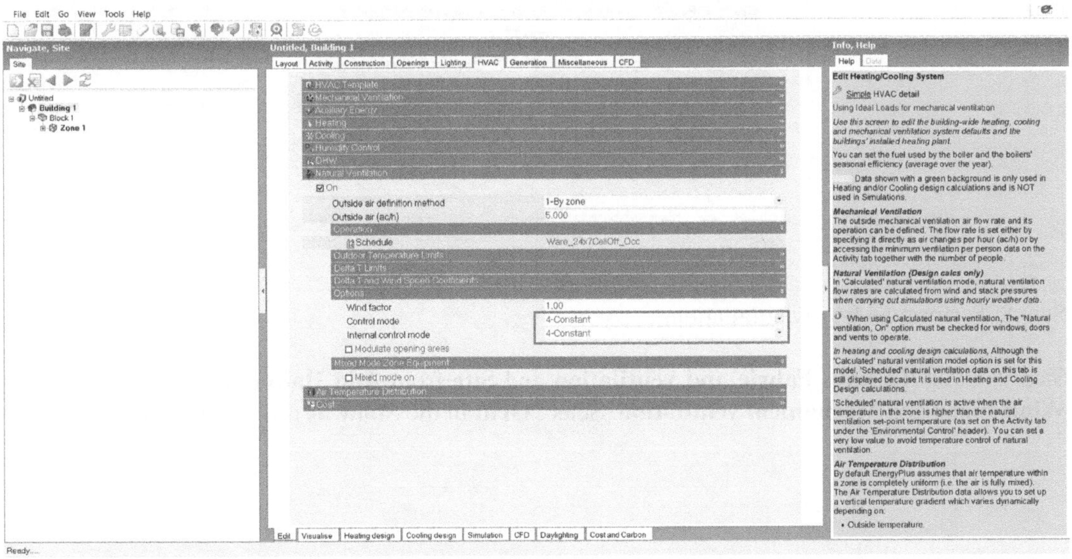

CONSTANT CONTROL MODE FOR NATURAL VENTILATION

Whenever an opening's operating schedule allows venting, all the zone's openable windows and doors are open, independent of indoor or outdoor conditions. Note that *constant* here means that the size of each opening is fixed while venting; the airflow through each opening can, of course, vary from time step to time step. This option allows modeling of a window that is opened for fresh air regardless of inside/outside temperature/enthalpy.

Source: www.designbuilder.co.uk/helpv7.3/Content/CalculatedNatVent.htm

Step 7: Select the **Opening** tab. Set **Window to wall %** to **20.00**. Click the **Free Aperture** section. Set **% Glazing area opens** to **100.0** and Select **Opening Position** as **1-Top** in the **Free Aperture** section.

Select **On 24/7** in the **Operation schedule**.

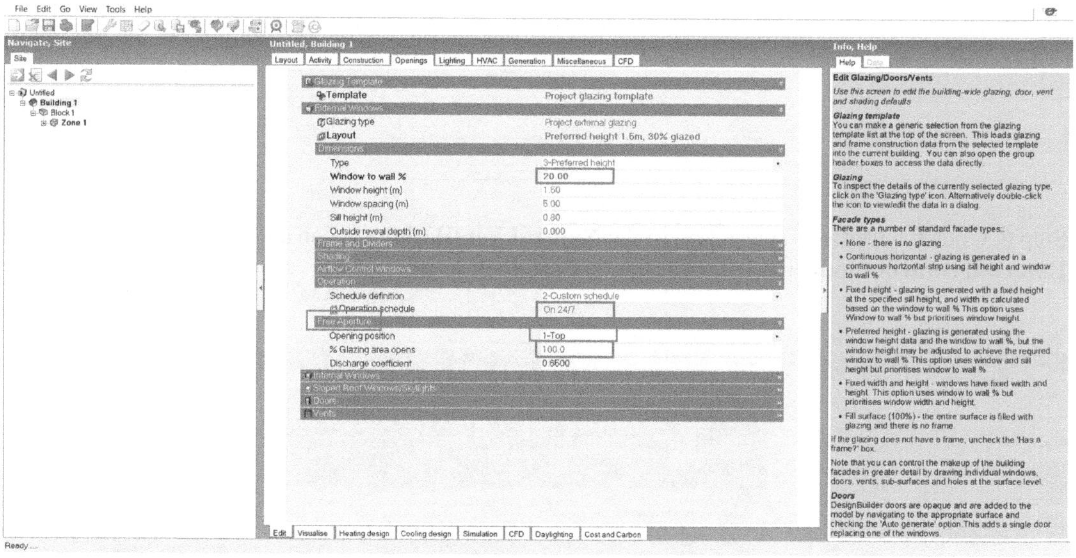

Step 8: Perform an hourly simulation.

Step 9: Click the links **Fabric and ventilation** and **Site** from **Add Data** to record the results for **Wind speed** and **Mechanical ventilation**. Select **Grid** in the **Show as** drop-down list.

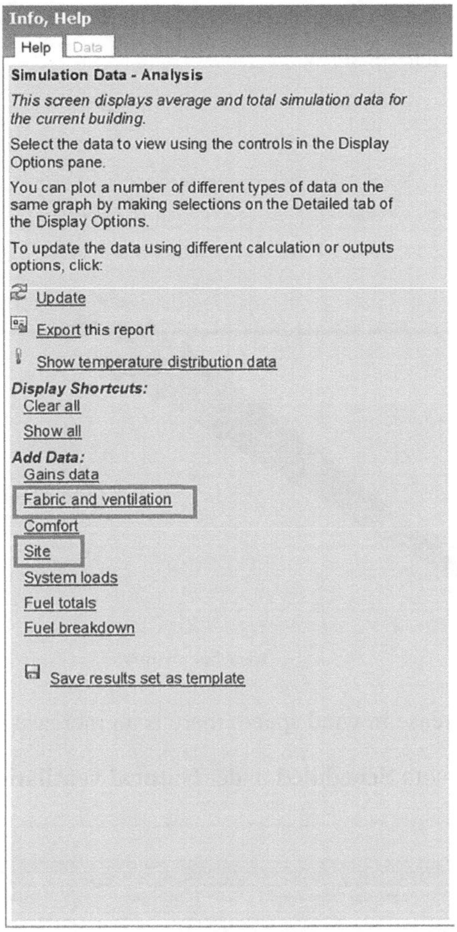

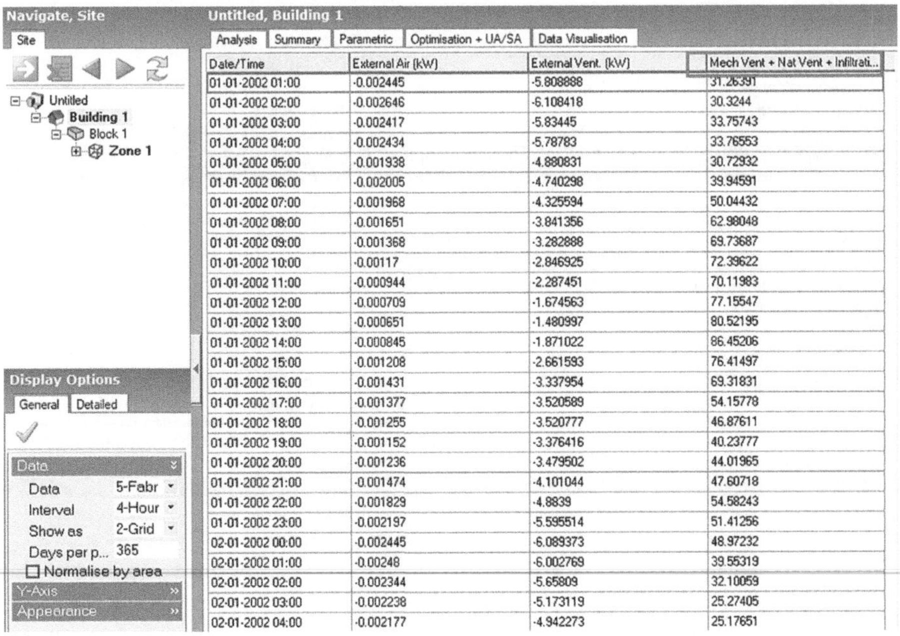

Date/Time	External Air (kW)	External Vent. (kW)	Mech Vent + Nat Vent + Infiltrati...
01-01-2002 01:00	-0.002445	-5.808888	31.26391
01-01-2002 02:00	-0.002646	-6.108418	30.3244
01-01-2002 03:00	-0.002417	-5.83445	33.75743
01-01-2002 04:00	-0.002434	-5.78783	33.76553
01-01-2002 05:00	-0.001938	-4.880831	30.72932
01-01-2002 06:00	-0.002005	-4.740298	39.94591
01-01-2002 07:00	-0.001968	-4.325594	50.04432
01-01-2002 08:00	-0.001651	-3.841356	62.98048
01-01-2002 09:00	-0.001368	-3.282888	69.73687
01-01-2002 10:00	-0.00117	-2.846925	72.39622
01-01-2002 11:00	-0.000944	-2.287451	70.11983
01-01-2002 12:00	-0.000709	-1.674563	77.15547
01-01-2002 13:00	-0.000651	-1.480997	80.52195
01-01-2002 14:00	-0.000845	-1.871022	86.45206
01-01-2002 15:00	-0.001208	-2.661593	76.41497
01-01-2002 16:00	-0.001431	-3.337954	69.31831
01-01-2002 17:00	-0.001377	-3.520589	54.15778
01-01-2002 18:00	-0.001255	-3.520777	46.87611
01-01-2002 19:00	-0.001152	-3.376416	40.23777
01-01-2002 20:00	-0.001236	-3.479502	44.01965
01-01-2002 21:00	-0.001474	-4.101044	47.60718
01-01-2002 22:00	-0.001829	-4.8839	54.58243
01-01-2002 23:00	-0.002197	-5.595514	51.41256
02-01-2002 00:00	-0.002445	-6.089373	48.97232
02-01-2002 01:00	-0.00248	-6.002769	39.55319
02-01-2002 02:00	-0.002344	-5.65809	32.10059
02-01-2002 03:00	-0.002238	-5.173119	25.27405
02-01-2002 04:00	-0.002177	-4.942273	25.17651

Export the results to a spreadsheet and plot a scatter graph between **Mech Vent + Nat Vent + Infiltration (ac/h)** and **Wind speed (m/s)**.

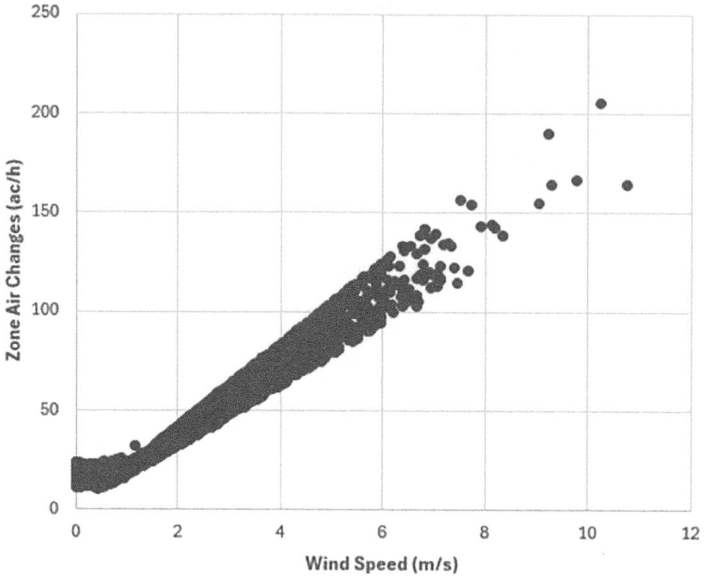

You can see that with an increase in wind speed, there is an increase in zone air changes.

Step 10: Repeat this tutorial with **Scheduled** under **Natural ventilation**.

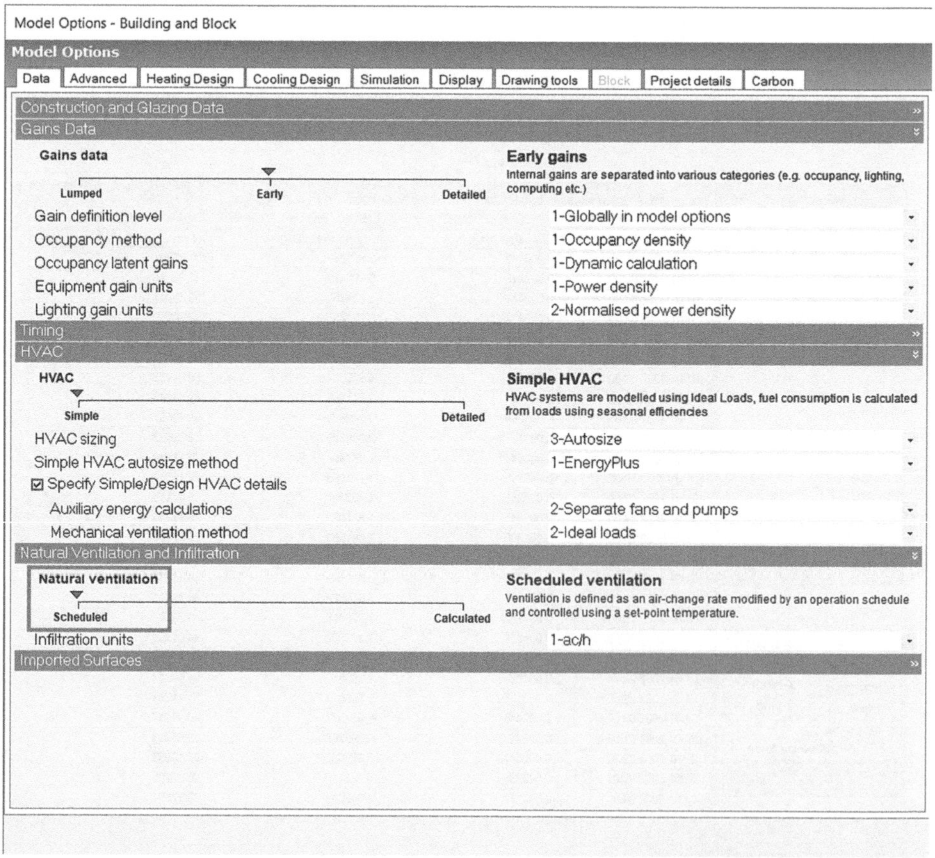

Step 11: Select the **Activity** tab. Clear both the **Indoor min and max temperature control** check-boxes in the **Natural Ventilation** section.

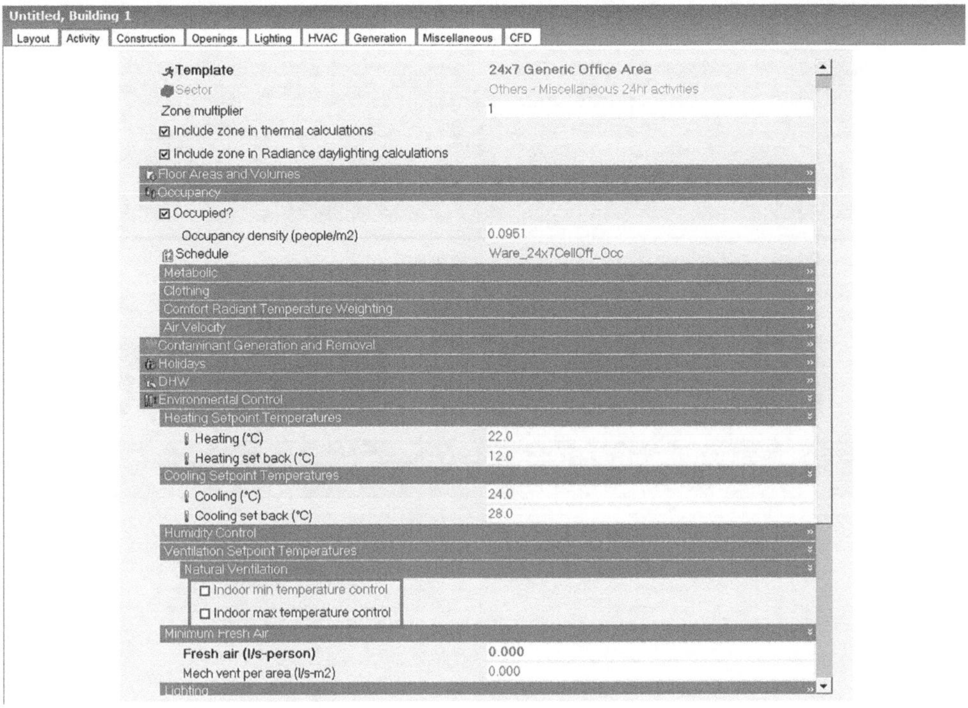

Step 12: Select the **Construction** tab. Clear **Model infiltration** in the **Airtightness** section.

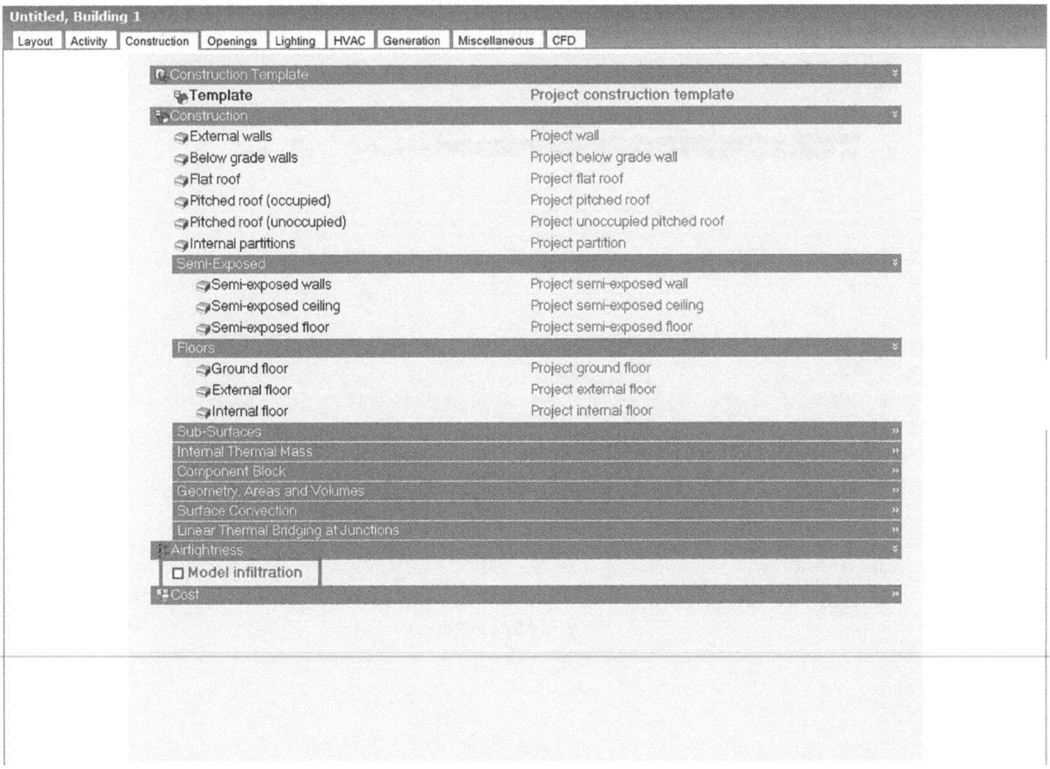

Step 13: Enter **3.000** for **Outside air (ac/h)** in the **Natural Ventilation** section and select **On 24/7** in the **Operation** section.

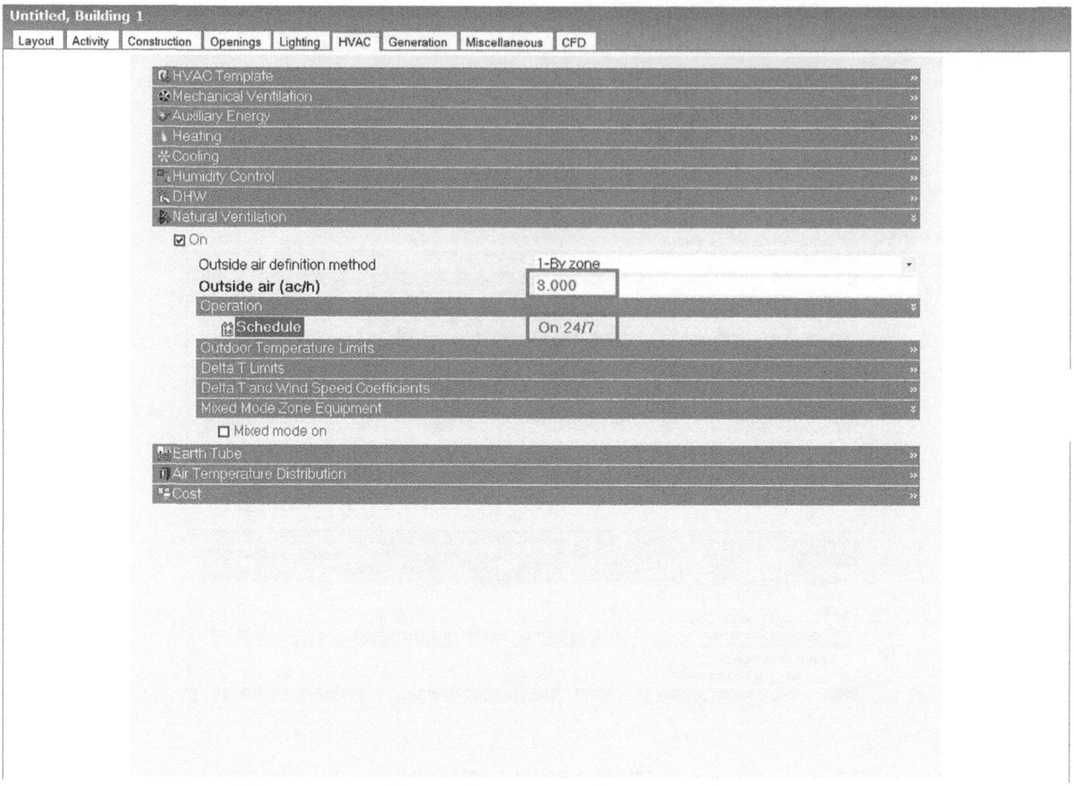

Perform an hourly simulation and record the results.

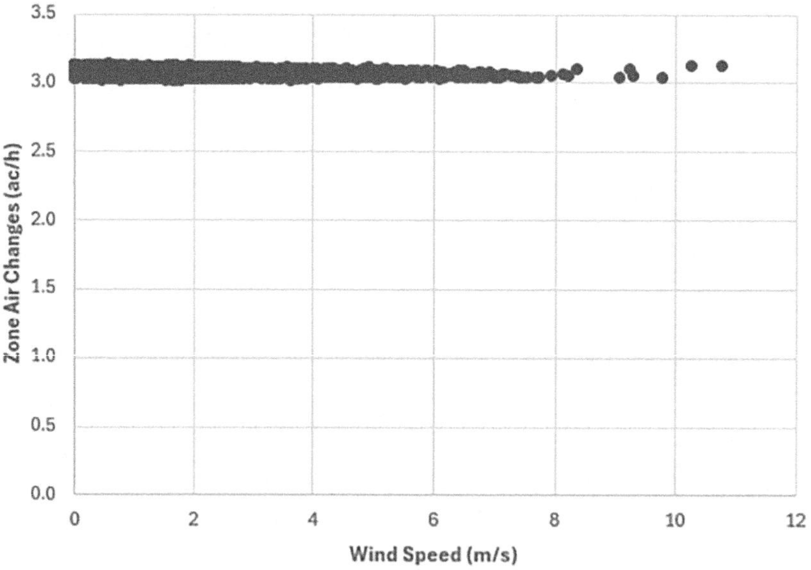

Save this model.

This tutorial illustrates the key differences between the **Calculated** and **Scheduled** approaches to natural ventilation in building energy modeling.

The calculated natural ventilation approach determines airflow rates dynamically based on real-time environmental parameters such as wind speed, wind direction, and temperature differences between indoor and outdoor air. Consequently, it is sensitive to external wind conditions and building orientation, often resulting in significant variability in ventilation performance throughout the day and across different seasons.

In contrast, the scheduled natural ventilation approach uses predefined airflow rates or user-specified schedules that remain constant regardless of actual weather or pressure conditions. While this method simplifies the modeling process and ensures predictable ventilation rates, it does not capture the dynamic interactions between the building and its external environment.

EXERCISE 6.1

Repeat this tutorial for **% Glazing area opens** to 50. Plot the chart between **Zone air changes (ac/h)** and **Wind speed (m/s)**. Observe the change in ACH.

TUTORIAL 6.2: EVALUATING THE IMPACT OF WIND SPEED ON NATURAL VENTILATION

Goal

To understand the impact of wind speed on natural ventilation

What Are You Going to Learn?

- How to change the weather data file for wind speed and direction

Problem Statement

In this tutorial, you are going to use a 10 m × 10 m single-zone model with operable windows on both the north and south façades. The objective is to assess the impact of wind speed on natural ventilation. To do this, you will modify the weather file to include two wind speed scenarios—1 m/s and 2 m/s—with a wind direction of 0°, meaning wind flows from north to south. The simulation will be run using the modified weather data to observe changes in ACH.

Use the **New Delhi/Safdarjung, India** location.

Solution

First, you are going to modify the weather data file for wind speed and direction.

Step 1: Download the weather data for New Delhi from the following link then download the .epw file or copy the file from the **Weather data** folder.

https://energyplus.net/weather-location/asia_wmo_region_2/IND/IND_New.Delhi.421820_ISHRAE

Step 2: Download the following **CSV editor** tool:

https://home.hccnet.nl/s.j.francke/csved/index.html (by Sam Francke).
(You can use any CSV editor; for this tutorial, we used CSVed 2.5.7.)

Step 3: Open the **CSV editor** tool and select the downloaded .epw file of New Delhi weather location.

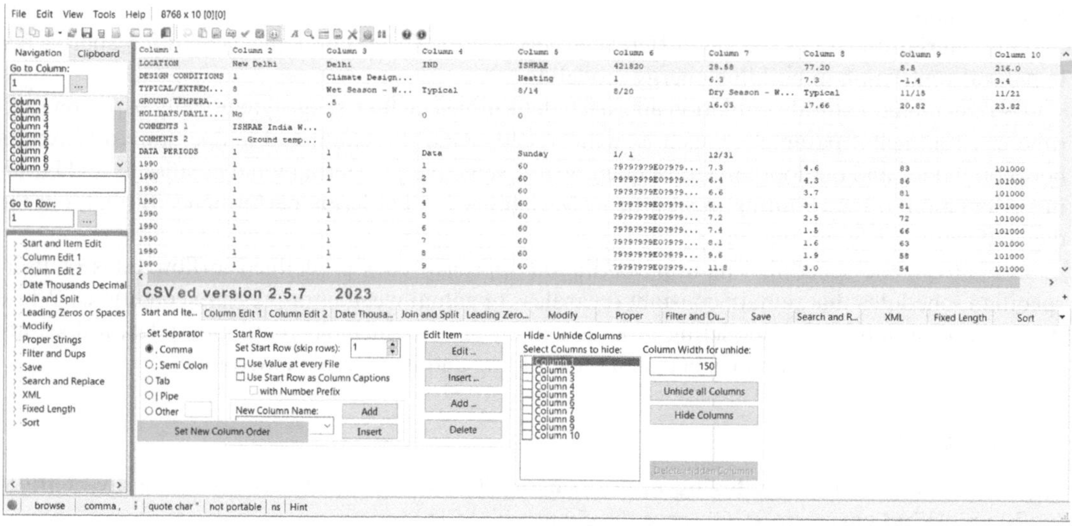

Step 4: After opening the file, go to **Set Start Row** in the **Start and Item Edit** tab. Set **Start Row** to **9**.

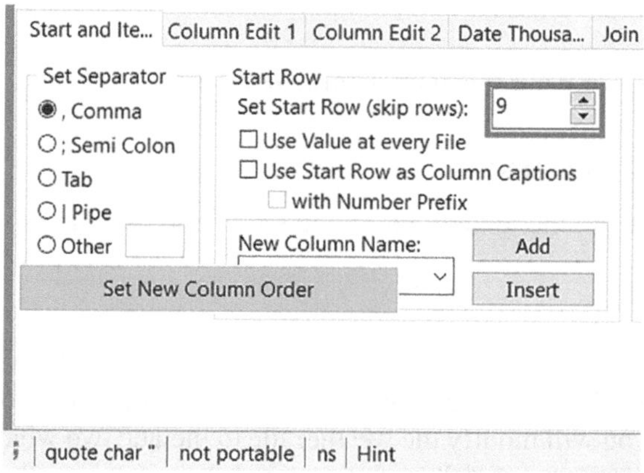

In the .epw file, columns 21–22 represent the wind direction and speed, respectively.

Column 17	Column 18	Column 19	Column 20	Column 21	Column 22	Column 23	Column 24	Column 25
0	0	0	0	289	1.4	0	5	1.3
0	0	0	0	270	1.7	0	5	1.3
0	0	0	0	265	2.0	0	1	1.3
0	0	0	0	248	1.9	0	7	1.3
0	0	0	0	270	1.7	0	5	1.3
0	0	0	0	276	2.6	1	7	1.3
0	0	0	0	280	3.3	2	8	1.3
0	0	0	0	292	3.9	3	4	1.3
0	0	0	0	293	3.9	5	9	1.3
0	0	0	0	297	3.9	6	0	1.3
0	0	0	0	292	3.9	6	8	1.3
0	0	0	0	294	4.4	6	4	1.3
0	0	0	0	290	4.7	6	0	1.3
0	0	0	0	292	5.0	5	8	1.3
0	0	0	0	287	4.5	3	2	1.3
0	0	0	0	291	3.9	2	4	1.3
0	0	0	0	292	2.8	1	6	1.3
0	0	0	0	304	2.3	1	4	1.3
0	0	0	0	308	1.9	1	2	1.3
0	0	0	0	315	2.2	2	9	1.3
0	0	0	0	306	2.4	2	2	1.3
0	0	0	0	302	2.9	2	1	1.3
0	0	0	0	292	2.8	1	4	1.3
0	0	0	0	294	2.7	1	8	1.3
0	0	0	0	290	2.1	0	6	1.3
0	0	0	0	292	1.7	0	2	1.3

Step 5: For changing the wind direction to 0°, select the **Column Edit 1** tab, select **Column 21** from the drop-down list and enter **0** in the **New Value** text box. Click **Edit Column**.

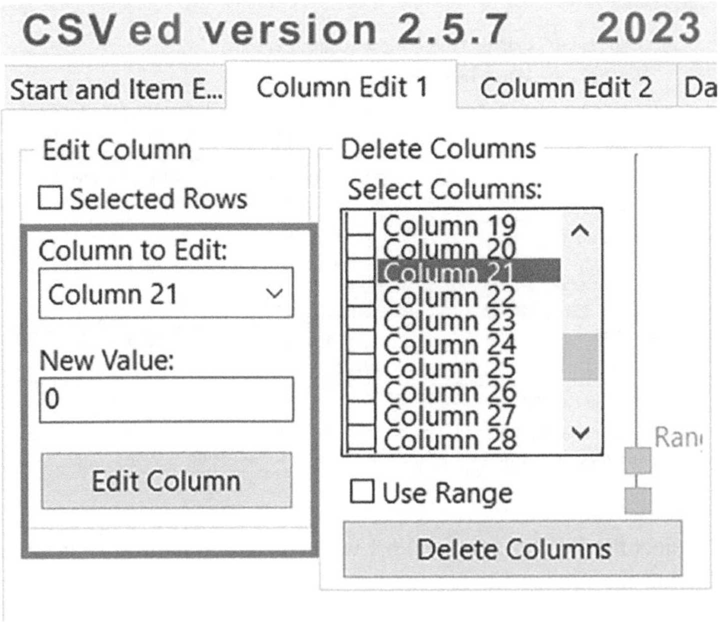

Step 6: Similarly, for changing the wind speed to 1 m/s, select **Column 22**, and enter **1** in the **New Value** text box. Click **Edit Column**.

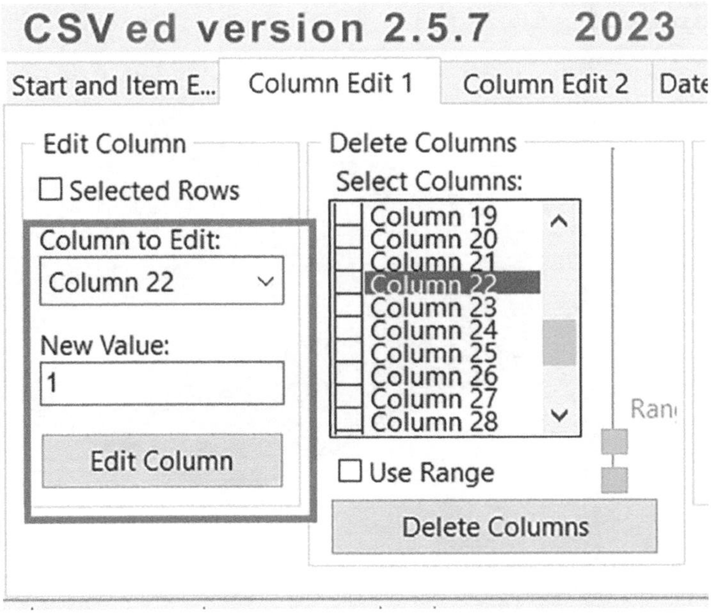

Step 7: Save the file with the name **Delhi Constant Wind.epw** on your desktop. Open **DesignBuilder**. Select the **File > Folders > Weather data** folder and paste it as the **Delhi Constant Wind.epw** weather file.

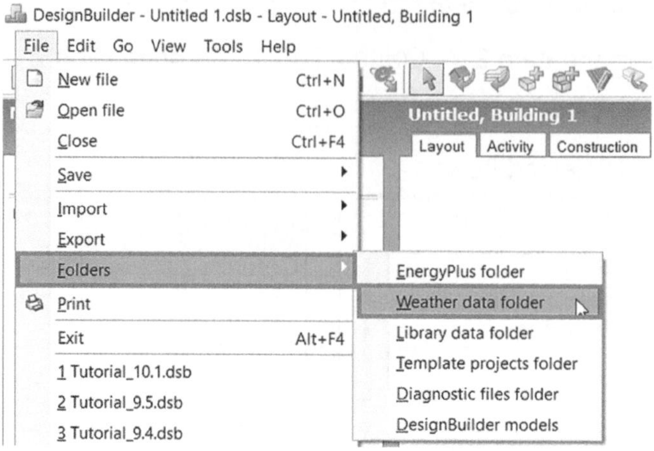

Step 8: Open the project file saved in Tutorial 6.1 with calculated natural ventilation and delete both the east and west façades.

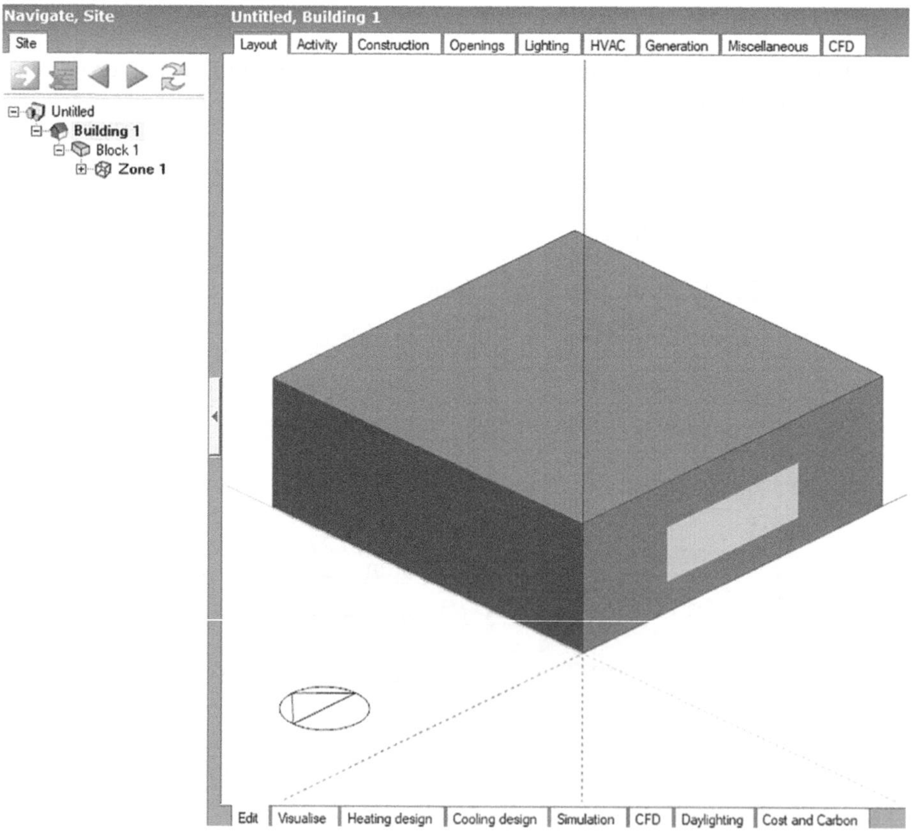

Step 9: Select the **Construction** tab. Select **Outer** for the **Zone volume calculation method** and **Zone floor area calculation method** in the **Geometry, Areas and Volumes** section.

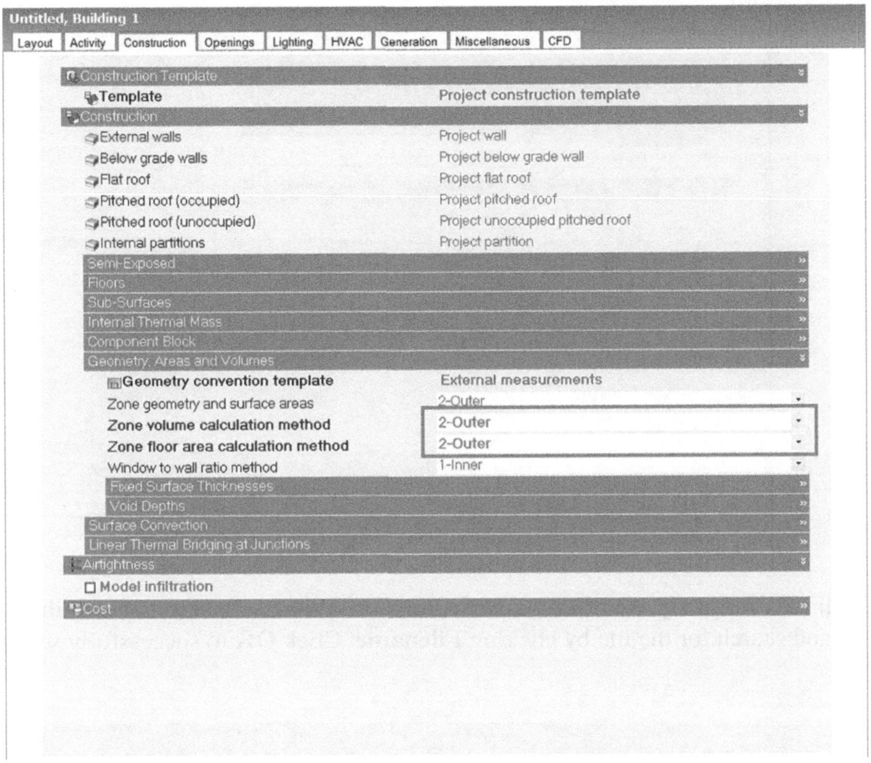

Step 10: Select the **Location** tab and select **Hourly weather data** under the **Simulation Weather Data** section.

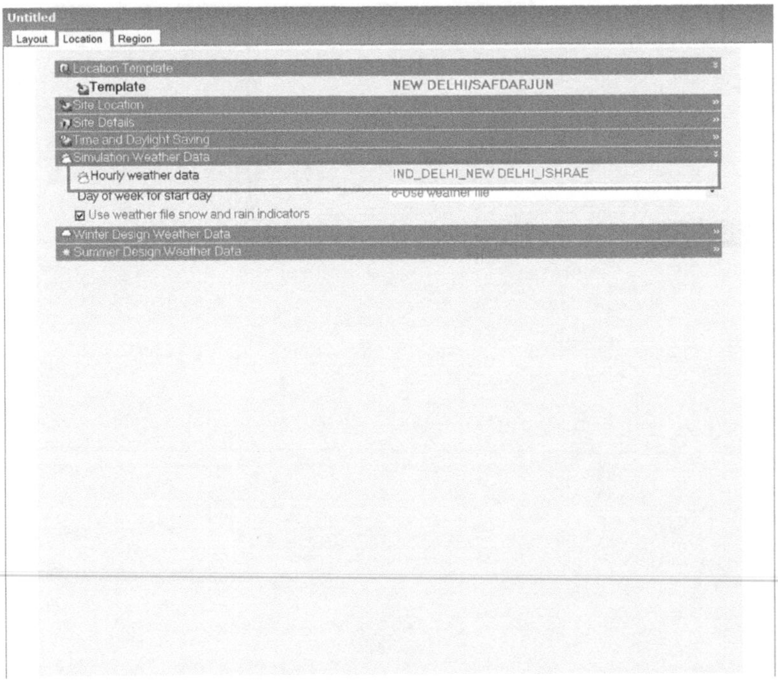

Step 11: Click the **Add new item** icon under the **Help** tab.

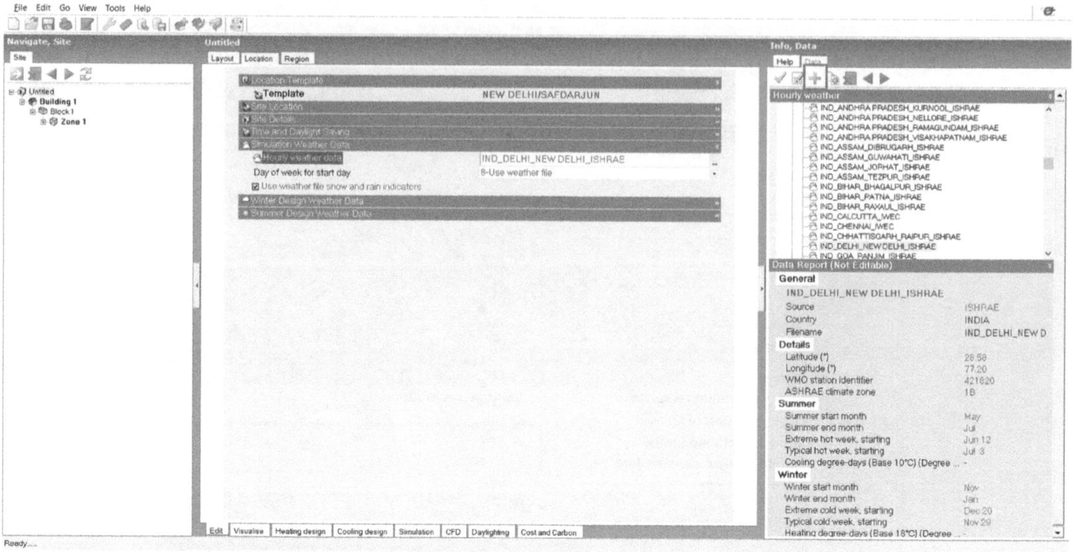

Step 12: In the **Edit hourly weather—Delhi Constant Wind** screen, rename with your custom weather file and search for the file by clicking **Filename**. Click **OK** to successfully select the file.

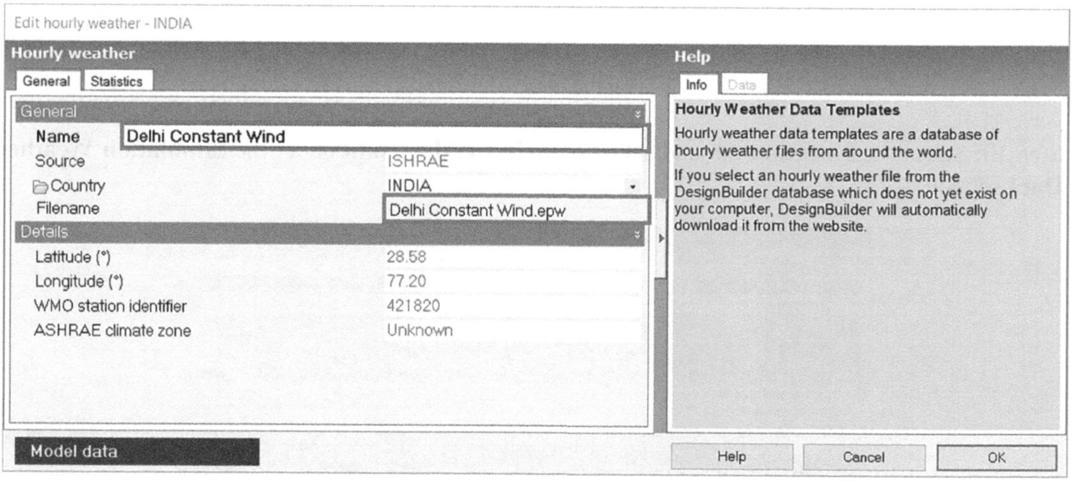

Step 13: Select **Exposed** from the **Exposure to wind** drop-down list in the **Site Details** section.

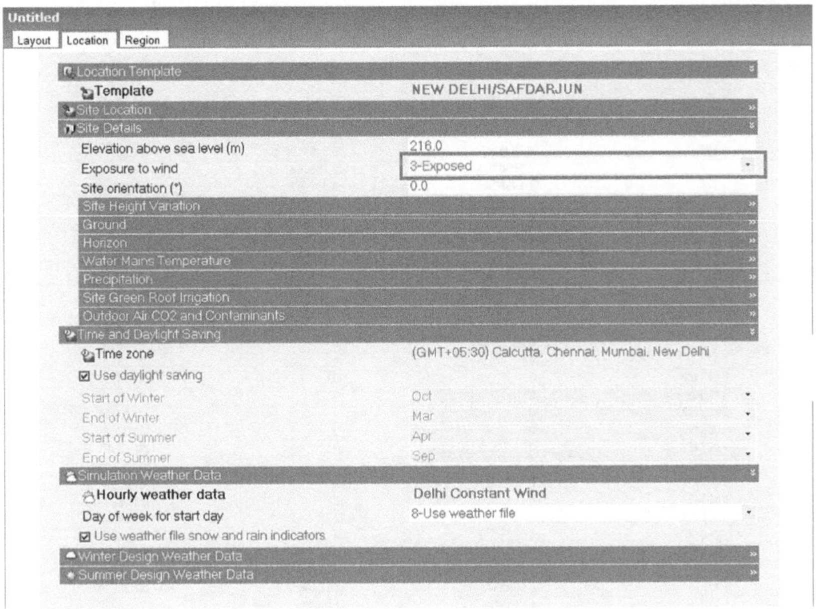

Step 14: Simulate the model and record the results. Observe the results for **Wind Speed**, **Wind Direction** and **Mech Vent + Nat Vent + Infiltration (ac/h)**.

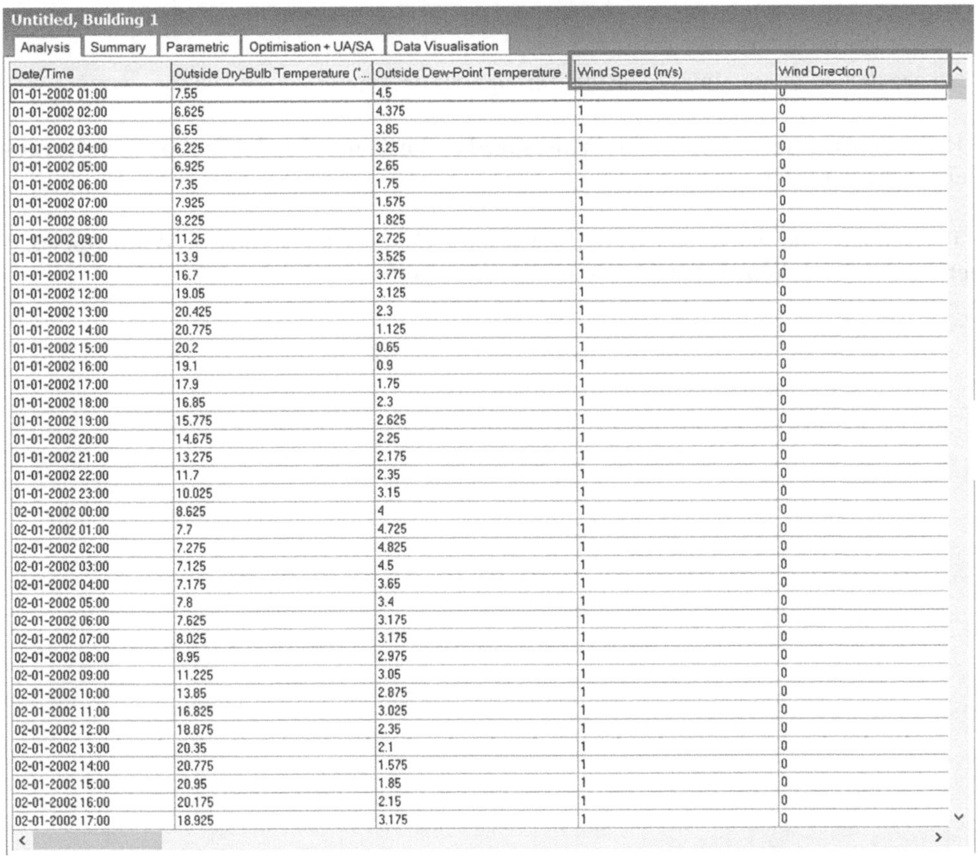
Untitled, Building 1

Analysis | Summary | Parametric | Optimisation + UA/SA | Data Visualisation

Date/Time	Outside Dry-Bulb Temperature (°...	Outside Dew-Point Temperature ...	Wind Speed (m/s)	Wind Direction (°)
01-01-2002 01:00	7.55	4.5	1	0
01-01-2002 02:00	6.625	4.375	1	0
01-01-2002 03:00	6.55	3.85	1	0
01-01-2002 04:00	6.225	3.25	1	0
01-01-2002 05:00	6.925	2.65	1	0
01-01-2002 06:00	7.35	1.75	1	0
01-01-2002 07:00	7.925	1.575	1	0
01-01-2002 08:00	9.225	1.825	1	0
01-01-2002 09:00	11.25	2.725	1	0
01-01-2002 10:00	13.9	3.525	1	0
01-01-2002 11:00	16.7	3.775	1	0
01-01-2002 12:00	19.05	3.125	1	0
01-01-2002 13:00	20.425	2.3	1	0
01-01-2002 14:00	20.775	1.125	1	0
01-01-2002 15:00	20.2	0.65	1	0
01-01-2002 16:00	19.1	0.9	1	0
01-01-2002 17:00	17.9	1.75	1	0
01-01-2002 18:00	16.85	2.3	1	0
01-01-2002 19:00	15.775	2.625	1	0
01-01-2002 20:00	14.675	2.25	1	0
01-01-2002 21:00	13.275	2.175	1	0
01-01-2002 22:00	11.7	2.35	1	0
01-01-2002 23:00	10.025	3.15	1	0
02-01-2002 00:00	8.625	4	1	0
02-01-2002 01:00	7.7	4.725	1	0
02-01-2002 02:00	7.275	4.825	1	0
02-01-2002 03:00	7.125	4.5	1	0
02-01-2002 04:00	7.175	3.65	1	0
02-01-2002 05:00	7.8	3.4	1	0
02-01-2002 06:00	7.625	3.175	1	0
02-01-2002 07:00	8.025	3.175	1	0
02-01-2002 08:00	8.95	2.975	1	0
02-01-2002 09:00	11.225	3.05	1	0
02-01-2002 10:00	13.85	2.875	1	0
02-01-2002 11:00	16.825	3.025	1	0
02-01-2002 12:00	18.875	2.35	1	0
02-01-2002 13:00	20.35	2.1	1	0
02-01-2002 14:00	20.775	1.575	1	0
02-01-2002 15:00	20.95	1.85	1	0
02-01-2002 16:00	20.175	2.15	1	0
02-01-2002 17:00	18.925	3.175	1	0

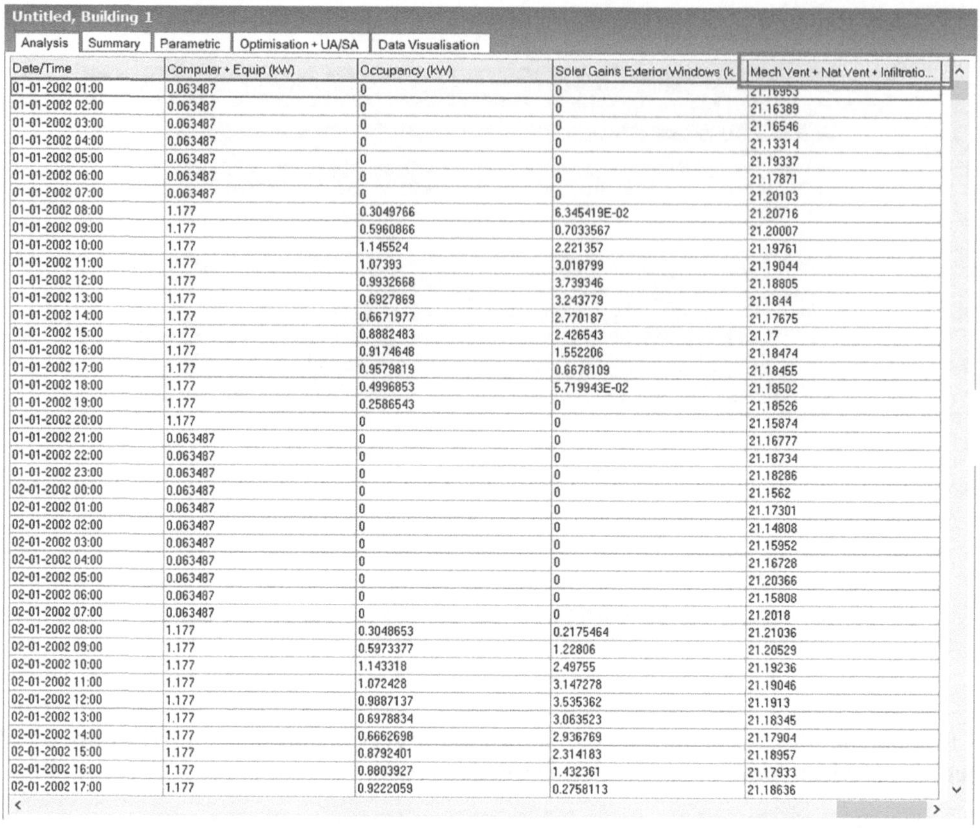

Date/Time	Computer + Equip (kW)	Occupancy (kW)	Solar Gains Exterior Windows (k	Mech Vent + Nat Vent + Infiltratio...
01-01-2002 01:00	0.063487	0	0	21.16953
01-01-2002 02:00	0.063487	0	0	21.16389
01-01-2002 03:00	0.063487	0	0	21.16546
01-01-2002 04:00	0.063487	0	0	21.13314
01-01-2002 05:00	0.063487	0	0	21.19337
01-01-2002 06:00	0.063487	0	0	21.17871
01-01-2002 07:00	0.063487	0	0	21.20103
01-01-2002 08:00	1.177	0.3049766	6.345419E-02	21.20716
01-01-2002 09:00	1.177	0.5960866	0.7033567	21.20007
01-01-2002 10:00	1.177	1.145524	2.221357	21.19761
01-01-2002 11:00	1.177	1.07393	3.018799	21.19044
01-01-2002 12:00	1.177	0.9932668	3.739346	21.18805
01-01-2002 13:00	1.177	0.6927869	3.243779	21.1844
01-01-2002 14:00	1.177	0.6671977	2.770187	21.17675
01-01-2002 15:00	1.177	0.8882483	2.426543	21.17
01-01-2002 16:00	1.177	0.9174648	1.552206	21.18474
01-01-2002 17:00	1.177	0.9579819	0.6678109	21.18455
01-01-2002 18:00	1.177	0.4996853	5.719943E-02	21.18502
01-01-2002 19:00	1.177	0.2586543	0	21.18526
01-01-2002 20:00	1.177	0	0	21.15874
01-01-2002 21:00	0.063487	0	0	21.16777
01-01-2002 22:00	0.063487	0	0	21.18734
01-01-2002 23:00	0.063487	0	0	21.18286
02-01-2002 00:00	0.063487	0	0	21.1562
02-01-2002 01:00	0.063487	0	0	21.17301
02-01-2002 02:00	0.063487	0	0	21.14808
02-01-2002 03:00	0.063487	0	0	21.15952
02-01-2002 04:00	0.063487	0	0	21.16728
02-01-2002 05:00	0.063487	0	0	21.20366
02-01-2002 06:00	0.063487	0	0	21.15808
02-01-2002 07:00	0.063487	0	0	21.2018
02-01-2002 08:00	1.177	0.3048653	0.2175464	21.21036
02-01-2002 09:00	1.177	0.5973377	1.22806	21.20529
02-01-2002 10:00	1.177	1.143318	2.49755	21.19236
02-01-2002 11:00	1.177	1.072428	3.147278	21.19046
02-01-2002 12:00	1.177	0.9887137	3.535362	21.1913
02-01-2002 13:00	1.177	0.6978834	3.063523	21.18345
02-01-2002 14:00	1.177	0.6662698	2.936769	21.17904
02-01-2002 15:00	1.177	0.8792401	2.314183	21.18957
02-01-2002 16:00	1.177	0.8803927	1.432361	21.17933
02-01-2002 17:00	1.177	0.9222059	0.2758113	21.18636

You can see that ACH are almost constant throughout the year.

Step 15: Modify the weather data with **Wind speed** as **2 m/s** and **Wind direction** as **0 degrees** with the help of previous steps.

Step 16: Simulate the model and record the results. Observe the results in **Wind Speed**, **Wind Direction** and **Mech Vent + Nat Vent + Infiltration (ac/h)**.

Untitled, Building 1

Analysis | Summary | Parametric | Optimisation + UA/SA | Data Visualisation

Date/Time	Outside Dry-Bulb Temperature (°...	Outside Dew-Point Temperature ...	Wind Speed (m/s)	Wind Direction (°)
01-01-2002 01:00	7.55	4.5	2	0
01-01-2002 02:00	6.625	4.375	2	0
01-01-2002 03:00	6.55	3.85	2	0
01-01-2002 04:00	6.225	3.25	2	0
01-01-2002 05:00	6.925	2.65	2	0
01-01-2002 06:00	7.35	1.75	2	0
01-01-2002 07:00	7.925	1.575	2	0
01-01-2002 08:00	9.225	1.825	2	0
01-01-2002 09:00	11.25	2.725	2	0
01-01-2002 10:00	13.9	3.525	2	0
01-01-2002 11:00	16.7	3.775	2	0
01-01-2002 12:00	19.05	3.125	2	0
01-01-2002 13:00	20.425	2.3	2	0
01-01-2002 14:00	20.775	1.125	2	0
01-01-2002 15:00	20.2	0.65	2	0
01-01-2002 16:00	19.1	0.9	2	0
01-01-2002 17:00	17.9	1.75	2	0
01-01-2002 18:00	16.85	2.3	2	0
01-01-2002 19:00	15.775	2.625	2	0
01-01-2002 20:00	14.675	2.25	2	0
01-01-2002 21:00	13.275	2.175	2	0
01-01-2002 22:00	11.7	2.35	2	0
01-01-2002 23:00	10.025	3.15	2	0
02-01-2002 00:00	8.625	4	2	0
02-01-2002 01:00	7.7	4.725	2	0
02-01-2002 02:00	7.275	4.825	2	0
02-01-2002 03:00	7.125	4.5	2	0
02-01-2002 04:00	7.175	3.65	2	0
02-01-2002 05:00	7.8	3.4	2	0
02-01-2002 06:00	7.625	3.175	2	0
02-01-2002 07:00	8.025	3.175	2	0
02-01-2002 08:00	8.95	2.975	2	0
02-01-2002 09:00	11.225	3.05	2	0
02-01-2002 10:00	13.85	2.875	2	0
02-01-2002 11:00	16.825	3.025	2	0
02-01-2002 12:00	18.875	2.35	2	0
02-01-2002 13:00	20.35	2.1	2	0
02-01-2002 14:00	20.775	1.575	2	0
02-01-2002 15:00	20.95	1.85	2	0
02-01-2002 16:00	20.175	2.15	2	0
02-01-2002 17:00	18.925	3.175	2	0

Untitled, Building 1

Analysis | Summary | Parametric | Optimisation + UA/SA | Data Visualisation

Date/Time	Computer + Equip (kW)	Occupancy (kW)	Solar Gains Exterior Windows (k...	Mech Vent + Nat Vent + Infiltratio...
01-01-2002 01:00	0.063487	0	0	42.41993
01-01-2002 02:00	0.063487	0	0	42.4282
01-01-2002 03:00	0.063487	0	0	42.40636
01-01-2002 04:00	0.063487	0	0	42.39238
01-01-2002 05:00	0.063487	0	0	42.40321
01-01-2002 06:00	0.063487	0	0	42.37769
01-01-2002 07:00	0.063487	0	0	42.38144
01-01-2002 08:00	1.177	0.3062825	6.345419E-02	42.40791
01-01-2002 09:00	1.177	0.6013043	0.7033567	42.39231
01-01-2002 10:00	1.177	1.157779	2.221357	42.38909
01-01-2002 11:00	1.177	1.087486	3.018799	42.36862
01-01-2002 12:00	1.177	1.006327	3.739346	42.36006
01-01-2002 13:00	1.177	0.7017935	3.243779	42.34918
01-01-2002 14:00	1.177	0.6759262	2.770187	42.33932
01-01-2002 15:00	1.177	0.9028008	2.426543	42.35122
01-01-2002 16:00	1.177	0.9359501	1.552206	42.37861
01-01-2002 17:00	1.177	0.9785101	0.6678109	42.38833
01-01-2002 18:00	1.177	0.5095469	5.719943E-02	42.38671
01-01-2002 19:00	1.177	0.2632762	0	42.39115
01-01-2002 20:00	1.177	0	0	42.37344
01-01-2002 21:00	0.063487	0	0	42.37552
01-01-2002 22:00	0.063487	0	0	42.39511
01-01-2002 23:00	0.063487	0	0	42.41157
02-01-2002 00:00	0.063487	0	0	42.42316
02-01-2002 01:00	0.063487	0	0	42.41878
02-01-2002 02:00	0.063487	0	0	42.39798
02-01-2002 03:00	0.063487	0	0	42.39544
02-01-2002 04:00	0.063487	0	0	42.38965
02-01-2002 05:00	0.063487	0	0	42.39429
02-01-2002 06:00	0.063487	0	0	42.36761
02-01-2002 07:00	0.063487	0	0	42.38354
02-01-2002 08:00	1.177	0.3064344	0.2175464	42.41422
02-01-2002 09:00	1.177	0.6028138	1.22806	42.39683
02-01-2002 10:00	1.177	1.15605	2.49755	42.38729
02-01-2002 11:00	1.177	1.086955	3.147278	42.36979
02-01-2002 12:00	1.177	1.002675	3.535362	42.36298
02-01-2002 13:00	1.177	0.7077262	3.063523	42.34998
02-01-2002 14:00	1.177	0.6755928	2.936769	42.34246
02-01-2002 15:00	1.177	0.8933061	2.314183	42.35477
02-01-2002 16:00	1.177	0.8966584	1.432361	42.36456
02-01-2002 17:00	1.177	0.9402581	0.2758113	42.3815

You can calculate the ACH for two windows with the following equation (from *ASHRAE Handbook of Fundamentals*, 2015. © 2015 ASHRAE, www.ashrae.org):

$$Q = U_w \times \sqrt{\frac{C_{p1} - C_{p2}}{\left(\frac{1}{A_1 \times C_1}\right)^2 + \left(\frac{1}{A_2 \times C_2}\right)^2}} \tag{6.1}$$

where C_p is the pressure drag coefficient, C is the discharge coefficient, U is the wind speed, and A is the area of the opening

WIND PRESSURE COEFFICIENT DATA

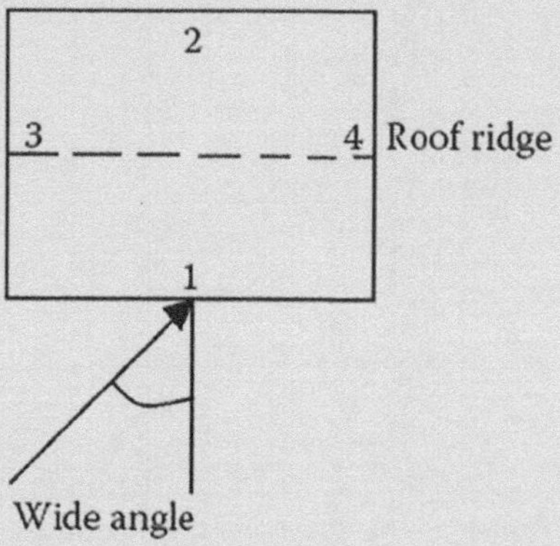

Low-rise buildings (up to three stories)
Length-to-width ratio: 1:1
Shielding condition: exposed
Wind speed reference level: building height
Wind angle (Table 6.1)

TABLE 6.1
Wind Pressure Coefficient Data

Location	0°	45°	90°	135°	180°	225°	270°	315°
Face 1	0.70	0.35	−0.50	−0.40	−0.20	−0.40	−0.50	0.35
Face 2	−0.20	−0.40	−0.50	0.35	0.70	0.35	−0.50	−0.40
Face 3	−0.50	0.35	0.70	0.35	−0.50	−0.40	−0.20	−0.40
Face 4	−0.50	−0.40	−0.20	−0.40	−0.50	0.35	0.70	0.35

Sources: www.designbuilder.co.uk/downloads/AIVCWindPressureCoefficientData.pdf and www.designbuilder.co.uk/helpv7.3/Content/Pressure_Coefficients_Data.htm

WIND PRESSURE COEFFICIENT TEMPLATES

Wind pressure coefficients are used when the **Natural ventilation model** option is set to **Calculated**. The EnergyPlus **Airflow network** calculations use pressure coefficients when calculating wind-induced pressure on each surface during simulations when the **Calculated Natural ventilation** option is selected.

DesignBuilder is supplied with a database of wind pressure coefficients based on the data from Martin Liddament, *Air Infiltration Calculation Techniques: An Applications Guide* (Paris: Air Infiltration and Ventilation Centre, International Energy Agency, 1986). The C_p data are buildings of three stories or fewer with square surfaces and for three levels of site exposure. The data are given in 45° increments.

Sources: https://designbuilder.co.uk/downloads/AIVCWindPressureCoefficientData.pdf and https://designbuilder.co.uk/helpv7.3/Content/Pressure_Coefficients_Data.htm

DISCHARGE COEFFICIENT

The *discharge coefficient*, which ranges from 0–1, is a measure of the amount of pressure that the air loses as it goes through every opening of flow resistance. The default of 0.65 is a good general-purpose setting.

Source: https://designbuilder.co.uk/helpv7.3/Content/Calculated_Natural_Ventilation_Options.htm

Wind speed provided in the weather file is for 10 m height. You can use the following formula to calculate the wind speed at 2 m height.

$$u = u_r \left(\frac{z}{z_r} \right)^{\alpha} \tag{6.2}$$

where α is approximately 1/7, or 0.143 (it varies with local topography)

Table 6.2 and Table 6.3 give the calculation for natural ventilation in the zone.

TABLE 6.2
Calculation of Zone Air Changes per Hour at a Wind Speed of 1 m/s at 10 m Height

Wind speed at 1 m	0.72 m/s
Area of window	6.59 m^2
C_{p1}	0.70
C_{p2}	–0.20
C_1	0.65
C_2	0.65
Q	2.07 m^3/s
Q	7,442 m^3/h
Zone volume	350 m^3
Air changes	21.26 ac/h

TABLE 6.3
Calculation of Zone Air Changes per Hour at a Wind Speed of 2 m/s at 10 m Height

Wind speed at 2 m	1.44 m/s
Area of window	6.59 m^2
C_{p1}	0.70
C_{p2}	−0.20
C_1	0.65
C_2	0.65
Q	4.13 m^3/s
Q	14,885 m^3/h
Zone volume	350 m^3
Air changes	42.53 ac/h

Refer to Table 6.1 for wind pressure coefficients.

Air change values calculated from the formula are matched with DesignBuilder simulated values.

The analysis confirms that an increase in wind speed leads to a rise in ACH, demonstrating the influence of wind-driven forces on natural ventilation performance. This relationship aligns with theoretical expectations, as greater wind speeds generate higher pressure differentials across openings, enhancing airflow through the space.

EXERCISE 6.2

Repeat this tutorial for a window area of 16 m^2.

TUTORIAL 6.3: EVALUATING THE IMPACT OF WINDOW OPENING AND CLOSING SCHEDULE

Goal

To evaluate the impact of window opening and closing schedule

What Are You Going to Learn?

- How to define a window opening and closing schedule for natural ventilation

Problem Statement

In this tutorial, you are going to use the model saved in Tutorial 6.2 and add windows in the remaining two directions. You need to simulate the model for the following two window operation schedules.

1. On (windows are always open)
2. 00:00–18:00 h Mon–Sat

Use the **New Delhi/Safdarjung, India** location.

Solution

Step 1: Open the project saved in Tutorial 6.2 and use the weather file IND_DELHI_NEW DELHI_ISHRAE.

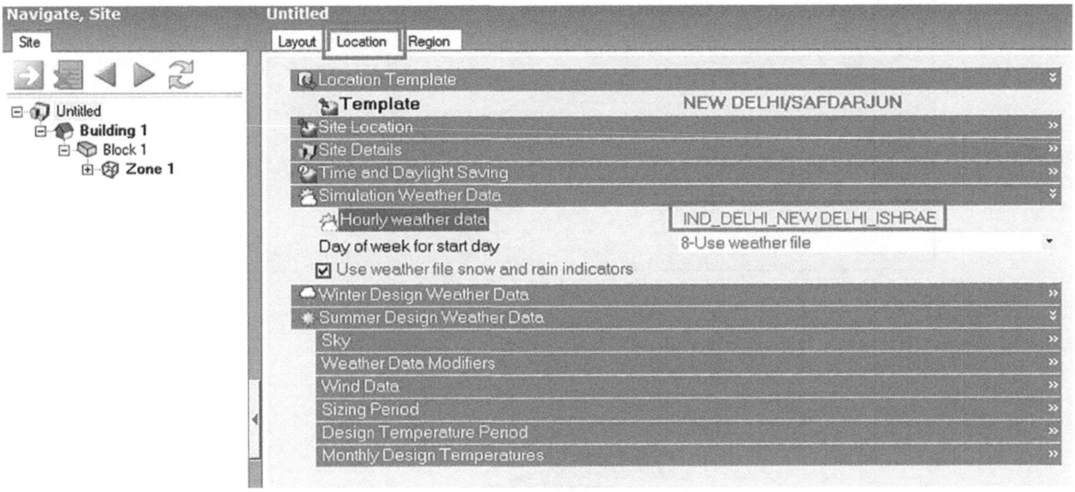

Step 2: In the navigation tree, select **East facing wall** by selecting **Wall—35.000 m²–90.0°**.

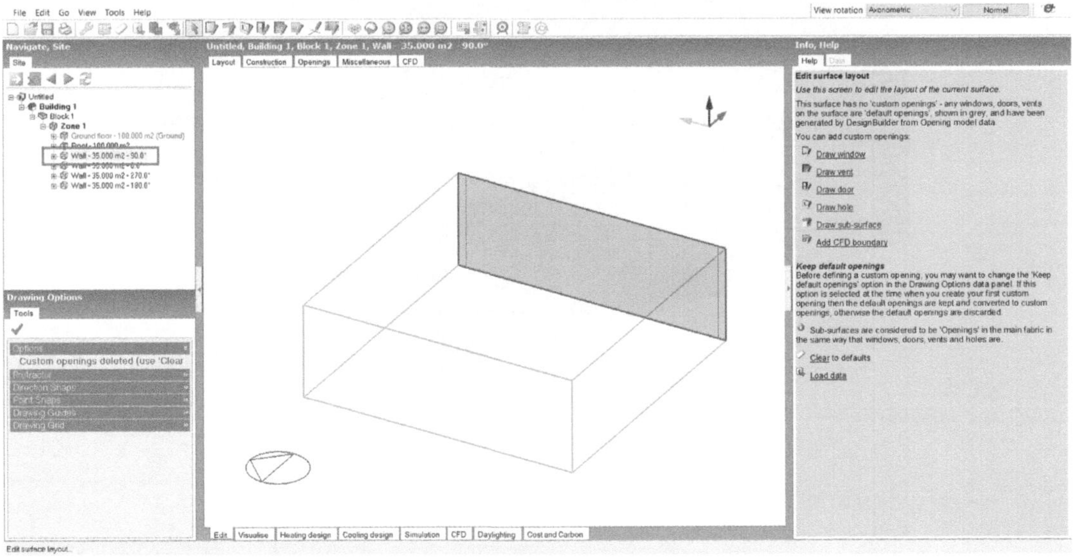

Step 3: Select **Left** in the **View rotation** drop-down list, and then select **Draw window** in the **Help** tab to draw a window with the same dimensions (sill height 0.8 m, 2.78 m from left, the height of window 1.5 m, width of window 4.4 m) as that on the north or south side.

Step 4: Similarly, draw the west window.

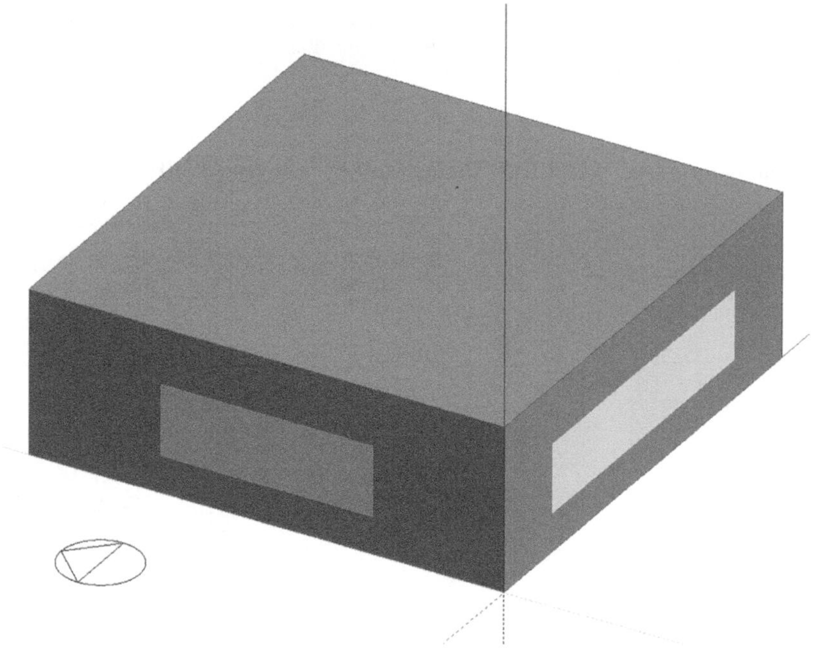

Step 5: Select the **Openings** tab. Click the **Operation** section. Make sure that **Operation schedule** is **On 24/7**.

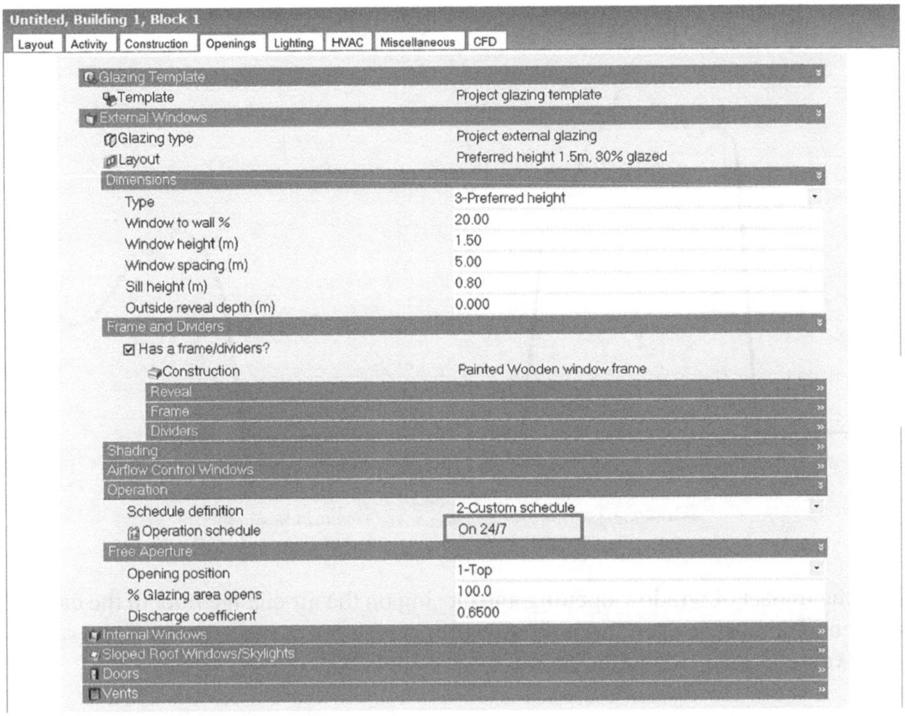

Simulate the model and record the results by exporting them to a spreadsheet.

Step 6: Repeat the preceding steps for an **Operation schedule** of **8:00–18:00 Mon–Sat**.

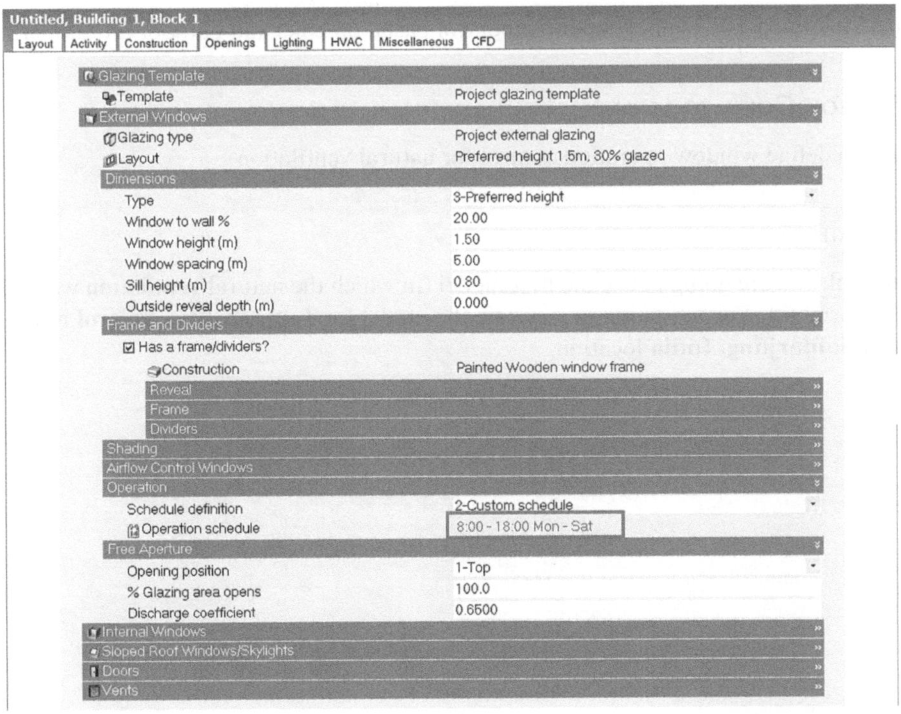

Simulate the model and record the results. Plot the graph for 1 January and 2 January for **Mech Vent + Nat Vent + Infiltration (Air changes per hour, ac/h)**.

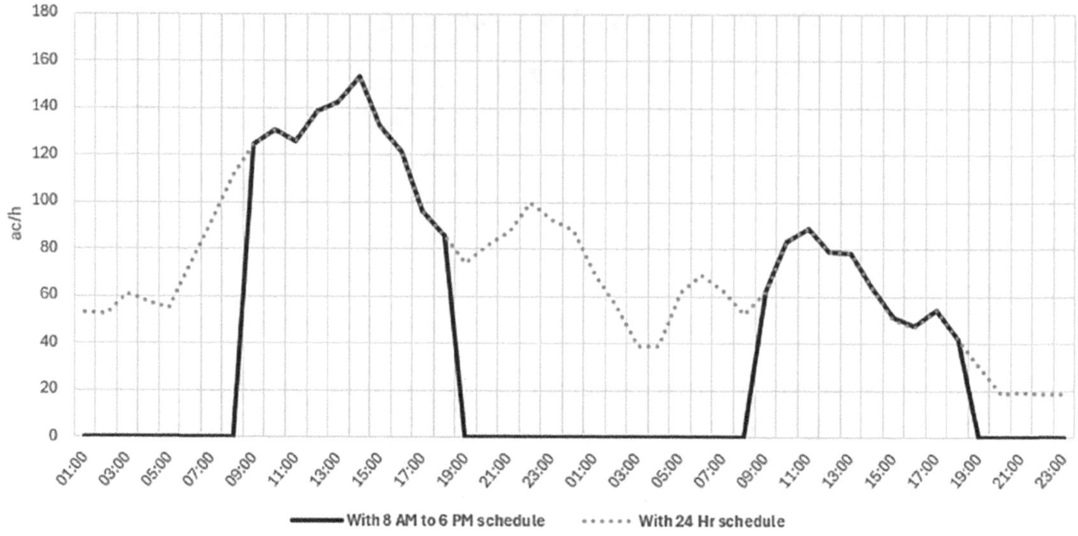

You can see the impact of window opening and closing on the air change rate, in the case of 24-hour schedule when windows are open throughout, while in the case of the 8:00–18:00 hh schedule, there is no natural ventilation during night.

TUTORIAL 6.4: EVALUATING THE IMPACT OF WINDOW OPENING CONTROL BASED ON OUTDOOR AND INDOOR TEMPERATURE

Goal

To evaluate the impact of window opening control based on zone and outdoor air temperature

What Are You Going to Learn?

- How to define window operation control for natural ventilation

Problem Statement

In this tutorial, you are going to use the first model (in which the natural ventilation was calculated) saved in Tutorial 6.1. You are going to simulate the model for **Temperature control mode**. Use the **New Delhi/Safdarjung, India** location.

Solution

Step 1: Open the model saved from Tutorial 6.1 that uses the *Calculated Natural Ventilation* option.

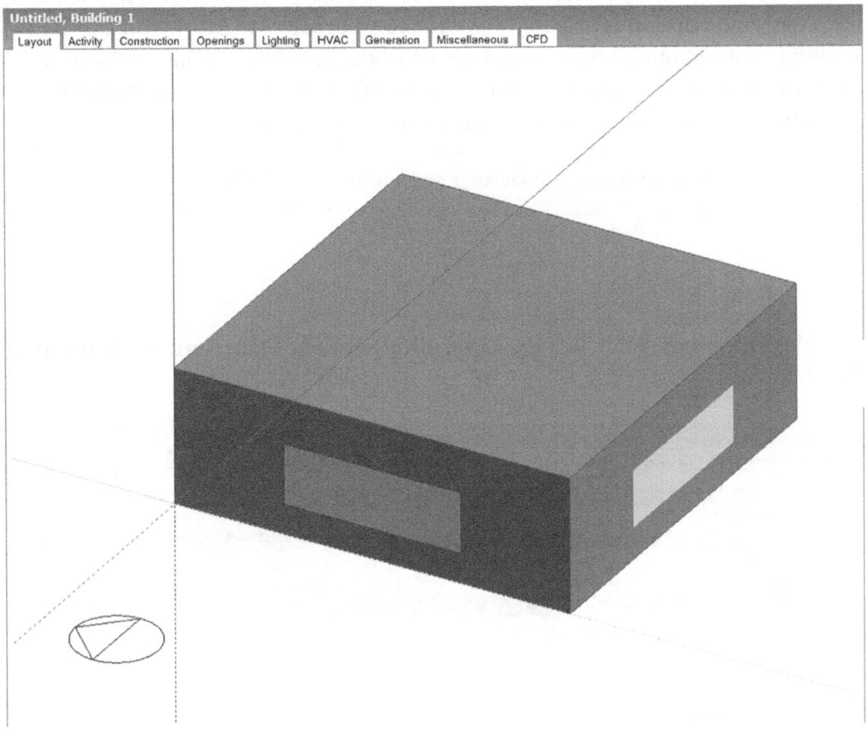

Step 2: Select the **Activity** tab. Set **Min temperature** to **20.0**.

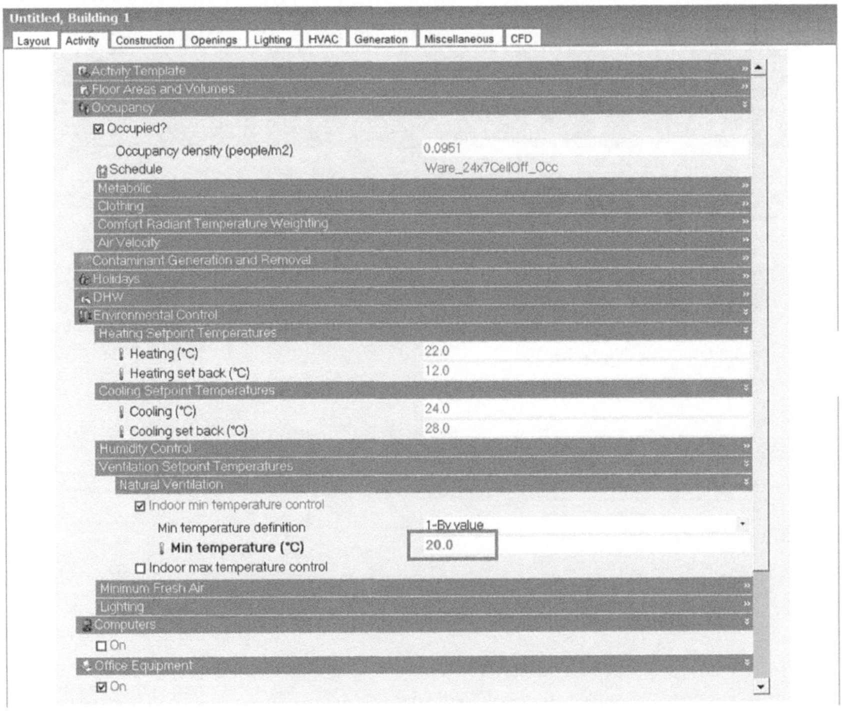

NATURAL VENTILATION MINIMUM TEMPERATURE

This is the fixed indoor temperature below which ventilation is shut off. The control is visible when the **1-By** value option is selected for **Min temperature** definition. It can be thought of as the cooling setpoint temperature which controls the activation of natural ventilation. If the inside air temperature is greater than this setpoint temperature (and the **Natural ventilation** operation schedule is on), then natural ventilation can take place.

Source: www.designbuilder.co.uk/helpv7.3/Content/_Environmental_comfort.htm

Step 3: Select the **Openings** tab. Click the **Operation** section. Make sure that **Operation schedule** is set to **On 24/7**.

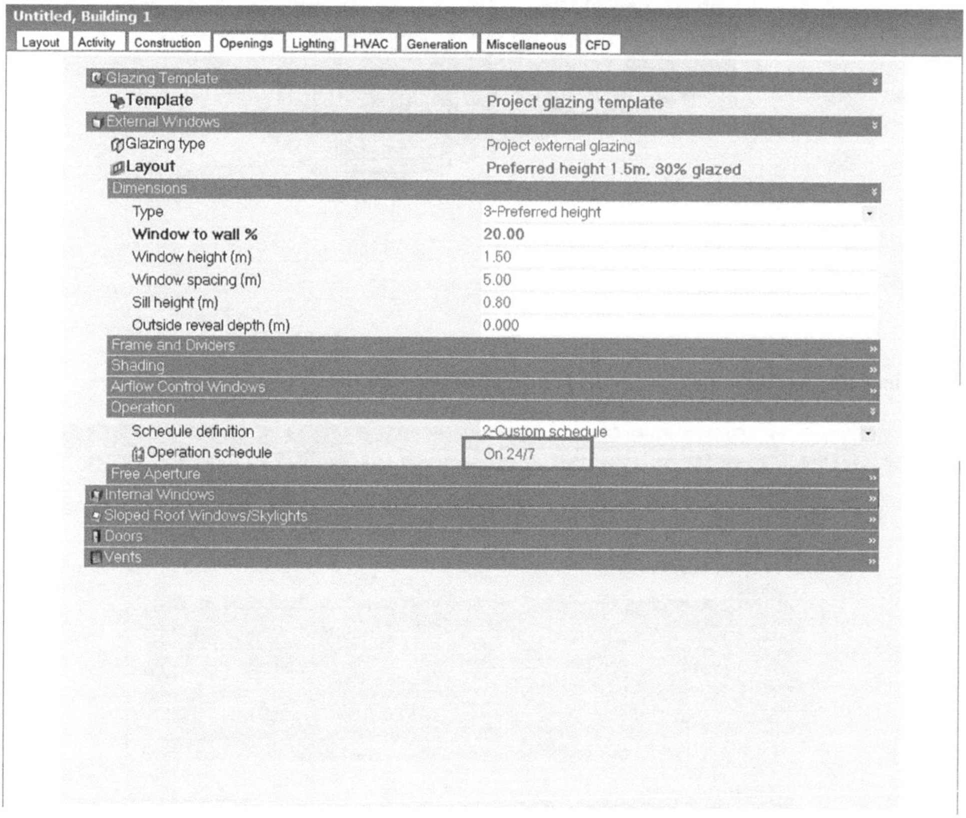

Step 4: Select the **HVAC** tab. Select **Temperature** from the **External** and **Internal Control mode** drop-down lists.

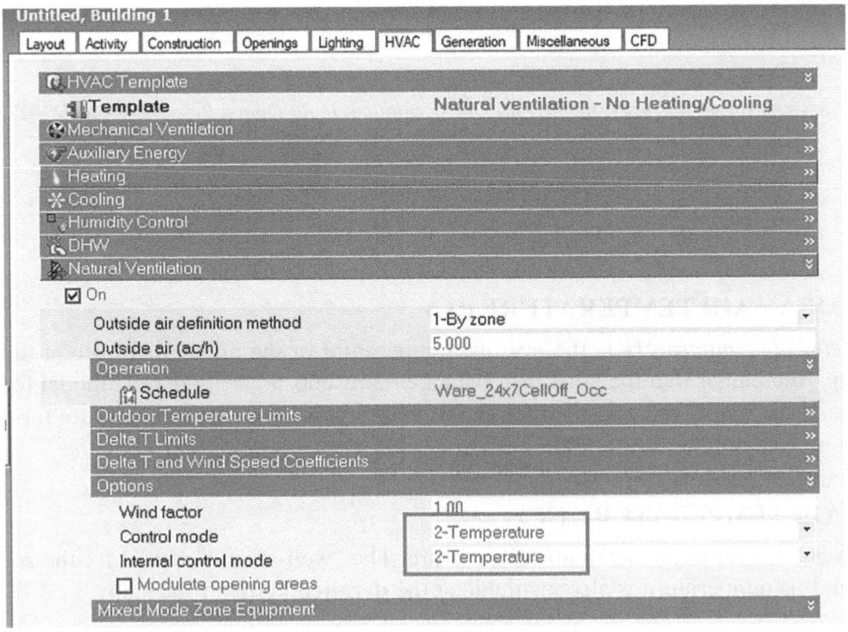

To see the effect of window opening controls, we need the following.

- Generate and view the results at system time steps instead of fixed user-defined time steps, which can be achieved by selecting the **detailed** frequency option on an HVAC output variable (e.g., **Zone Air Temperature**)
- Generate more output variables such as window opening factor, zone infiltration, and zone air temperature

FIELD: VENTILATION CONTROL MODE

This specifies the type of zone-level natural ventilation control. The windows are "open" when $T_{zone_air} > T_{setpoint}$ and $t_{zone_air} > T_{outside_air}$ and the schedule value = 1, whereby $T_{outside_air}$ is equal to the outdoor air temperature, T_{zone_air} is equal to the previous time step's zone air temperature, and $T_{setpoint}$ is equal to the **Vent Temperature Schedule** value.

Source: www.designbuilder.co.uk/helpv7.3/Content/_Operation2.htm

The *system time step* is a variable-length time step that governs the driving time step for HVAC and plant system modeling. The user cannot directly control the system time step (except by use of the **ConvergenceLimits** object). When the HVAC portion of the simulation begins its solution for the current zone time step, it uses the zone time step as its maximum length but can then reduce the time step, as necessary, to improve the solution. Users can see the system time step used if they select the **detailed** frequency option on an HVAC output variable (e.g., **Zone Air Temperature**). In contrast, the **Zone** variables will only be reported on the zone time step (e.g., **Zone Mean Air Temperature**). Note that hourly data (such as outdoor conditions expressed by **Design Days** or **Weather data**) are interpolated to the **Zone Time** step.

Source: U.S. Department of Energy, *EnergyPlus Documentation, Input-Output Reference*, March 2022; available at https://energyplus.net/documentation

ZONE MEAN AIR TEMPERATURE (°C)

Zone mean air temperature is the average temperature of the air temperature at the system time step. Remember that the zone heat balance represents a "well-stirred" model for a zone; therefore, there is only one mean air temperature to represent the air temperature for the zone. This is reported only at the zone time step.

ZONE AIR TEMPERATURE (°C)

This is very similar to the mean air temperature. The "well-stirred" model for the zone is the basis, but this temperature is also available at the **detailed** system time step.

Source: U.S. Department of Energy, *EnergyPlus Documentation, Input-Output Reference*, https://energyplus.net/documentation

Step 5: Minimize the **DesignBuilder** window. Create a new file named **output.idf**. You can use any text editor such as **Notepad++** or **Editplus**. Write additional output variables as follows:

Output:Variable, *, Zone Air Temperature, Detailed;
Output:Variable, *, AFN Surface Venting Window or Door Opening Factor, Detailed;
Output:Variable, *, AFN Zone Infiltration, Detailed;

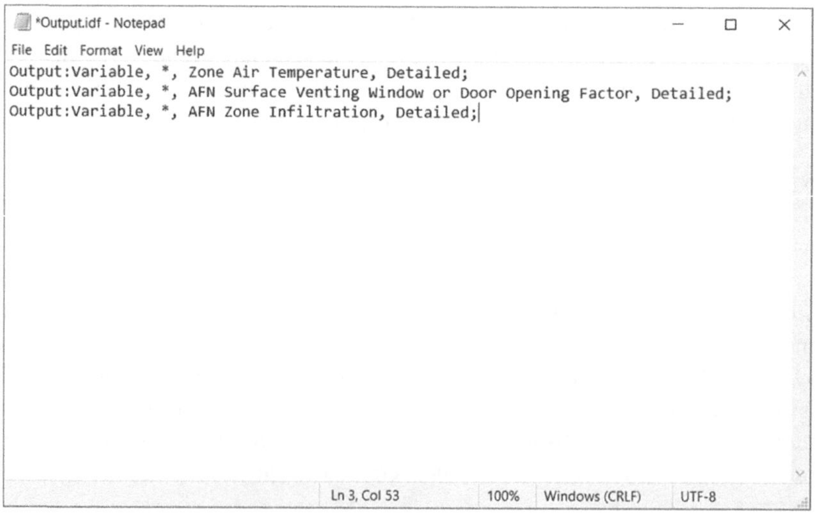

This .idf file must be placed in the path shown in the following figure.

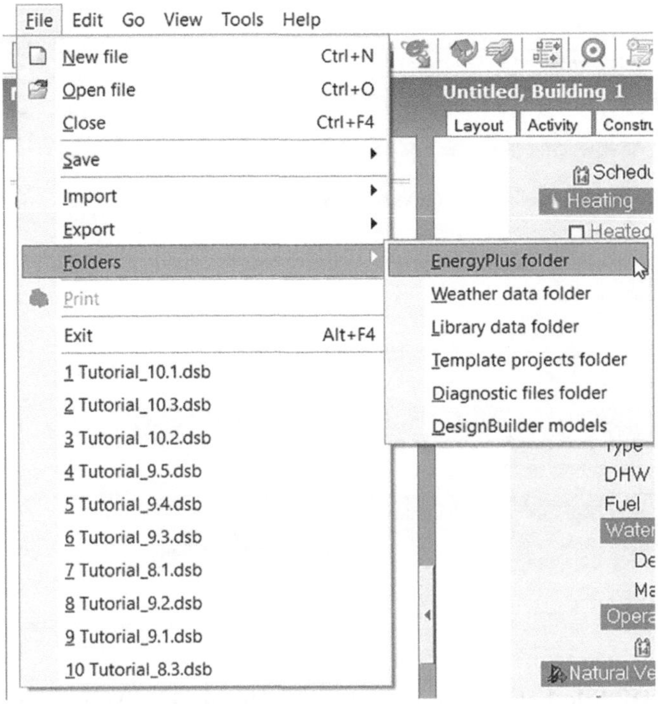

Step 6: Maximize the **DesignBuilder** window, and select the **HVAC** tab. Click **Simple** in the **Help** tab. The **Model Options—Building and Block** screen appears.

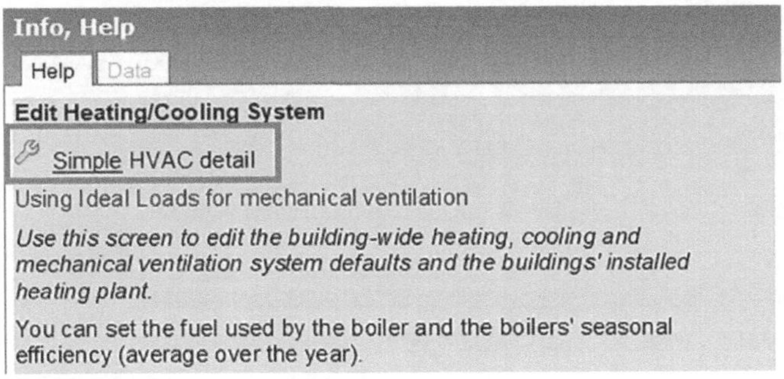

Step 7: Select the **Simulation** screen tab. Click the **Advanced** section.

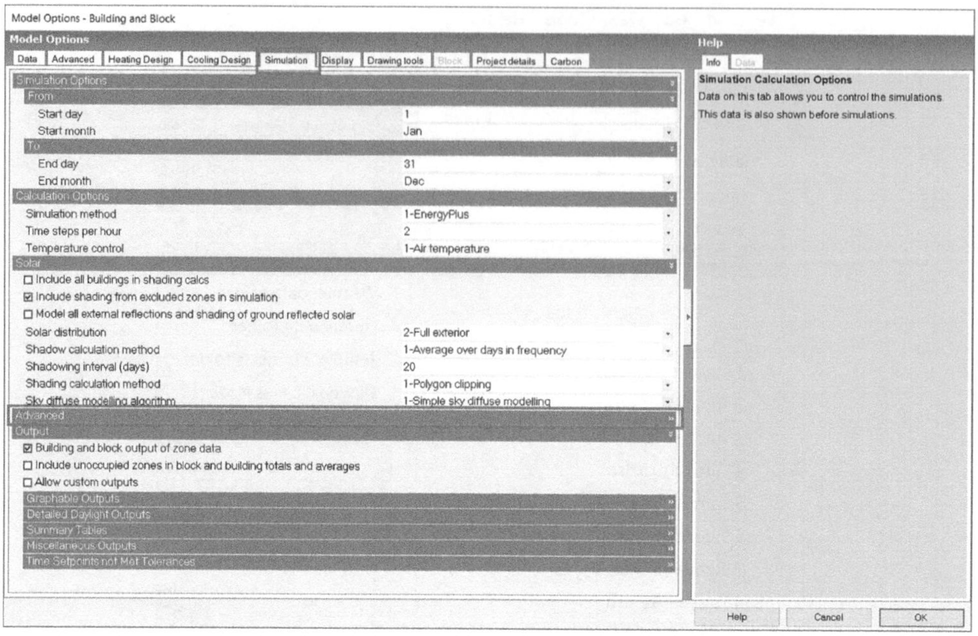

Step 8: Select the **IDF File 1** checkbox and select the **Output.idf** file from the **EnergyPlus** folder. Click **OK**.

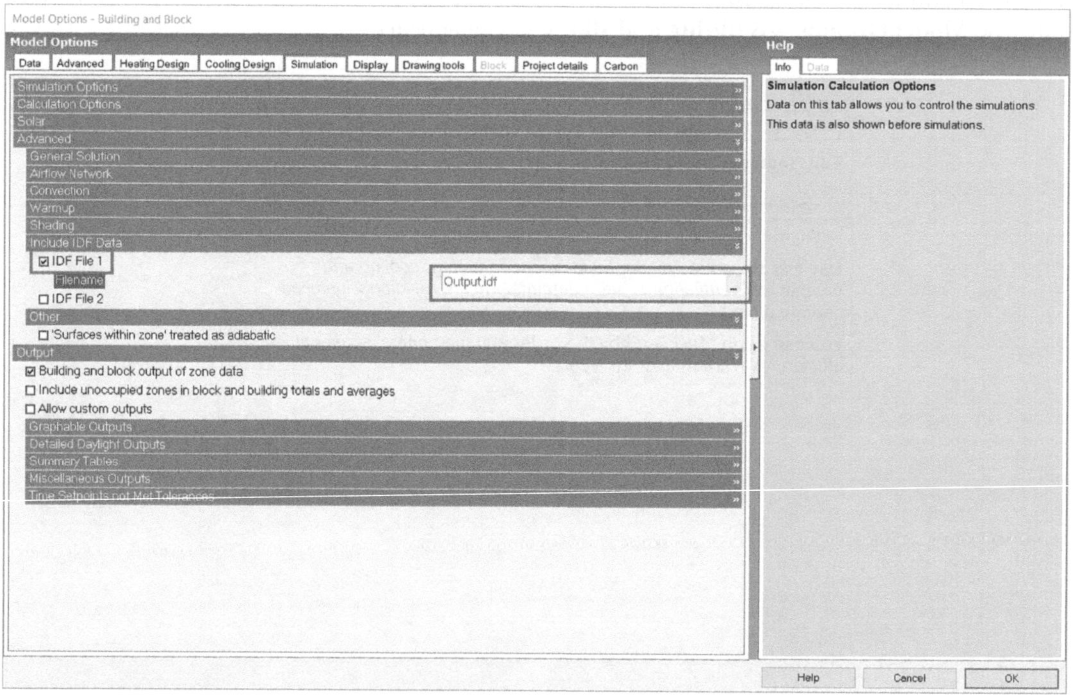

Step 9: Select the **Simulation** tab. The **Edit Calculation Options** screen appears. Click **OK**.

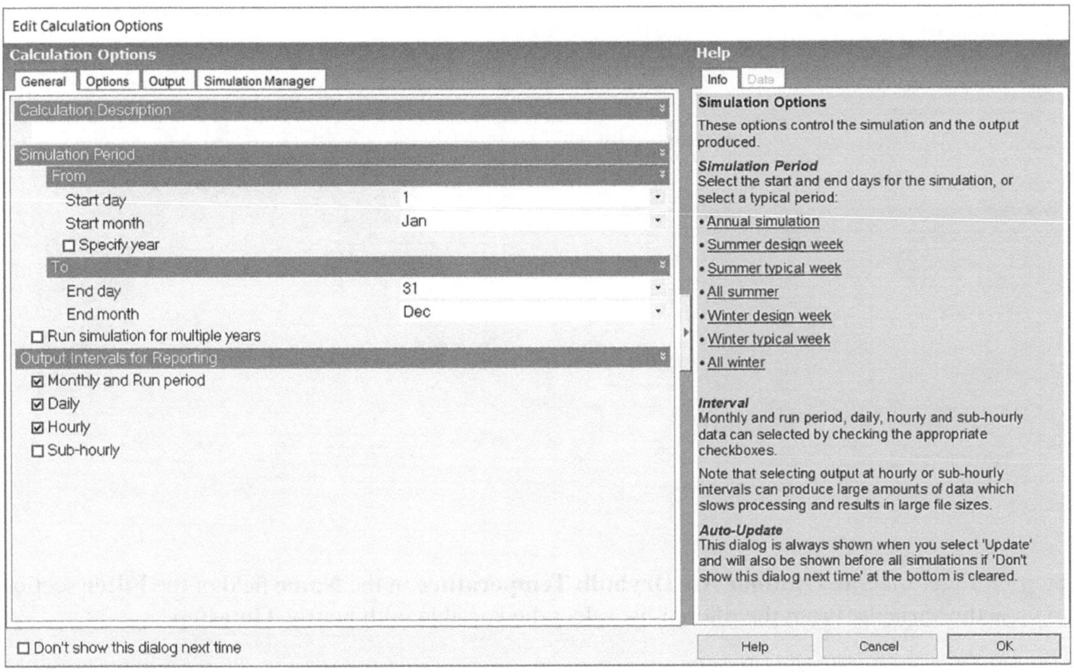

Save this simulation model.

After the simulation, the results appear on the screen, but DesignBuilder does not support viewing "Detailed" reporting frequency. After the simulation, an ESO file is generated in the **EnergyPlus** folder which contains results at detailed reporting frequency. You can view this in the **xESOView** tool, which can be downloaded from http://xesoview.sourceforge.net/.

Step 10: Click the **View EnergyPlus results** icon. From the **EnergyPlus** folder, select the **eplusout.eso** file. The **xESOView** tool will open.

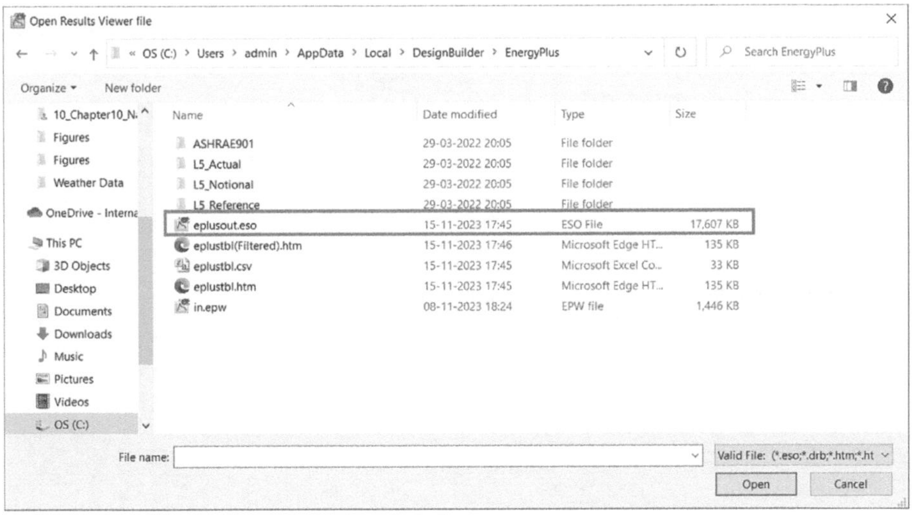

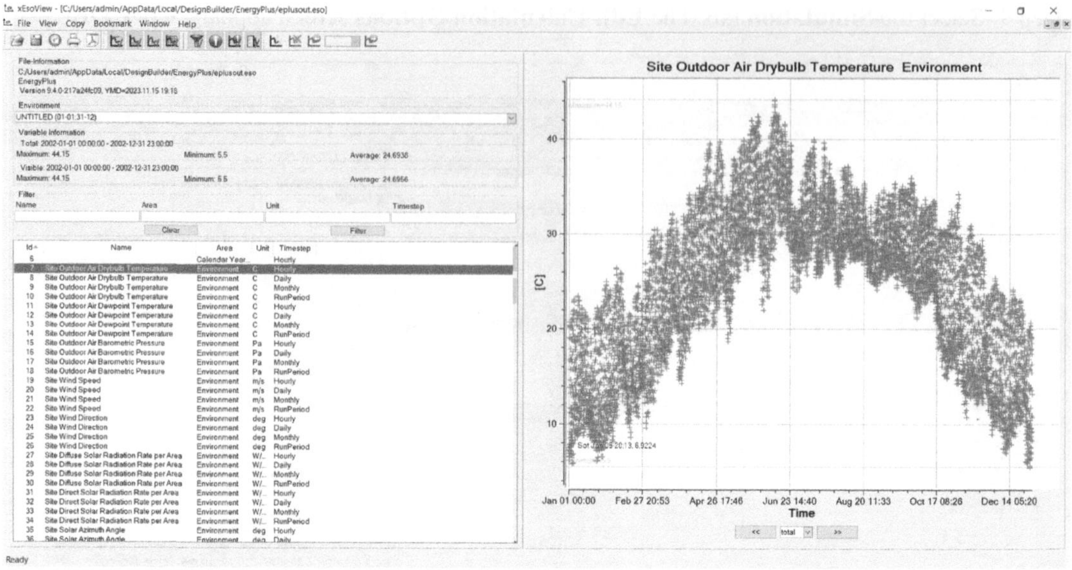

Step 11: Enter the **Site Outdoor Air Drybulb Temperature** in the **Name** field of the **Filter** section to view the variable. From the filtered list select the variable with hourly **Timestep**.

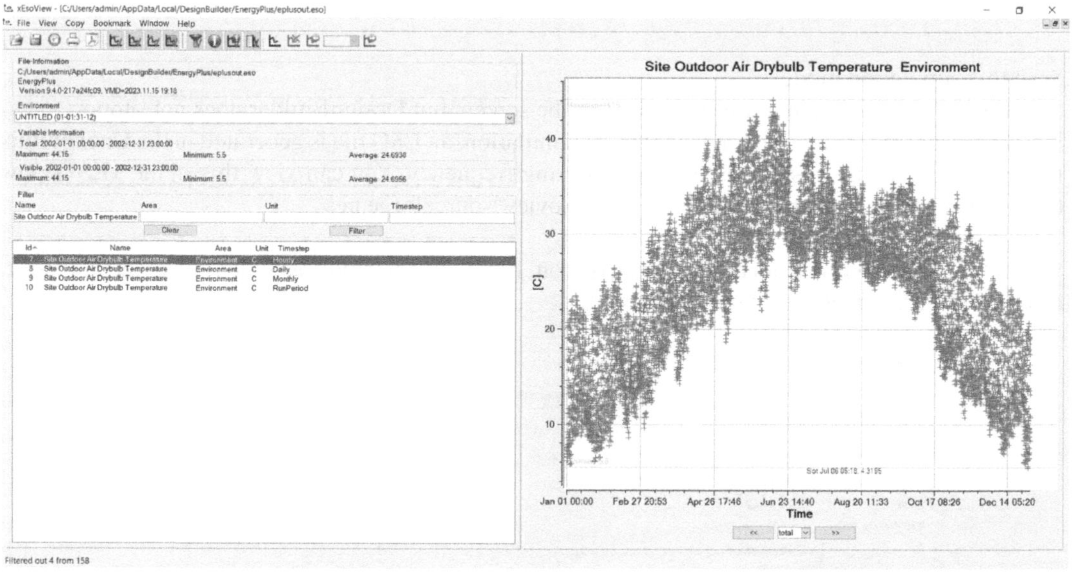

You can copy the variable data to view in a spreadsheet.

Step 12: Click **Copy** and select **Copy variable data**.

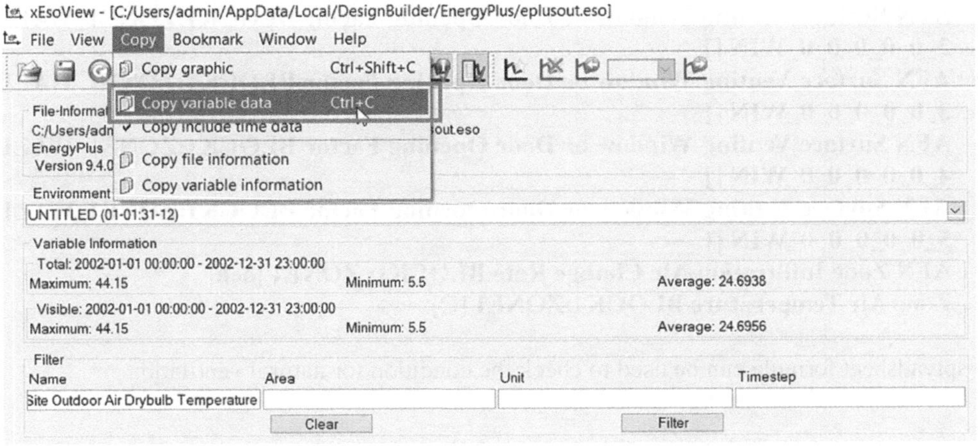

You can copy only one variable at a time.

Step 13: Open a spreadsheet program and paste the data into a new worksheet.

	A	B	C
1	Date/Time	Site Outdoor Air Drybulb Temperature Environment [C]	
2	01-01-2002 00:00	-1111110	
3	01-01-2002 00:30	-1111110	
4	01-01-2002 00:00	7.55	
5	01-01-2002 01:00	-1111110	
6	01-01-2002 01:15	-1111110	
7	01-01-2002 01:30	-1111110	
8	01-01-2002 01:45	-1111110	
9	01-01-2002 01:00	6.625	
10	01-01-2002 02:00	-1111110	
11	01-01-2002 02:30	-1111110	
12	01-01-2002 02:00	6.55	
13	01-01-2002 03:00	-1111110	
14	01-01-2002 03:30	-1111110	
15	01-01-2002 03:00	6.225	
16	01-01-2002 04:00	-1111110	
17	01-01-2002 04:15	-1111110	
18	01-01-2002 04:30	-1111110	
19	01-01-2002 04:45	-1111110	
20	01-01-2002 04:00	6.925	
21	01-01-2002 05:00	-1111110	
22	01-01-2002 05:30	-1111110	
23	01-01-2002 05:00	7.35	
24	01-01-2002 06:00	-1111110	
25	01-01-2002 06:30	-1111110	
26	01-01-2002 06:00	7.925	
27	01-01-2002 07:00	-1111110	
28	01-01-2002 07:10	-1111110	
29	01-01-2002 07:20	-1111110	
30	01-01-2002 07:30	-1111110	
31	01-01-2002 07:40	-1111110	
32	01-01-2002 07:50	-1111110	

Sheet1 Sheet2 Sheet3

Similarly, copy and paste the following variables with **Timestep** as **hourly:**

- **AFN Surface Venting Window or Door Opening Factor BLOCK1:ZONE1_WALL_2_0_0_0_0_0_WIN []**
- **AFN Surface Venting Window or Door Opening Factor BLOCK1:ZONE1_WALL_3_0_0_0_0_0_WIN []**
- **AFN Surface Venting Window or Door Opening Factor BLOCK1:ZONE1_WALL_4_0_0_0_0_0_WIN []**
- **AFN Surface Venting Window or Door Opening Factor BLOCK1:ZONE1_WALL_5_0_0_0_0_0_WIN []**
- **AFN Zone Infiltration Air Change Rate BLOCK1: ZONE1 [ach]**
- **Zone Air Temperature BLOCK1:ZONE1 [C]**

The spreadsheet formula can be used to check the condition for natural ventilation.

	A	B	C	D	E	F	G	H	I	J
1	Date/Time	Site Outdoor Air Drybulb Temperature Environment [C]	AFN Surface Venting Window or Door Opening Factor BLOCK1:ZONE1_WALL_2_0_0_0_0_0_WIN []	AFN Surface Venting Window or Door Opening Factor BLOCK1:ZONE1_WALL_3_0_0_0_0_0_WIN []	AFN Surface Venting Window or Door Opening Factor BLOCK1:ZONE1_WALL_4_0_0_0_0_0_WIN []	AFN Surface Venting Window or Door Opening Factor BLOCK1:ZONE1_WALL_5_0_0_0_0_0_WIN []	AFN Zone Infiltration Air Change Rate BLOCK1:ZONE1 [ach]	Zone Air Temperature BLOCK1:ZONE1 [C]	Check if Tzone>20 AND Tzone>Tout	
15927	03-02-2002 18:00	18.95	-	-	-	-		22.8989	TRUE	
15928	03-02-2002 18:15	18.95	0.95	0.95	0.95	0.95		20.3325	TRUE	
15929	03-02-2002 18:30	18.95	0.96	0.96	0.96	0.96		20.1382	TRUE	
15930	03-02-2002 18:34	18.95	0.97	0.97	0.97	0.97		19.2795	FALSE	
15931	03-02-2002 18:38	18.95	-	-	-	-		20.2343	TRUE	
15932	03-02-2002 18:42	18.95	0.97	0.97	0.97	0.97		20.0819	TRUE	
15933	03-02-2002 18:47	18.95	0.97	0.97	0.97	0.97		19.5355	FALSE	
15934	03-02-2002 18:51	18.95	-	-	-	-		20.4348	TRUE	
15935	03-02-2002 18:55	18.95	0.97	0.97	0.97	0.97		20.1002	TRUE	
15936	03-02-2002 18:00	17.78					15.16			
15937	03-02-2002 19:00	17.78	-	-	-	-		21.1479	TRUE	
15938	03-02-2002 19:05	17.78	0.96	0.96	0.96	0.96		19.5819	FALSE	
15939	03-02-2002 19:10	17.78	-	-	-	-		19.7995	FALSE	
15940	03-02-2002 19:15	17.78	-	-	-	-		21.2276	TRUE	
15941	03-02-2002 19:20	17.78	0.95	0.95	0.95	0.95		19.6818	FALSE	
15942	03-02-2002 19:25	17.78	-	-	-	-		19.8132	FALSE	
15943	03-02-2002 19:30	17.78	0.96	0.96	0.96	0.96		18.9367	FALSE	
15944	03-02-2002 19:32	17.78	-	-	-	-		18.9749	FALSE	
15945	03-02-2002 19:35	17.78	-	-	-	-		19.9281	FALSE	
15946	03-02-2002 19:38	17.78	-	-	-	-		21.0676	TRUE	
15947	03-02-2002 19:40	17.78	0.95	0.95	0.95	0.95		19.5908	FALSE	
15948	03-02-2002 19:43	17.78	-	-	-	-		19.2445	FALSE	
15949	03-02-2002 19:46	17.78	-	-	-	-		19.9864	FALSE	
15950	03-02-2002 19:49	17.78	-	-	-	-		21.0621	TRUE	
15951	03-02-2002 19:51	17.78	0.95	0.95	0.95	0.95		19.5848	FALSE	
15952	03-02-2002 19:54	17.78	-	-	-	-		19.2491	FALSE	
15953	03-02-2002 19:57	17.78	-	-	-	-		19.9964	FALSE	
15954	03-02-2002 19:00	16.23					8.50			
15955	03-02-2002 20:00	16.23	-	-	-	-		20.4595	TRUE	
15956	03-02-2002 20:03	16.23	0.95	0.95	0.95	0.95		18.8735	FALSE	
15957	03-02-2002 20:06	16.23	-	-	-	-		18.6685	FALSE	

You can see from the preceding that if the following conditions are true, then there is natural ventilation:

$$T_{zone_air} > T_{setpoint} \text{ and } T_{zone_air} > T_{outside_air} \quad (6.3)$$

In this tutorial, the zone setpoint is set to 20°C. The windows open in the timestep immediately following the point at which the specified condition is met. It is important to note that the zone infiltration air change rate (as shown in column H) is reported on an hourly basis, not at the system timestep. Consequently, even when the windows open, the associated ventilation will not be captured in that specific timestep but will appear in the subsequent hourly report.

The simulation results confirmed that window operation was controlled by a temperature-based logic: windows opened when the indoor temperature exceeded 20°C and was higher than the outdoor temperature, activating natural ventilation.

TUTORIAL 6.5: EVALUATING THE IMPACT OF WINDOW OPENING AREA MODULATION ON NATURAL VENTILATION

GOAL

To evaluate the impact of window opening area modulation based on zone and outdoor air temperature

WHAT ARE YOU GOING TO LEARN?

- How to define window opening area modulation

PROBLEM STATEMENT

In this tutorial, you are going to use the model saved in Tutorial 6.4. You are going to simulate the model for **Temperature control mode**. Use the **New Delhi/Safdarjung, India** weather location.

The modulation takes the following form when **Ventilation Control Mode = Temperature**:

$T_{zone} - T_{out}$ = (lower value on inside/outside temperature difference for modulating the venting open factor)

Then the multiplication factor is 1.0.

(Lower value on inside/outside temperature difference for modulating the venting open factor) < $T_{zone} - T_{out}$ < (upper value on inside/outside temperature difference for modulating the venting open factor)

Then the multiplication factor varies linearly from 1.0 to the limit value on multiplier for modulating the venting open factor.

$T_{zone} - T_{out}$ = (upper value on inside/outside temperature difference for modulating the venting open factor)

Then the multiplication factor equals the limit value on multiplier for modulating the venting open factor.

Source: U.S. Department of Energy, *EnergyPlus Documentation, Input-Output Reference*, March 2025; available at https://energyplus.net/documentation.

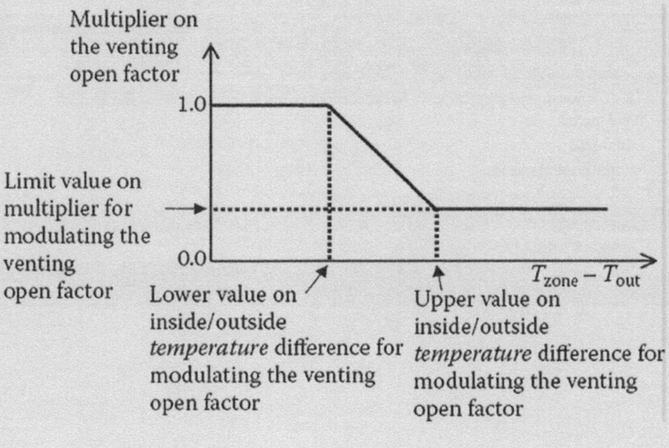

Solution

Step 1: Open the project saved in Tutorial 6.4.

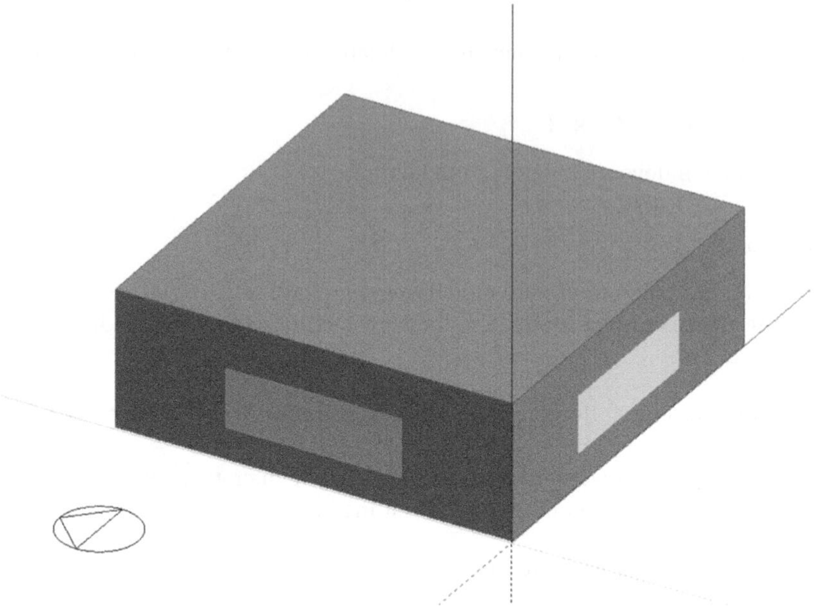

Step 2: Select the **HVAC** tab. Expand the **Natural Ventilation** section.

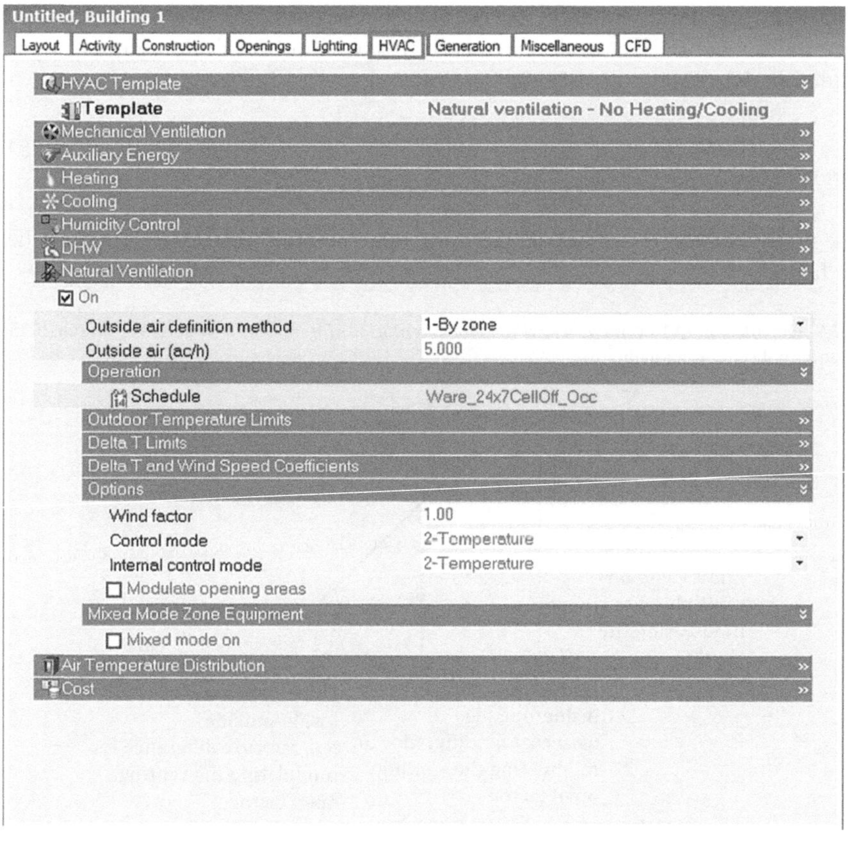

Step 3: Select the **Modulate opening areas** checkbox. Enter the lower value of $T_{in} - T_{out}$ as **5.0** and the upper value as **15**.

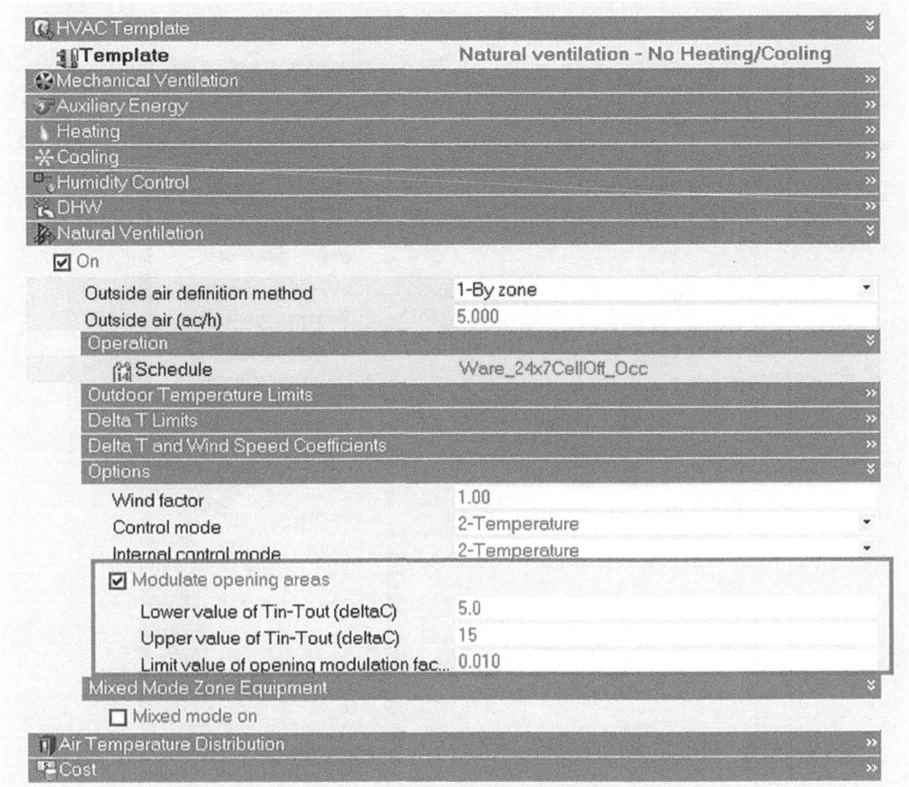

Step 4: Minimize the **DesignBuilder** window, create an **output.idf** file, and write the following additional output variables.

Output: Variable, *, Zone Air Temperature, Detailed;
Output:Variable, *, AFN Surface Venting Window or Door Opening Factor, Detailed;
Output: Variable, *, AFN Zone Infiltration, Detailed;

This .idf file needs to be placed in the path shown in the following figure.

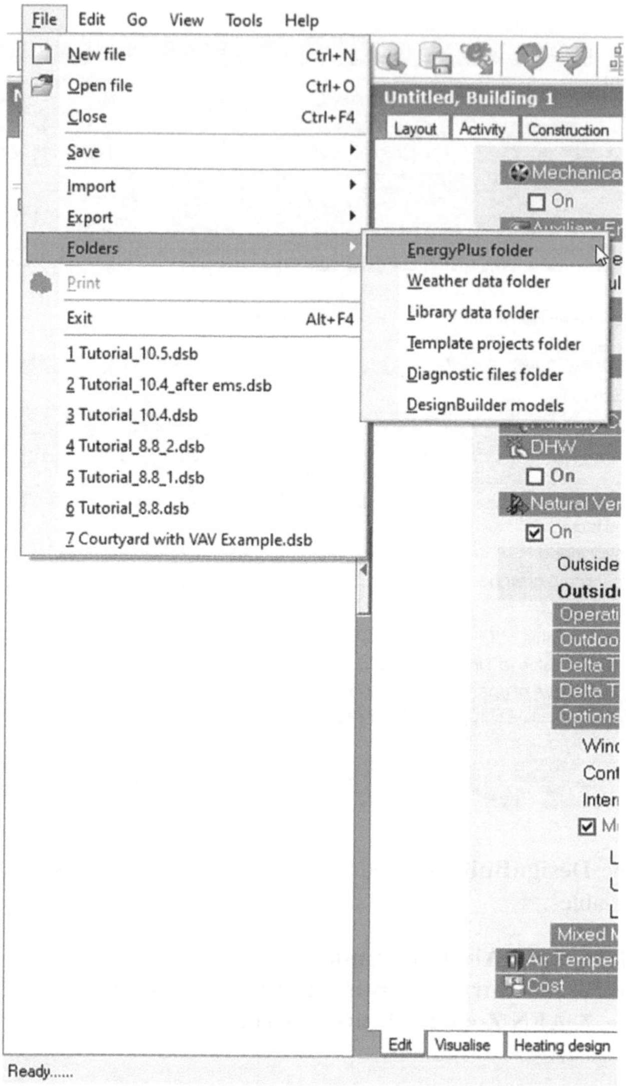

Step 5: Maximize the **DesignBuilder** window, and select the **HVAC** tab. Click **Simple** under the **Help** tab. The **Model Options—Building and Block** screen appears.

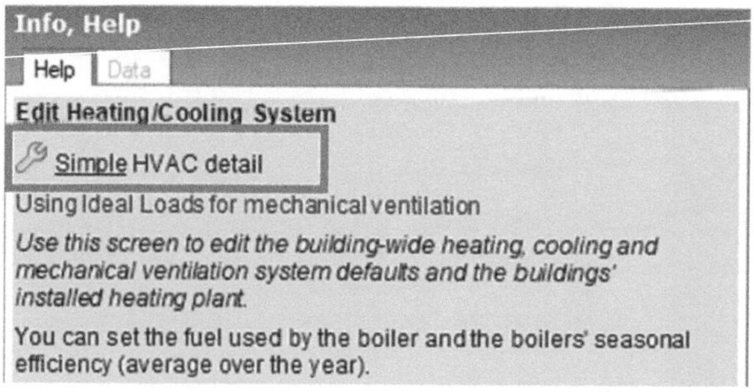

Step 6: Select the **Simulation** tab. Click the **Advanced** section.

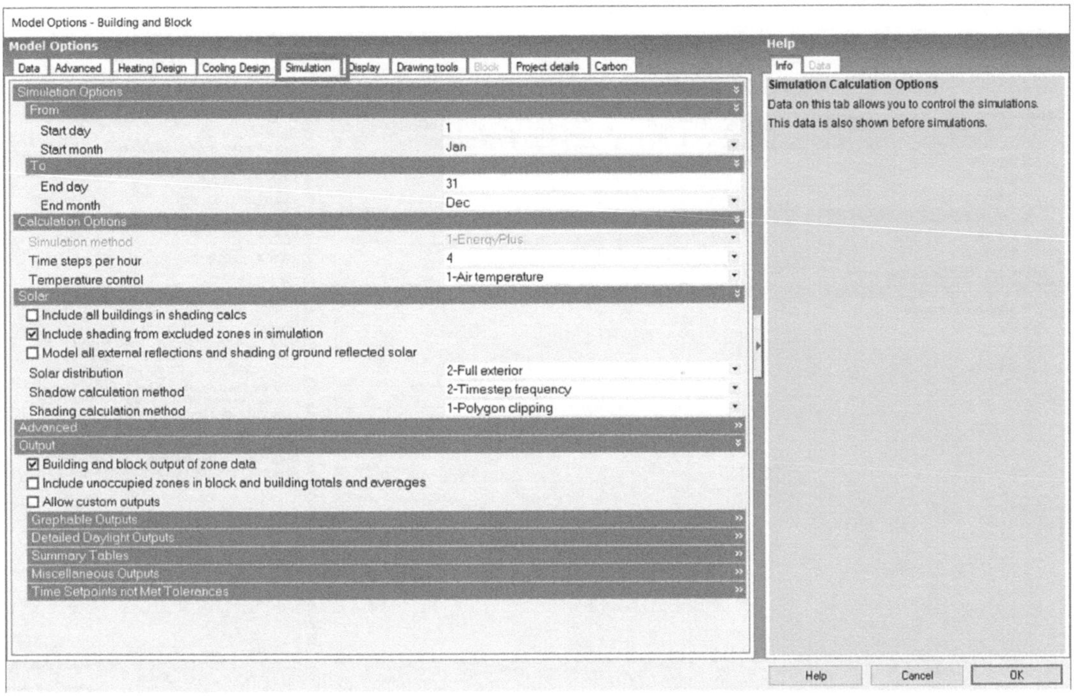

Step 7: Select the **IDF File 1** checkbox and select the **Output.idf** file from the **EnergyPlus** folder. Click **OK**.

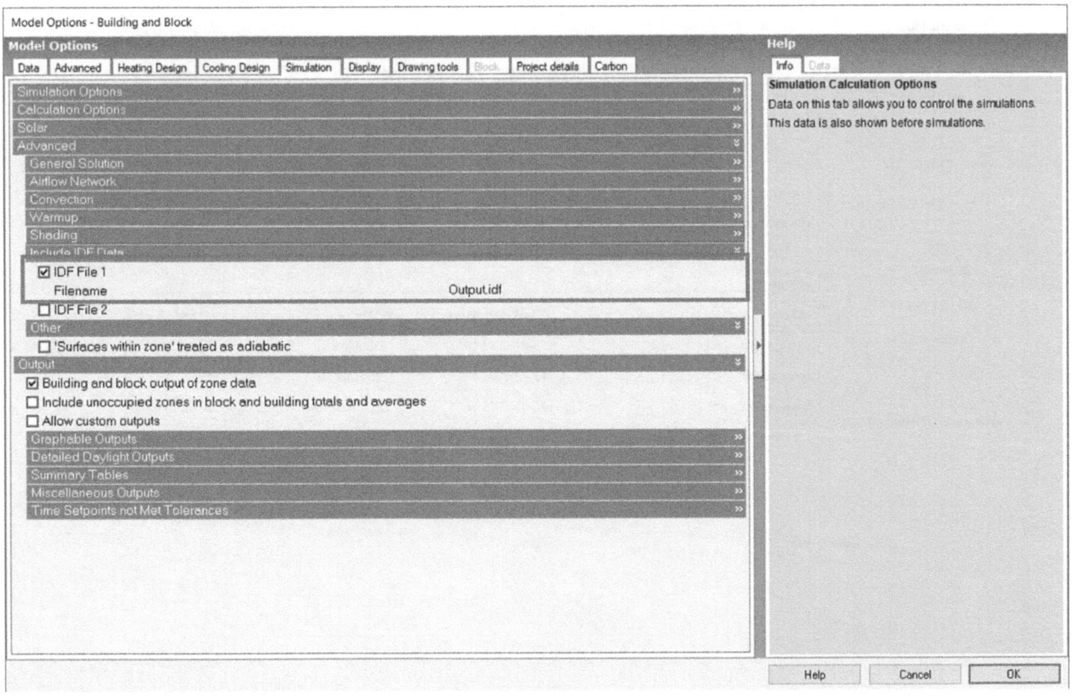

Step 8: Select the **Simulation** screen tab. The **Edit Calculation Options** screen appears. Click **OK**.

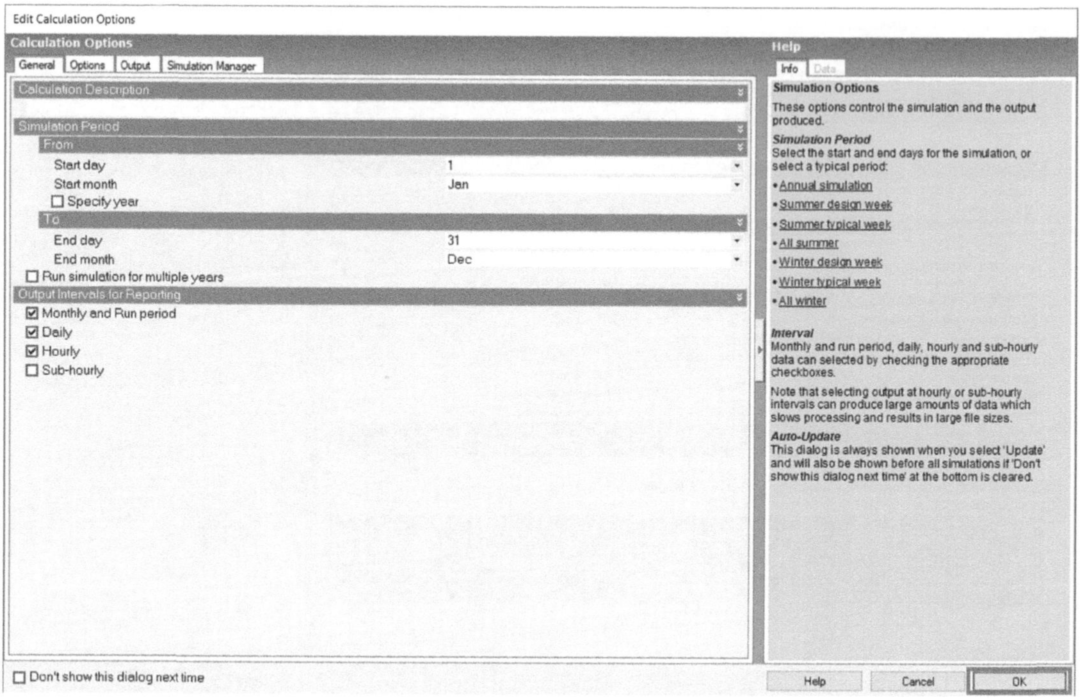

Step 9: Select the **EnergyPlus** results. Select the **eplusout** file so that the file will automatically open in the **xESOView** tool.

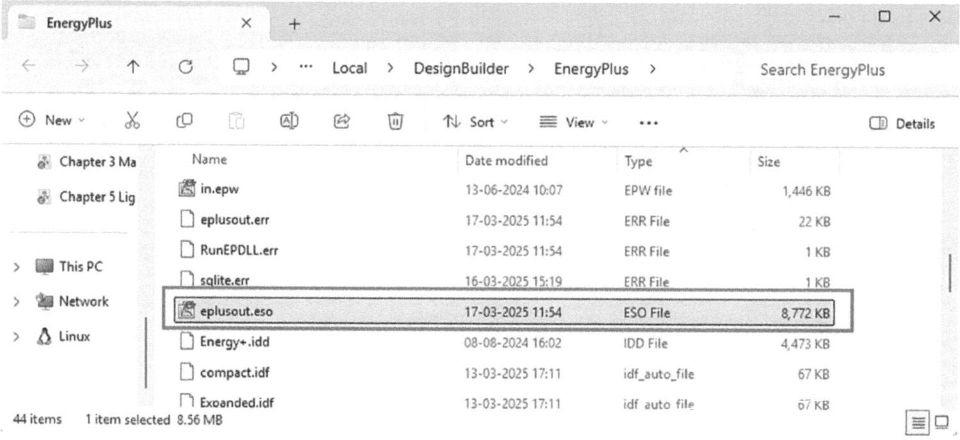

The **xEsoView** screen appears.

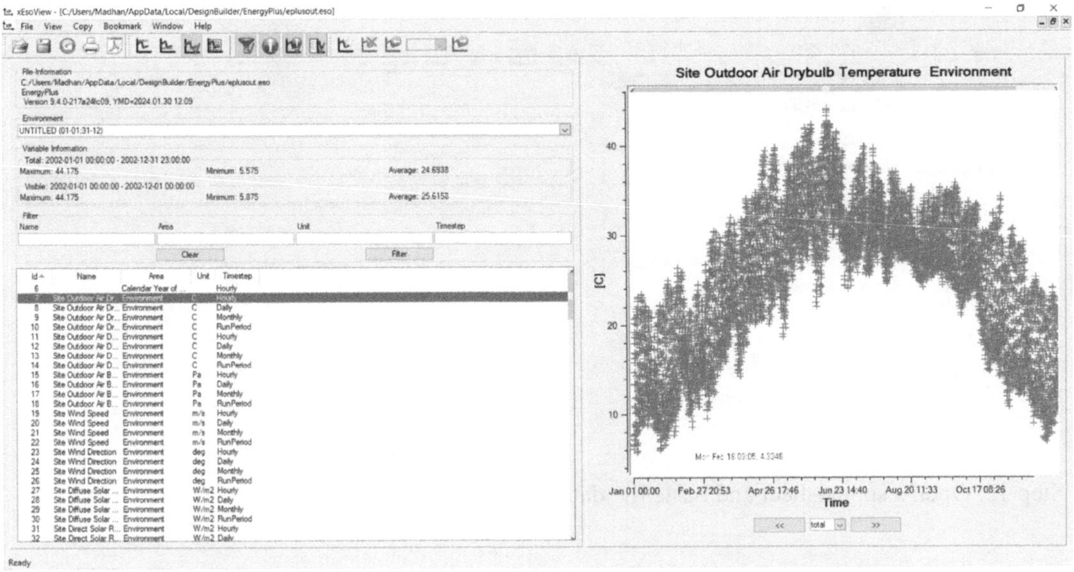

Step 10: Enter **Site Outdoor Air Drybulb Temperature** in the **Name** field of the **Filter** section to view the variable.

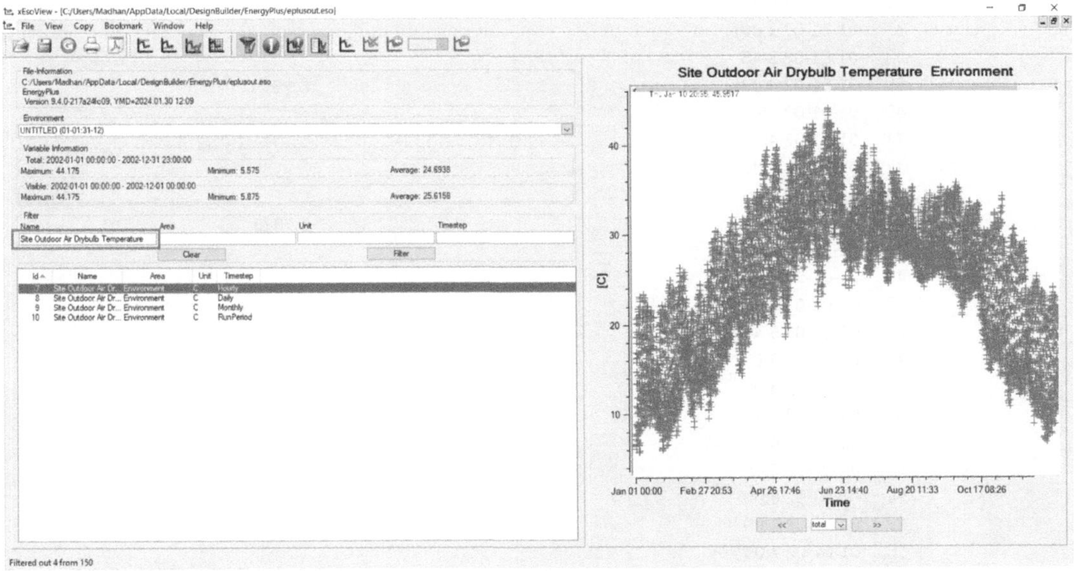

You can copy the variable data to view in a spreadsheet.

Step 11: Click **Copy** and select **Copy variable data**.

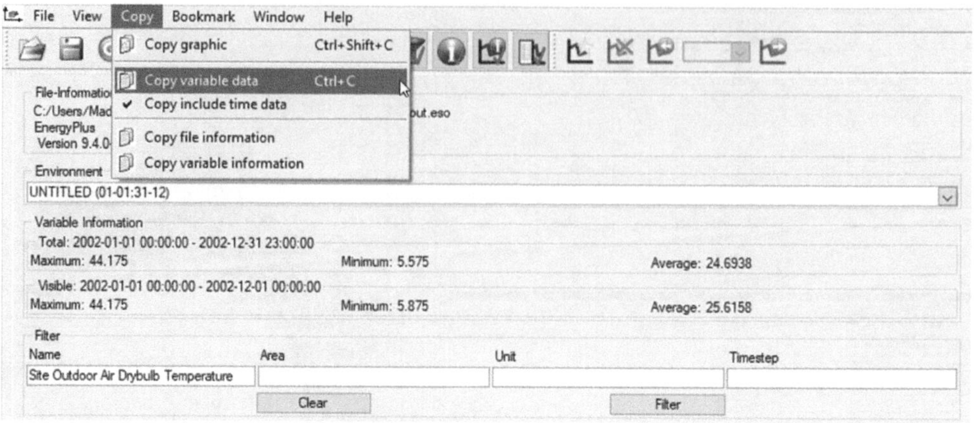

Step 12: Open a spreadsheet and paste the data.

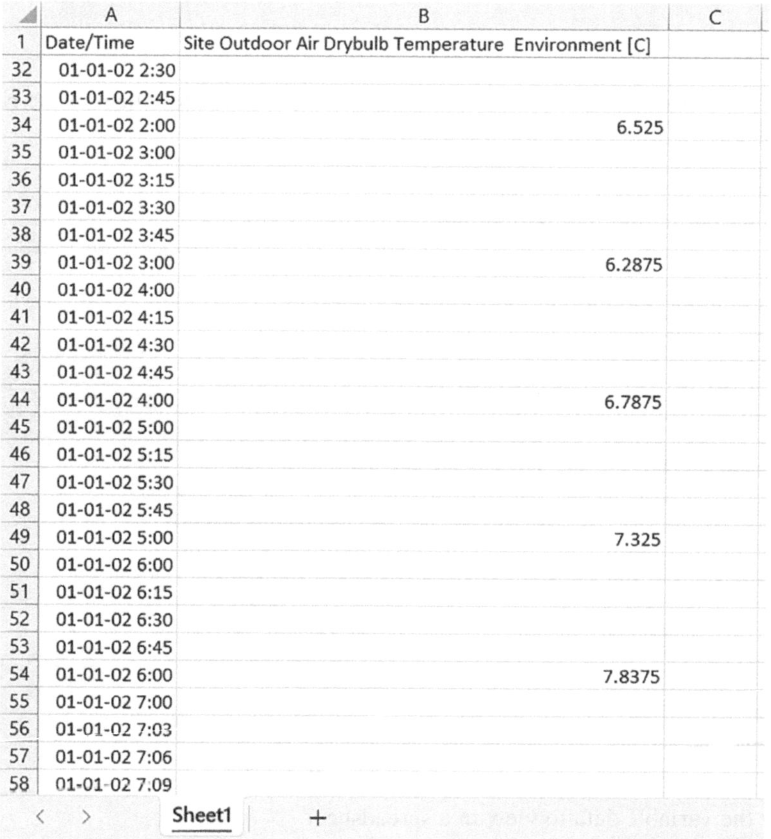

	A	B	C
1	Date/Time	Site Outdoor Air Drybulb Temperature Environment [C]	
32	01-01-02 2:30		
33	01-01-02 2:45		
34	01-01-02 2:00	6.525	
35	01-01-02 3:00		
36	01-01-02 3:15		
37	01-01-02 3:30		
38	01-01-02 3:45		
39	01-01-02 3:00	6.2875	
40	01-01-02 4:00		
41	01-01-02 4:15		
42	01-01-02 4:30		
43	01-01-02 4:45		
44	01-01-02 4:00	6.7875	
45	01-01-02 5:00		
46	01-01-02 5:15		
47	01-01-02 5:30		
48	01-01-02 5:45		
49	01-01-02 5:00	7.325	
50	01-01-02 6:00		
51	01-01-02 6:15		
52	01-01-02 6:30		
53	01-01-02 6:45		
54	01-01-02 6:00	7.8375	
55	01-01-02 7:00		
56	01-01-02 7:03		
57	01-01-02 7:06		
58	01-01-02 7:09		

Sheet1

You can check if natural ventilation is working as per the controls applied.

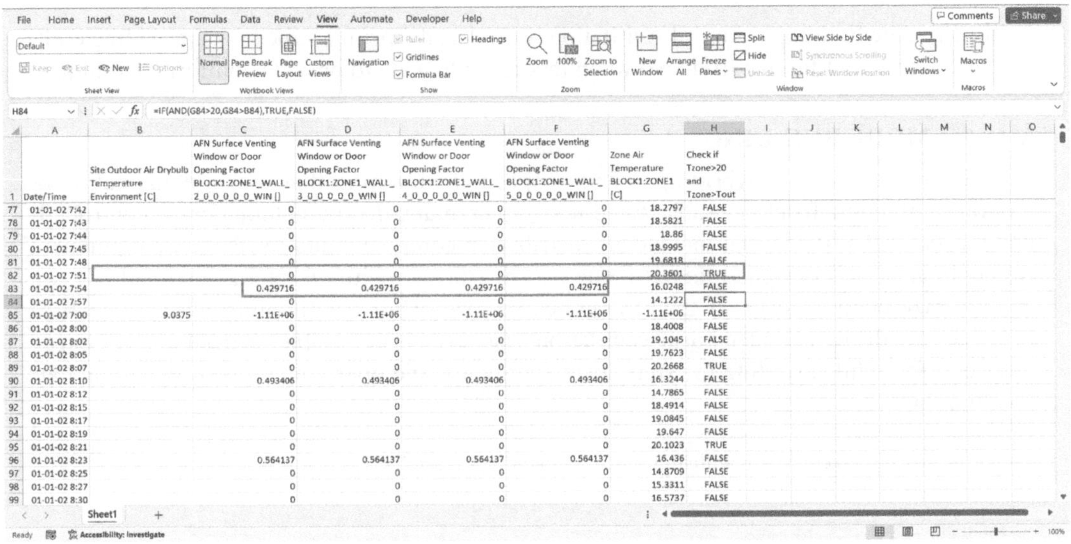

	Date/Time	Site Outdoor Air Drybulb Temperature Environment [C]	AFN Surface Venting Window or Door Opening Factor BLOCK1:ZONE1_WALL_2_0_0_0_0_0_WIN []	AFN Surface Venting Window or Door Opening Factor BLOCK1:ZONE1_WALL_3_0_0_0_0_0_WIN []	AFN Surface Venting Window or Door Opening Factor BLOCK1:ZONE1_WALL_4_0_0_0_0_0_WIN []	AFN Surface Venting Window or Door Opening Factor BLOCK1:ZONE1_WALL_5_0_0_0_0_0_WIN []	Zone Air Temperature BLOCK1:ZONE1 [C]	Check if Tzone>20 and Tzone>Tout
77	01-01-02 7:42		0	0	0	0	18.2797	FALSE
78	01-01-02 7:43		0	0	0	0	18.5821	FALSE
79	01-01-02 7:44		0	0	0	0	18.86	FALSE
80	01-01-02 7:45		0	0	0	0	18.9995	FALSE
81	01-01-02 7:48		0	0	0	0	19.6818	FALSE
82	01-01-02 7:51		0	0	0	0	20.3601	TRUE
83	01-01-02 7:54		0.429716	0.429716	0.429716	0.429716	16.0248	FALSE
84	01-01-02 7:57		0	0	0	0	14.1222	FALSE
85	01-01-02 7:00	9.0375	-1.11E+06	-1.11E+06	-1.11E+06	-1.11E+06	-1.11E+06	FALSE
86	01-01-02 8:00		0	0	0	0	18.4008	FALSE
87	01-01-02 8:02		0	0	0	0	19.1045	FALSE
88	01-01-02 8:05		0	0	0	0	19.7623	FALSE
89	01-01-02 8:07		0	0	0	0	20.2668	TRUE
90	01-01-02 8:10		0.493406	0.493406	0.493406	0.493406	16.3244	FALSE
91	01-01-02 8:12		0	0	0	0	14.7865	FALSE
92	01-01-02 8:15		0	0	0	0	18.4914	FALSE
93	01-01-02 8:17		0	0	0	0	19.0845	FALSE
94	01-01-02 8:19		0	0	0	0	19.647	FALSE
95	01-01-02 8:21		0	0	0	0	20.1023	TRUE
96	01-01-02 8:23		0.564137	0.564137	0.564137	0.564137	16.436	FALSE
97	01-01-02 8:25		0	0	0	0	14.8709	FALSE
98	01-01-02 8:27		0	0	0	0	15.3311	FALSE
99	01-01-02 8:30		0	0	0	0	16.5737	FALSE

You can see from the preceding that if the following conditions are true, then there is natural ventilation:

$$T_{zone_air} > T_{setpoint} \; and \, T_{zone_air} > T_{outside_air} \tag{6.4}$$

and **Window opening fraction** depends on the $T_{zone} - T_{out}$.

The simulation results highlight the effectiveness of temperature-based modulation of window openings in regulating natural ventilation. When the indoor/outdoor temperature difference was less than 5°C, windows opened fully, allowing maximum airflow, while differences greater than 15°C reduced openings to a predefined minimum. Between these thresholds, the opening fraction varied linearly, enabling a controlled response to changing thermal conditions.

TUTORIAL 6.6: EVALUATING THE IMPACT OF MIXED-MODE VENTILATION

Goal

To evaluate the impact of mixed-mode ventilation on energy performance

What Are You Going to Learn?

- How to model mixed-mode ventilation

Problem Statement

In this tutorial, you are going to use the model saved in Tutorial 6.4. Determine the energy consumption for the following two cases.

1. With mixed mode
2. Fully air-conditioned

Use the **New Delhi/Safdarjung—India** location.

MIXED-MODE COOLING

In mixed-mode buildings, natural ventilation is used as the primary means of providing cooling, and active cooling is introduced when this is inadequate to provide comfort conditions.

MIXED-MODE BUILDING

Mixed mode refers to a hybrid approach to space conditioning that uses a combination of natural ventilation from operable windows (either manually or automatically controlled) and mechanical systems that include air distribution equipment and refrigeration equipment for cooling. A well-designed mixed-mode building begins with an intelligent façade design to minimize cooling loads. It then integrates the use of air conditioning when and where necessary, with the use of natural ventilation whenever it is feasible or desirable, to maximize comfort while avoiding the significant energy use and operating costs of year-round air conditioning.

Source: https://cbe.berkeley.edu/research/mixed-mode-building-research

Solution

Step 1: Open the file saved in Tutorial 6.4.

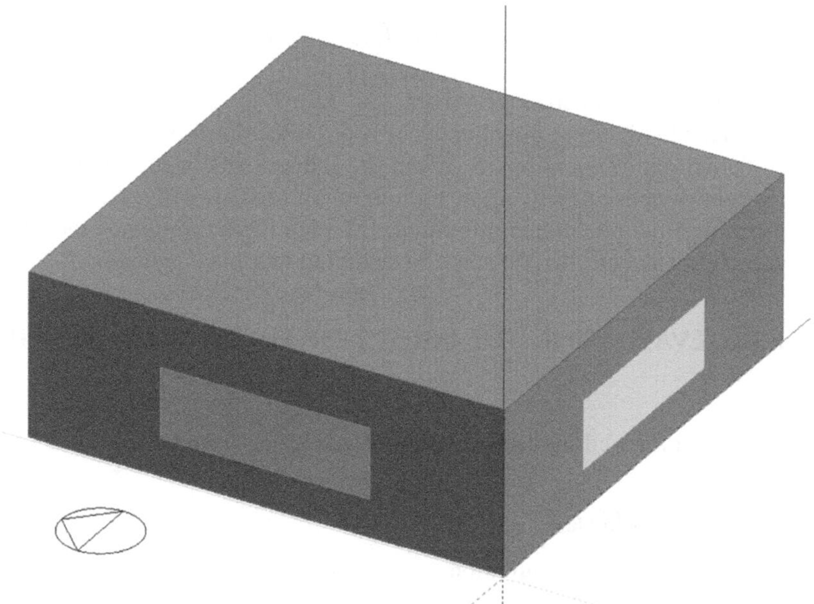

Step 2: Select the **HVAC** tab. Select the **HVAC template** as **Radiator heating, Boiler HW, Mixed mode Nat Vent, Local comfort cooling**.

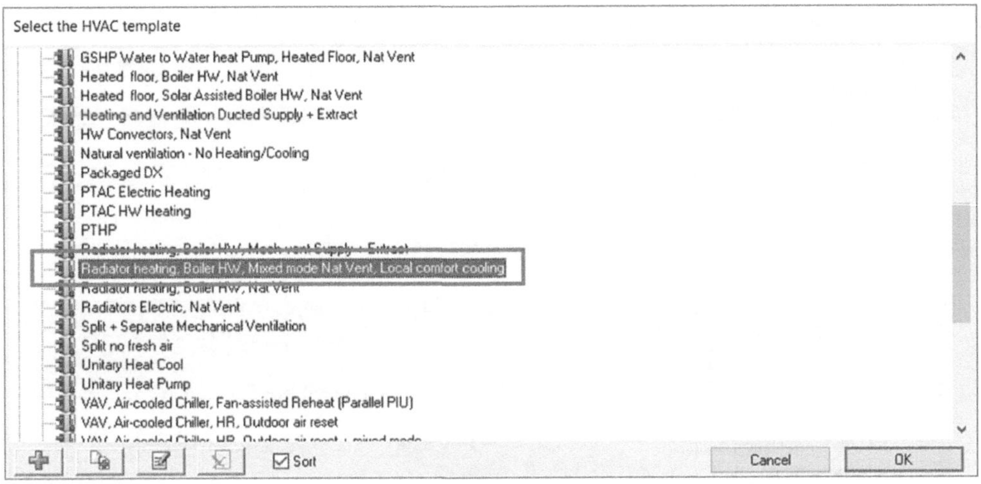

Step 3: Clear the **Heated** checkbox in the **Heating** section.

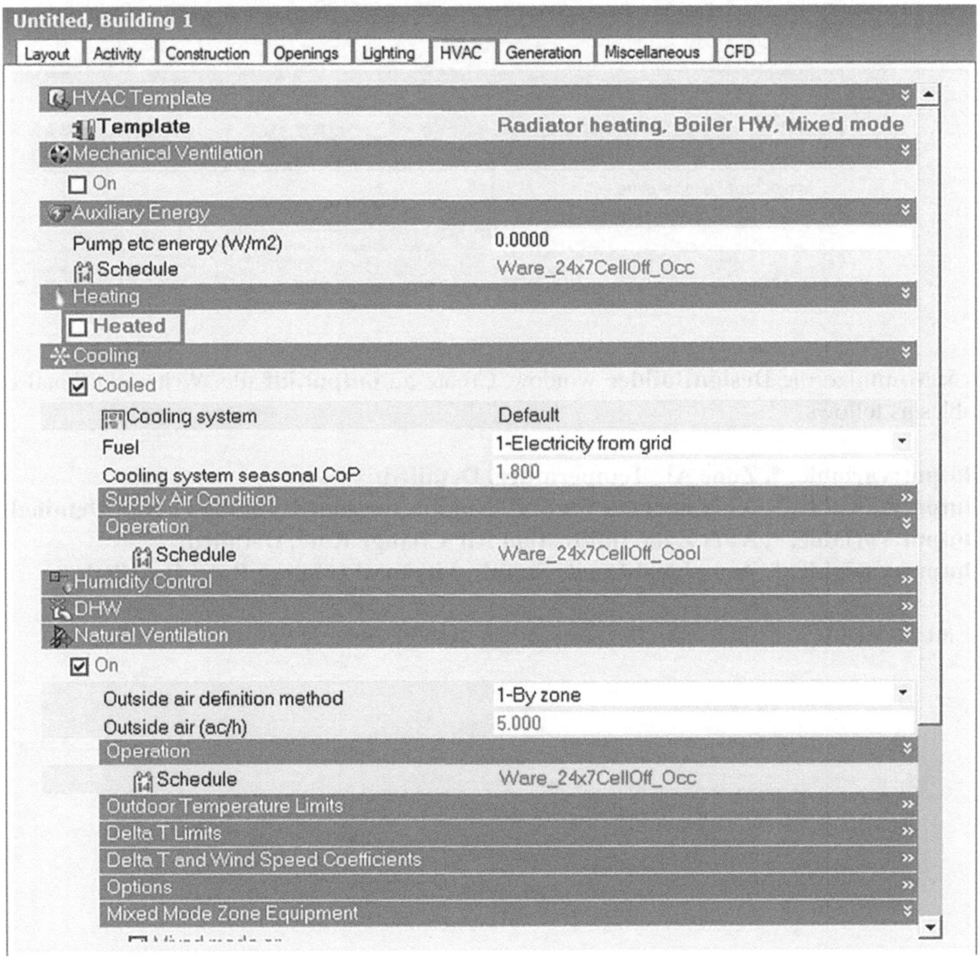

Step 4: Ensure that the **Mixed mode on** checkbox is selected.

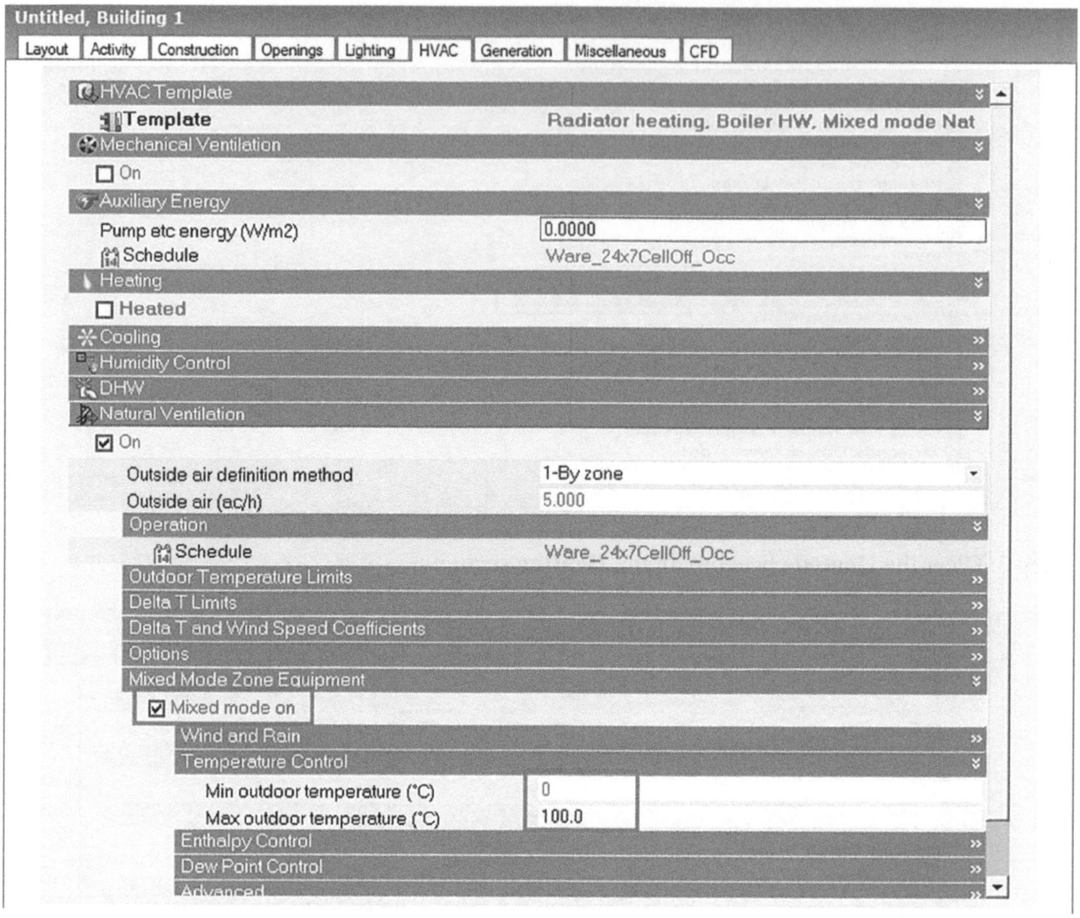

Step 5: Minimize the **DesignBuilder** window. Create an **output.idf** file. Write additional output variables as follows.

Output:Variable, *, Zone Air Temperature, Detailed;
Output:Variable, *, AFN Surface Venting Window or Door Opening Factor, Detailed;
Output:Variable, *, AFN Zone Infiltration Air Change Rate, Detailed;
Output:Variable, *, Zone Ideal Loads Supply Air Total Cooling Rate, Detailed;

This .idf file needs to be placed in the path shown in the following figures.

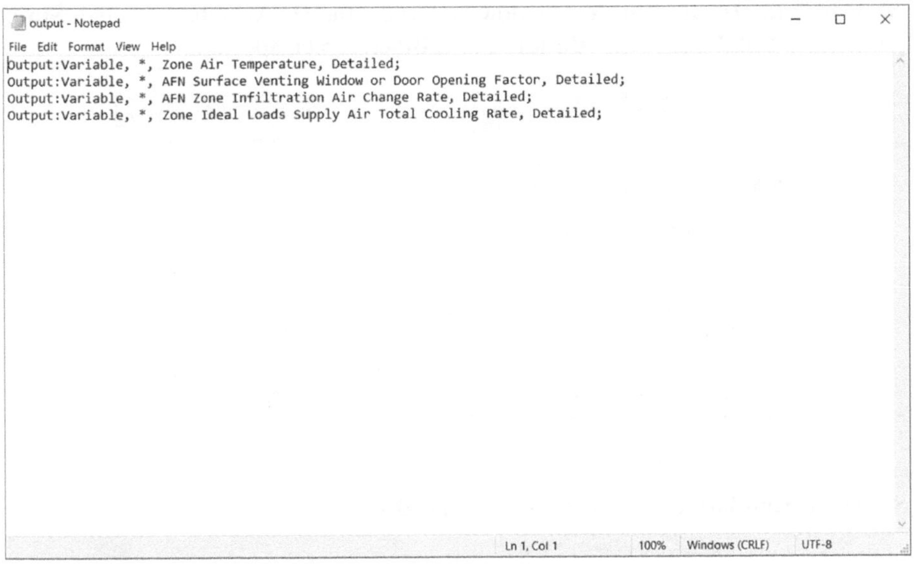
output - Notepad
File Edit Format View Help
Output:Variable, *, Zone Air Temperature, Detailed;
Output:Variable, *, AFN Surface Venting Window or Door Opening Factor, Detailed;
Output:Variable, *, AFN Zone Infiltration Air Change Rate, Detailed;
Output:Variable, *, Zone Ideal Loads Supply Air Total Cooling Rate, Detailed;
Ln 1, Col 1
100%
Windows (CRLF)
UTF-8

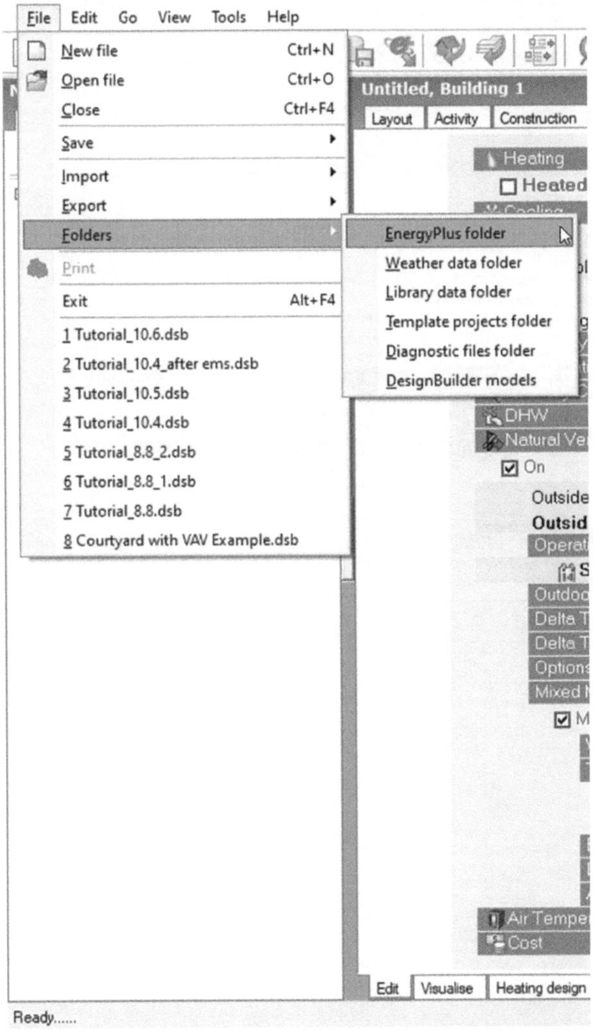
File Edit Go View Tools Help
New file Ctrl+N
Open file Ctrl+O
Close Ctrl+F4
Save
Import
Export
Folders
Print
Exit Alt+F4
1 Tutorial_10.6.dsb
2 Tutorial_10.4_after ems.dsb
3 Tutorial_10.5.dsb
4 Tutorial_10.4.dsb
5 Tutorial_8.8_2.dsb
6 Tutorial_8.8_1.dsb
7 Tutorial_8.8.dsb
8 Courtyard with VAV Example.dsb
EnergyPlus folder
Weather data folder
Library data folder
Template projects folder
Diagnostic files folder
DesignBuilder models
Untitled, Building 1
Layout
Activity
Construction
Heating
Heated
DHW
On
Cost
Edit
Visualise
Heating design
Ready......

Step 6: Maximize the **DesignBuilder** window and select the **HVAC** tab, then click **Simple** under the **Help** tab. The **Model Options – Building and Block** screen appears.

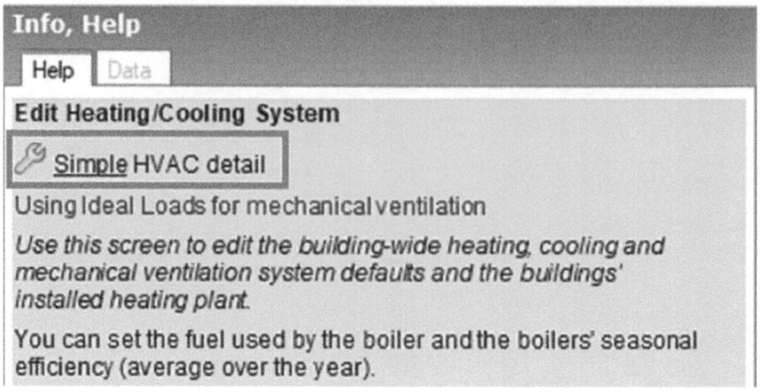

Step 7: Select the **Simulation** tab. Click the **Advanced** section.

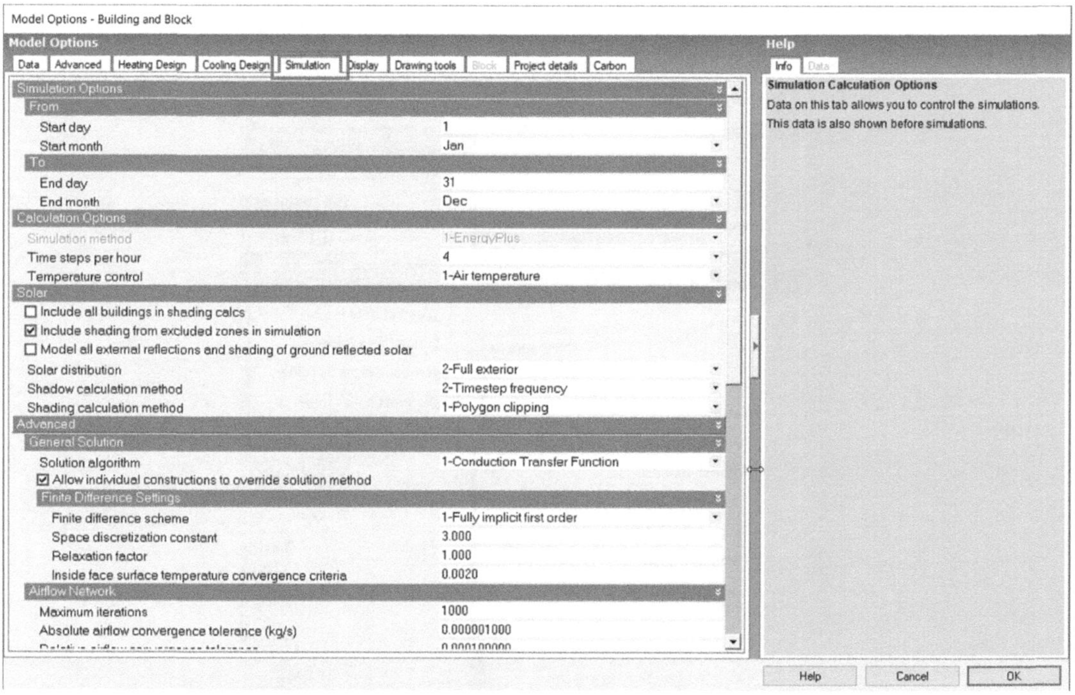

Step 8: Select the **IDF File 1** checkbox and select the **output.idf** file from the **EnergyPlus** folder. Click **OK**.

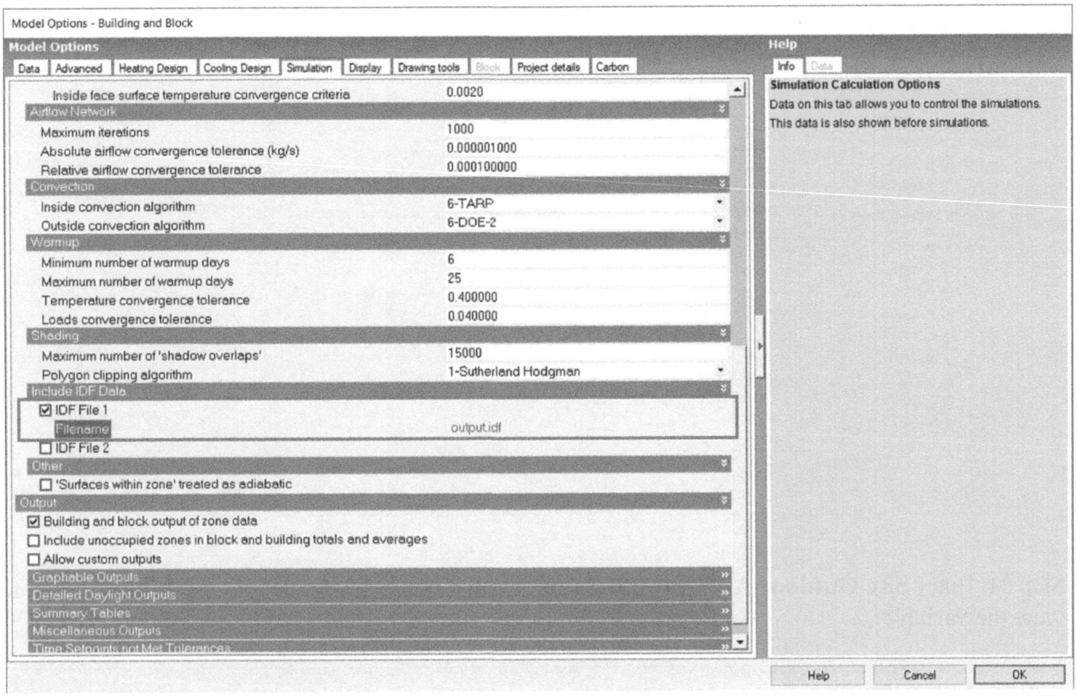

Step 9: Select the **Simulation** screen tab. The **Edit Calculation Options** screen appears. Click **OK**.

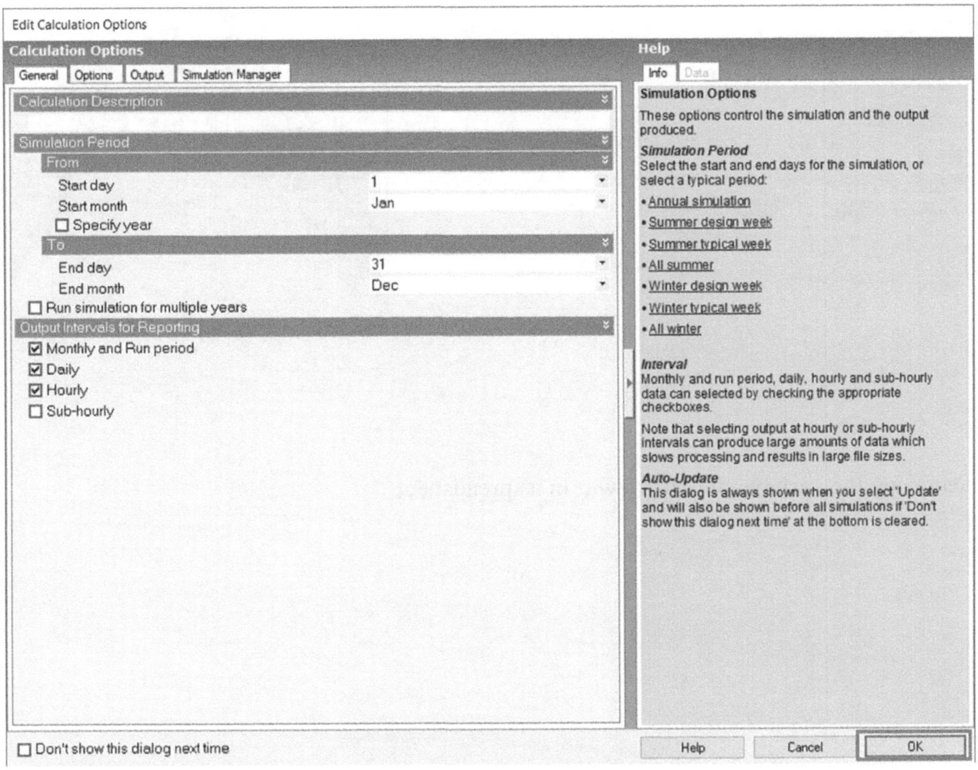

Step 10: Select the **EnergyPlus** results. Select the **eplusout.eso**. This will automatically launch the file in the **xESOView** tool.

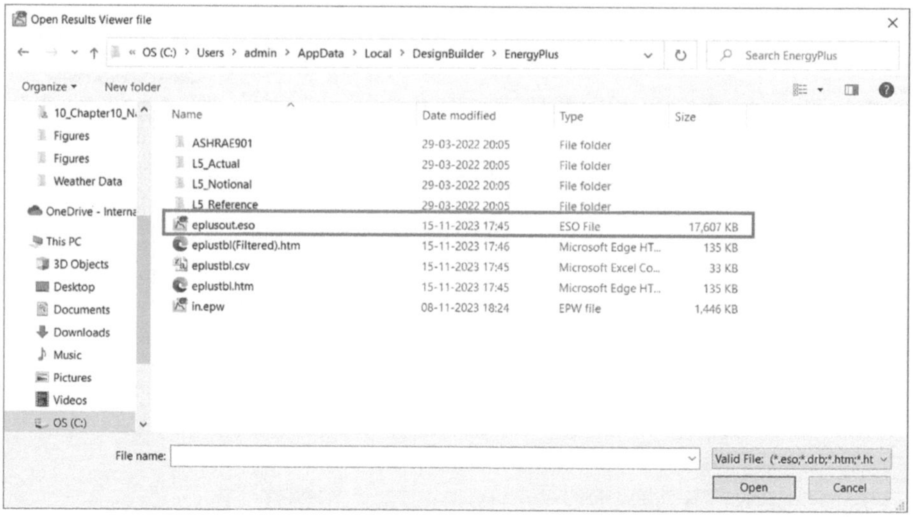

Step 11: Enter **Site Outdoor Air Drybulb Temperature** in the **Name** field of the **Filter** section to view the variable.

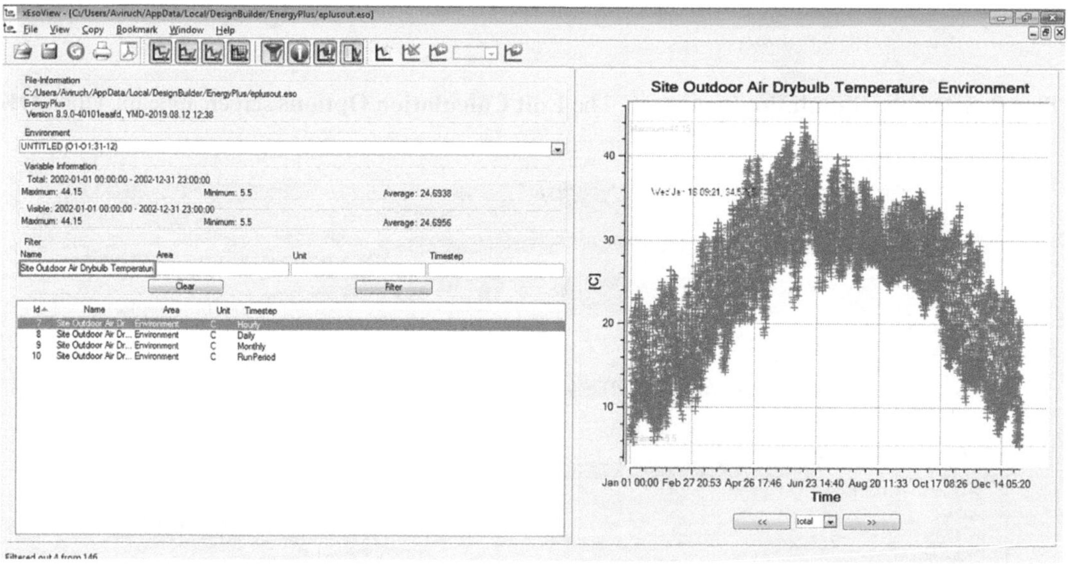

You can copy the variable data to view it in a spreadsheet.

Step 12: Click **Copy** then select **Copy variable data**.

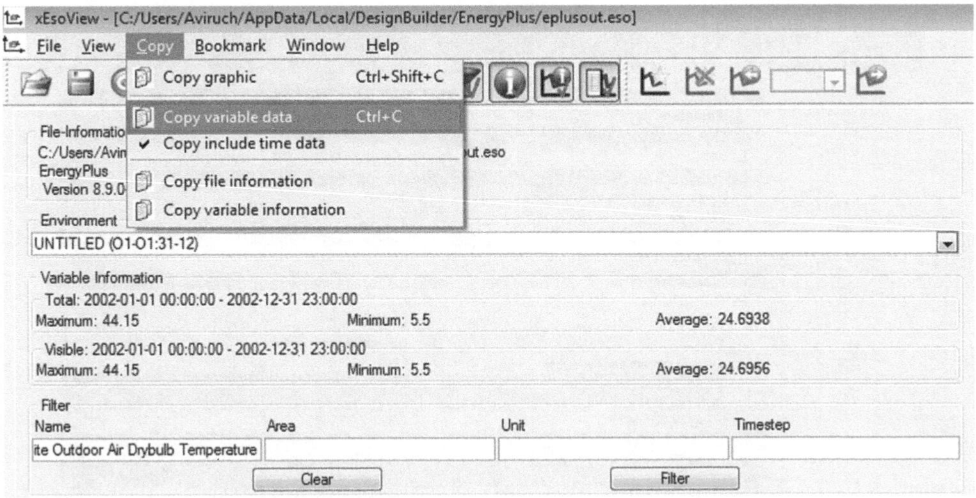

Step 13: Open a spreadsheet and paste the data.

Date/Time	Site Outdoor Air Drybulb Temperature Environment [C]	AFN Surface Venting Window or Door Opening Factor BLOCK1:MYZONE_WALL_2_0_0_0_0_WIN []	AFN Surface Venting Window or Door Opening Factor BLOCK1:MYZONE_WALL_3_0_0_0_0_WIN []	AFN Surface Venting Window or Door Opening Factor BLOCK1:MYZONE_WALL_4_0_0_0_0_WIN []	AFN Surface Venting Window or Door Opening Factor BLOCK1:MYZONE_WALL_5_0_0_0_0_WIN []	AFN Zone Infiltration Air Change Rate BLOCK1:MYZONE [ach]	Zone Air Temperature BLOCK1:MYZONE [C]	Zone Ideal Loads Supply Air Total Cooling Rate BLOCK1:MYZONE IDEAL LOADS AIR [W]
17/11/2002 9:46		1.00	1.00	1.00	1.00		23.3292	
17/11/2002 9:49		1.00	1.00	1.00	1.00		23.3389	
17/11/2002 9:51		1.00	1.00	1.00	1.00		23.3397	
17/11/2002 9:54		1.00	1.00	1.00	1.00		23.3377	
17/11/2002 9:57		1.00	1.00	1.00	1.00		23.3367	
17/11/2002 9:00	22.23					31.00		0.00
17/11/2002 10:00		0.00	0.00	0.00	0.00		23.6863	
17/11/2002 10:03		0.00	0.00	0.00	0.00		24.3996	
17/11/2002 10:06		0.00	0.00	0.00	0.00		24.0324	
17/11/2002 10:09		0.00	0.00	0.00	0.00		24.0003	
17/11/2002 10:12		0.00	0.00	0.00	0.00		24	
17/11/2002 10:15		0.00	0.00	0.00	0.00		24	
17/11/2002 10:18		0.00	0.00	0.00	0.00		24	
17/11/2002 10:21		0.00	0.00	0.00	0.00		24	
17/11/2002 10:24		0.00	0.00	0.00	0.00		24	
17/11/2002 10:27		0.00	0.00	0.00	0.00		24	
17/11/2002 10:30		0.00	0.00	0.00	0.00		24	
17/11/2002 10:00	24.93					0.00		1918.43
17/11/2002 11:00		0.00	0.00	0.00	0.00		24	
17/11/2002 11:30		0.00	0.00	0.00	0.00		24	
17/11/2002 11:00	26.65					0.00		2236.20
17/11/2002 12:00		0.00	0.00	0.00	0.00		24	
17/11/2002 12:30		0.00	0.00	0.00	0.00		24	
17/11/2002 12:00	27.83					0.00		2618.66
17/11/2002 13:00		0.00	0.00	0.00	0.00		24	
17/11/2002 13:30		0.00	0.00	0.00	0.00		24	

You can see from the preceding that if the following conditions are true, then there is natural ventilation:

$$T_{zone_air} > T_{setpoint} \text{ and } T_{zone_air} > T_{outside_air} \quad (6.5)$$

When the outside temperature is more than the zone temperature, natural ventilation is off and mechanical cooling takes place.

Step 14: Clear the **On** checkbox to switch off the natural ventilation so that the zone runs in the **Air condition** mode.

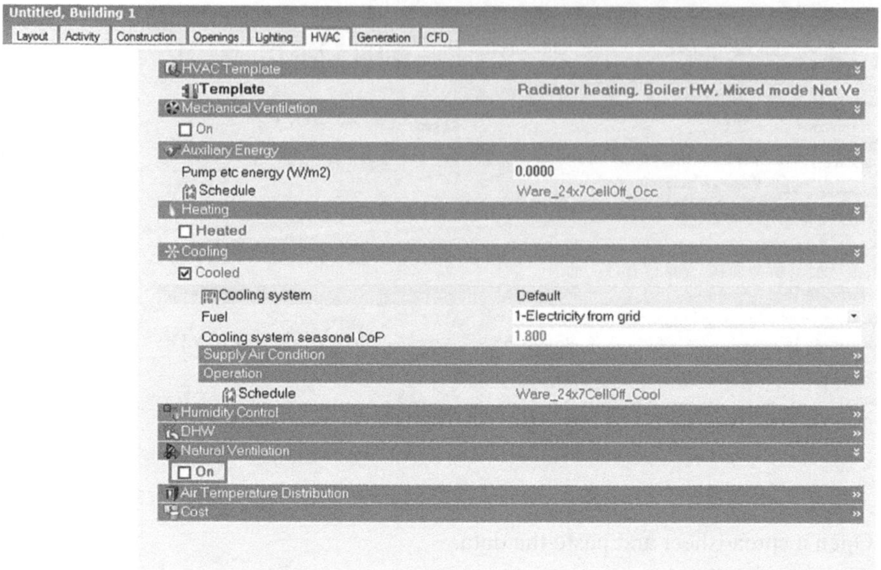

Step 15: Simulate the model and record the monthly energy consumption results (make sure to clear the **IDF file 1** check box). Compare the results in both cases.

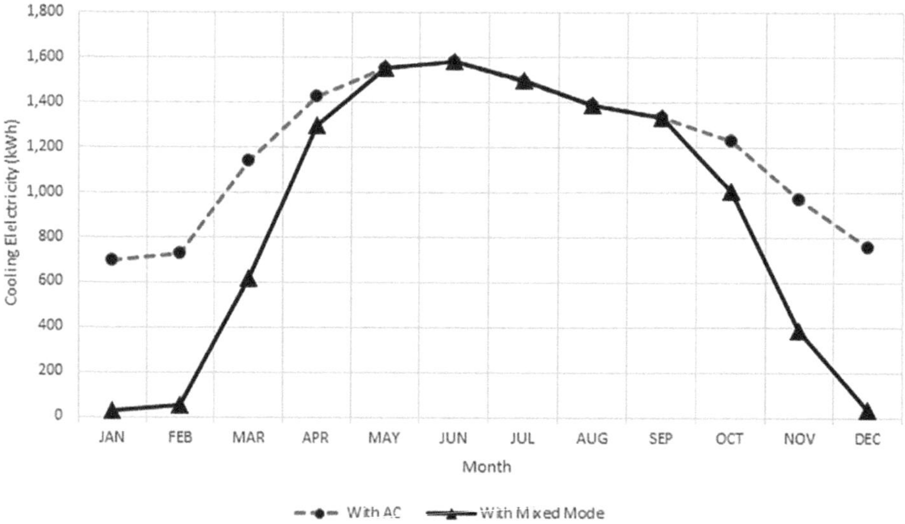

Step 16: Open the monthly outdoor temperature plot.

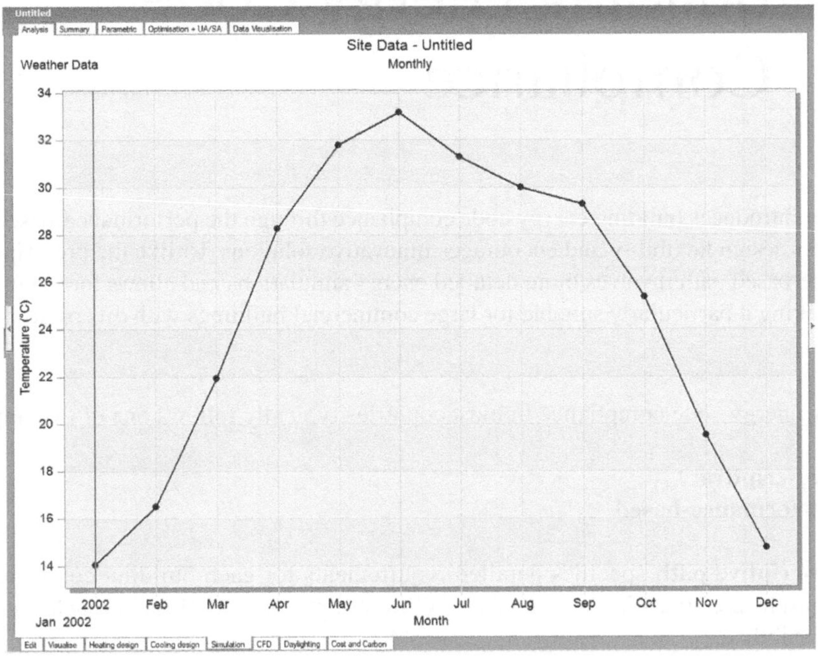

The simulation results demonstrate the energy-saving potential of mixed-mode ventilation systems. In this approach, natural ventilation is prioritized, with mechanical cooling used only when outdoor conditions cannot maintain indoor comfort. During cooler months, particularly from January–March and October–December, natural ventilation proves effective, reducing the need for mechanical cooling. This translates into lower energy consumption compared to the fully air-conditioned scenario, underscoring the value of leveraging favorable climatic conditions for passive cooling.

In contrast, the fully air-conditioned operation led to consistently higher energy use, as mechanical systems were solely responsible for meeting the building's cooling demands. This comparison highlights the advantages of integrating natural ventilation into building design—especially in location like New Delhi, where certain periods of the year allow for effective passive cooling.

EXERCISE 6.6

Repeat this tutorial for window area modulation based on zone and outdoor air temperature.

7 Building Energy Code Compliance

This chapter introduces building energy code compliance through the performance-based path, which offers greater design flexibility and encourages innovative solutions. Unlike the prescriptive path, the performance-based path involves more detailed energy simulations and allows for trade-offs between systems, making it particularly suitable for large commercial buildings with diverse uses and loads.

Building energy code compliance in most countries typically follows one of two approaches.

1. **Prescriptive**
2. **Performance-based**

The **prescriptive path** specifies detailed requirements for each building component, all of which must be met to achieve compliance. While straightforward, this approach can constrain design flexibility.

In contrast, the **performance-based path** allows for greater design freedom and enables innovative solutions by permitting trade-offs among systems. However, it requires advanced energy simulations and a deeper understanding of system interactions.

Source: USDOE, Building Energy Code Enforcement Toolkit 2012

ASHRAE Standard 90.1 is a widely recognized standard developed by the American Society of Heating, Refrigerating and Air-Conditioning Engineers (ASHRAE). It sets forth the minimum requirements for the energy-efficient design of buildings, excluding low-rise residential buildings. The standard is extensively used as a basis for commercial building energy codes across the globe.

A central feature of the standard is Appendix G, the Performance Rating Method (PRM), which forms the foundation of the performance-based compliance path. Appendix G provides a structured methodology to evaluate a building's energy performance by comparing two simulation models: a baseline building and a proposed design. The baseline model is defined using standardized characteristics based on the building type, size, and climate zone, serving as a reference for comparison. The proposed design model, on the other hand, represents the actual building design with all intended energy efficiency measures.

Both models are simulated under identical conditions using hourly energy modeling tools. The annual energy cost of the proposed building is then compared against that of the baseline to assess compliance and potential savings. Table G3.1 in the standard outlines the modeling requirements for both models, covering aspects such as space use classification, occupancy and operation schedules, building envelope characteristics, lighting systems, HVAC configurations, service hot water systems, and receptacle loads.

This chapter explains some of the baseline building requirements related to the building envelope and lighting through four tutorials. These include the modeling of building orientation (rotation), external wall construction, vertical fenestration assemblies, and lighting power density. The tutorials in this chapter focus only on the modeling requirements. There are no simulations performed in this chapter.

DOI: 10.1201/9781003563013-7

TUTORIAL 7.1: MODELING BUILDING PERFORMANCE IN FOUR ORIENTATIONS

Goal

To model the baseline building orientation in accordance with building energy code requirements

What Are You Going to Learn?

- How to use the Site Orientation input to rotate the building to 90°, 180°, and 270° from its original orientation

In most of the building energy codes, the baseline building performance is generated by simulating the building with its actual orientation and again after rotating the entire building by 90°, 180°, and 270°, and then deriving the average from the results. This is illustrated by the following example. The field **Site orientation** can be used to achieve the desired orientation.

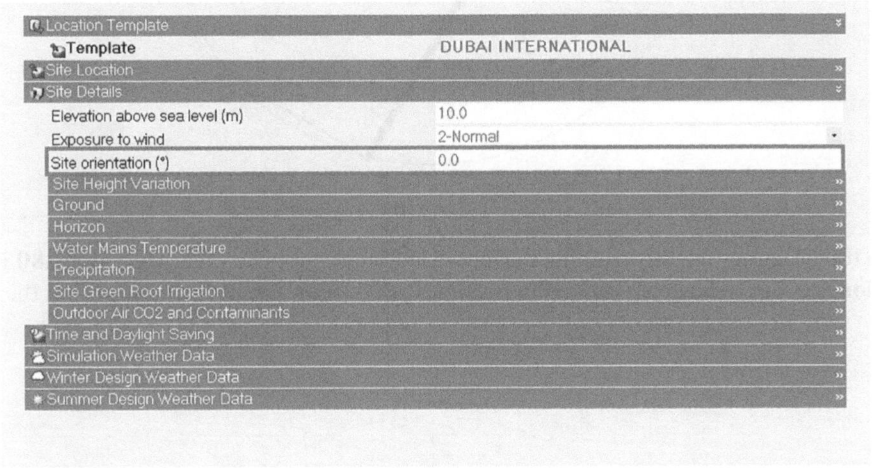

Let us consider the building illustrated in the following figure. The rectangle represents the building. The bold line indicates the primary façade, which is oriented 150° from true north.

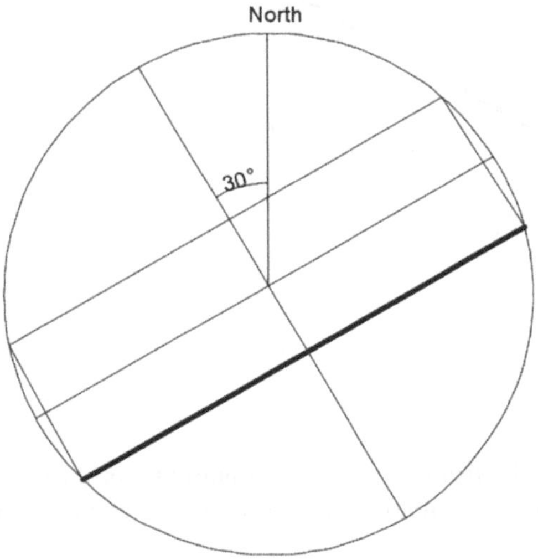

To rotate the building by 90° from its current orientation, you can enter the value **90** in the **Site orientation** text box. The new orientation becomes 240° from true north, as shown in the following figure.

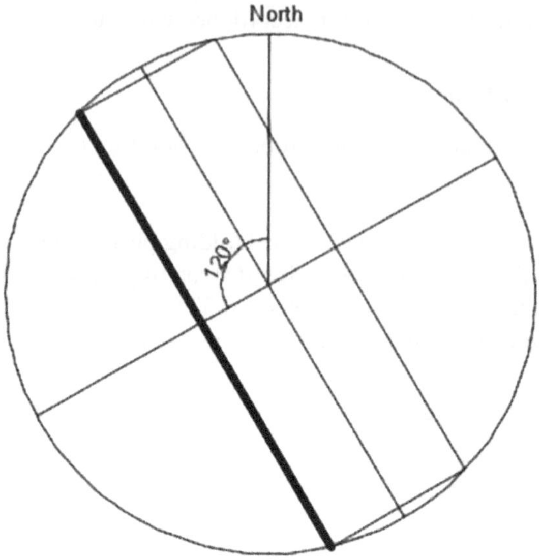

To rotate the building by 180° from its current orientation, you can enter the value **180** in the **Site orientation** text box. The new orientation becomes 330° from true north, as shown in the following figure.

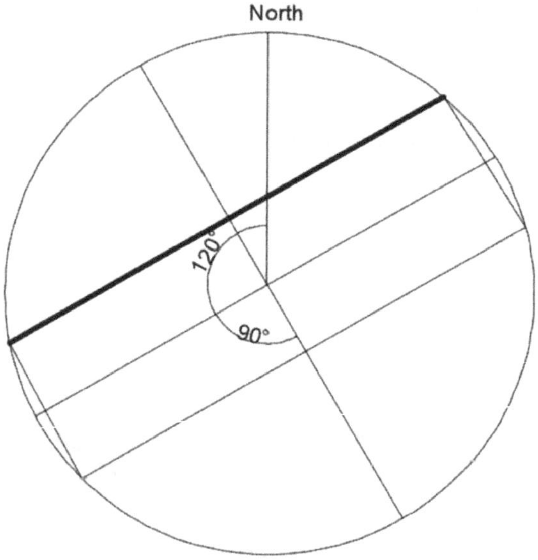

To rotate the building by 270° from its current orientation, you can enter the value **270** in the **Site orientation** text box. The new orientation becomes 60° from true north, as shown in the following figure.

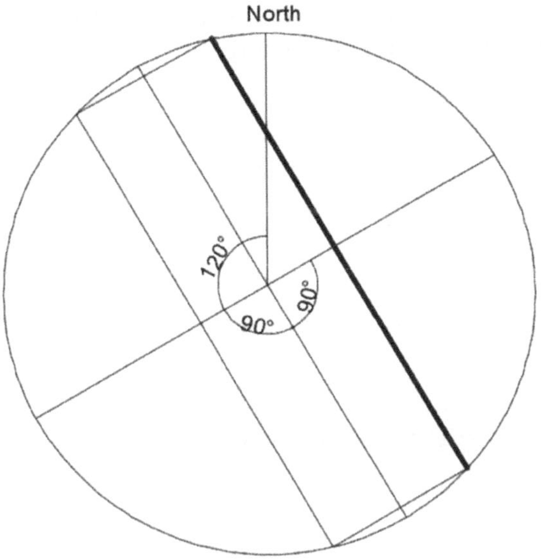

TUTORIAL 7.2: CREATING THE BASELINE BUILDING EXTERNAL WALL FOR ASHRAE STANDARD 90.1–2016, APPENDIX G

Goal

To select a steel-framed external wall construction that meets the baseline requirements specified in ASHRAE Standard 90.1–2016, Appendix G

What Are You Going to Learn?

- How to select a steel-framed wall assembly that complies with the required U-value
- How to inspect and verify wall properties

According to ASHRAE Standard 90.1–2016, Appendix G, the baseline external wall construction is required to be steel-framed. The applicable U-factor for the baseline wall must be selected from Tables G3.4.1–G3.4.8 of the standard, based on the building's climate zone. For example, if the building is located in Climate Zone 1, a wall with a U-value of 0.70 W/m²-K must be used to comply with the requirement.

Step 1: Create a new external wall, and select Wall, Steel-Framed, R-13 (2.3), U-0.124 (0.70)—16 in. (400mm) On Centre, 3.5 in. (89mm) Depth Framing.

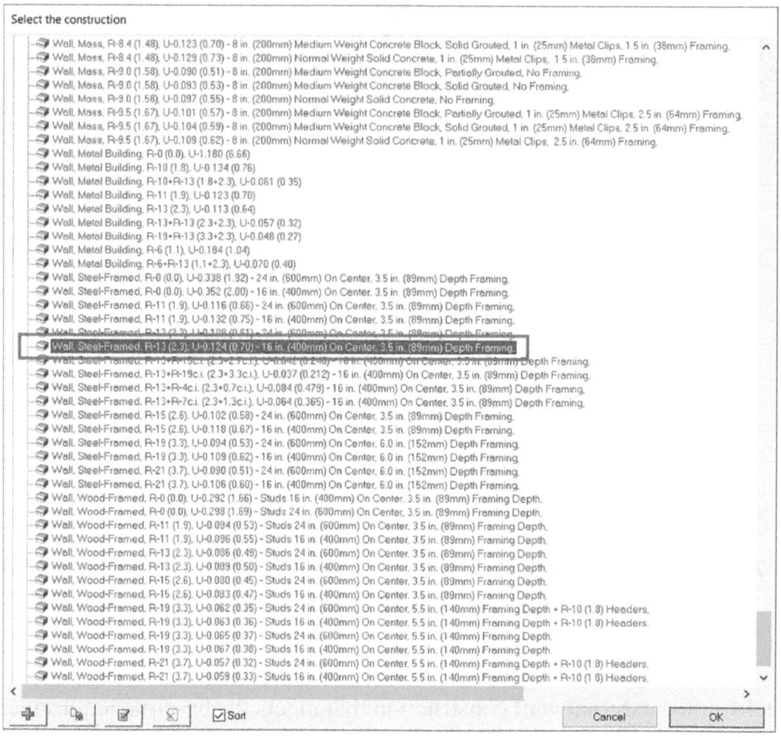

Step 2: Edit the construction to view properties.

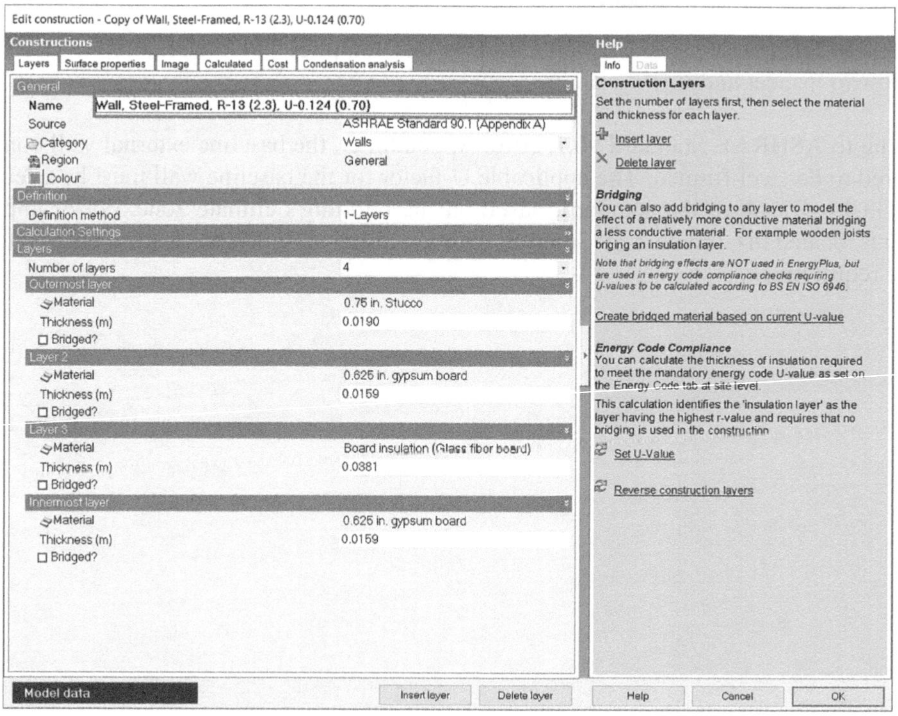

Step 3: Select the **Calculated** tab. It displays all construction properties.

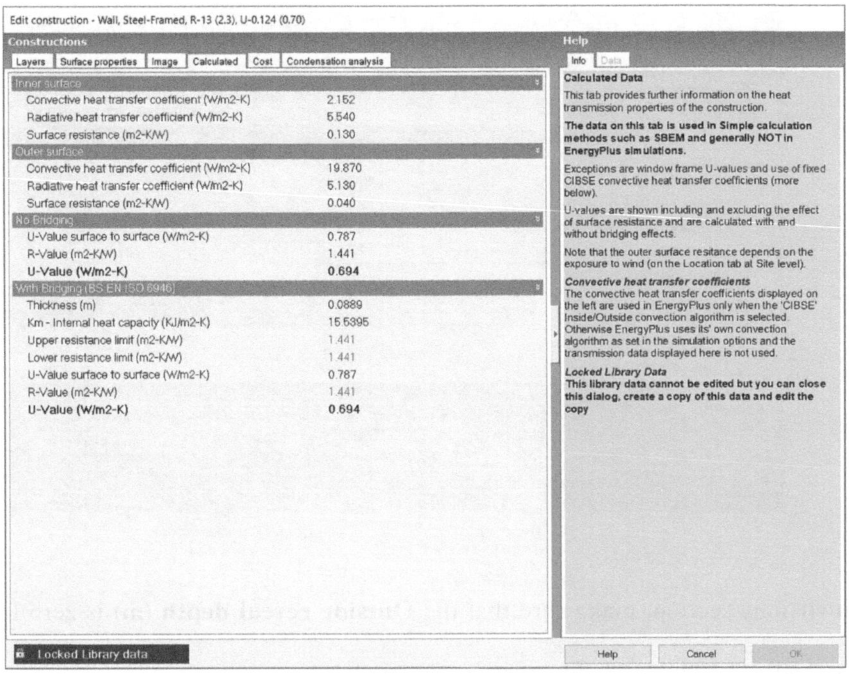

TUTORIAL 7.3: MODELING FLUSH WINDOWS FOR THE BASE CASE

Goal

To model flush windows for the baseline building in compliance with ASHRAE Standard 90.1–2016, Appendix G

What Are You Going to Learn?

- How to ensure that shading projections are not modeled on fenestrations in the base case building
- How to set the **Outside Reveal Depth** to zero, so windows are flush with the exterior surface

ASHRAE Standard 90.1–2016, Appendix G, Table G3.1, requires that shading projections on the fenestrations of the baseline building must not be modeled. The fenestrations need to be modeled as flush with the exterior wall or roof.

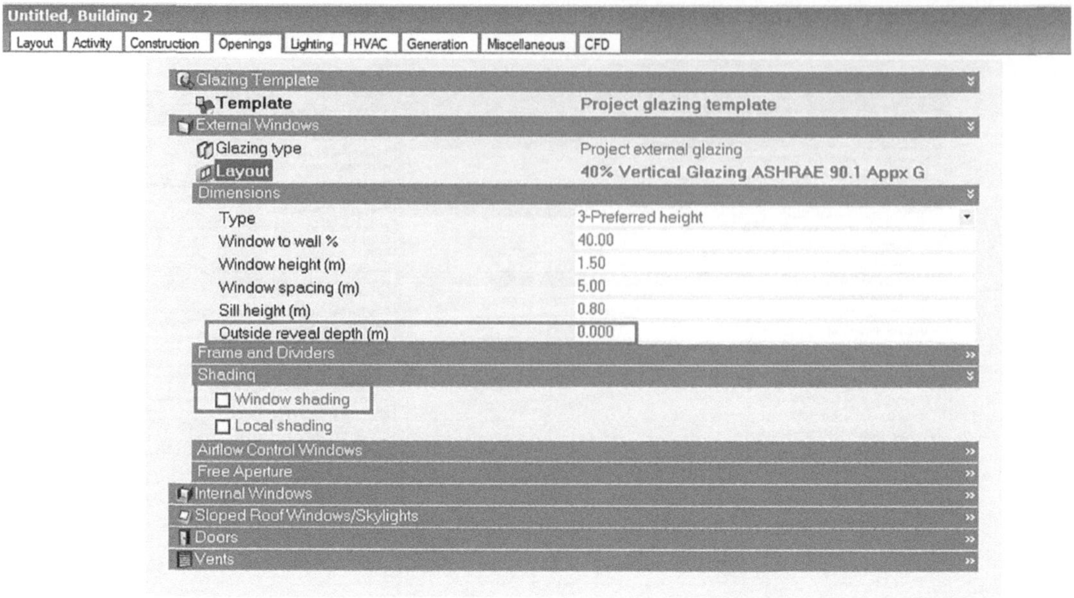

In the **Dimensions** section, make sure that the **Outside reveal depth (m)** is zero for the base case.

Window shading and local shading must not be modeled.

TUTORIAL 7.4: MODELING BASELINE BUILDING INTERIOR LIGHTING POWER

Goal

To model interior lighting power for the baseline building as required by ASHRAE Standard 90.1–2013, Appendix G

What Are You Going to Learn?

- How to set lighting power density (LPD) for the baseline building using either the Building Area Method or the Space-by-Space Method

As per ASHRAE Standard 90.1–2013, Appendix G, interior lighting power in the baseline building design shall be determined using the same categorization procedure (building area or space-by-space method) and categories as the proposed design with lighting power set equal to the maximum allowed for the corresponding method and category in 9.2 of the standard.

To determine lighting power allowances, ASHRAE Standard 90.1-2013 provides the following two calculation methods.

- The **Building Area Method** assigns a single **lighting power density (LPD)** value to the entire building based on its **primary function or occupancy type**
- The **Space-by-Space Method** assigns specific LPD values to each **space type** within the building based on its function

For example, in the case of an office building where the building area method is employed, the baseline building lighting power density should be set at 8.8 W/m², as specified in Table 9.5.1 of the standard.

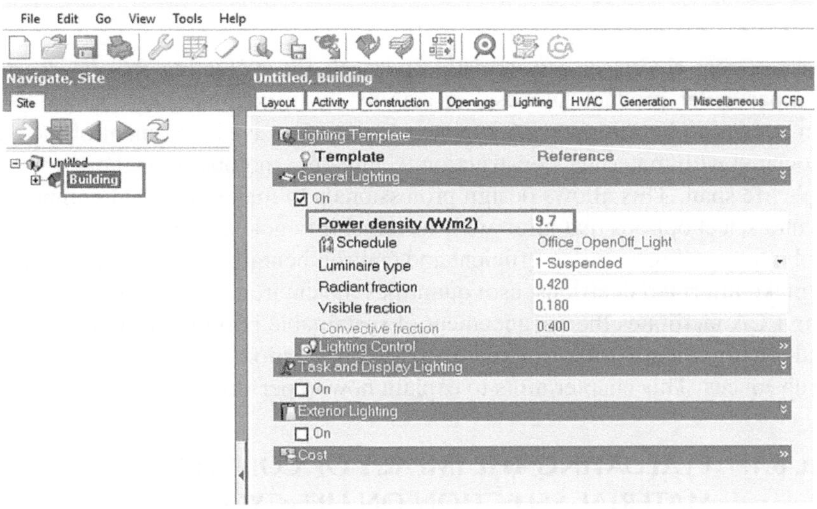

REFERENCES

American Society of Heating, Refrigerating and Air-Conditioning Engineers (ASHRAE), ASHRAE Standard 90.1–2013: *Energy Standard for Buildings Except Low-Rise Residential Buildings*. Atlanta, GA: ASHRAE, www.ashrae.org.

American Society of Heating, Refrigerating and Air-Conditioning Engineers (ASHRAE), ASHRAE Standard 90.1–2016: *Energy Standard for Buildings Except Low-Rise Residential Buildings*. Atlanta, GA: ASHRAE, www.ashrae.org.

8 Life-Cycle Assessment

Life-cycle assessment (LCA) is a systematic approach for evaluating the environmental consequences of a particular product, process, or activity, from acquiring raw materials to final disposal. When applied to buildings, LCA enables a comprehensive assessment of the environmental impacts associated with materials, construction techniques, and operational energy use throughout the building's life span. This allows design professionals to make informed comparisons between alternatives and select options that align with sustainability goals. As a result, LCA has become an essential tool in promoting resource-efficient and environmentally responsible building design. It is essential to make informed decisions, as it quantifies the environmental impacts of a specific product. Utilizing LCA facilitates the advancement of sustainable building design, resulting in a minimized overall environmental footprint and decreased utilization and emission of non-renewable or hazardous substances. This chapter aims to explain how to perform LCA for buildings.

TUTORIAL 8.1: EVALUATING THE IMPACT OF CONSTRUCTION MATERIAL SELECTION ON LIFE-CYCLE EMISSIONS

Goal

To evaluate the impact of construction material on life-cycle emissions

What Are You Going to Learn?

- Calculating life-cycle emissions

Problem Statement

In this tutorial, you are going to use a 50 m × 25 m five-zone model with a 5 m perimeter depth. You are going to evaluate the following materials for external walls with 200 mm Aerated concrete block to get the same R-value of 3 m^2-K/W of insulation.

1. Foam—Polyurethane
2. Glass fiber insulation board

TABLE 8.1
Properties of the Foam—Polyurethane and Glass Fiber Insulation Board

Serial No.	Property	Foam—Polyurethane	Board insulation (Glass Fiber Board)
1	Thermal conductivity (W/m-K)	0.028	0.036
2	Density (kg/m^3)	30	160
3	Specific Heat (J/kg-K)	1,470	840
4	Thickness (mm) required for R-3 m^2-K/W	84	108

Perform calculations for life-cycle emissions using the **One Click LCA** tool.
Use weather file for **London Gatwick** location.

DOI: 10.1201/9781003563013-8

SOLUTION

Step 1: Open a new project and create a **50 m × 25 m** building with a **5 m** perimeter depth. Select the **Construction** tab.

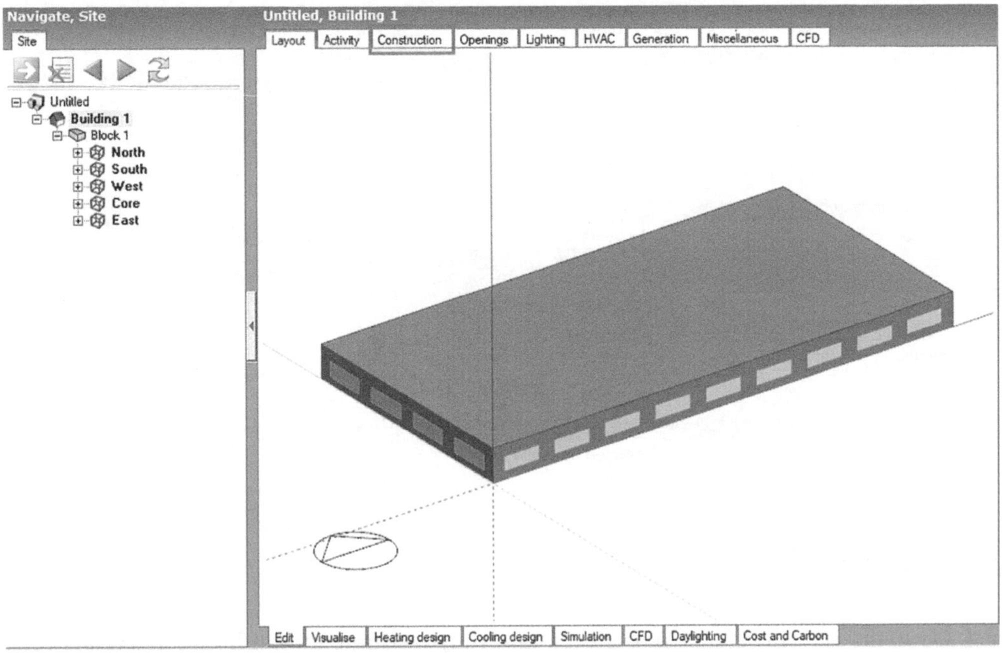

Step 2: In the Construction section, click **Project wall. External walls** text gets highlighted. Click the **Add new item** icon. The **Edit construction-Wall** screen appears.

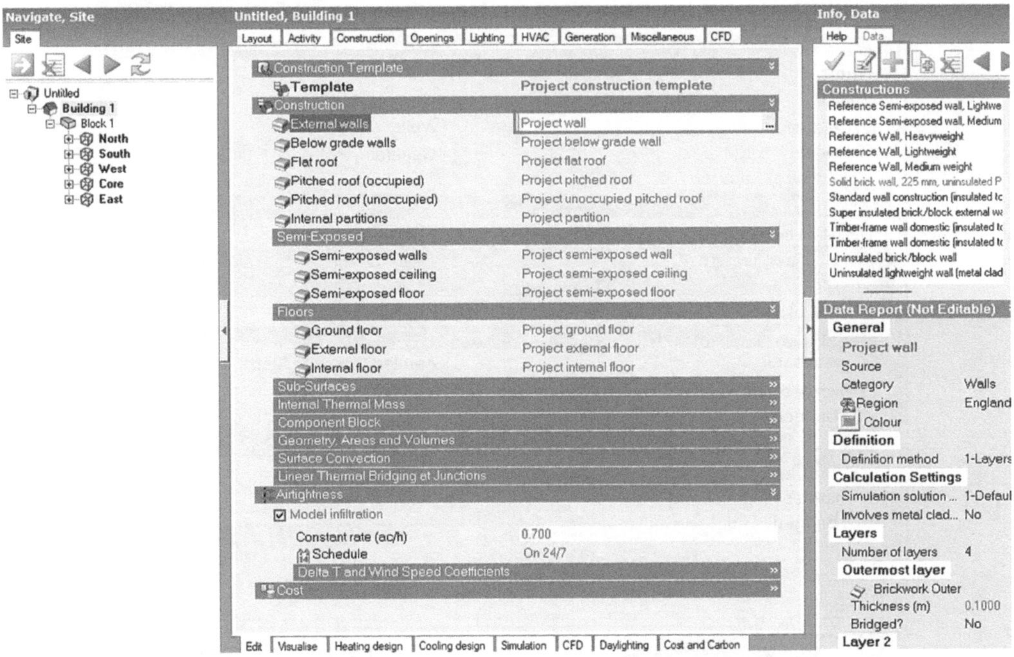

Step 3: Enter **Project External Wall with PUF** in the **Name** textbox. Go to the **Number of Layers** drop-down list and select **2**.

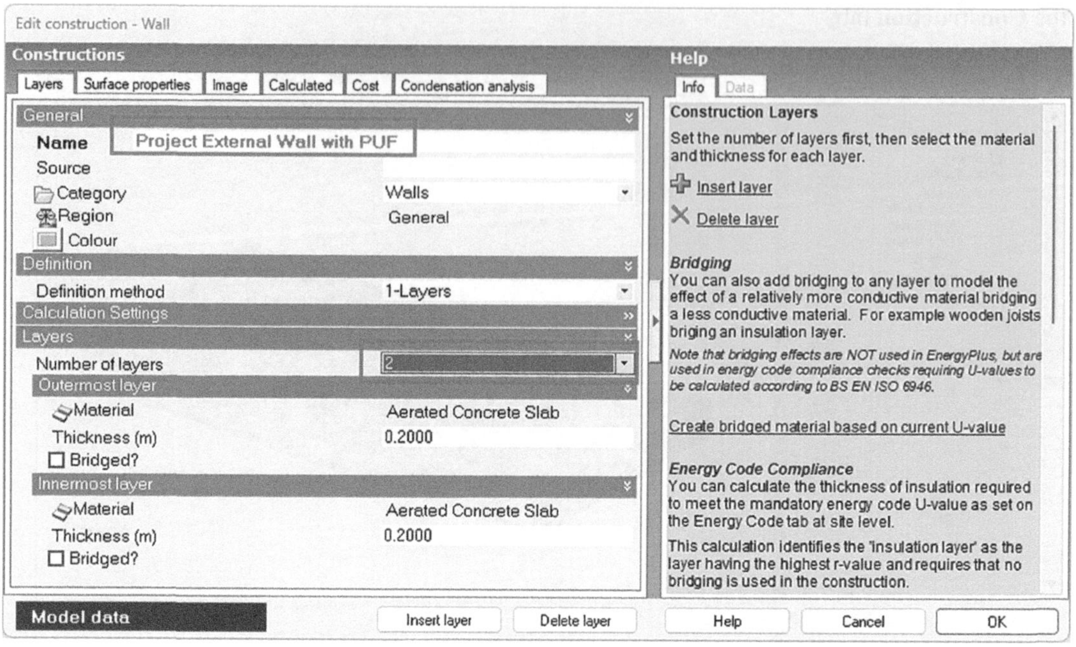

Step 4: Go to the **Layers** section and select **Foam polyurethane** in **Insulating materials** category as the innermost layer. Set the thickness to **0.0840**.

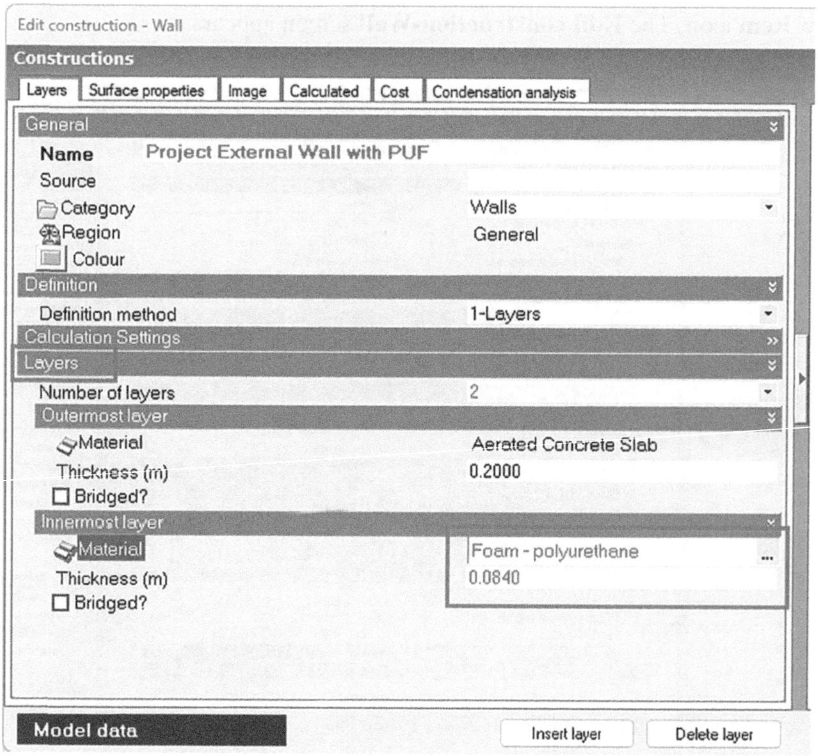

Step 5: Click on **Create copy of highlighted item** and then click on **Edit/View highlighted item** to edit the material properties of the copied item.

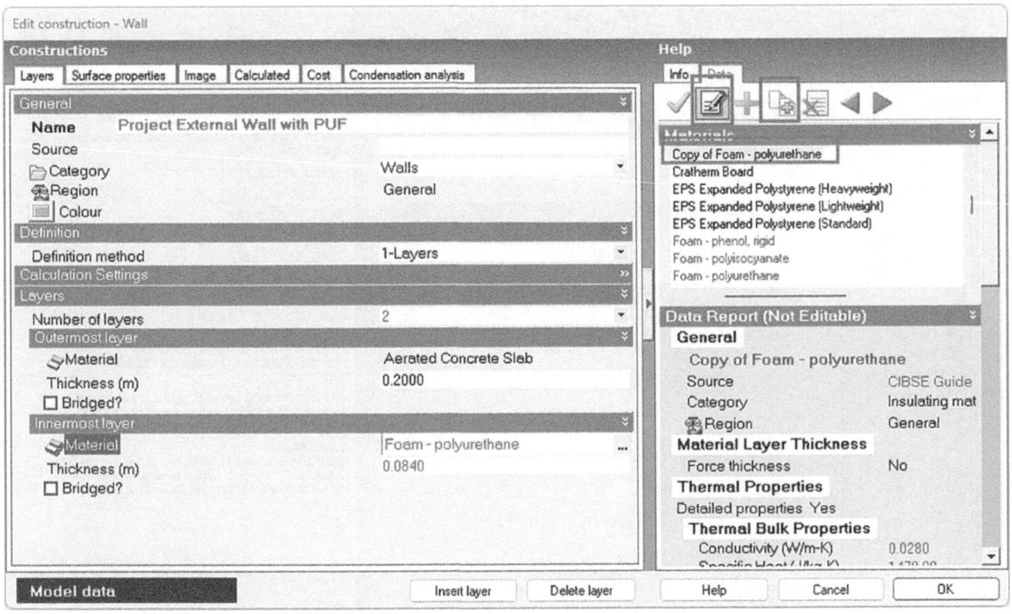

Step 6: Enter **Foam—polyurethane_project** in the **Name** textbox.

Step 7: Click on the **Simulation** screen tab. **Edit Calculation Options** screen appears.

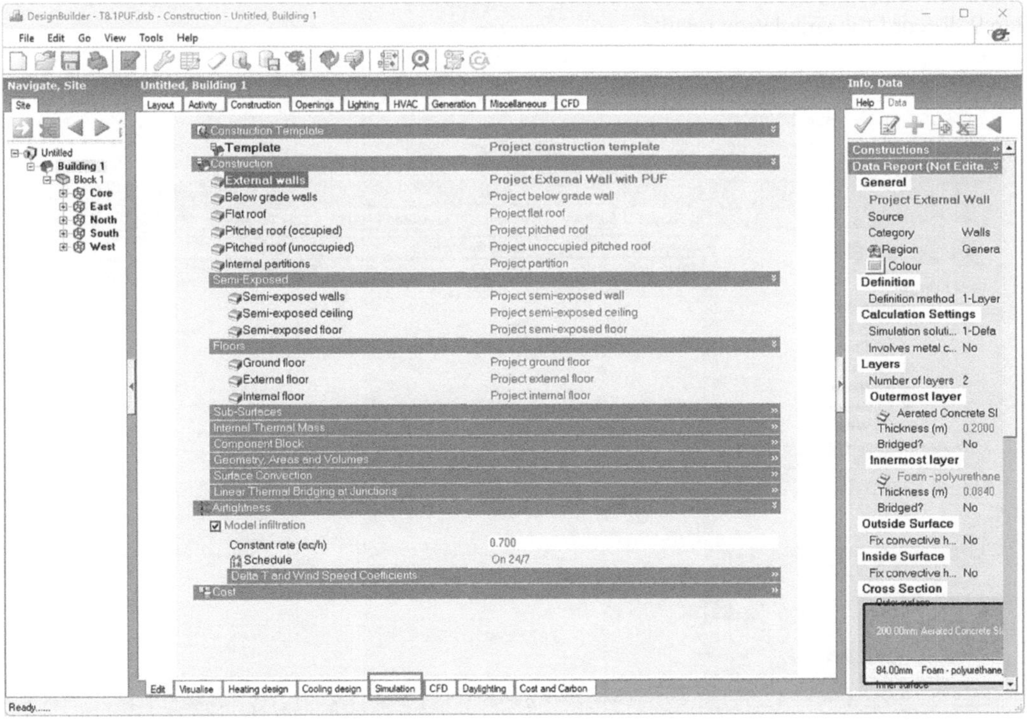

Step 8: Click on **Annual Simulation**.

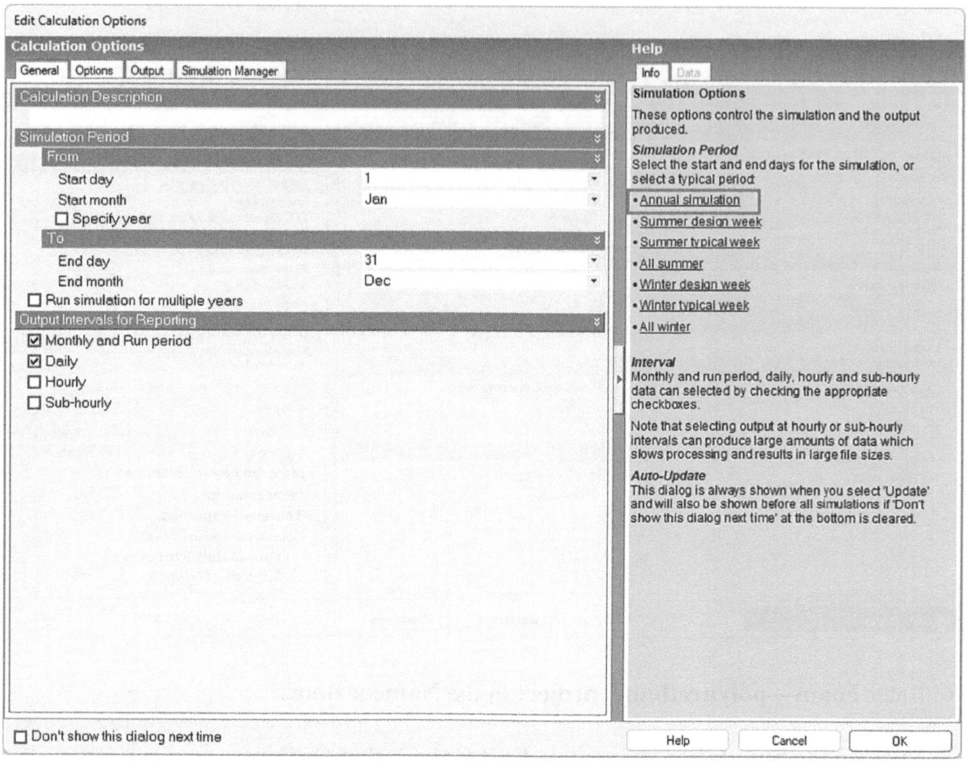

Step 9: Record the simulation results.

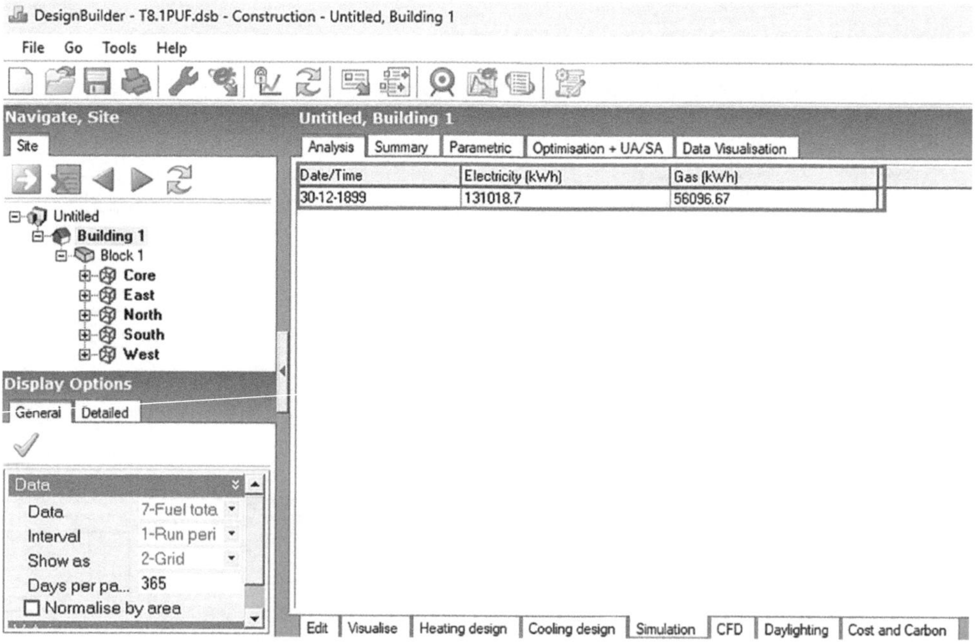

Save this model as "Tutorial 8.1_PUF".

In the following steps, you are going to replace **Foam polyurethane** with **Board insulation** in the external wall construction.

Step 10: Copy the model "Tutorial 8.1_PUF" and rename it as "Tutorial 8.1_GlassBoradInsu". Open this model and click the **Edit** screen tab. The **Construction Template** screen appears. Click **External walls** under the **Construction** section. Click on **Create copy of highlighted item** and then click on **Edit/View highlighted item**. The **Edit construction** screen appears.

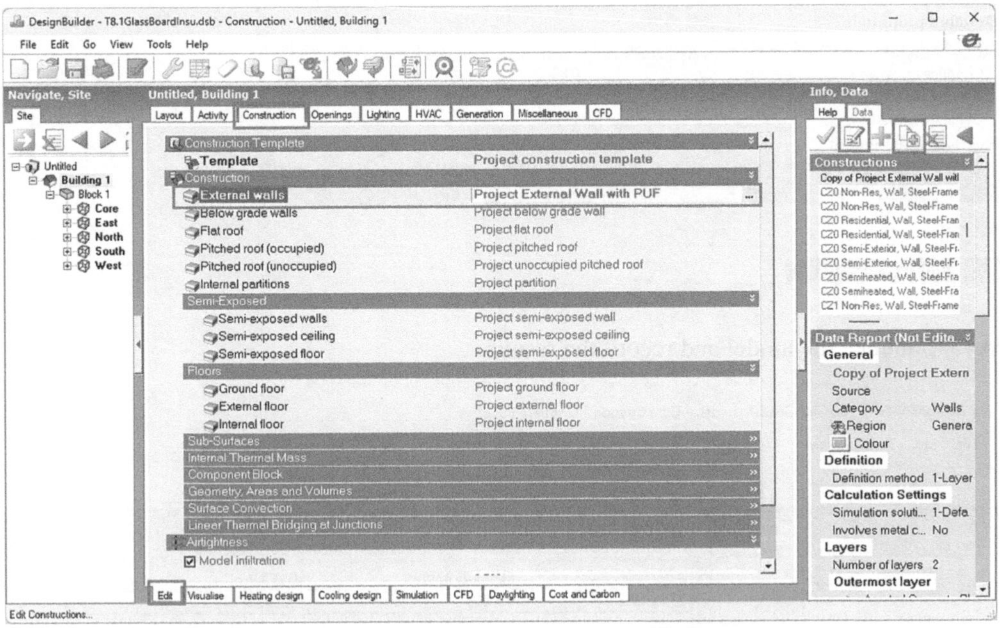

Step 11: Change the name to **Project External Wall with Board Insulation**. Go to the **Layers** section and select **Board insulation (Glass fiber board)** as the innermost layer. Set the thickness to **0.1080.**

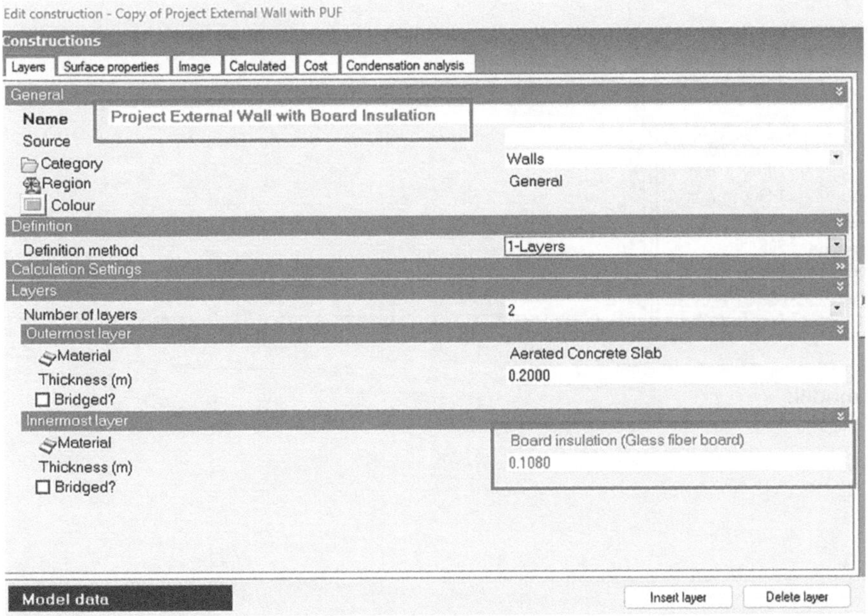

Step 12: Copy the existing material. Enter **Name** as **Board insulation (Glass fiber board)** **_project**.

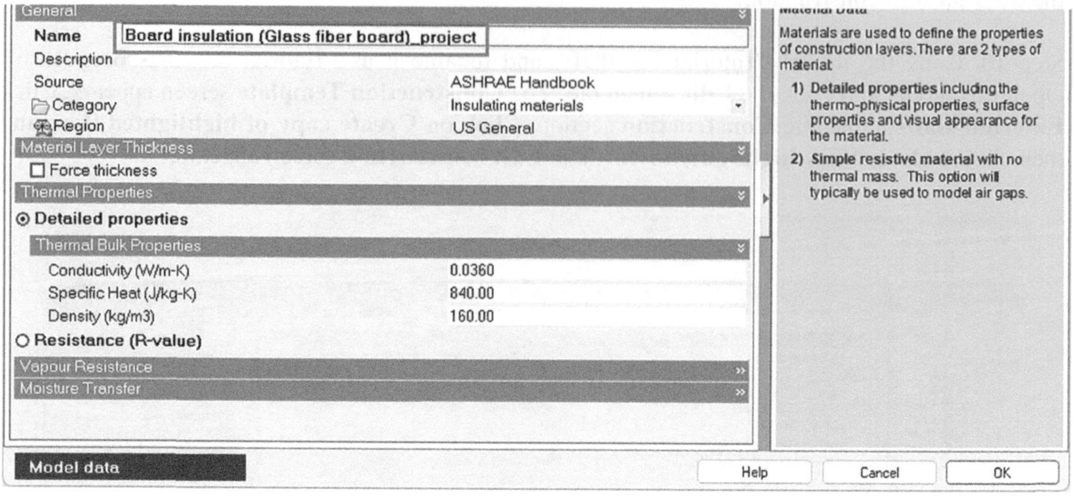

Step 13: Simulate the model and record the results.

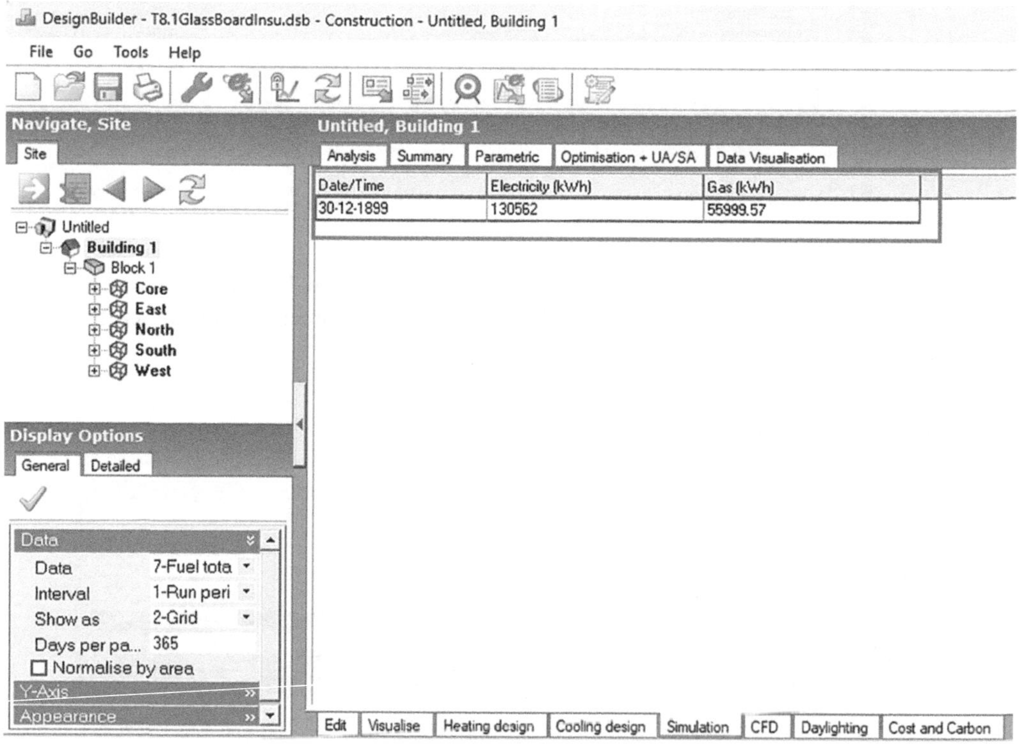

Save this model.

Step 14: Open https://oneclicklca.com in the browser.

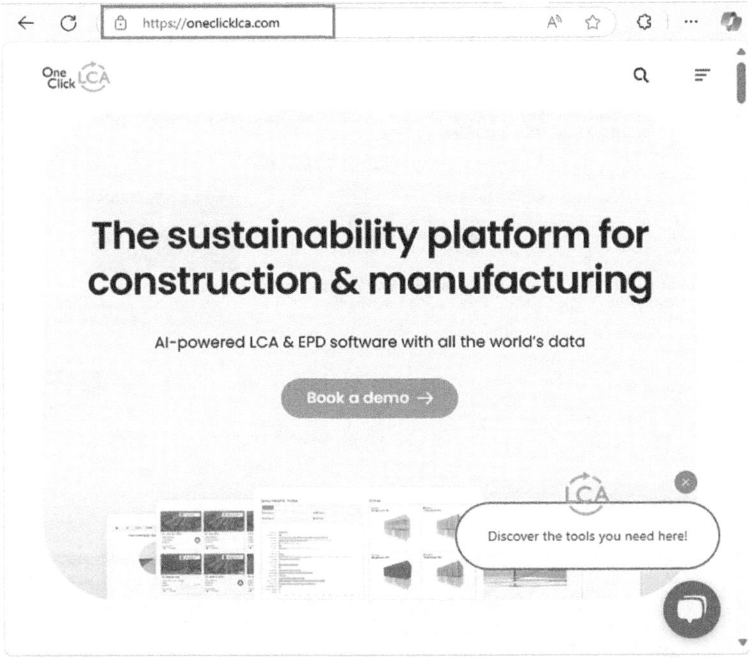

Step 15: Click on **Log in.** Enter your **username and password.** If you do not have a OneClick LCA account, click on **Register** to create a new one.

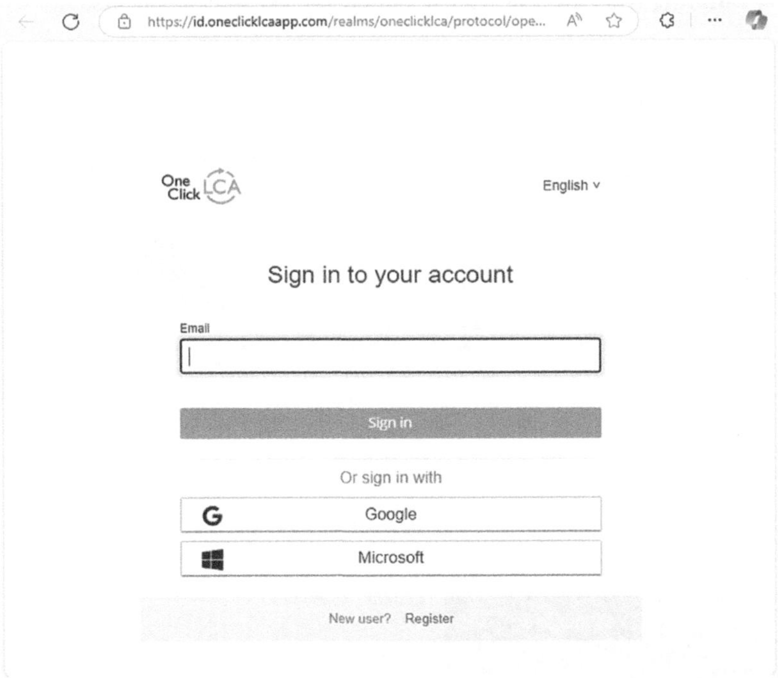

Step 16: Once you are logged in, click on **+ Add**.

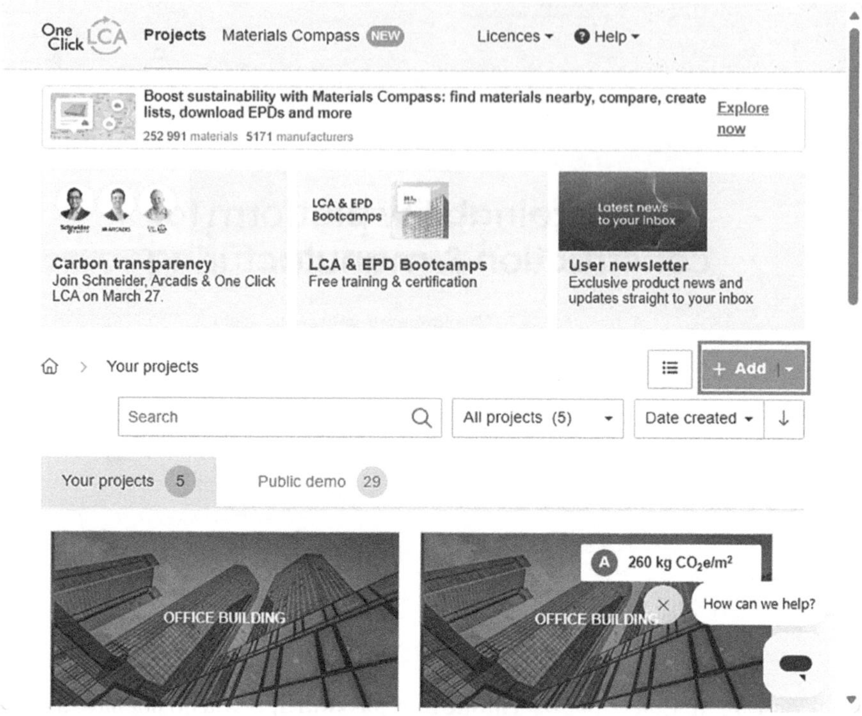

Step 17: Select **Building** from drop-down list. A new project screen appears.

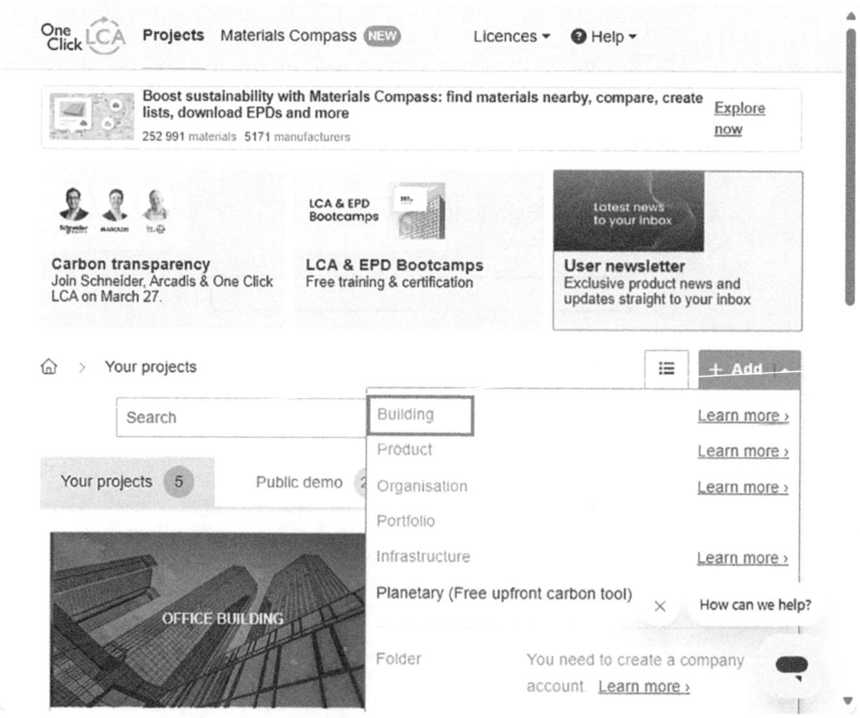

Step 18: Enter **Tutorial_8_2** in the **Name (mandatory)** text box. Select **Office buildings** from **Type (mandatory)** drop-down list. Select **United Kingdom** from **Country (mandatory)** drop-down list. Click on **Next**.

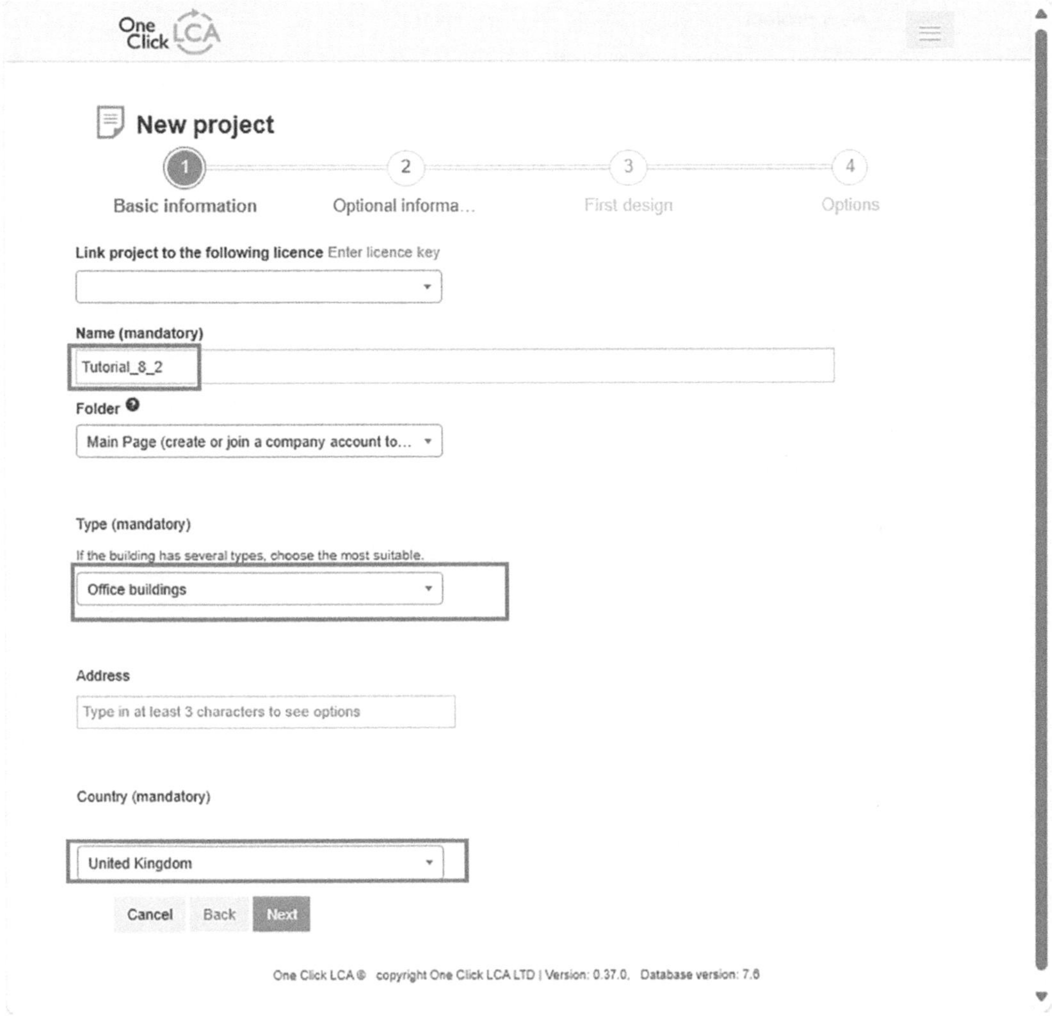

Step 19: Enter **1200** in **Gross Floor Area (m^2)**. Enter **1** in **Number of above ground floors**. Select **Concrete frame** from **Frame type** drop-down list. Click on **Next**.

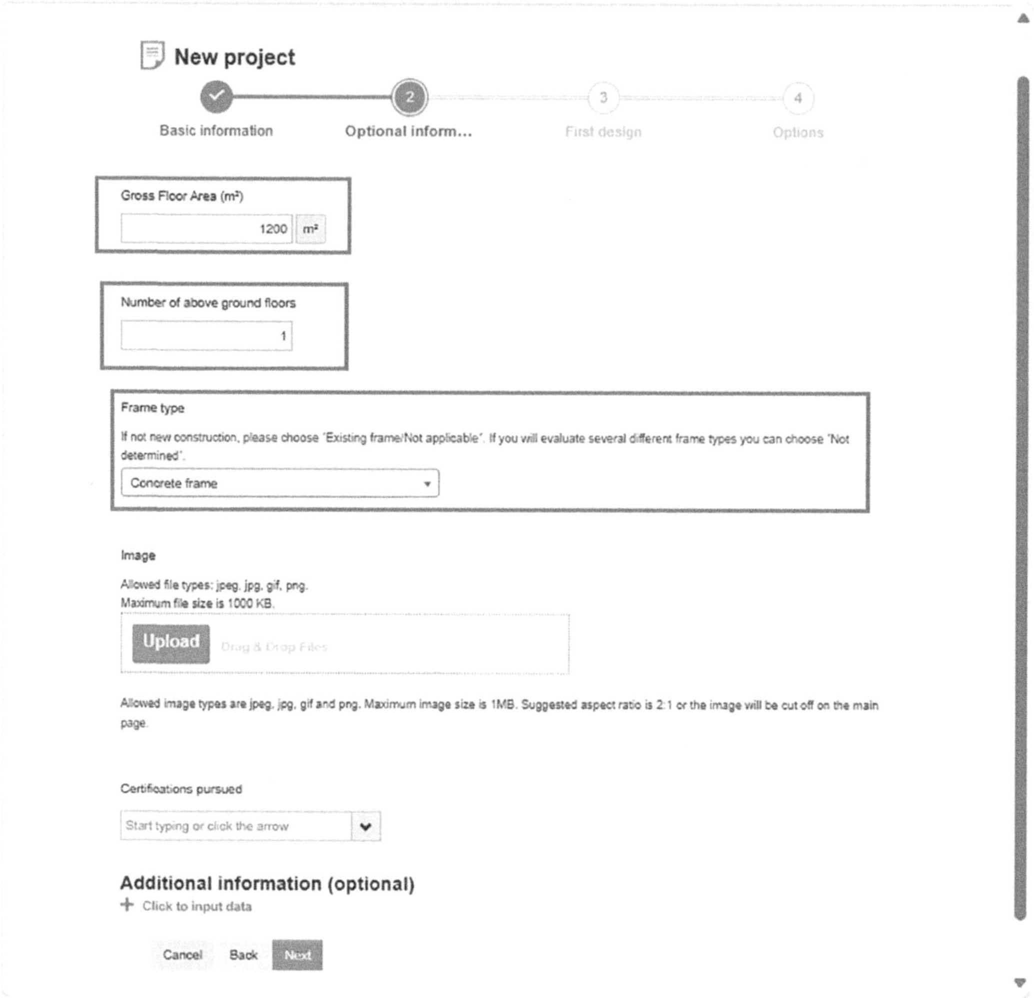

Step 20: Click on **Enter license key**. A text box appears, enter your license key in text box.

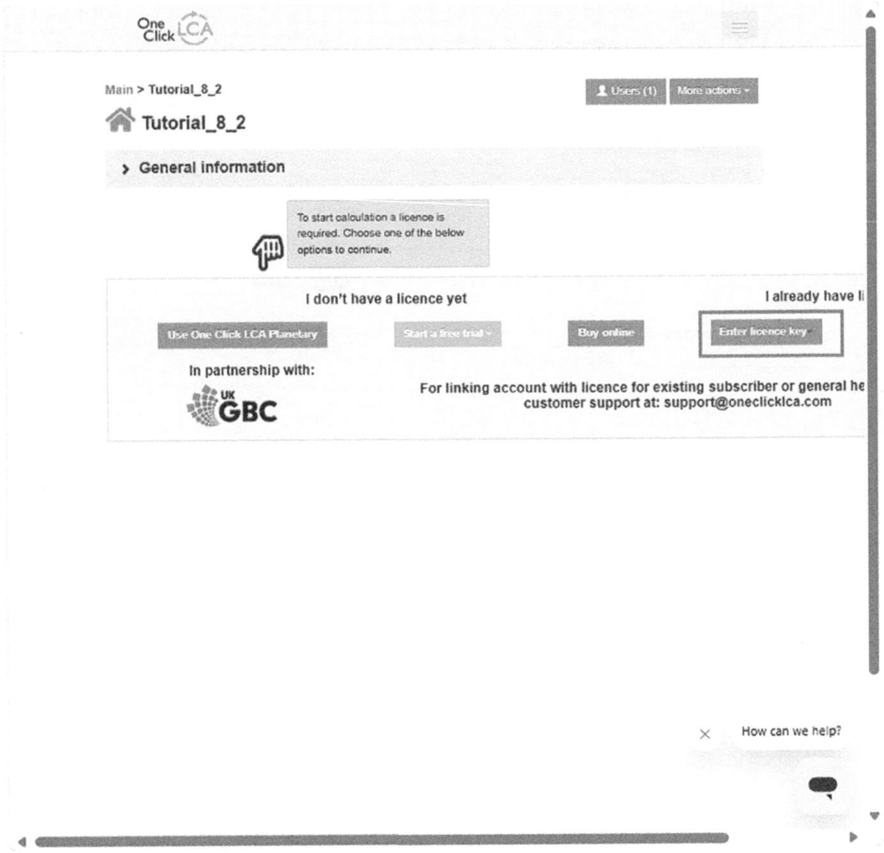

It is important to note that the activation method shown in step 20 is limited to educational users. Commercial One Click LCA subscriptions use alternative activation processes.

Sources: https://help.oneclicklca.com/en/collections/647509-product-guide-licensing and https://help.oneclicklca.com/en/articles/275601-licensing-your-project

Step 21: Click on **Get started**. The available tools screen appears.

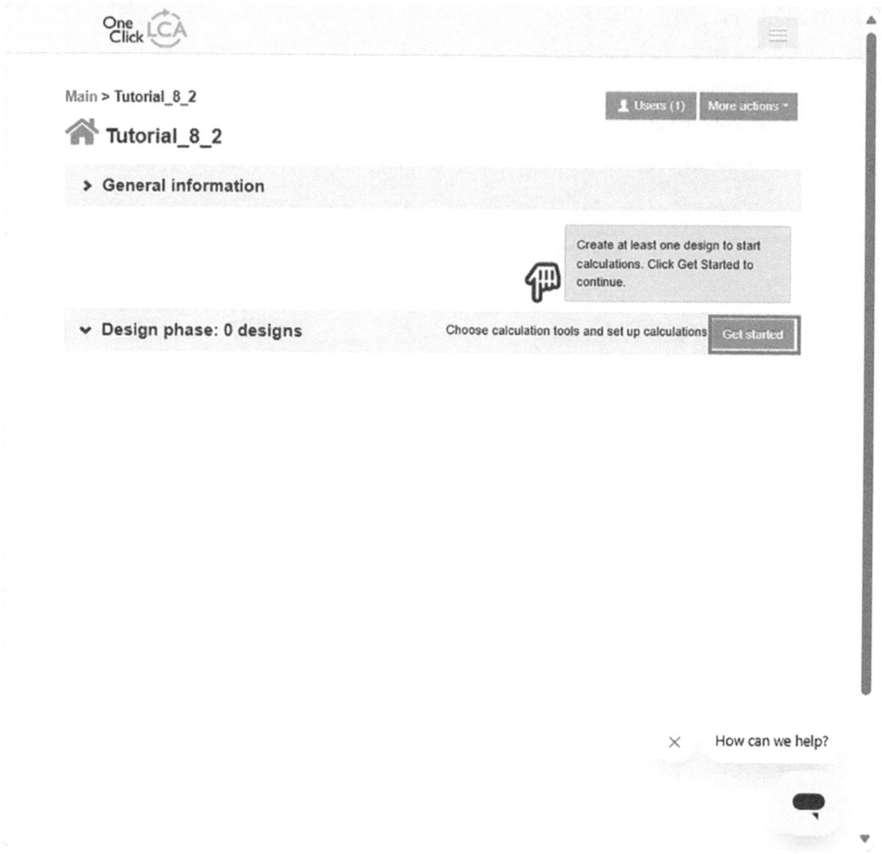

Step 22: Select **Level(s) life-cycle assessment (EN15804 +A1) Life-cycle assessment for Level(s)—overarching assessment tool 7: Cradle to cradle Life Cycle Asse** check box. Click on **Next**.

In case EN15804+A1 is not available, you can choose an equivalent calculation tool available with your license.

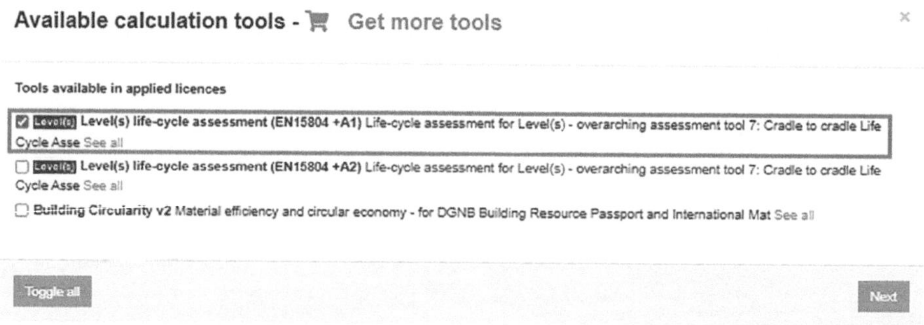

Step 23: Enter **Case-I PUF** in **Name** text box. Click on **Next**.
Life-Cycle Assessment Parameters setup screen appears.

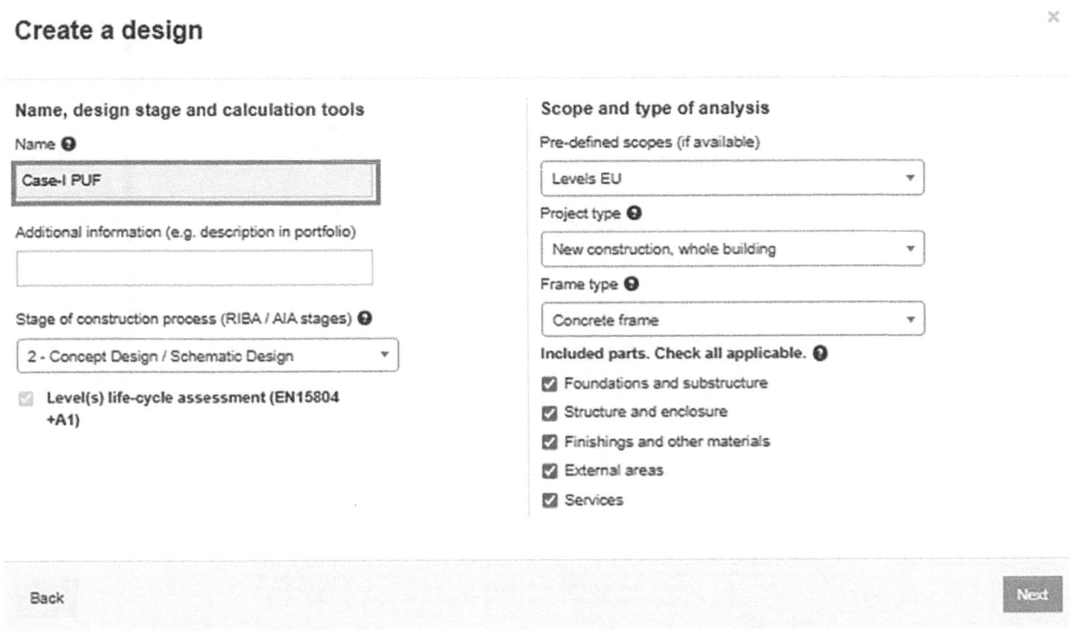

Step 24: Click on **Review and adjust LCA parameters**.

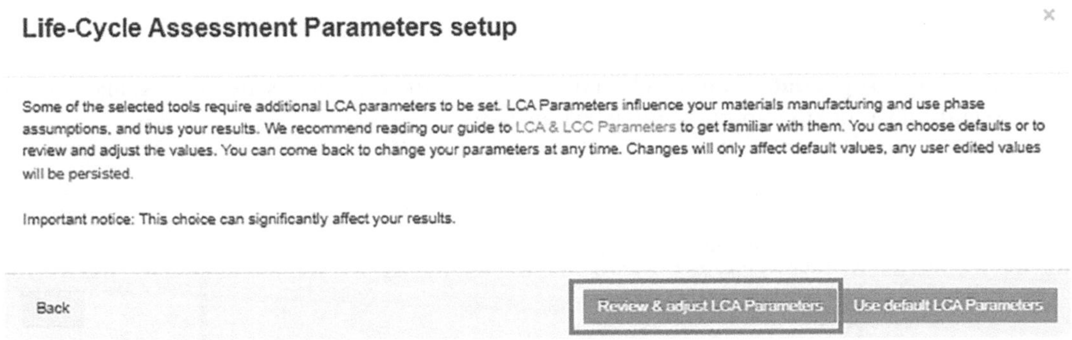

Step 25: Click on **Save**.

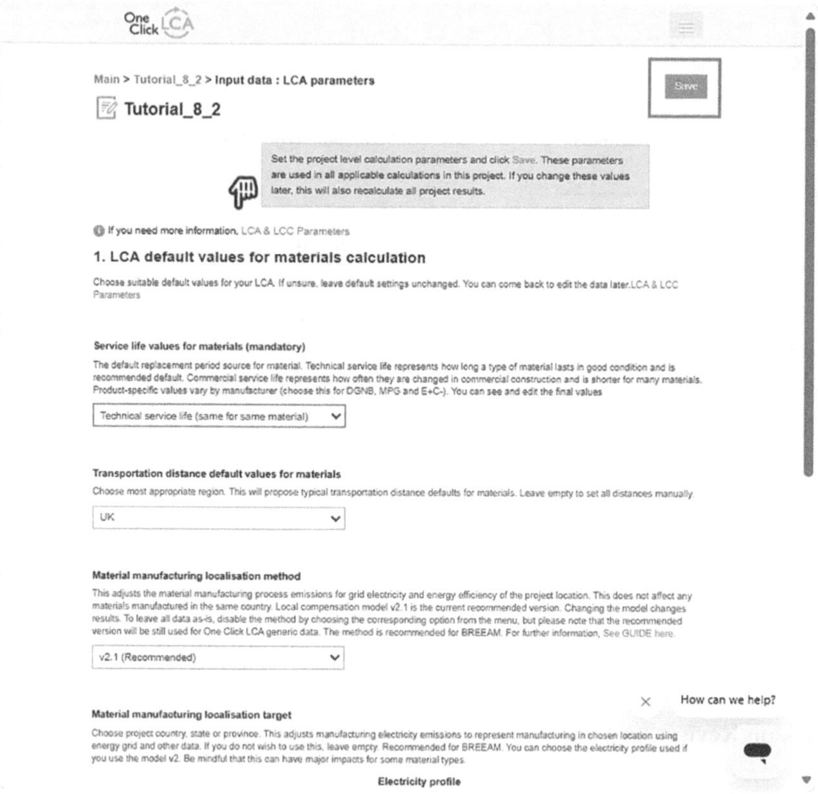

Step 26: Click on **Input data**. You can observe that most of the Data inputs are missing, and you can use DesignBuilder tool to import the model parameters.

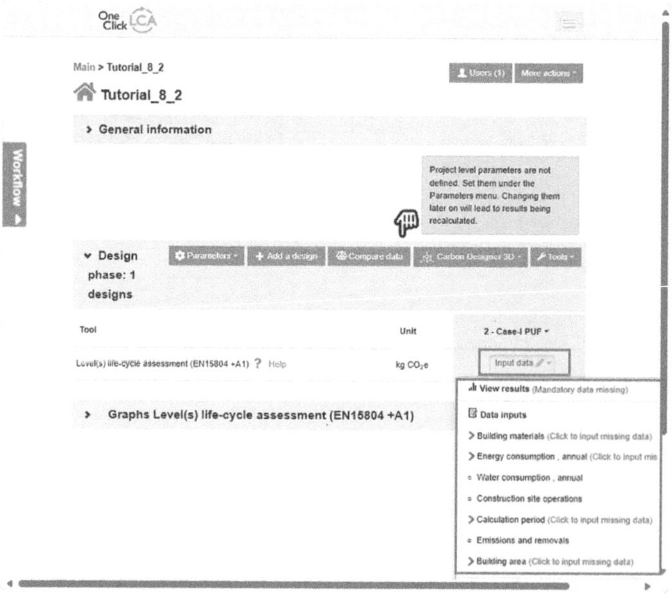

Step 27: Open the DesignBuilder model saved with polyurethane insulation foam.

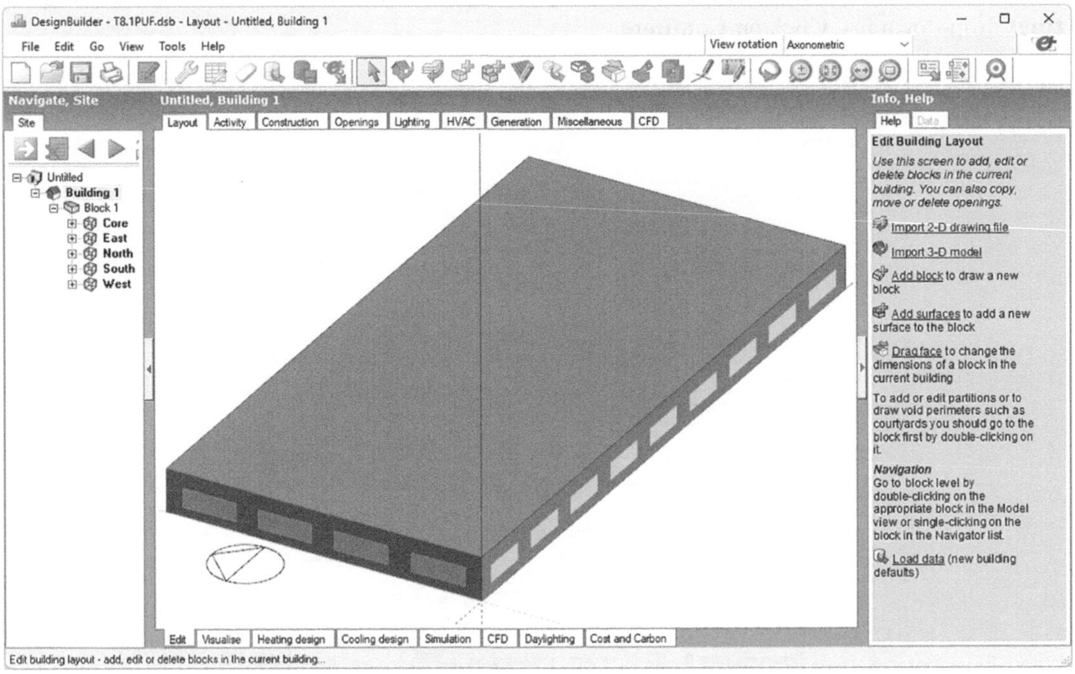

Step 28: Click on **Edit** Screen tab. Click on **LCA** button. A new screen appears on the web browser.

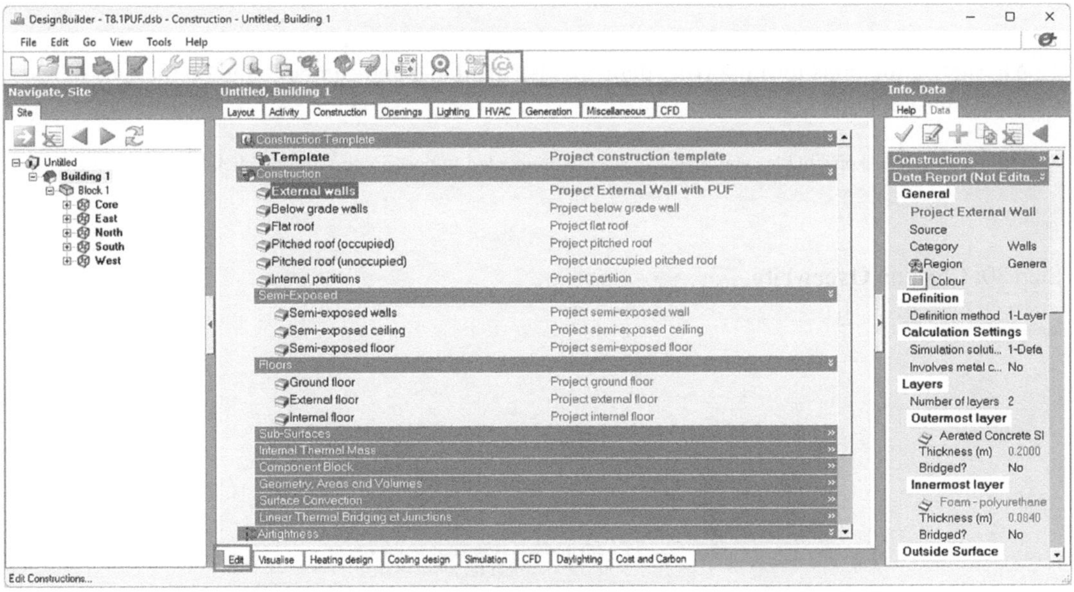

Step 29: Select **Tutorial_8_2** from the **Choose the project or entity** drop-down list. Select **Case-I PUF** from the **Choose or create a design** drop-down list. Select **All data** from the **Filtering settings** drop-down list. Click on **Continue**.

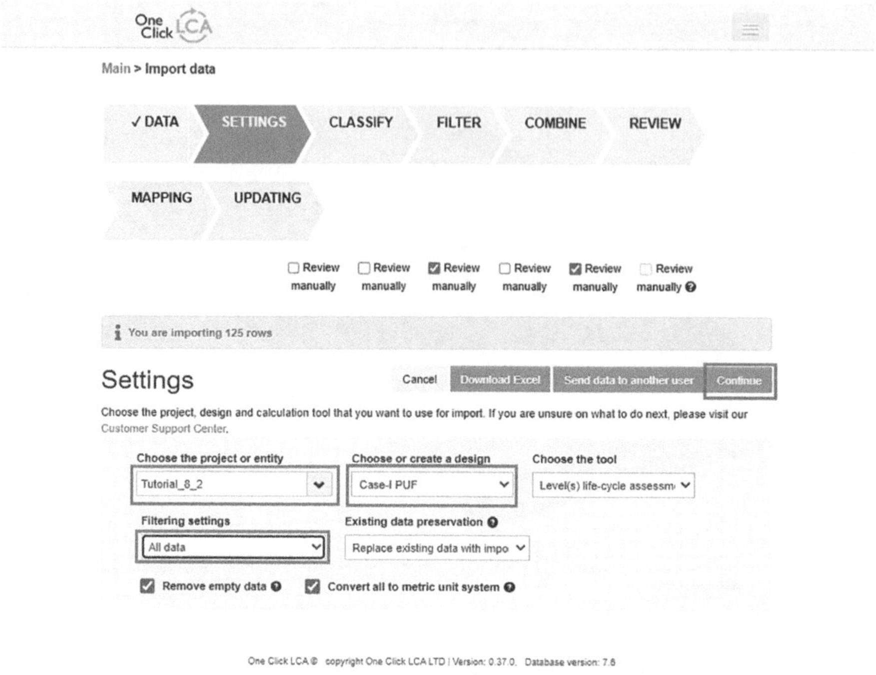

Alternative methods for importing data are also available and can be found on the One Click LCA website.

Source: https://help.oneclicklca.com/en/articles/275869-updated-import-process-in-one-click-lca

Step 30: Click on **Overwrite**.

Step 31: Click on **Continue**.

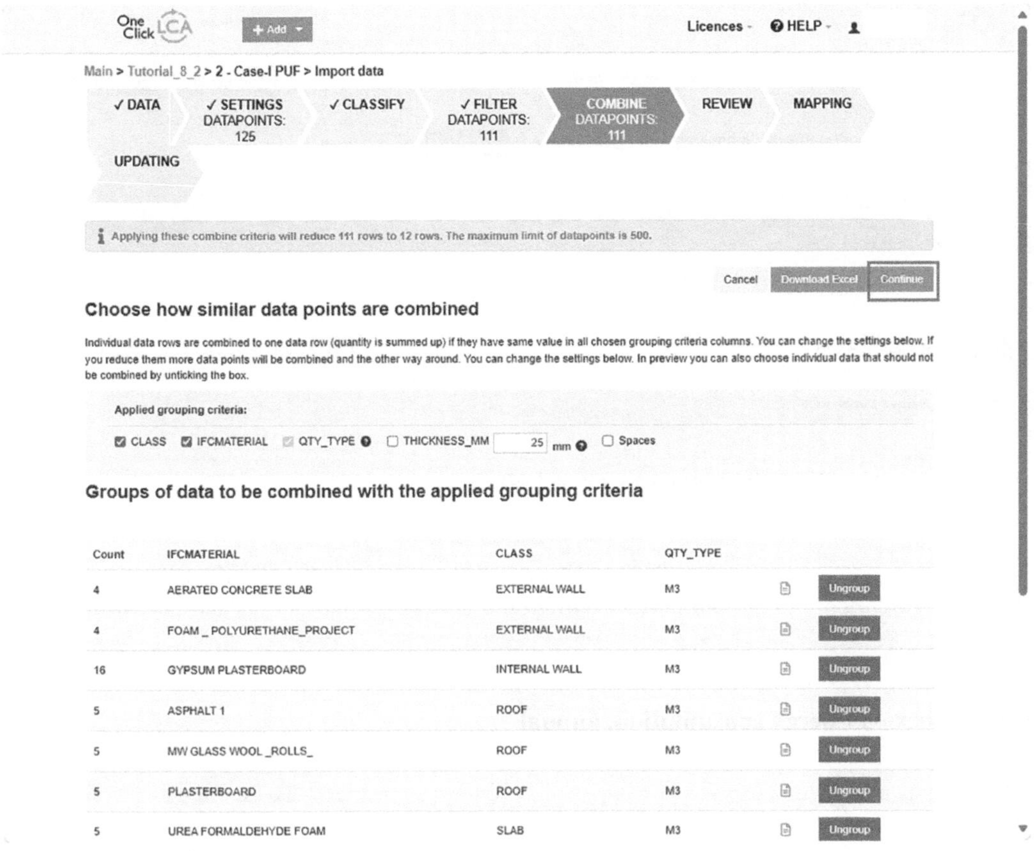

Step 32: Click on **Continue**.

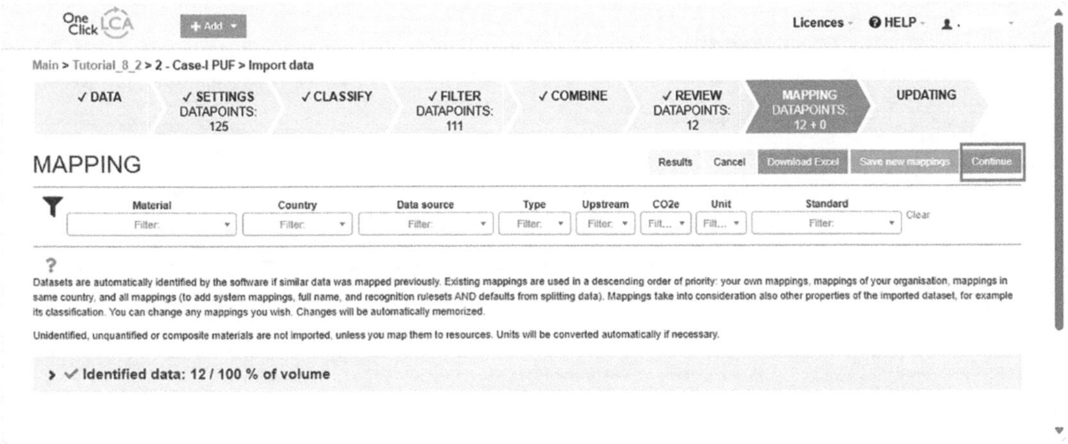

Step 33: Click on **Energy consumption—Click to input missing data**.

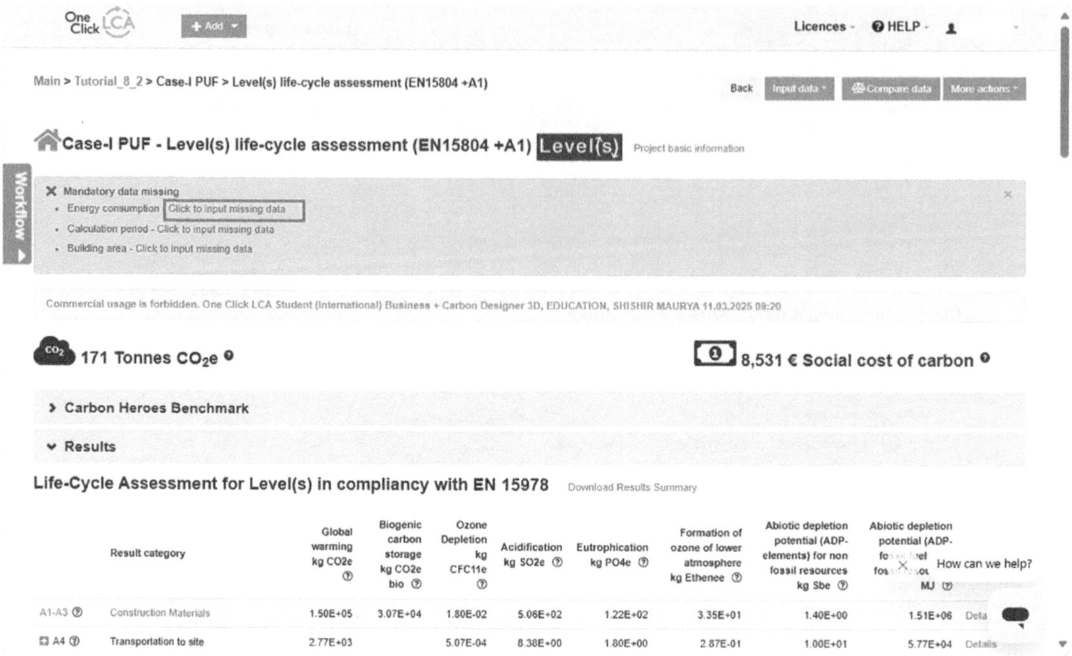

Step 34: Click on **Energy consumption, annual**.

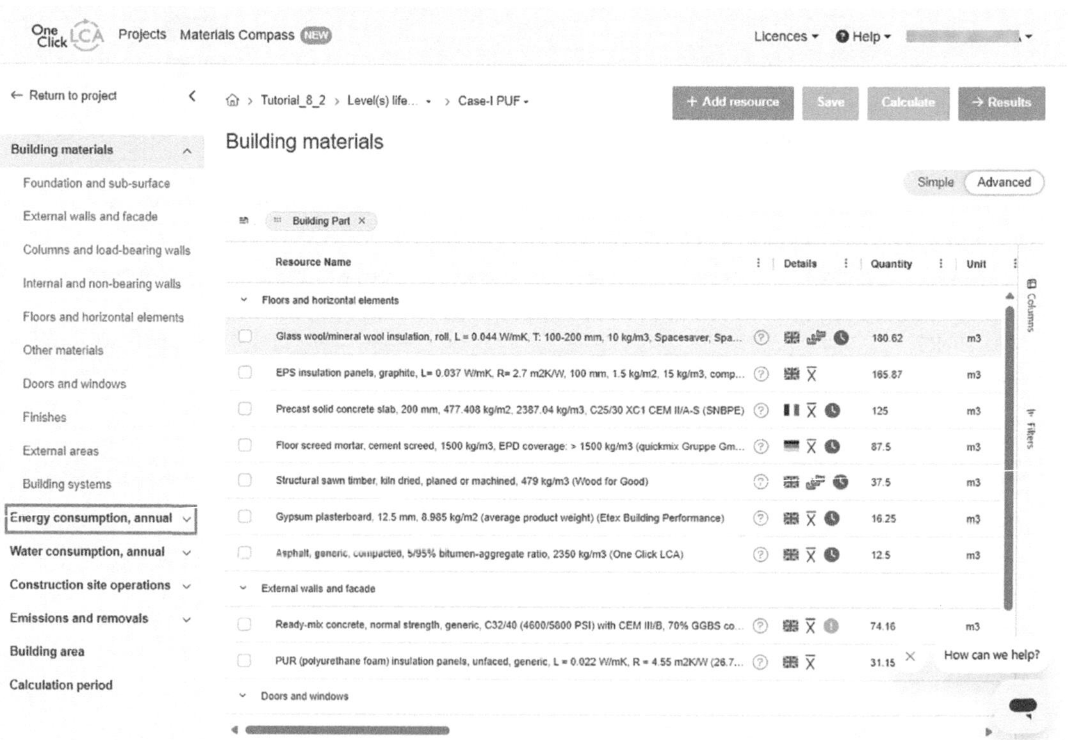

Step 35: Click on **Electricity use**. Click on **Add resource**.

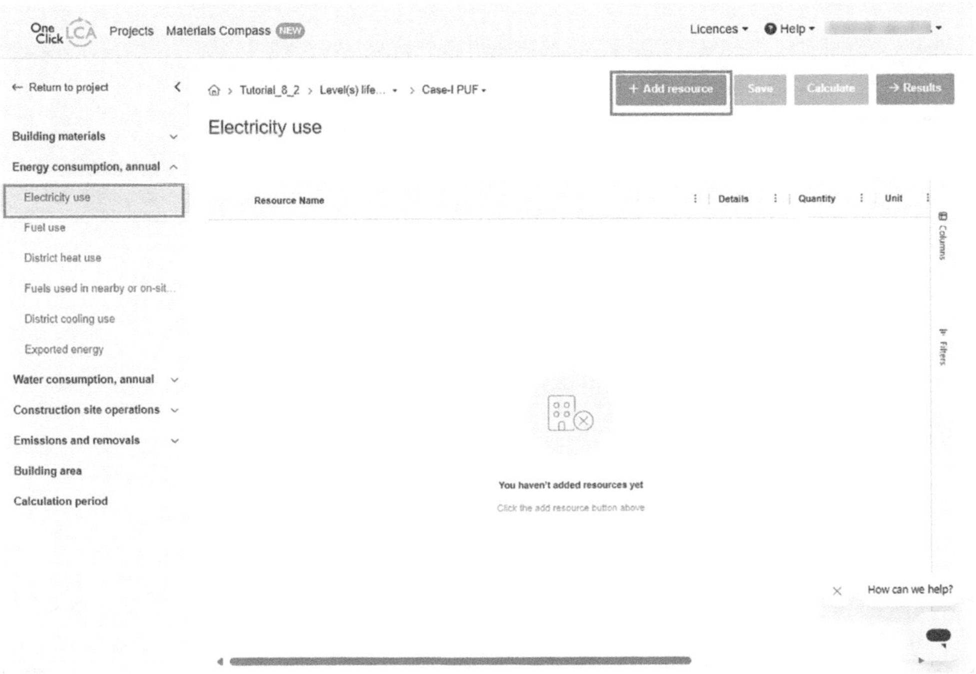

Step 36: Select **United Kingdom** check box. Select **Electricity, United Kingdom, SAP 10.1 and 10.2**. Click on **Add resources (1)**.

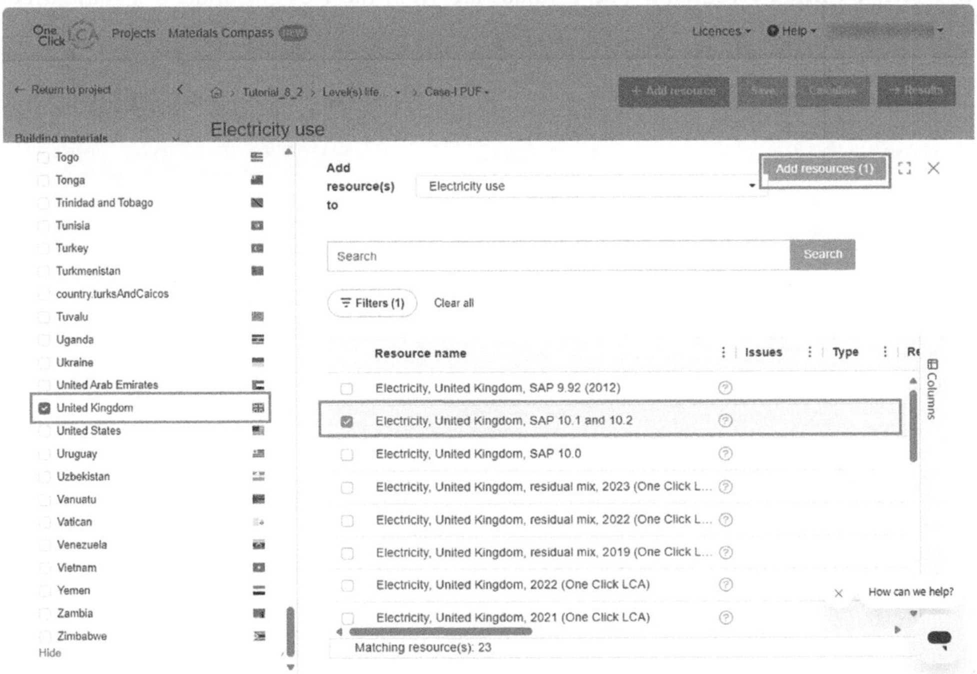

Step 37: Select **Fuel use** from **Add resource(s) to** drop-down. Select **Natural gas SAP 10.1 and 10.2**. Click on **Add resources (1)**.

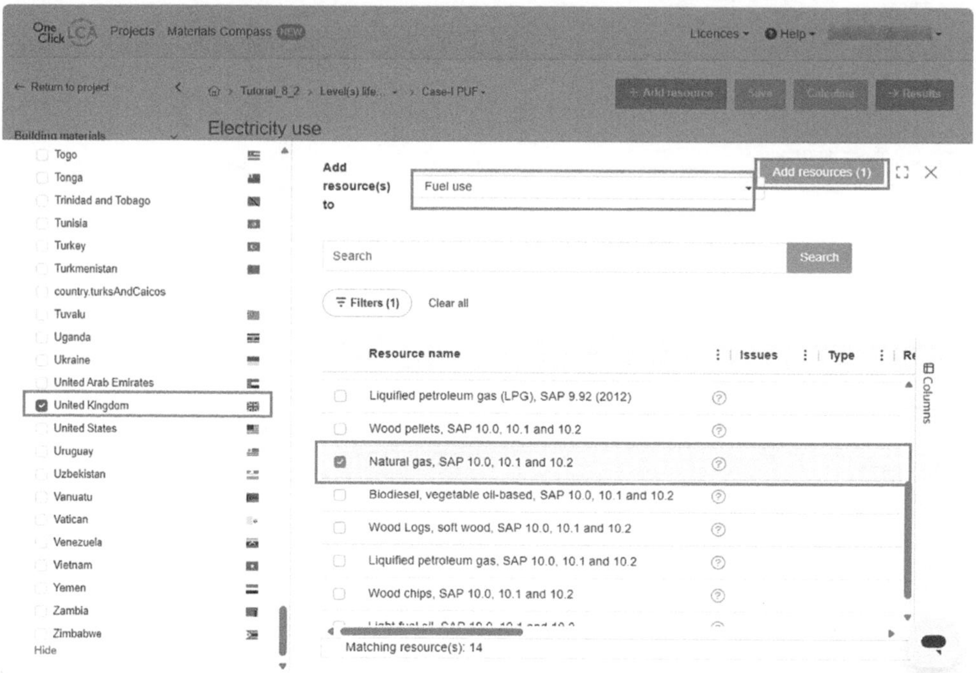

Step 38: Enter **56096.67** kWh **in Natural Gas, SAP 10.0, 10.1, and 10. Enter 131018.7 kWh in Electricity, United Kingdom.** These values are from the DesignBuilder simulation recorded earlier.

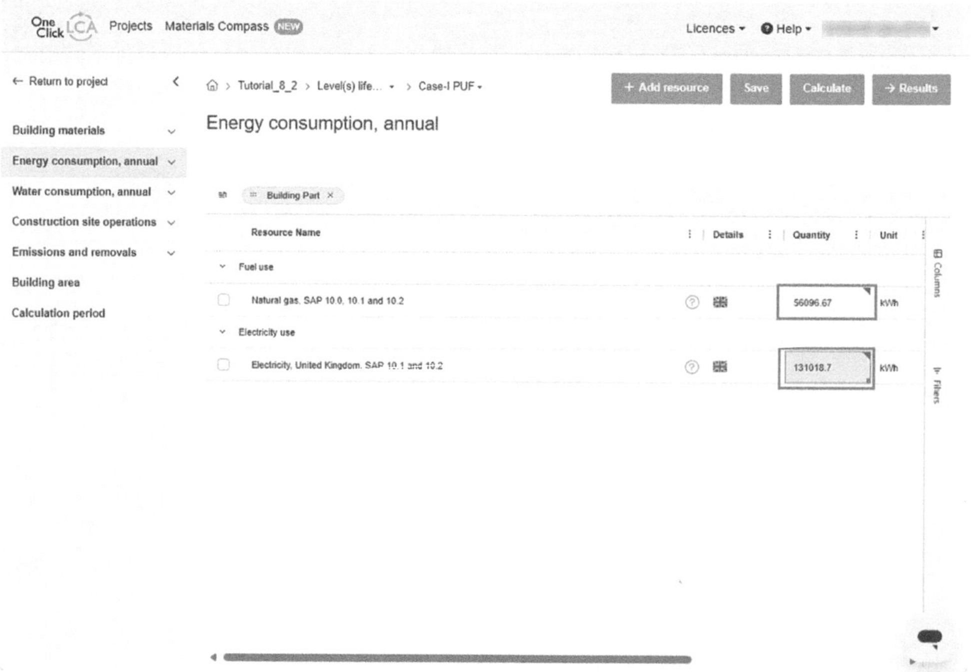

Step 39: Click **Building area**. Enter **1200** in the **Gross Internal Floor Area** textbox.

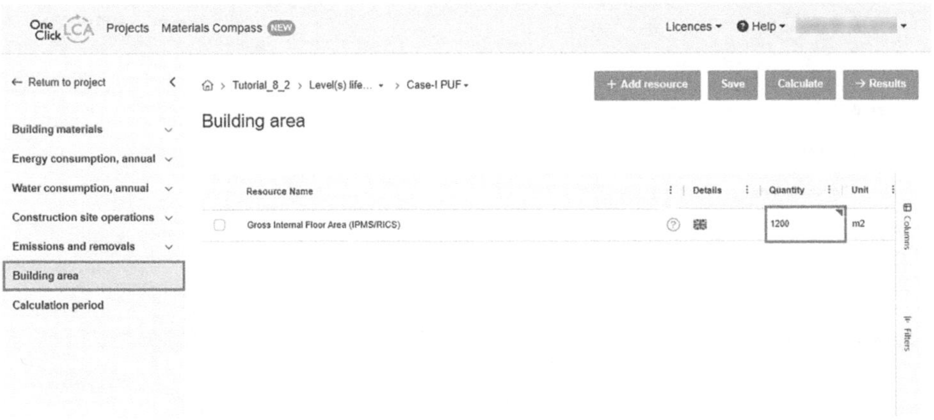

Step 40: Click on **Calculation period** and Enter **60** in the **years** textbox. Click on **Results.**

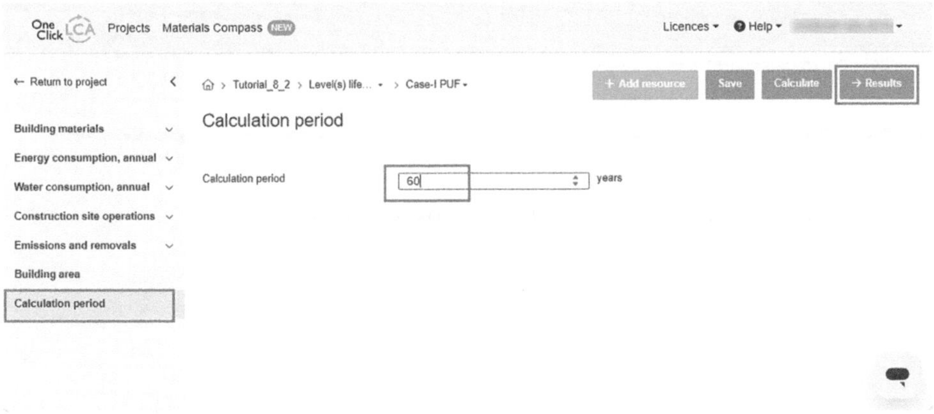

Step 41: Record the results. Click **Tutorial_8_2**.

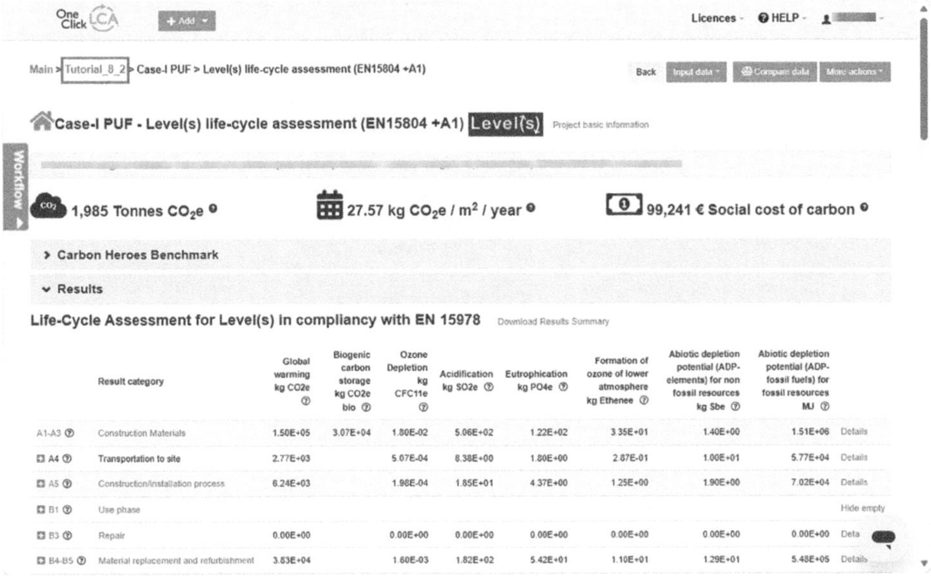

Step 42: Click **2-Case-I PUF** and select **Copy**.

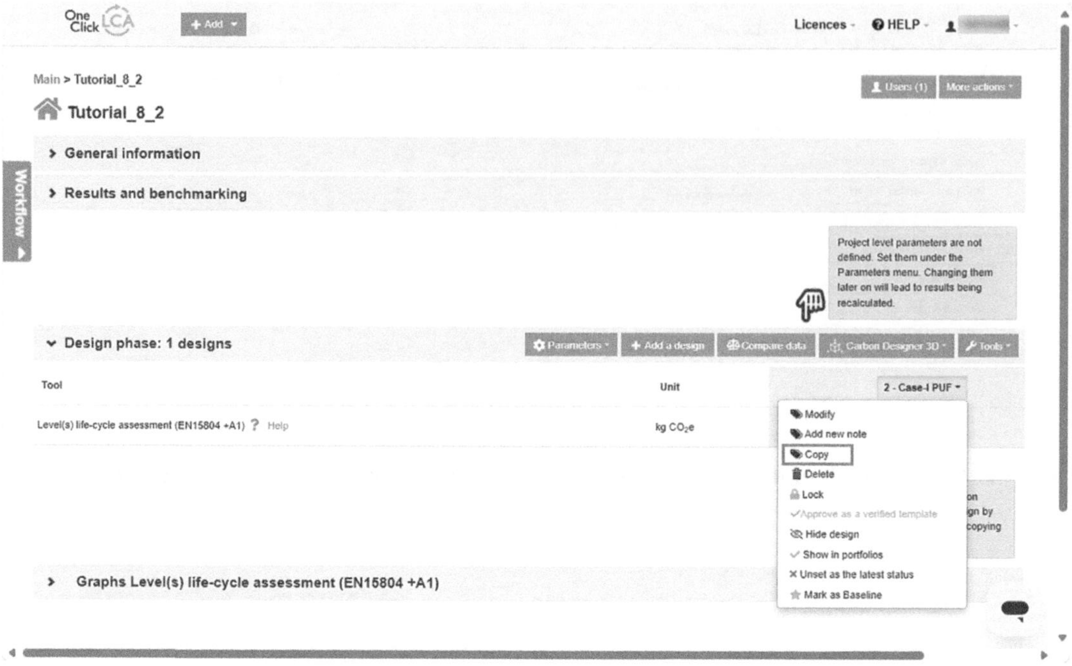

Step 43: Enter **Case II Glass Board** in **Name** textbox. Click **Add**.

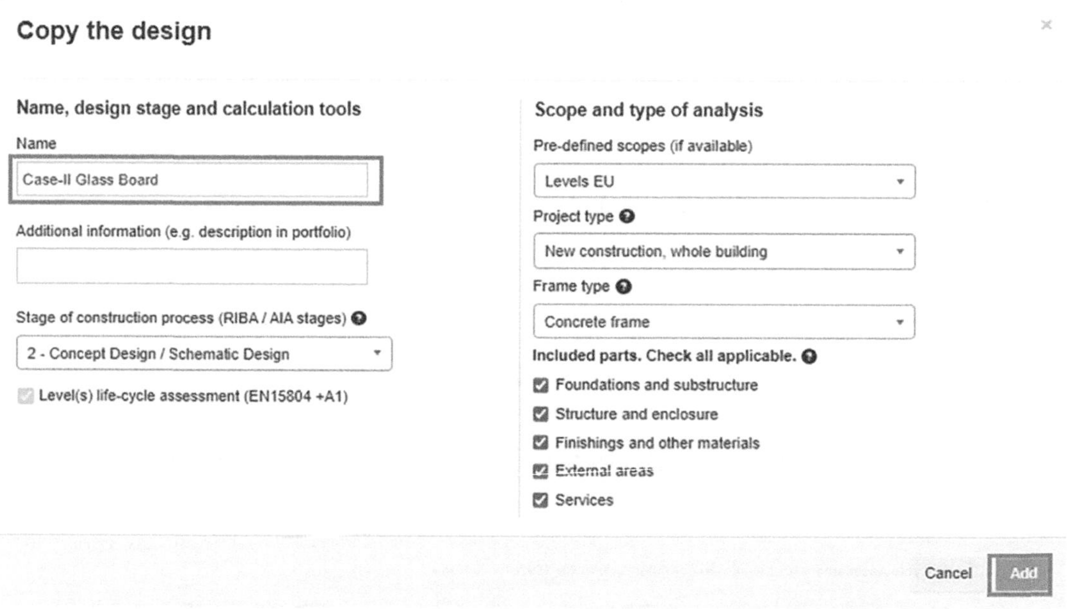

Step 44: Observe that in both cases, the values are the same.

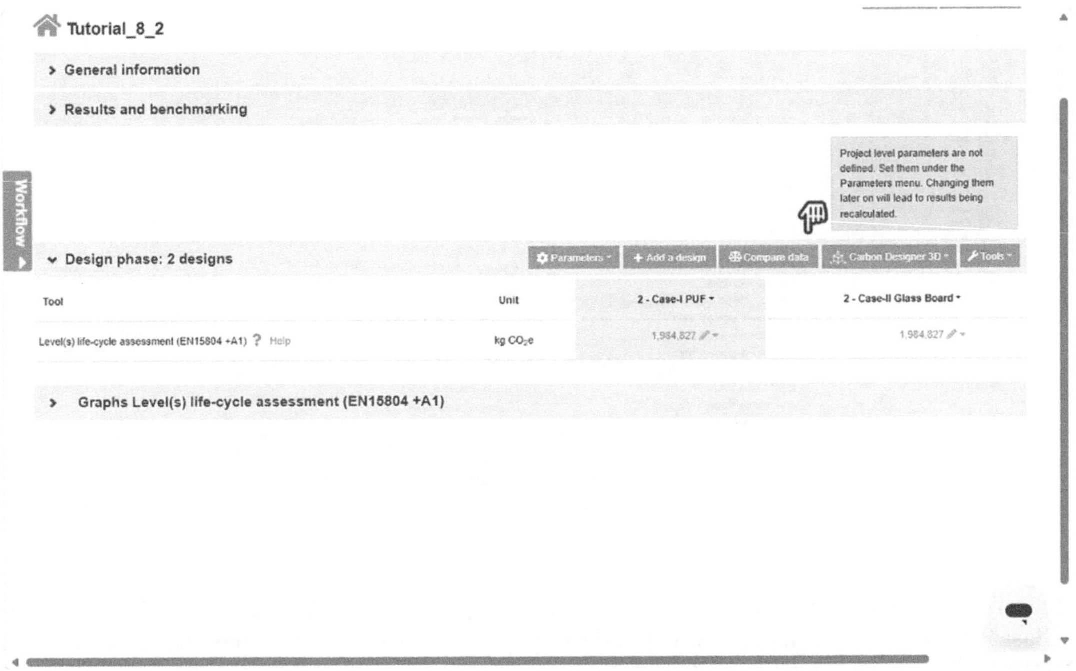

Step 45: Now open the model created in Tutorial 8.1 with **Board insulation (Glass Fiber Board)**.

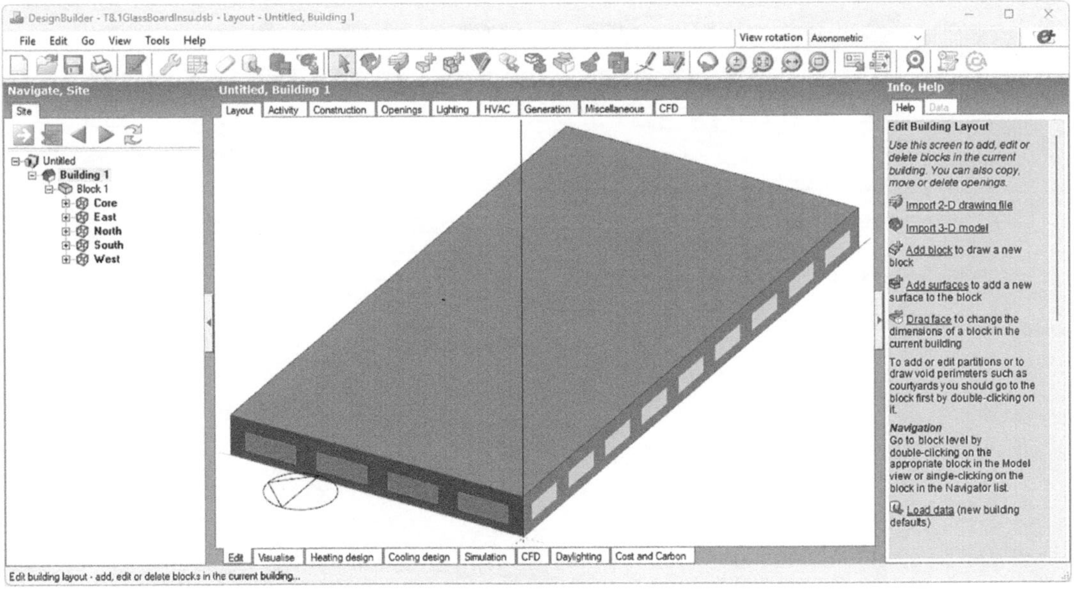

Step 46: Click on **LCA** button. A new screen appears on the web browser.

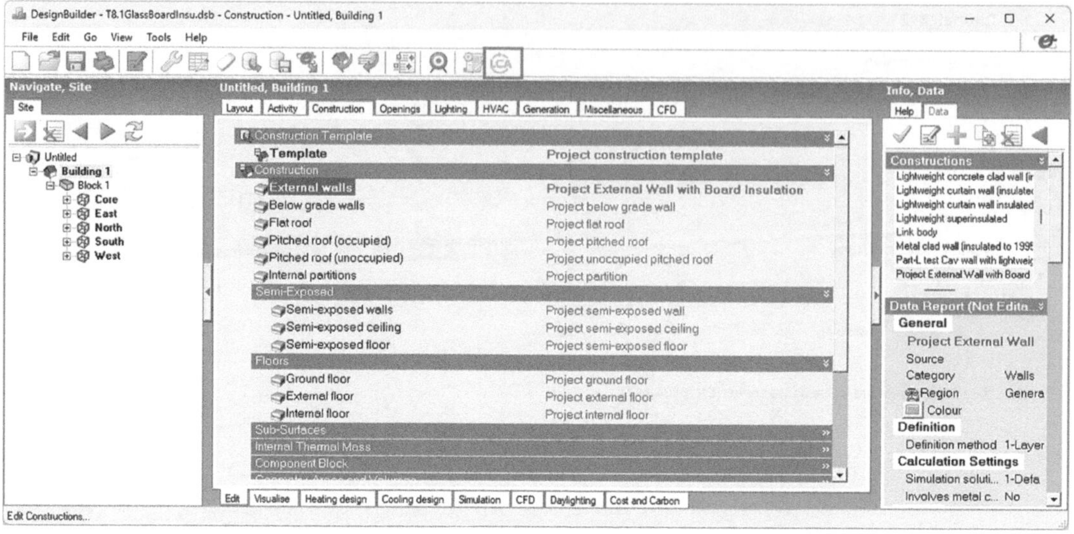

Step 47: Select **Tutorial_8_2** from the **Choose the project or entity** drop-down list. Select **Case-II Glass Board** from the **Choose or create a design** drop-down list. Select **All data** from the **Filtering settings** drop-down list. Click on **Continue**.

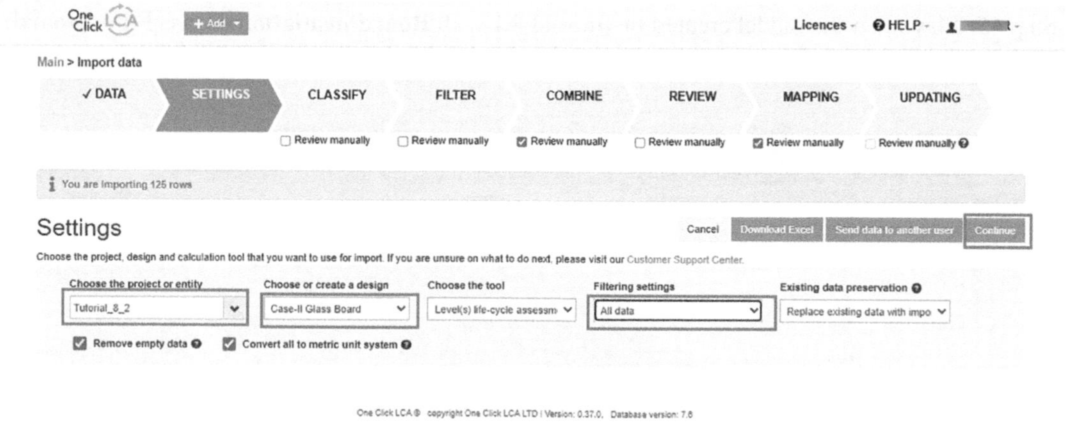

Step 48: Enter **130562** kWh **Quantity** in **Electricity, United Kingdon**. Enter **55999.57** kWh **Quantity** in **Natural Gas, SAP 10.0, 10.1 and 10**. These values are from DesignBuilder annual simulation. (Please refer to steps 21–31 in this tutorial)

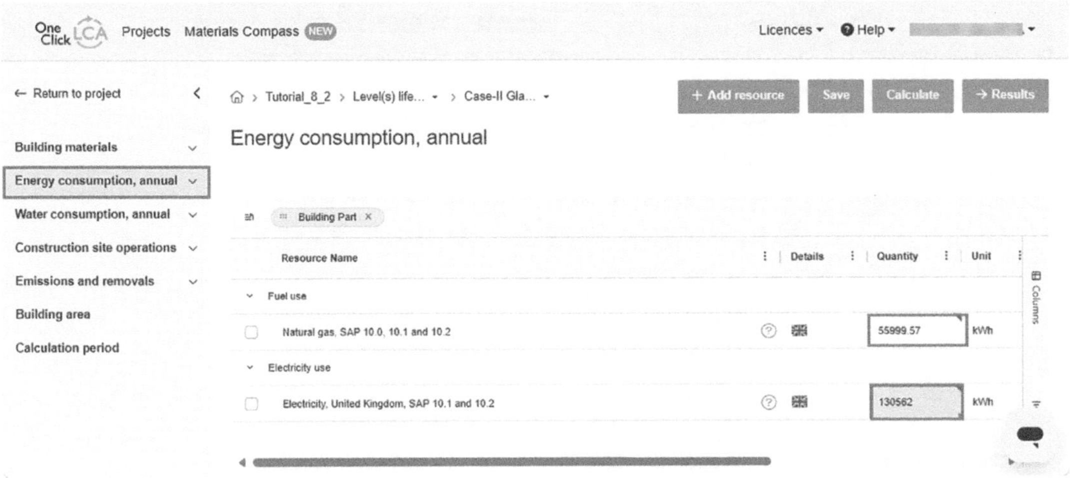

Step 49: Record the results. Click **Tutorial_8_2**.

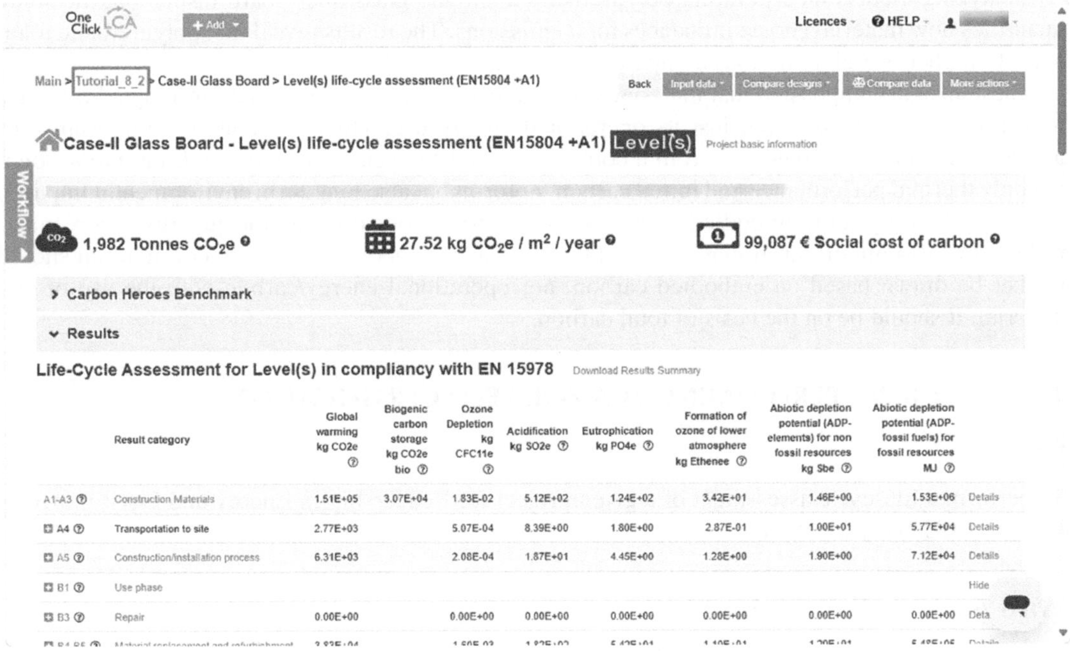

Step 50: Click **Graphs Level(s) life-cycle assessment (EN15804+A1)** and compare the results.

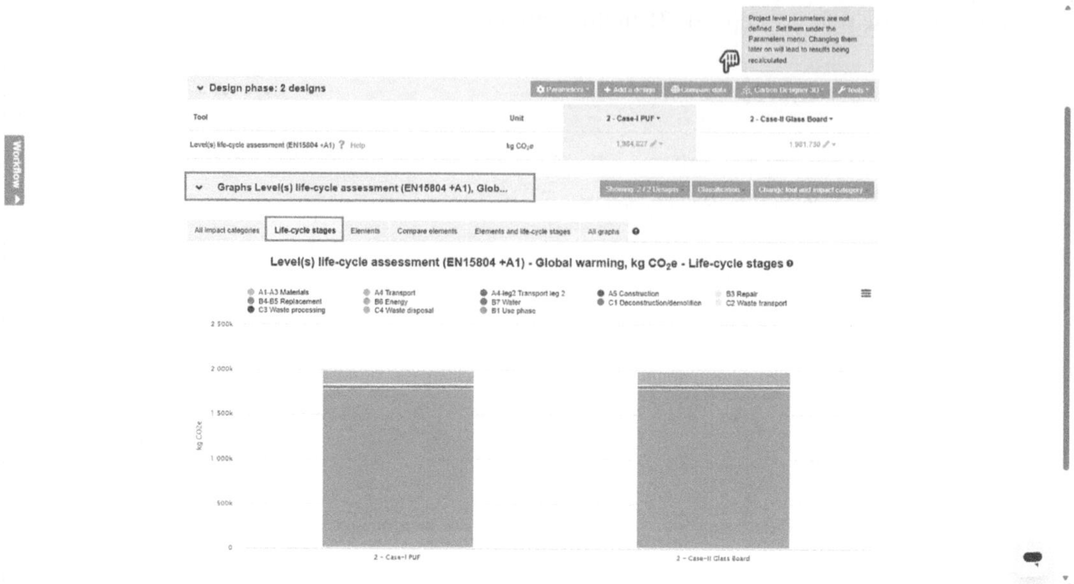

By analyzing models incorporating polyurethane foam and glass fiber board insulation, the study quantifies how material choice influences total emissions. The results reveal that polyurethane foam leads to higher overall emissions compared to glass fiber board.

These insights emphasize that material selection plays a crucial role in determining a building's total life-cycle emissions, not just its operational energy use. This highlights the importance of assessing construction materials from a comprehensive life-cycle perspective, taking into account not only thermal performance and initial carbon footprint but also long-term environmental impacts.

Merely looking at the embodied energy per unit mass is not sufficient; the quantity of any material used is also important. It may be noted that conclusions about the selection of material should neither be drawn based on embodied carbon, nor operational energy/carbon with the use of any material. It should be on the basis of total carbon.

TUTORIAL 8.2: PERFORMING LCA FOR LEED CERTIFICATION

Goal

To perform a life-cycle assessment of a given project for Leadership in Energy and Environmental Design (LEED) certification

What Are You Going to Learn?

- Performing Whole-Building Life-Cycle Assessment for LEED certification

Problem Statement

In tutorial 8.1, 50 m × 25 m five-zone models with a 5 m perimeter depth were simulated. Use the same building model prepared with the polyurethane insulation board.

Prepare baseline and proposed cases as shown in Table 8.2.

TABLE 8.2
Baseline and Proposed Case Parameters

Parameter	Baseline	Proposed Design
Floor-to-floor slab height	3.5 m	2.8 m

Calculate reduction and compare with the baseline building in the following seven impact categories.

- Global warming kg CO_2e
- Ozone Depletion kg CFC-11e
- Acidification kg SO_2e
- Eutrophication kg PO_4e
- Formation of ozone of lower-atmosphere kg Ethene
- Abiotic depletion potential (ADP-elements) for non-fossil resources kg Sb_e
- Abiotic depletion potential (ADP-fossil fuels) for fossil resources MJ

Solution

Step 1: Open https://oneclicklca.com in the browser.

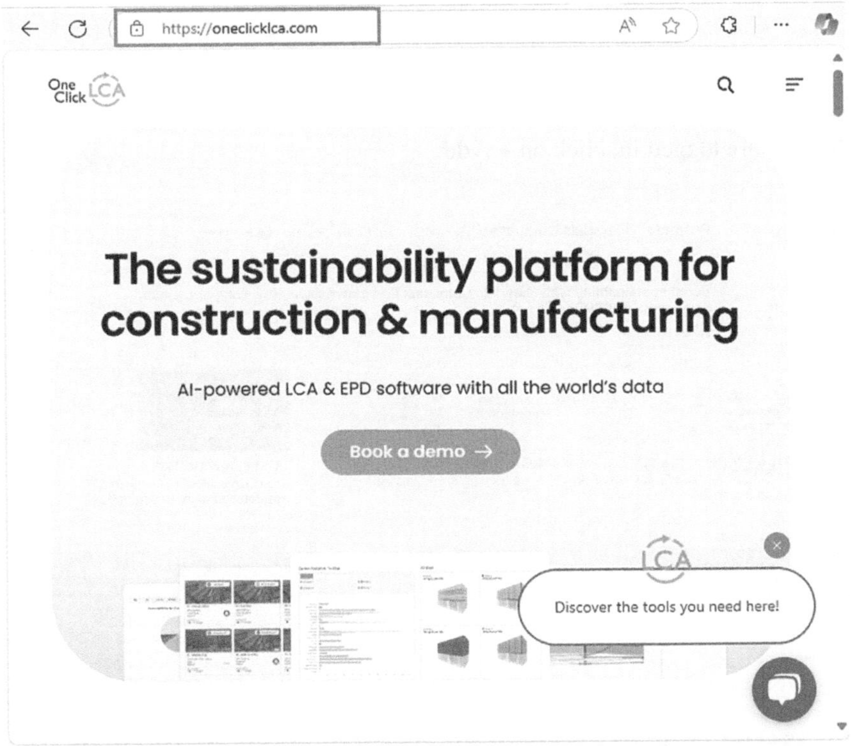

Step 2: Click on **Log in.** Enter your **username and password.** If you do not have a OneClick LCA account, click on **Register** to create a new one.

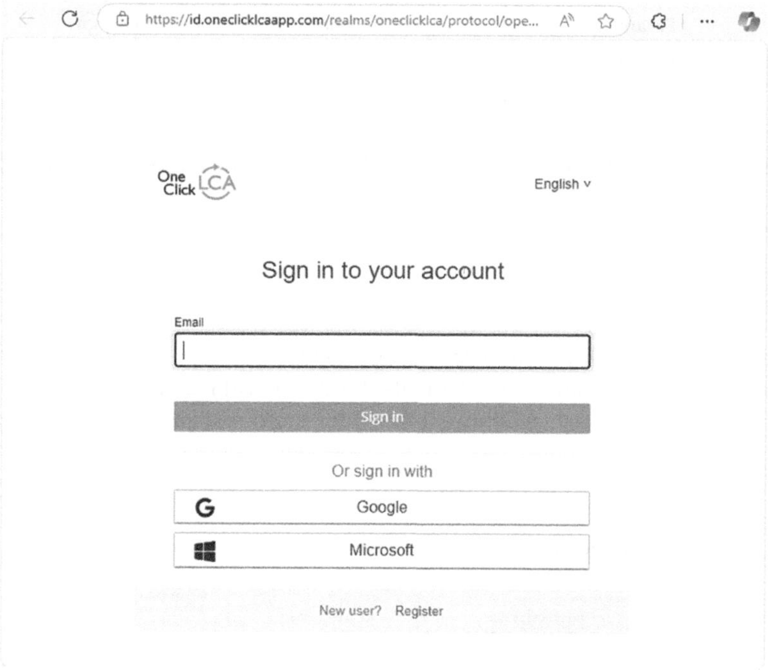

Step 3: Once you are logged in, click on **+ Add**.

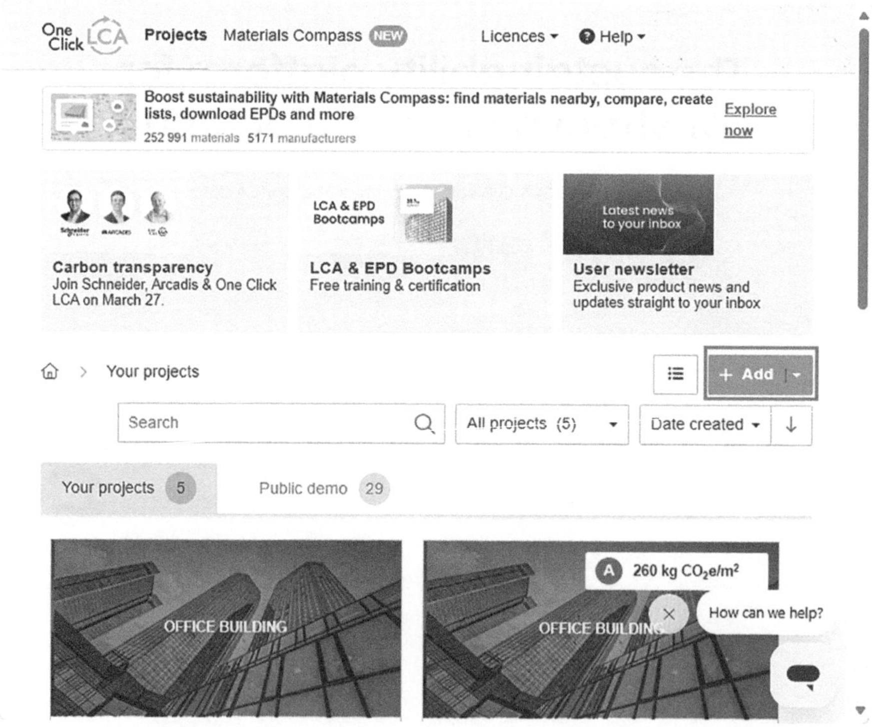

Step 4: Select **Building** from drop-down list. A new project screen appears.

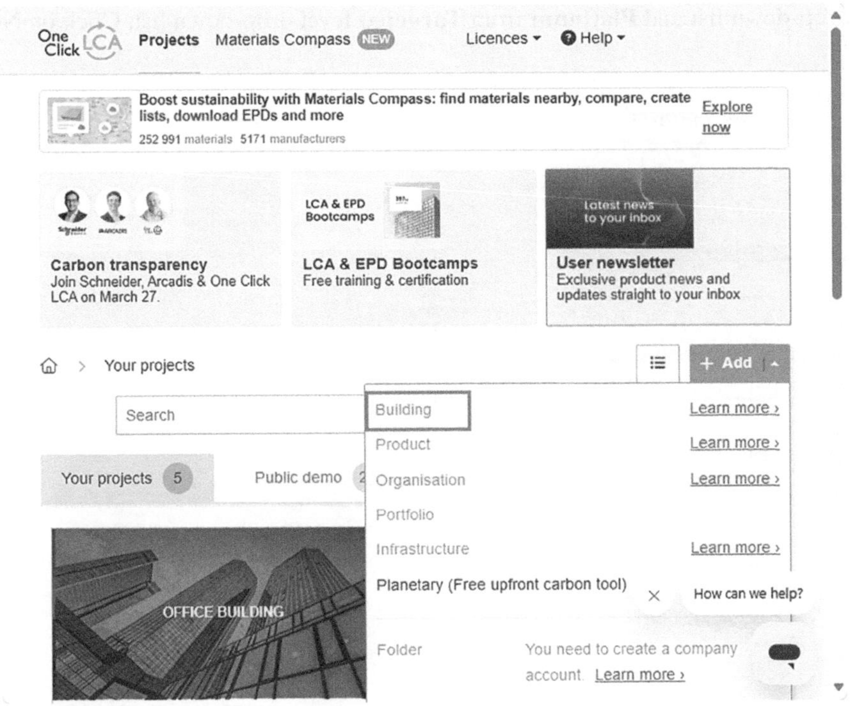

Step 5: Enter **Tutorial_8_3** in the **Name (mandatory)** text box. Select **Office buildings** from **Type (mandatory)** drop-down list. Select **United Kingdom** from **Country (mandatory)** drop-down list. Click on **Next**.

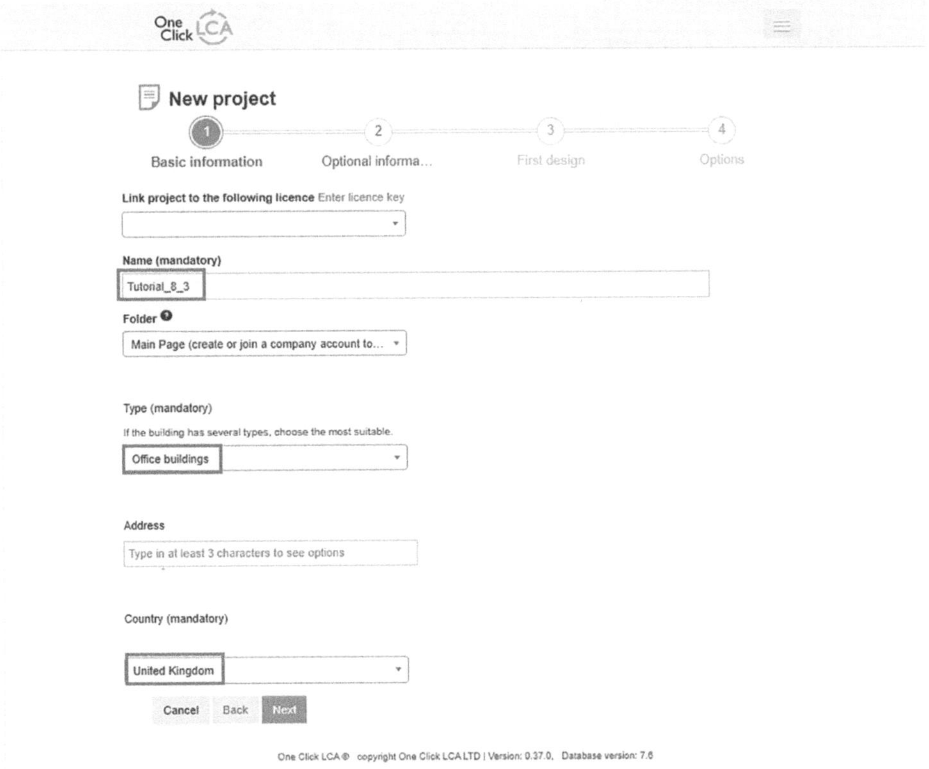

Step 6: Enter **1200** in **Gross Floor Area (m^2)**. Enter **1** in **Number of above ground floors**. Select **Concrete frame** from **Frame type** drop-down list. Select **LEED V4.1 BD+C** from **Certifications pursued** drop-down list and **Platinum** from **Targeted level** drop-down list. Click on **Next**.

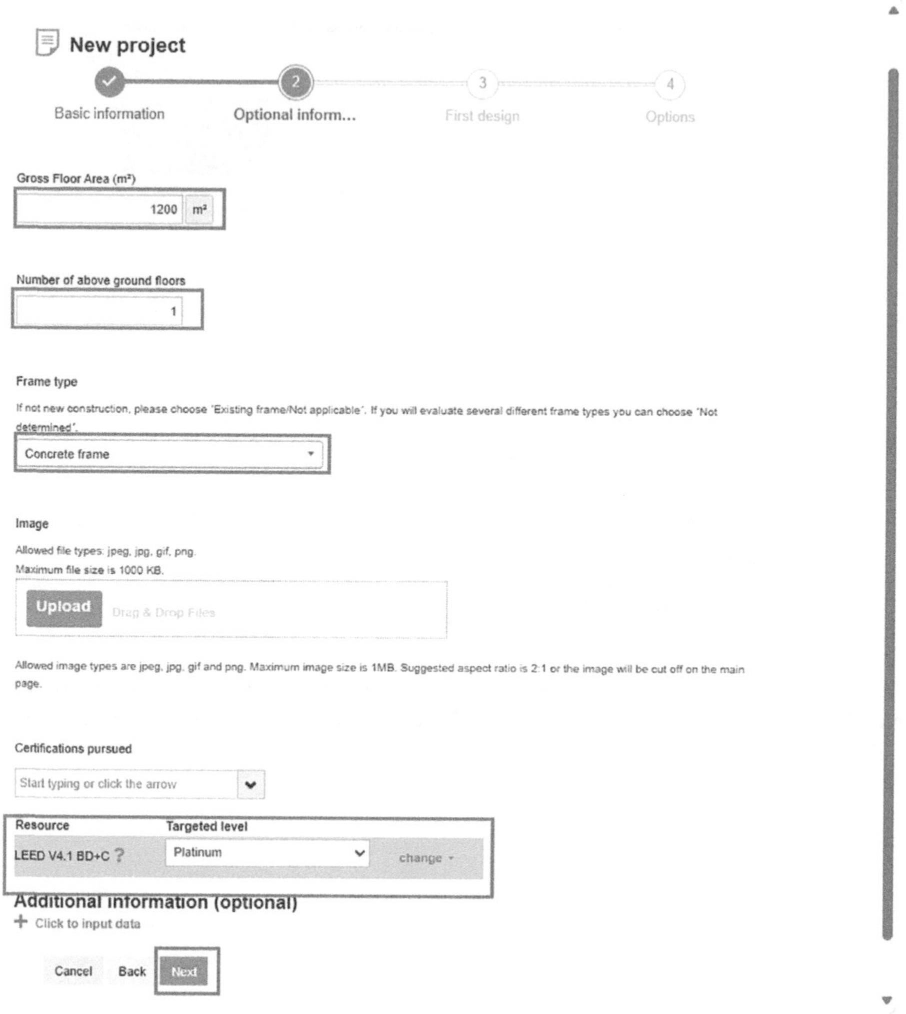

Step 7: Click on **Enter license key**. A text box appears, enter your license key in text box.

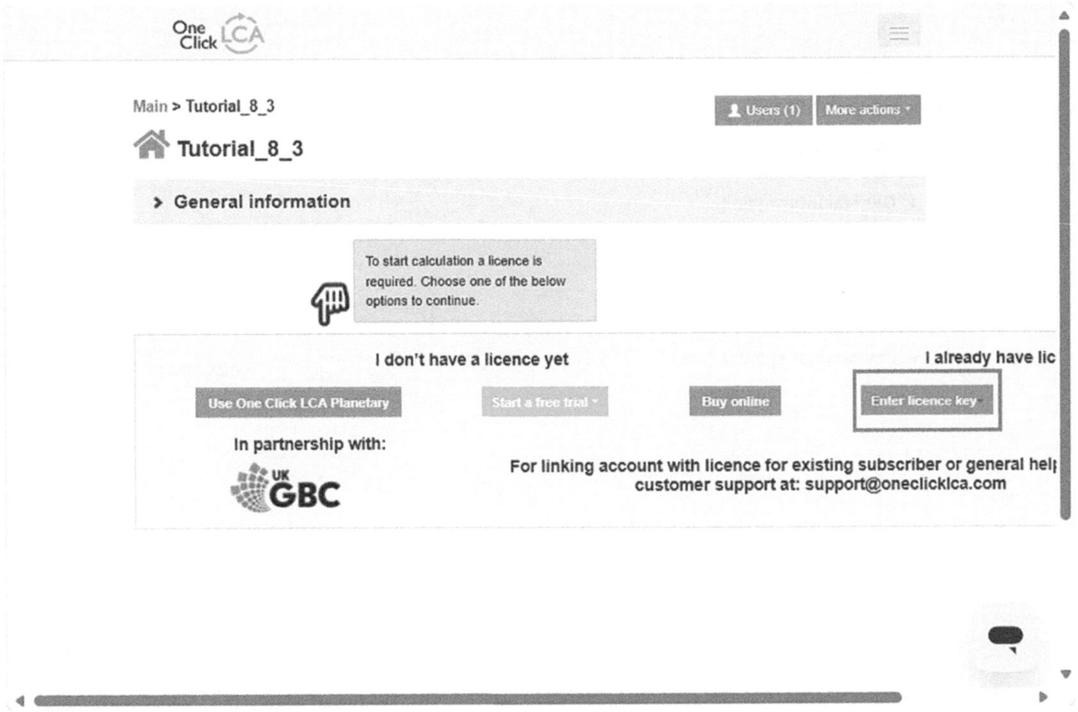

It is important to note that the activation method shown in step 7 is limited to educational users. Commercial One Click LCA subscriptions use alternative activation processes.

Sources: https://help.oneclicklca.com/en/collections/647509-product-guide-licensing and https://help.oneclicklca.com/en/articles/275601-licensing-your-project

Step 8: Click on **Get started**. Available tools screen appears.

Step 9: Select **Level(s) life-cycle assessment (EN15804 +A1) Life-cycle assessment for Level(s)—overarching assessment tool 7: Cradle to cradle Life Cycle Asse** check box. Click on **Next**.

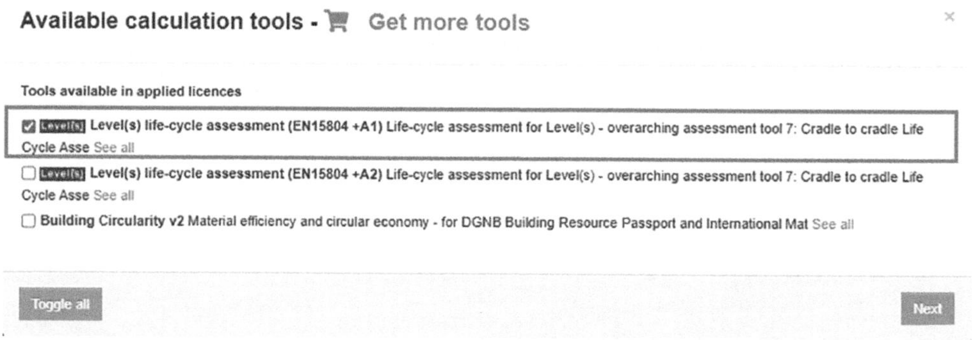

In case EN15804+A1 tool is not available, you can choose an equivalent calculation tool available with your license.

Step 10: Enter **Case-I PUF** in **Name** text box. Click on **Next**. **Life-Cycle Assessment Parameters** setup screen appears.

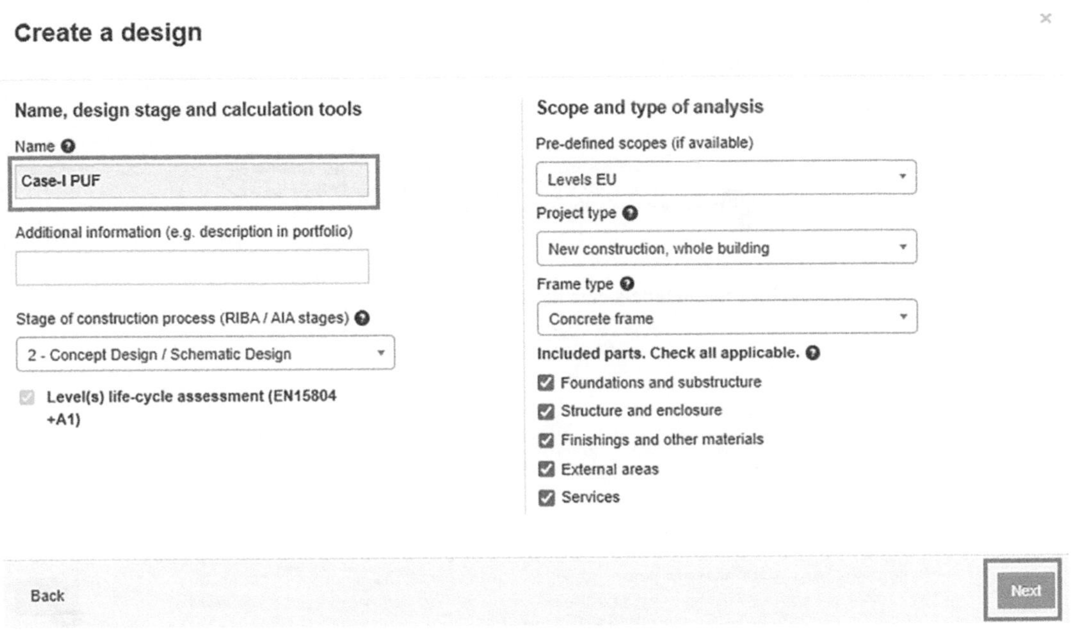

Step 11: Click on **Review and adjust LCA parameters**.

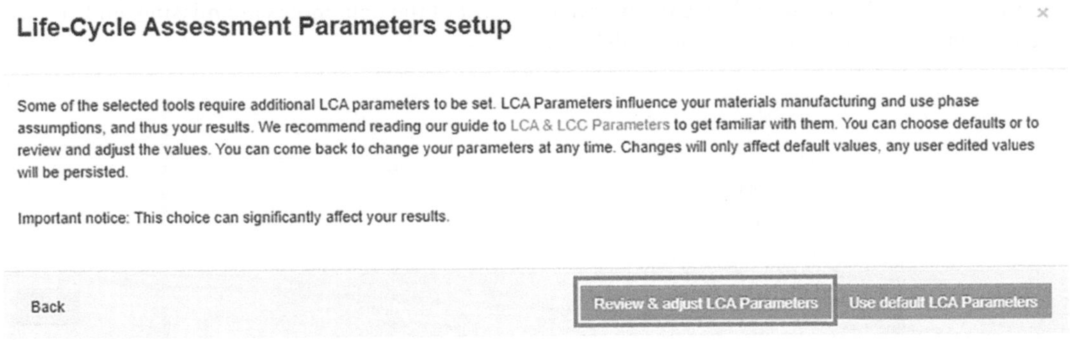

Step 12: Click on **Save**.

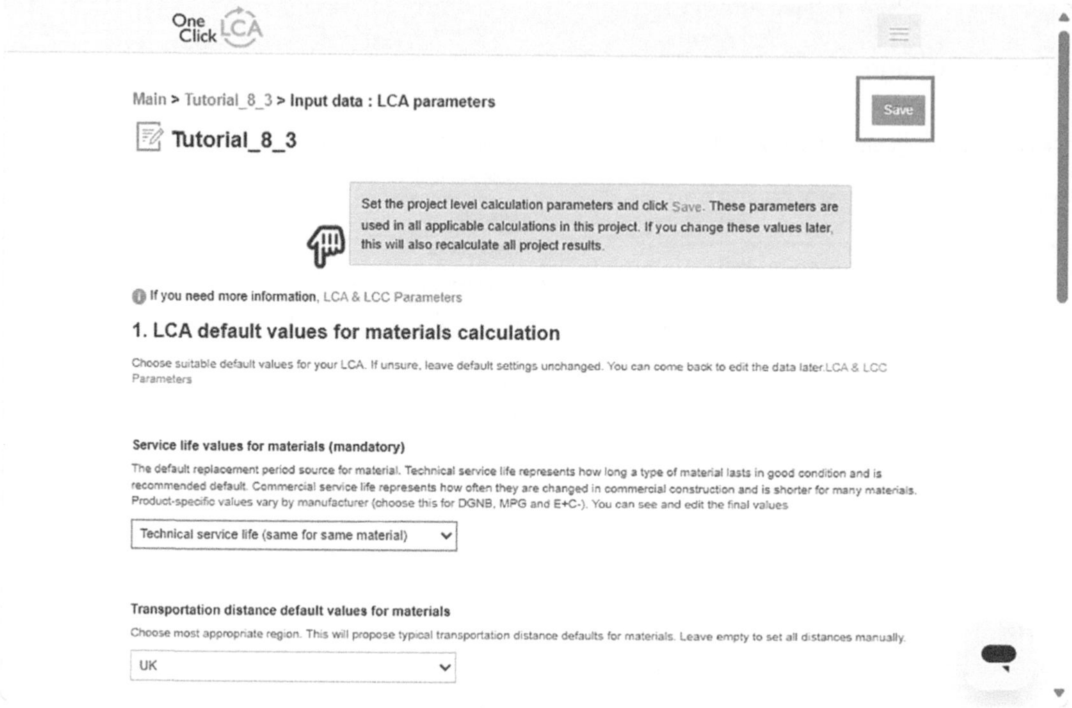

Step 13: Click on **Input data**. You can observe that most of the data inputs are missing, and you can use DesignBuilder tool to import the model parameters.

Step 14: Open the DesignBuilder model saved in Tutorial 8.1 prepared with the polyurethane insulation board.

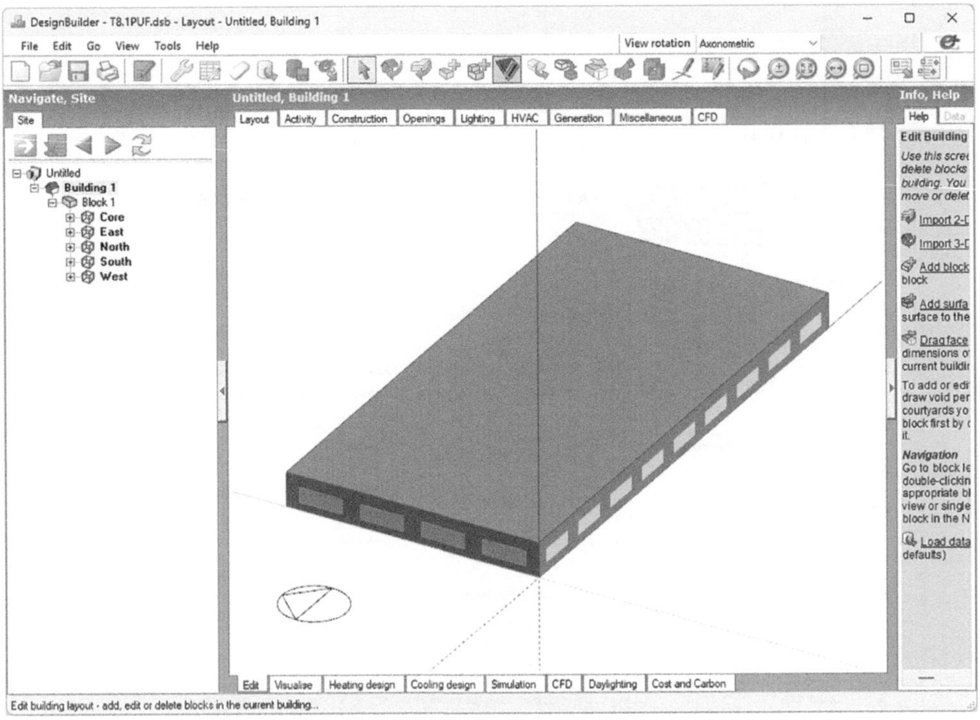

Step 15: Click on the **LCA** button. A new screen on the web browser appears.

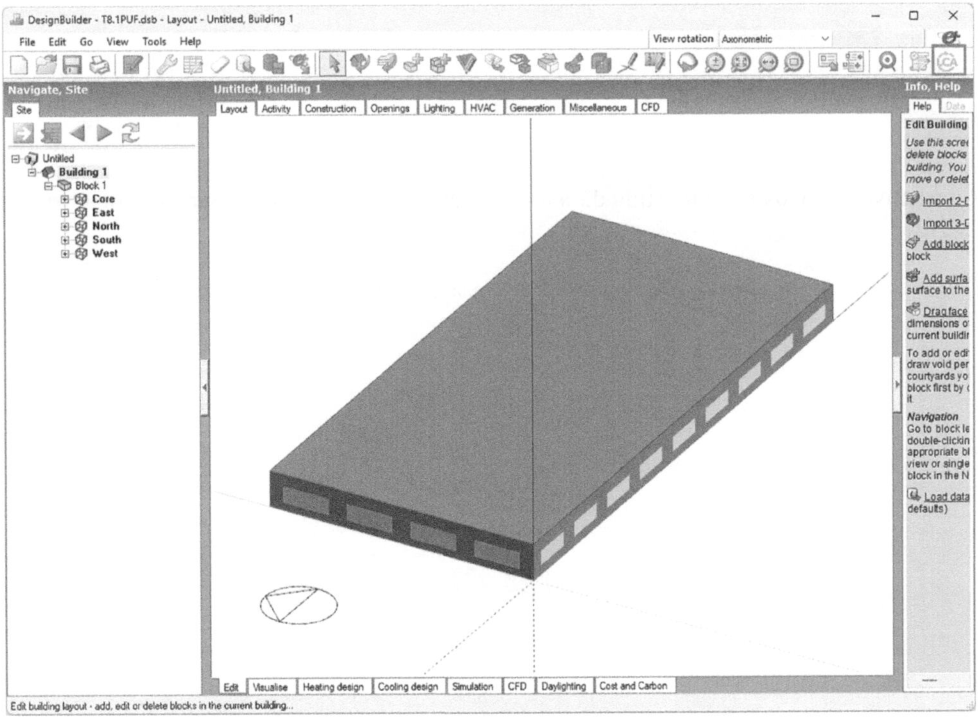

Step 16: Select **Tutorial_8_3** from the **Choose the project or entity** drop-down list. Select **Case-I Base** from the **Choose or create a design** drop-down list. Select **All data** from the **Filtering settings** drop-down list. Click on **Continue.**

Alternative methods for importing data are also available and can be found on the One Click LCA website.

Source: https://help.oneclicklca.com/en/articles/275869-updated-import-process-in-one-click-lca

Step 17: Click on **Overwrite**.

Step 18: Click on **Continue.**

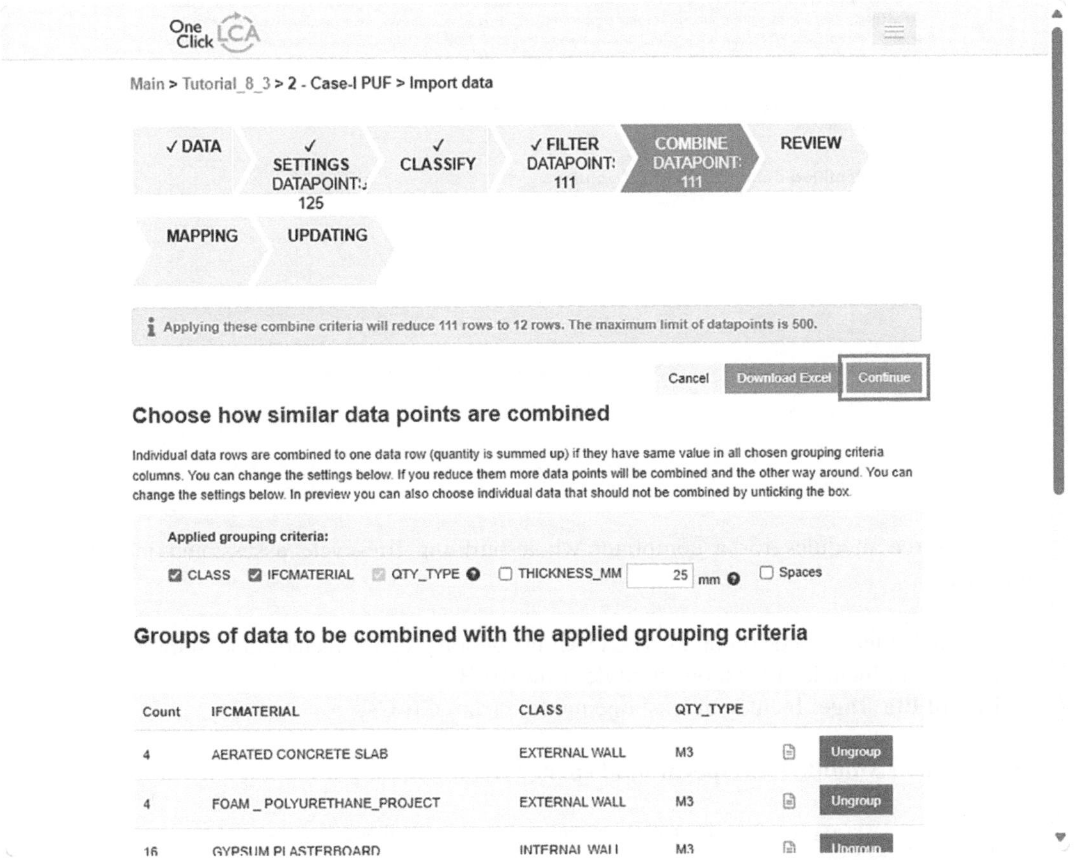

Step 19: Click on **Continue.**

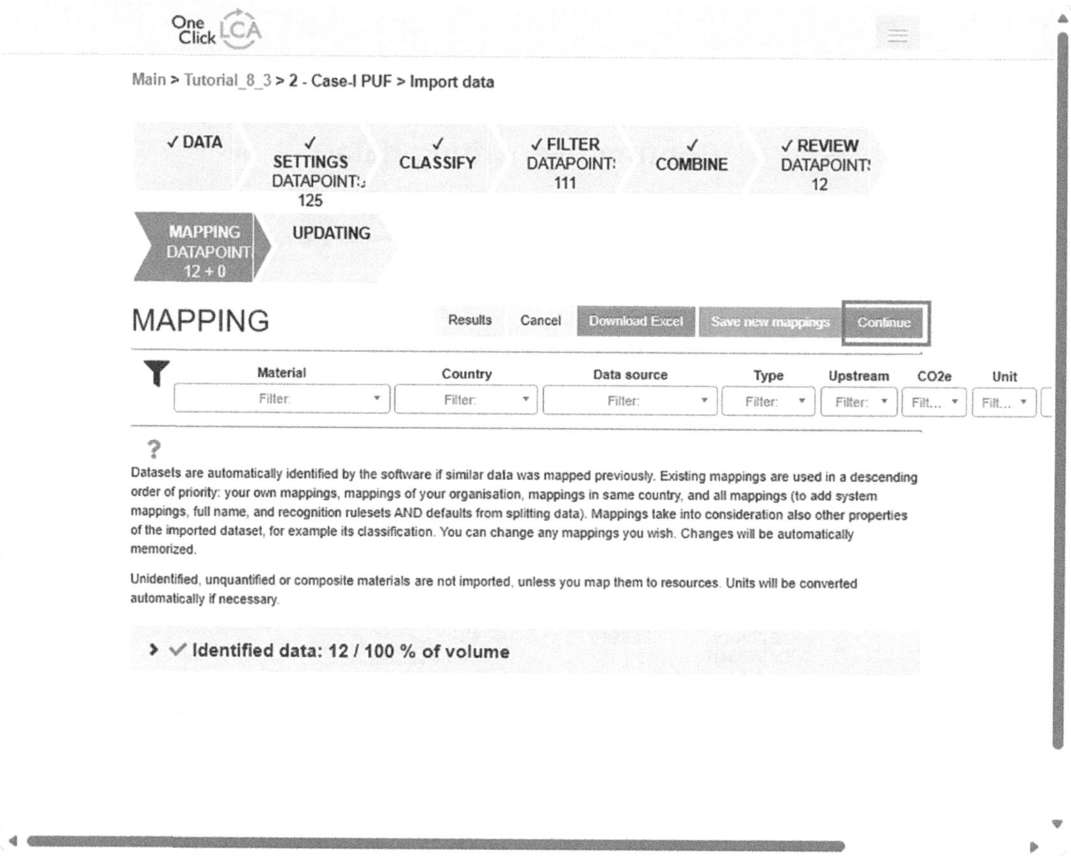

The required modules for a compliant whole-building life-cycle assessment include the following.

Product stage: include modules A1–A3. Construction process: include at least module A4.
Use stage: Include at least one module from B1–B5.
End of life stage: Include at least one module from C1–C4.

Do not report **Module B6: Operational energy use.**

Source: https://www.usgbc.org/credits/new-construction-core-and-shell-schools-new-construction-retail-new-construction-data-27

Step 20: Click on **Calculation period—Click to input missing data**.

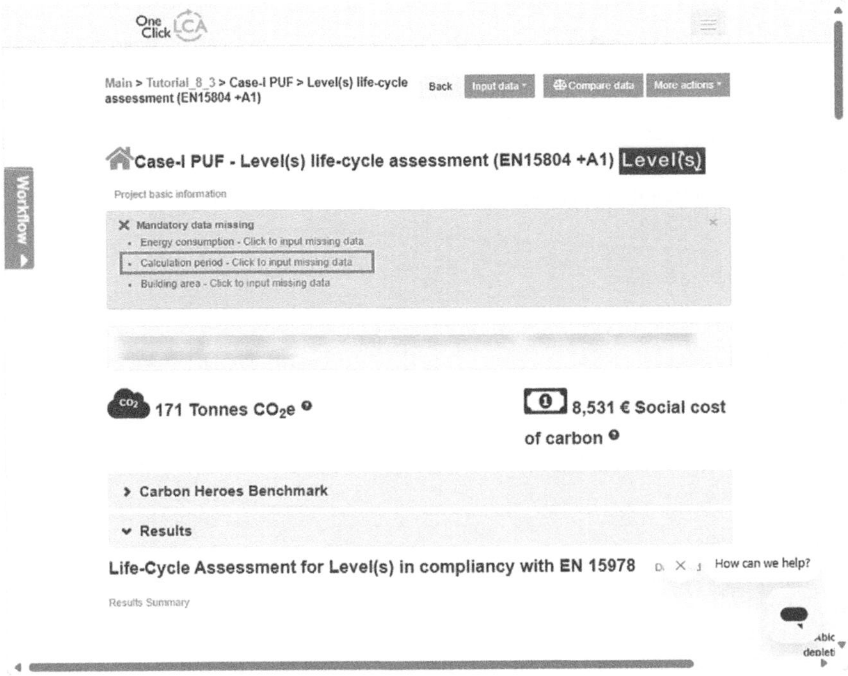

Step 21: Click on **Energy consumption, annual**. Click on **Electricity use**. Click on **Add resource**. Select **United Kingdom** check box. Select **Electricity, United Kingdom, SAP 10.1 and 10.2**. Click on **Add resources** (1). Enter **0** kWh **in Electricity, United Kingdom.**

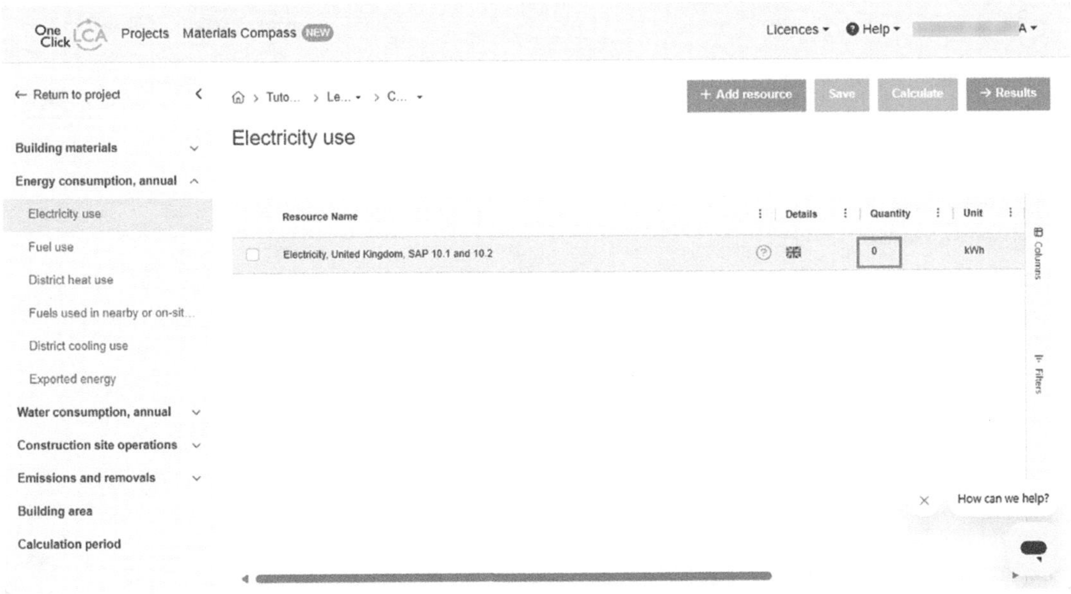

Step 22: Click on **Calculation period** and Enter **60** in **years** textbox.

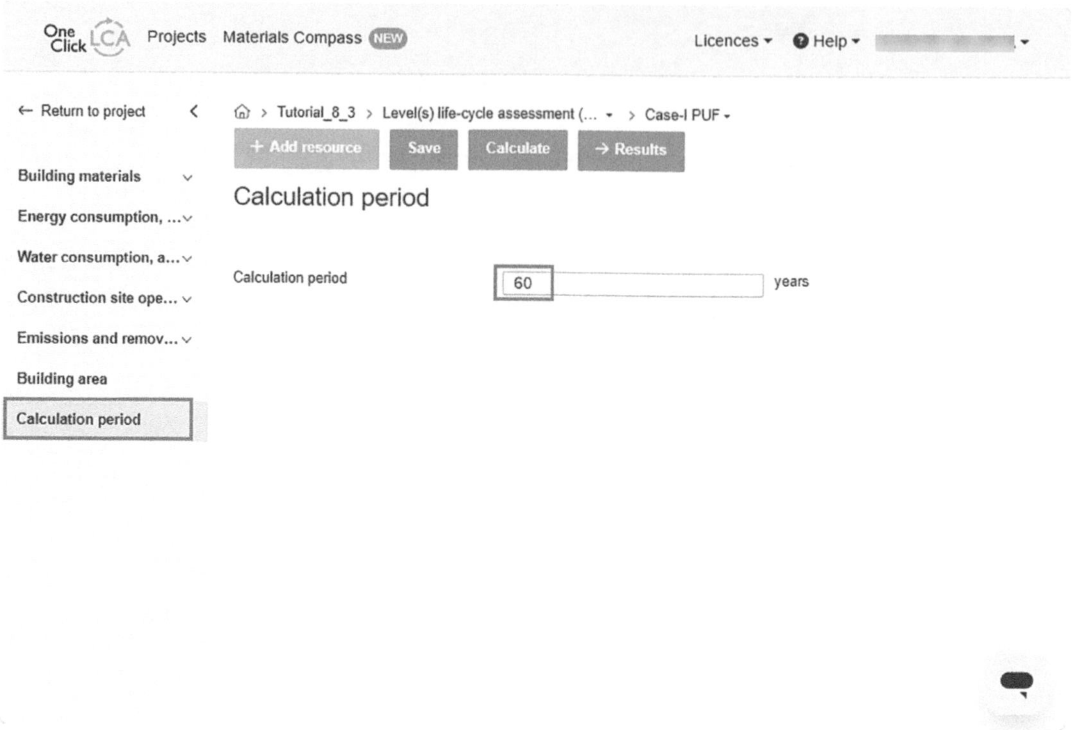

Step 23: Click **Building area**. Enter **1200** in **Gross Internal Floor Area** textbox. Click on **Results.**

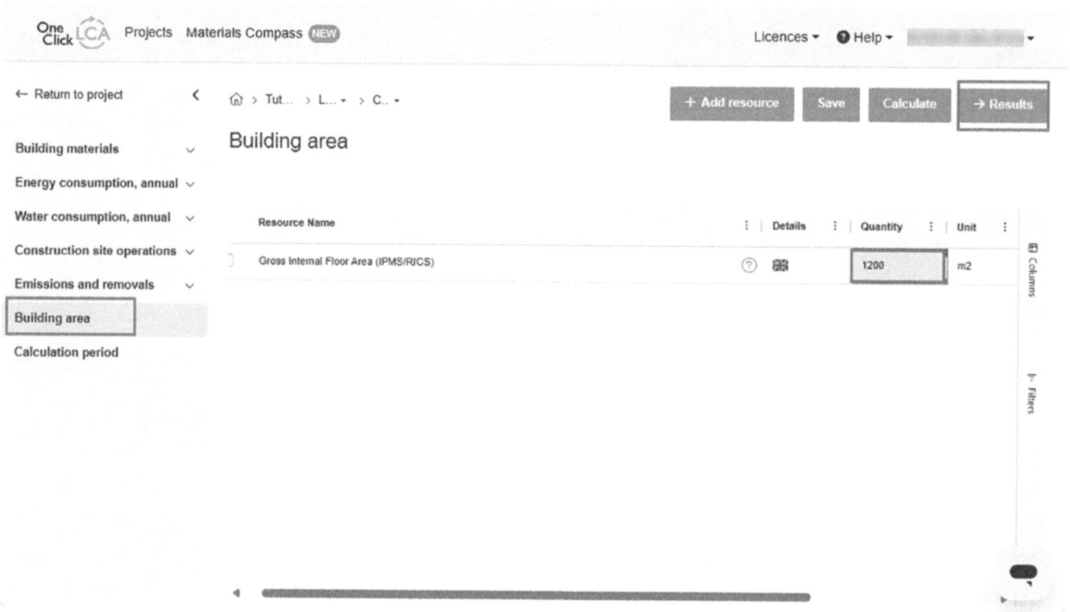

Step 24: Record the results.

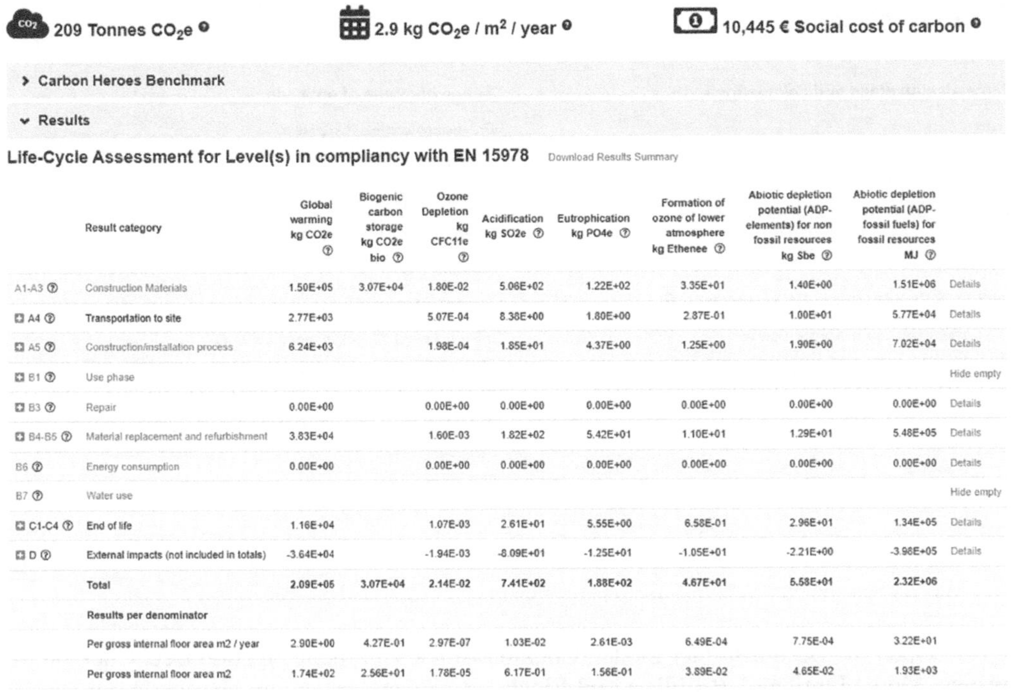

209 Tonnes CO_2e | 2.9 kg CO_2e / m^2 / year | 10,445 € Social cost of carbon

> Carbon Heroes Benchmark

Results

Life-Cycle Assessment for Level(s) in compliancy with EN 15978   Download Results Summary

	Result category	Global warming kg CO2e	Biogenic carbon storage kg CO2e bio	Ozone Depletion kg CFC11e	Acidification kg SO2e	Eutrophication kg PO4e	Formation of ozone of lower atmosphere kg Ethenee	Abiotic depletion potential (ADP-elements) for non fossil resources kg Sbe	Abiotic depletion potential (ADP-fossil fuels) for fossil resources MJ	
A1-A3	Construction Materials	1.50E+05	3.07E+04	1.80E-02	5.06E+02	1.22E+02	3.35E+01	1.40E+00	1.51E+06	Details
A4	Transportation to site	2.77E+03		5.07E-04	8.38E+00	1.80E+00	2.87E-01	1.00E+01	5.77E+04	Details
A5	Construction/installation process	6.24E+03		1.98E-04	1.85E+01	4.37E+00	1.25E+00	1.90E+00	7.02E+04	Details
B1	Use phase									Hide empty
B3	Repair	0.00E+00		0.00E+00	0.00E+00	0.00E+00	0.00E+00	0.00E+00	0.00E+00	Details
B4-B5	Material replacement and refurbishment	3.83E+04		1.60E-03	1.82E+02	5.42E+01	1.10E+01	1.29E+01	5.48E+05	Details
B6	Energy consumption	0.00E+00		0.00E+00	0.00E+00	0.00E+00	0.00E+00	0.00E+00	0.00E+00	Details
B7	Water use									Hide empty
C1-C4	End of life	1.16E+04		1.07E-03	2.61E+01	5.55E+00	6.58E-01	2.96E+01	1.34E+05	Details
D	External impacts (not included in totals)	-3.64E+04		-1.94E-03	-8.09E+01	-1.25E+01	-1.05E+01	-2.21E+00	-3.98E+05	Details
	Total	2.09E+05	3.07E+04	2.14E-02	7.41E+02	1.88E+02	4.67E+01	5.58E+01	2.32E+06	
	Results per denominator									
	Per gross internal floor area m2 / year	2.90E+00	4.27E-01	2.97E-07	1.03E-02	2.61E-03	6.49E-04	7.75E-04	3.22E+01	
	Per gross internal floor area m2	1.74E+02	2.56E+01	1.78E-05	6.17E-01	1.56E-01	3.89E-02	4.65E-02	1.93E+03	

The next step is to create a proposed building.

Proposed Case

Step 25: Click **Tutorial_8_3**. Click **2-Case-I PUF** and select **Copy**.

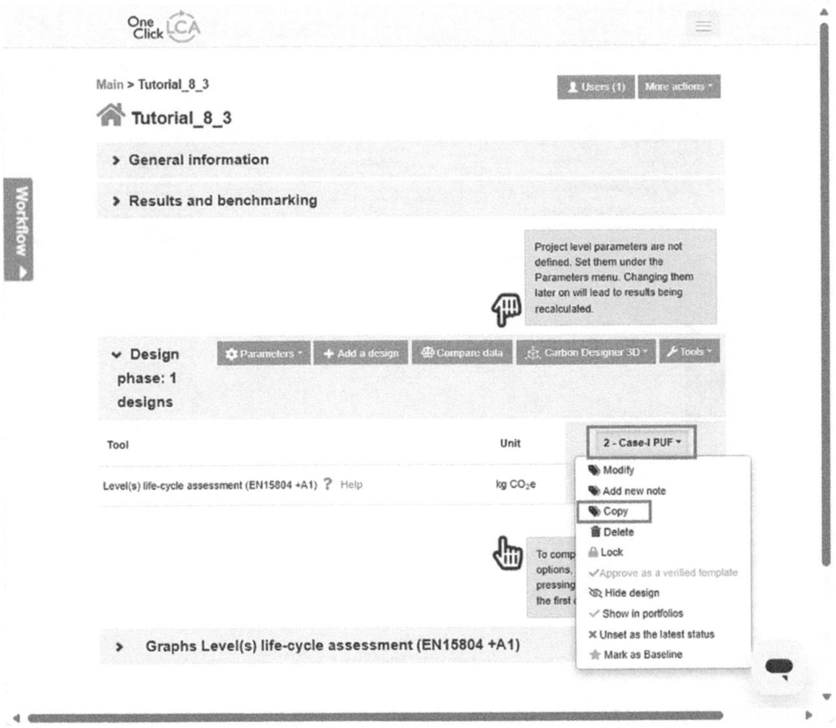

Step 26: Enter **Case-2 Proposed** in **Name** textbox. Click **Add**.

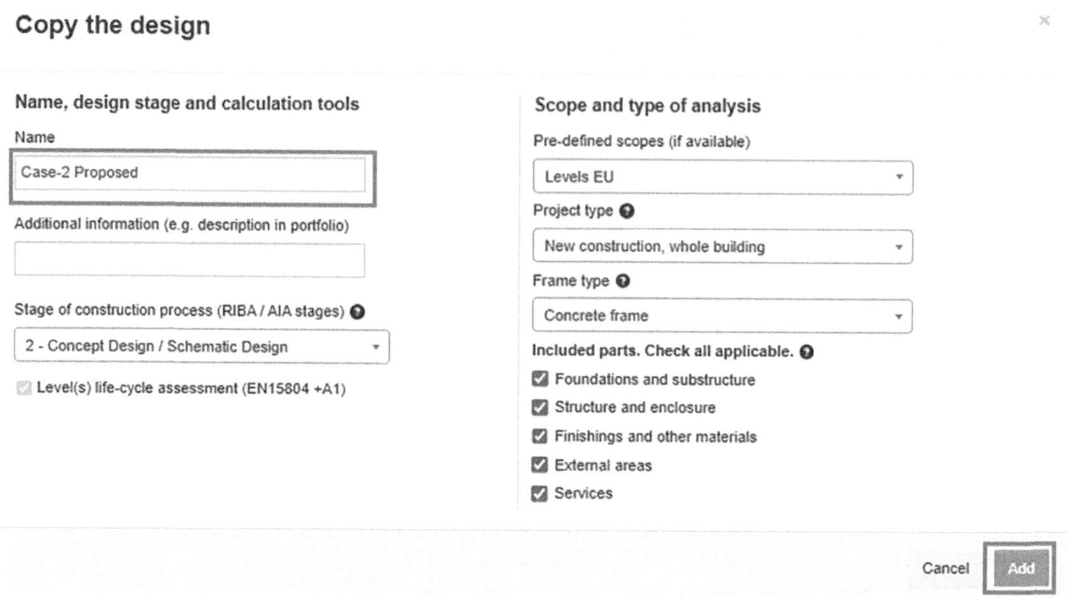

Step 27: Open the DesignBuilder model. Click on **Block 1** in the navigation tree. Click on **Model Options**. **Model Options—Building and Block** screen appears.

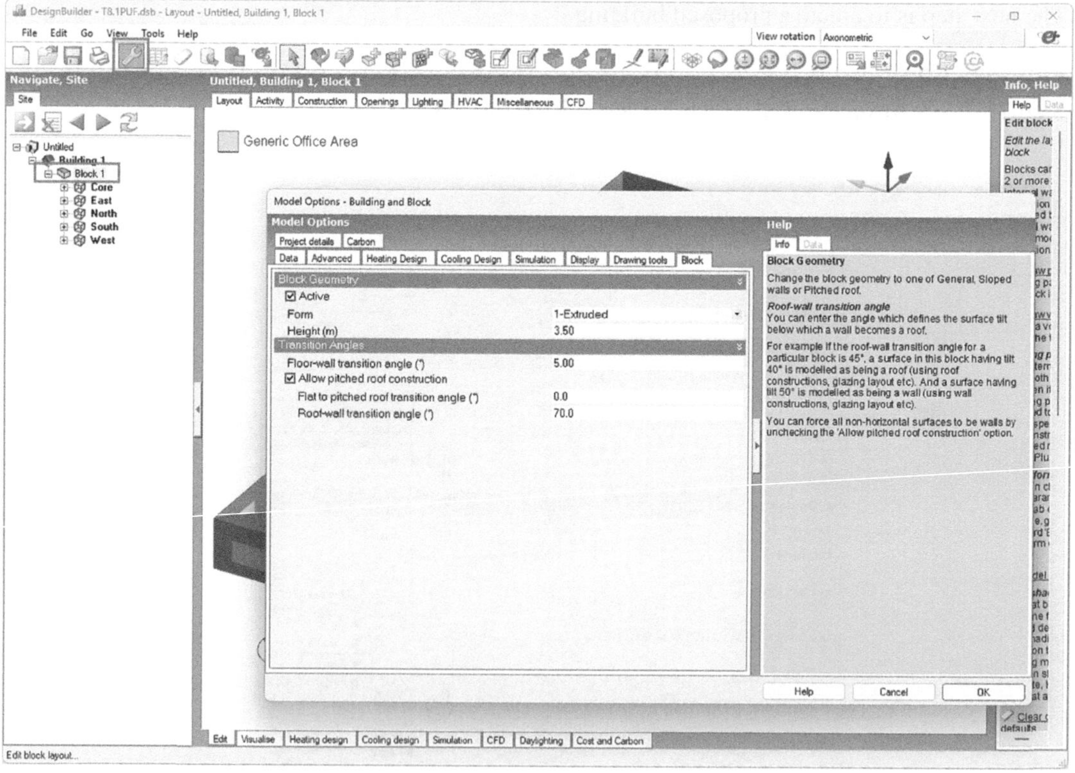

Step 28: In **Block Geometry** section, enter **2.8** in **Height (m)** text box.

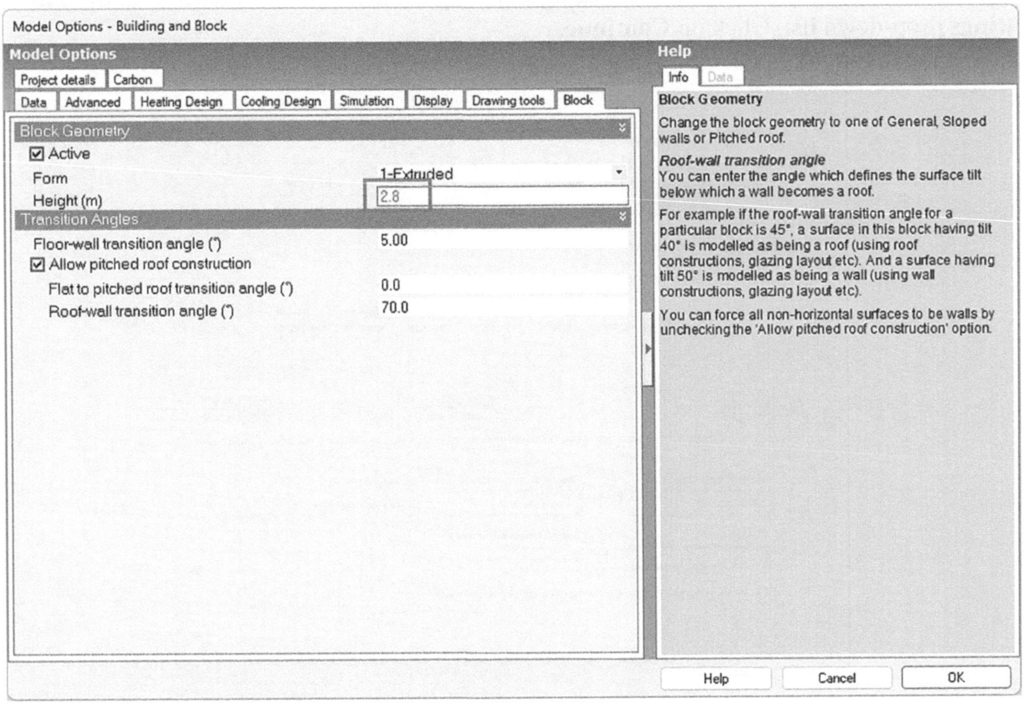

Step 29: Click on **LCA** button. A new screen appears on the web browser.

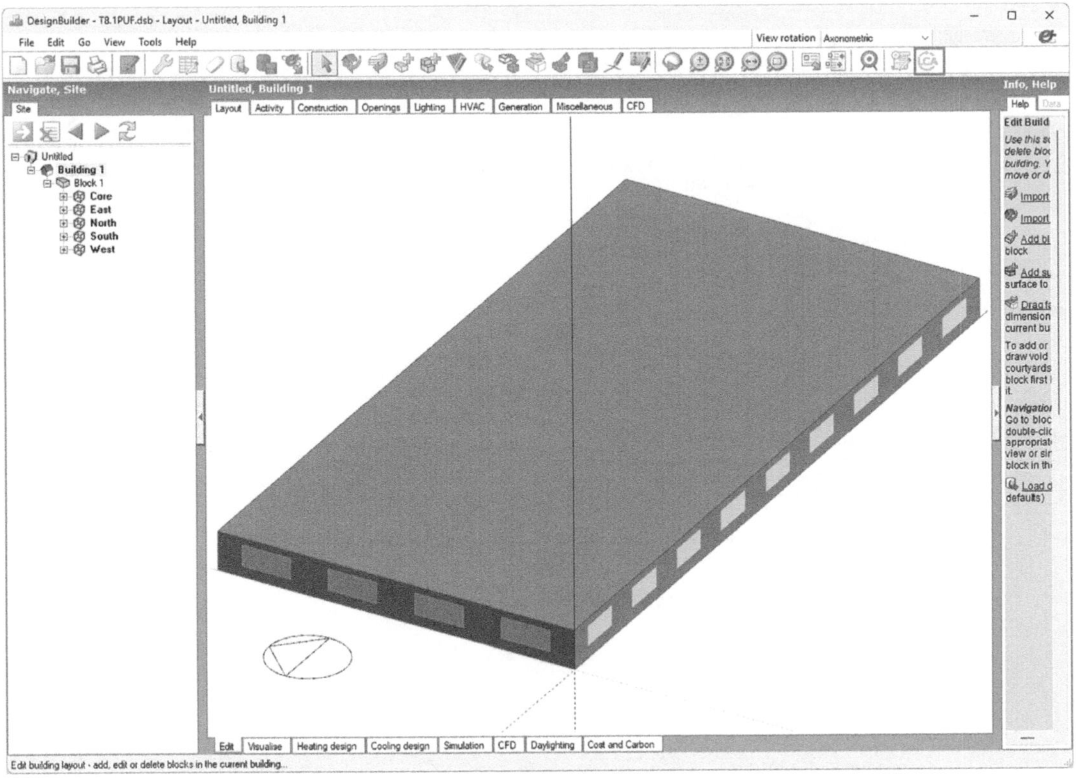

Step 30: Select **Tutorial_8_3** from the **Choose the project or entity** drop-down list. Select **Case-2 Proposed** from the **Choose or create a design** drop-down list. Select **All data** from the **Filtering settings** drop-down list. Click on **Continue**.

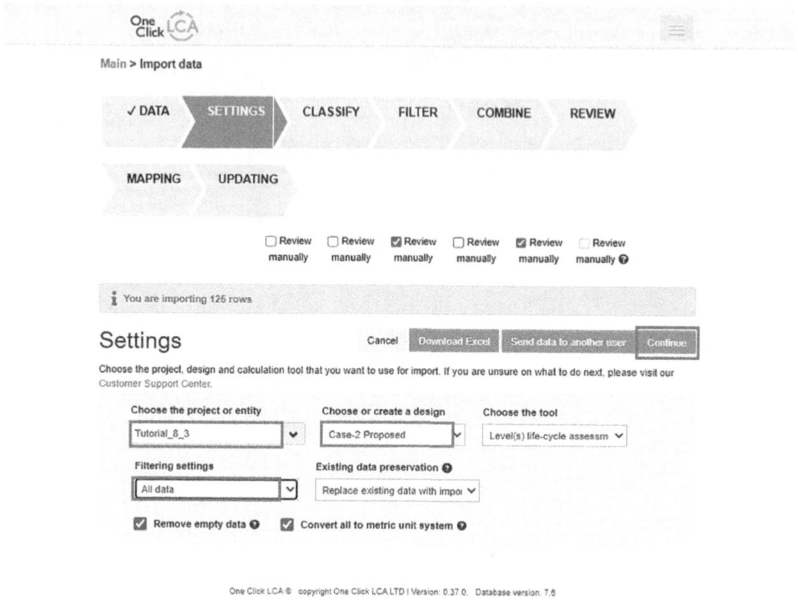

Step 31: Click on **Continue**.

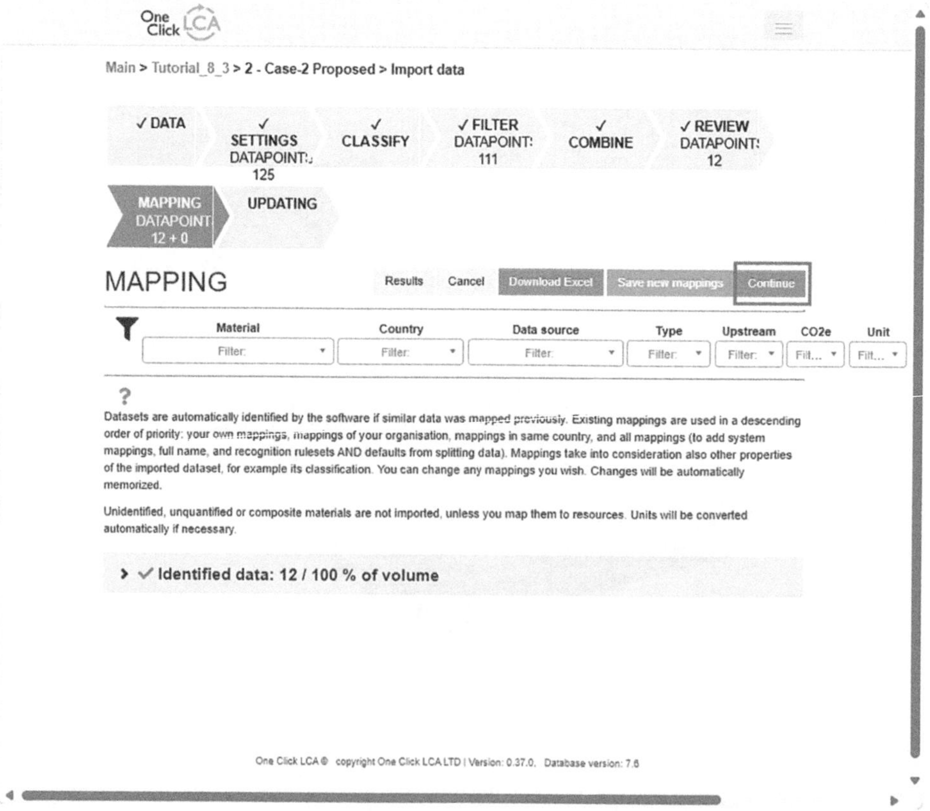

Step 32: Record the results. Click **Tutorial_8_3**.

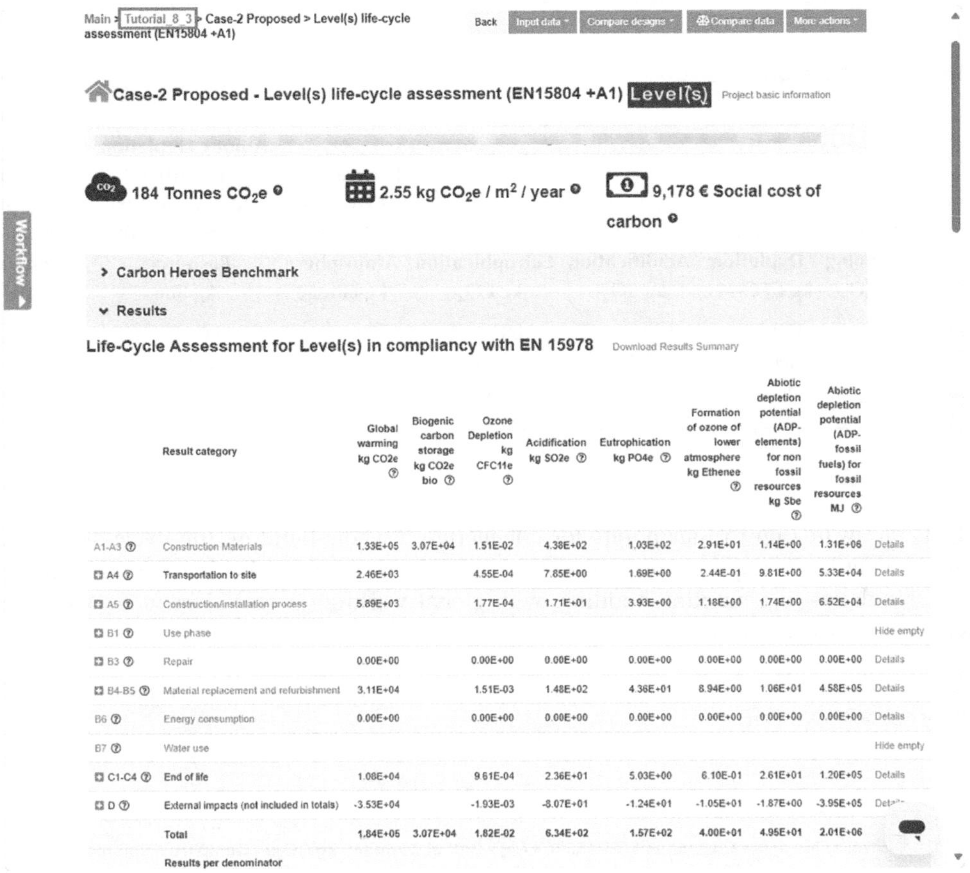

	Result category	Global warming kg CO2e	Biogenic carbon storage kg CO2e bio	Ozone Depletion kg CFC11e	Acidification kg SO2e	Eutrophication kg PO4e	Formation of ozone of lower atmosphere kg Ethenee	Abiotic depletion potential (ADP-elements) for non fossil resources kg Sbe	Abiotic depletion potential (ADP-fossil fuels) for fossil resources MJ	
A1-A3	Construction Materials	1.33E+05	3.07E+04	1.51E-02	4.38E+02	1.03E+02	2.91E+01	1.14E+00	1.31E+06	Details
A4	Transportation to site	2.46E+03		4.55E-04	7.85E+00	1.69E+00	2.44E-01	9.81E+00	5.33E+04	Details
A5	Construction/installation process	5.89E+03		1.77E-04	1.71E+01	3.93E+00	1.18E+00	1.74E+00	6.52E+04	Details
B1	Use phase									Hide empty
B3	Repair	0.00E+00		0.00E+00	0.00E+00	0.00E+00	0.00E+00	0.00E+00	0.00E+00	Details
B4-B5	Material replacement and refurbishment	3.11E+04		1.51E-03	1.48E+02	4.36E+01	8.94E+00	1.06E+01	4.58E+05	Details
B6	Energy consumption	0.00E+00		0.00E+00	0.00E+00	0.00E+00	0.00E+00	0.00E+00	0.00E+00	Details
B7	Water use									Hide empty
C1-C4	End of life	1.08E+04		9.61E-04	2.36E+01	5.03E+00	6.10E-01	2.61E+01	1.20E+05	Details
D	External impacts (not included in totals)	-3.53E+04		-1.93E-03	-8.07E+01	-1.24E+01	-1.05E+01	-1.87E+00	-3.95E+05	Det...
	Total	1.84E+05	3.07E+04	1.82E-02	6.34E+02	1.57E+02	4.00E+01	4.95E+01	2.01E+06	
	Results per denominator									

Step 33: Click **Graphs Level(s) life-cycle assessment (EN15804+A1)** and compare the results.

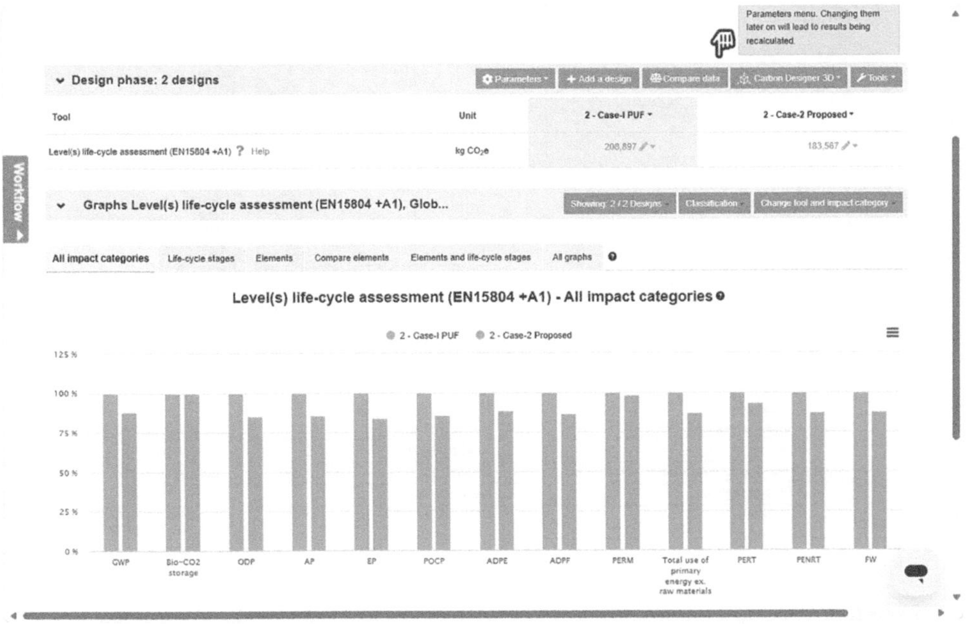

Step 34: Compare the results.

TABLE 8.3
Comparison of Results

Case	Global Warming kg CO_2e	Ozone Depletion kg CFC-11e	Acidification kg SO_2e	Eutrophication kg PO_4e	Formation of Ozone of Lower Atmosphere kg Ethene	Abiotic Depletion Potential (ADP-Elements) for Non-Fossil Resources kg Sbe	Abiotic Depletion Potential (ADP-Fossil Fuels) For fossil Resources MJ
Baseline	100	100	100	100	100	100	100
Proposed	87.87	84.98	85.55	83.64	85.73	88.59	86.76
Reduction	12.13%	15.02%	14.45%	16.36%	14.27%	11.41%	13.24%

LEED v4 awards up to four points for conducting a whole-building life-cycle assessment. Higher point levels require demonstrating progressively greater reductions in environmental impacts relative to a baseline building, with global warming potential included among the assessed impact categories.

Source: https://www.usgbc.org/credits/new-construction-core-and-shell-schools-new-construction-retail-new-construction-data-27

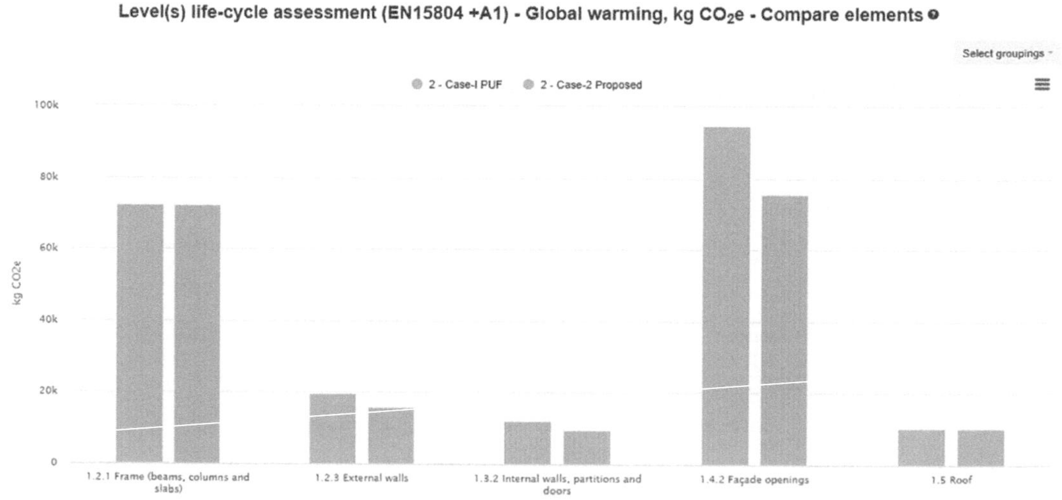

Chart

Category	2 - Case-I PUF	2 - Case-2 Proposed
1.2.1 Frame (beams, columns and slabs)	72245.41923333809	72245.41923333809
1.2.3 External walls	19642.71308970459	15712.74684947177
1.3.2 Internal walls, partitions and doors	12082.96759245514	9666.374073964114
1.4.2 Façade openings	94622.79516007507	75639.21983170242
1.5 Roof	10302.795768685719	10302.795768685719

TABLE 8.4
Comparison of Global Warming for Both Cases

Category	Global Warming, kg CO_2e 3.5 m	2.8 m	Reduction
1.2.1 Frame (beams, columns and slabs)	72,245	72,245	0%
1.2.3 External walls	19,643	15,713	20%
1.3.2 Internal walls, partitions and doors	12,083	9,666	20%
1.4.2 Façade openings	94,623	75,639	20%
1.5 Roof	10,303	10,303	0%
Total	**2,08,897**	**1,83,567**	**12%**

TABLE 8.5
Comparison of Ozone Depletion for Both Cases

Category	Ozone Depletion, kg CFC-11e 3.5 m	2.8 m	Reduction
1.2.1 Frame (beams, columns and slabs)	0.003	0.003	0%
1.2.3 External walls	0.002	0.002	20%
1.3.2 Internal walls, partitions and doors	0.001	0.000	20%
1.4.2 Façade openings	0.014	0.011	20%
1.5 Roof	0.003	0.003	0%
Total	**0.0214**	**0.0182**	**15%**

You can also generate a comparative report between the baseline and proposed designs. This feature is available in the following tools.

- LCA for LEED Int'l
- LCA for LEED, US (TRACI)
- LCA for LEED, Canada (TRACI)
- LCA for CalGreen (US)

Source: https://help.oneclicklca.com/en/articles/275626-comparative-report

The results show that the proposed design yielded reductions across all impact categories compared to the baseline. Specifically, there was a 12% reduction in global warming potential (kg CO_2e), 14.45% in acidification potential (kg SO_2e), and 16.36% in eutrophication potential (kg PO_4e). Other impact reductions included 15.02% in ozone depletion, 14.27% in the formation of lower-atmosphere ozone, 11.41% in abiotic depletion of non-fossil resources, and 13.24% in abiotic depletion of fossil fuels. These reductions were achieved primarily through reducing the vertical height and thereby the material quantities used in structural and envelope components.

These outcomes align well with the LEED v4.1 Materials and Resources credit for Building Life-Cycle Impact Reduction. According to LEED, achieving at least a 10% reduction in three or more impact categories—including global warming potential—qualifies the project for three points under Path 3 of the credit. Since the proposed design meets this criterion, it not only advances sustainability objectives but also contributes to the project's LEED certification potential.

Index

F

G

H

I

L

M

N

O

P

R

S

T

U

V

W

Z